Advances in Industrial Control

Other titles published in this series:

Digital Controller Implementation and Fragility
Robert S.H. Istepanian and James F. Whidborne (Eds.)

Optimisation of Industrial Processes at Supervisory Level
Doris Sáez, Aldo Cipriano and Andrzej W. Ordys

Robust Control of Diesel Ship Propulsion
Nikolaos Xiros

Hydraulic Servo-systems
Mohieddine Jelali and Andreas Kroll

Model-based Fault Diagnosis in Dynamic Systems Using Identification Techniques
Silvio Simani, Cesare Fantuzzi and Ron J. Patton

Strategies for Feedback Linearisation
Freddy Garces, Victor M. Becerra, Chandrasekhar Kambhampati and Kevin Warwick

Robust Autonomous Guidance
Alberto Isidori, Lorenzo Marconi and Andrea Serrani

Dynamic Modelling of Gas Turbines
Gennady G. Kulikov and Haydn A. Thompson (Eds.)

Control of Fuel Cell Power Systems
Jay T. Pukrushpan, Anna G. Stefanopoulou and Huei Peng

Fuzzy Logic, Identification and Predictive Control
Jairo Espinosa, Joos Vandewalle and Vincent Wertz

Optimal Real-time Control of Sewer Networks
Magdalene Marinaki and Markos Papageorgiou

Process Modelling for Control
Benoît Codrons

Computational Intelligence in Time Series Forecasting
Ajoy K. Palit and Dobrivoje Popovic

Modelling and Control of Mini-Flying Machines
Pedro Castillo, Rogelio Lozano and Alejandro Dzul

Ship Motion Control
Tristan Perez

Hard Disk Drive Servo Systems (2nd Ed.)
Ben M. Chen, Tong H. Lee, Kemao Peng and Venkatakrishnan Venkataramanan

Measurement, Control, and Communication Using IEEE 1588
John C. Eidson

Piezoelectric Transducers for Vibration Control and Damping
S.O. Reza Moheimani and Andrew J. Fleming

Manufacturing Systems Control Design
Stjepan Bogdan, Frank L. Lewis, Zdenko Kovačić and José Mireles Jr.

Windup in Control
Peter Hippe

Nonlinear H_2/H_∞ Constrained Feedback Control
Murad Abu-Khalaf, Jie Huang and Frank L. Lewis

Practical Grey-box Process Identification
Torsten Bohlin

Control of Traffic Systems in Buildings
Sandor Markon, Hajime Kita, Hiroshi Kise and Thomas Bartz-Beielstein

Wind Turbine Control Systems
Fernando D. Bianchi, Hernán De Battista and Ricardo J. Mantz

Advanced Fuzzy Logic Technologies in Industrial Applications
Ying Bai, Hanqi Zhuang and Dali Wang (Eds.)

Practical PID Control
Antonio Visioli

(continued after Index)

Riccardo Marino · Patrizio Tomei
Cristiano M. Verrelli

Induction Motor Control Design

Springer

Riccardo Marino
Patrizio Tomei
Cristiano M. Verrelli

Università di Roma
Tor Vergata
Dipto. Ingegneria Elettronica
Via del Politecnico 1
00133 Roma
Italy

marino@ing.uniroma2.it
tomei@ing.uniroma2.it
verrelli@ing.uniroma2.it

ISSN 1430-9491
ISBN 978-1-4471-5708-3 ISBN 978-1-84996-284-1 (eBook)
DOI 10.1007/978-1-84996-284-1
Springer London Dordrecht Heidelberg New York

British Library Cataloguing in Publication Data
A catalogue record for this book is available from the British Library

Cover design: eStudioCalamar, Figueres/Berlin

Springer is part of Springer Science+Business Media (www.springer.com)

Advances in Industrial Control

Professor (Emeritus) O.P. Malik
Department of Electrical and Computer Engineering
University of Calgary
2500, University Drive, NW
Calgary, Alberta
T2N 1N4
Canada

Professor K.-F. Man
Electronic Engineering Department
City University of Hong Kong
Tat Chee Avenue
Kowloon
Hong Kong

Professor G. Olsson
Department of Industrial Electrical Engineering and Automation
Lund Institute of Technology
Box 118
S-221 00 Lund
Sweden

Professor A. Ray
Department of Mechanical Engineering
Pennsylvania State University
0329 Reber Building
University Park
PA 16802
USA

Professor D.E. Seborg
Chemical Engineering
3335 Engineering II
University of California Santa Barbara
Santa Barbara
CA 93106
USA

Doctor K.K. Tan
Department of Electrical and Computer Engineering
National University of Singapore
4 Engineering Drive 3
Singapore 117576

Professor I. Yamamoto
Department of Mechanical Systems and Environmental Engineering
The University of Kitakyushu
Faculty of Environmental Engineering
1-1, Hibikino,Wakamatsu-ku, Kitakyushu, Fukuoka, 808-0135
Japan

To Rebecca and Rosanna

R. Marino

To my family and the memory of my parents

P. Tomei

To Paola, my family and the memory of my grandparents

C.M. Verrelli

Series Editors' Foreword

The series *Advances in Industrial Control* aims to report and encourage technology transfer in control engineering. The rapid development of control technology has an impact on all areas of the control discipline. New theory, new controllers, actuators, sensors, new industrial processes, computer methods, new applications, new philosophies..., new challenges. Much of this development work resides in industrial reports, feasibility study papers and the reports of advanced collaborative projects. The series offers an opportunity for researchers to present an extended exposition of such new work in all aspects of industrial control for wider and rapid dissemination.

Over recent years there has been considerable interest in trying to understand and quantify the potential benefits that nonlinear control could bring to industrial applications. One obstacle to the widespread use of nonlinear control has been the issue of finding appropriate nonlinear system models easily. This obstacle is commonly avoided by finding a linear model of limited validity and then designing a robust control able to deliver satisfactory system performance for a wider range of system parameter variations. Finding analytical dynamical nonlinear models for routine industrial application to allow the straightforward development of nonlinear control designs has been a little more problematic; however, there are some industrial areas, such as electrical machines, marine systems, and chemical processes, where nonlinear system models are more readily available for use in nonlinear control designs.

In the field of chemical processes, K.M. Hangos, J. Bokor, and G. Szederkényi exploited nonlinear system models and used nonlinear control techniques in their textbook *Analysis and Control of Nonlinear Process Systems* (ISBN 978-1-85233-600-4, 2004) that was published in our related series: *Advanced Textbooks in Control and Signal Processing*. Marine systems is another field where there are well-known nonlinear models, and researchers K. D. Do and J. Pan have recently developed a whole series of new nonlinear control algorithms for the different control tasks demanded of such systems. A comprehensive presentation of their work can be found in the *Advances in Industrial Control* monograph *Control of Ships and Underwater Vehicles* (ISBN 978-1-84882-729-5, 2009).

This *Advances in Industrial Control* monograph by R. Marino, P. Tomei and G.M. Verrelli is devoted to the control of induction motors from the industrial field of electrical machines. In the monograph the authors report the systematic application of nonlinear control techniques to develop a sequence of sophisticated control algorithms. The key facilitator in this development is the availability of a set of analytical dynamical state space models for induction motor behavior. The authors then exploit the structure of these models in a variety of ingenious ways to develop the increasingly complex nonlinear control algorithms. Despite the closely argued theoretical presentation in the monograph, the basic outline of the approach taken should be easily recognizable to any industrial engineer familiar with the modern control paradigm, namely:

- modelling in this case basic electrical equations leading to nonlinear state space models (Chapter 1);
- open-loop inverse model-based control (Chapter 1);
- feedback control based on a full state vector, including states that are unmeasurable (Chapter 2);
- observers to reconstruct unmeasurable states and observers to estimate uncertain system parameter values (Chapter 3). State observers driven by measurable outputs will facilitate output feedback designs and parameter observers will facilitate adaptive control designs;
- general output feedback control designs (Chapter 4); and
- specialized output feedback designs in this case speed sensorless feedback control (Chapter 5).

The authors use this agenda for induction motor control, carefully absorbing more and more realistic practical assumptions to develop increasingly general control algorithms. At each step of the way, useful validating simulation results are presented. These use the same induction motor parameters so it is possible to compare results within and indeed across chapters as the various control schemes evolve. Surprisingly, not too many different analysis tools from Lyapunov and nonlinear control theory are used in the development and the context and explanation of those that are used can be found in two useful reference appendices.

The *Advances in Industrial Control* monograph series has not seen many entries that present a wholly nonlinear viewpoint for industrial control system design so it is a pleasure to welcome this exhaustive volume by R. Marino, P. Tomei and G.M. Verrelli to the series. The authors have stated that one of their aims in writing this monograph was to give a unified presentation of these nonlinear induction motor controls that subsumes and archives the last thirty or more years of development since the engineers at Siemens (1971) and Toshiba (1980) first developed the nonlinear control method called the direct field-oriented control algorithm. Also threaded into the volume is their own significant research contribution to the field. Clearly, the monograph will be of great interest to electrical and control engineers working in the electrical machines field. Academics, postgraduate students and researchers working in the nonlinear control paradigm will also find new inspiration from the work of the authors and much transferable knowledge for tackling nonlinear control problems in other industrial applications fields.

Industrial Control Centre
Glasgow
Scotland, UK
2008

M.J. Grimble
M.A. Johnson

Preface

The control of induction motors has attracted much attention from researchers and engineers since 1971. More than 4,000 journal papers have been published on induction motor control and more than 500 specifically on the adaptive control of induction motors: it is still a very active research area since more than 300 journal papers appeared in 2008. The industrial interest in induction motor control is documented by over 80,000 patents on this subject. The availability of low cost powerful digital signal processors and significant advances in power electronics motivated the design of complex induction motor controls. The aim is to achieve the same, or even superior, performance on speed tracking and power efficiency which are obtained by more sophisticated and expensive, but less reliable, electric motors such as direct current or permanent magnet ones. Direct current motors are extensively used in variable speed applications since their flux and torque are independently controlled by the field and the armature current. However, they have disadvantages due to the mechanical commutator and the brushes so that they are limited in high-speed, high-voltage operating conditions. Induction motors are much more difficult to control but have definite advantages since they have no commutator, no brushes, no rotor windings in squirrel cage motors, they have a simple rugged structure, can tolerate heavy overloading, and can produce higher torques with a lower weight, smaller size, and lower rotating mass than direct current motors.

The design of control algorithms for induction motors is, however, very complex for many reasons. It is a multivariable control problem since there are two independent control inputs and two outputs to be controlled: the primary output is the rotor speed to achieve the required dynamic performance, while the secondary output is the rotor flux modulus for power efficiency maximization. It is an intrinsic nonlinear problem since the electromagnetic torque, which controls the rotor speed, is a nonlinear function of stator currents and rotor fluxes, and the operating conditions of interest are away from the equilibrium points so that linear approximation techniques do not apply. It involves parameters such as load torque and rotor resistance which may vary widely during operation; they are critical in the control design and should be identified online to maximize power efficiency. The control design cannot rely on state variable feedback since rotor flux measurements are not easily available. If rotor speed is not measured in order to reduce costs or due to sensor failures and only stator currents and voltages are available from measurements, the control problem is called speed-sensorless. In this case the desired reference signals for rotor speed and flux modulus are to be tracked in spite of parameter perturbations, while both tracking errors are not available for feedback to the controller. The feedforward control which solves the tracking problem in open-loop may be explicitly obtained by computing the induction motor nonlinear inverse dynamics. The stability of the resulting open-loop controlled motor is, however, not always guaranteed since it depends both on the reference trajectories to be tracked and on motor parameters. Even in stable operating conditions the dynamic responses may be unsatisfactory and poorly damped. Hence, feedback from stator currents, and from rotor speed when available, has the goal of enhancing both stability and robustness with respect to parameter perturbations; moreover, it should improve transient behaviors and power efficiency. This book is focused on the nonlinear feedback control design techniques, including adaptive ones, which are required to achieve high speed tracking performance along with high power efficiency in induction motor control.

Besides its technological motivations for electric traction and electric drives, the control of induction motors has an intrinsic interest from the view point of nonlinear control theory, since it involves clearly modeled nonlinear terms such as electromagnetic torque and two critical parameters; the appropriate tools belong to the theory of adaptive output feedback for multi-input, multi-output nonlinear systems. Such a theory started to be developed in 1992 for special classes of single-input, single-output nonlinear systems but it does not encompass the induction motor models. Hence, the control of induction motors constitutes a very interesting case study which evolved into a benchmark nonlinear control problem. In fact, most of the fundamental concepts of nonlinear control theory apply in a nontrivial way. Induction motors are not feedback linearizable by static state feedback but they are feedback linearizable by a dynamic state feedback. It is enough to add one integrator to achieve feedback linearization; this can be done in many different ways even though they all lead to singularities that make it inadvisable to render the closed-loop linear in all operating conditions. Induction motors are input–output feedback linearizable but the input–output feedback linearizing control, which makes the rotor flux angle unobservable, is singular when the rotor flux is zero and it is not power efficient at low rotor flux levels. Induction motors are observable from rotor speed and stator current measurements so that flux observers, including adaptive ones, can be designed. Observer-based output feedback controls can also be designed using Lyapunov techniques. The steady-state dynamics of induction motors are very intriguing: in the case of constant reference signals for rotor speed and rotor flux modulus, they constitute a limit cycle in the state space where the rotating speed of the flux vector is equal to the sum of the desired rotor speed and the so-called slip speed, which depends on the load torque, the rotor resistance, and the flux modulus. In the more general case of bounded reference signals the steady-state dynamics may be very complex. They remain bounded but their stability and attractivity are in general difficult to study. In many cases the attractivity is not global and the stability is not exponential, depending on the reference signals and physical parameters, and instabilities or poor dynamic responses may arise.

Engineers at Siemens and Toshiba developed *ante litteram* in 1971 and 1980, respectively, nonlinear feedback control algorithms which are now called direct field-oriented control and indirect field-oriented control. At that time nonlinear control theory was just at its beginning: researchers were investigating basic controllability (1972) and observability (1977) properties. In fact, the proof that indirect field-oriented control is globally stable was published in 1996. Using today's terminology we can say that direct field-oriented control is an asymptotic state feedback linearizing control which has a singularity when rotor flux is zero, while indirect field-oriented control is a global dynamic output feedback control which has no singularities and allows the motor to start from any initial condition. Field-oriented controls were originally conceived for current-fed motors in which the stator currents can be controlled very rapidly by stator voltages, so that they may be considered control inputs by neglecting the stator currents dynamics; they were then extended to general induction motor models. During the 1980s new important tools in nonlinear state feedback design were developed: feedback linearization and input–output decoupling along with their adaptive generalizations. Good theories proved once again very useful in applications since they led to very innovative control algorithms for induction motors, with superior performance with respect to field-oriented controls. An adaptive feedback linearizing control with online identification of load torque and rotor resistance was developed in 1991. The goals of field-oriented controls and feedback linearizing controls are indeed very similar: they both use nonlinear feedback and nonlinear change of coordinates so that the feedback systems have a simpler structure.

Since both direct field-oriented control and feedback linearizing control make use of rotor flux signals there was a strong motivation to design rotor flux observers. At that time nonlinear observer theory was not fully developed. Nevertheless nonlinear observers for induction motors were designed in 1978: they were called bilinear observers. A complete theory for rotor flux observers, including observers with arbitrary rate of convergence, was successively developed. Adaptive flux observers were also designed which are adaptive with respect to rotor resistance, since rotor flux observers were found to be very sensitive with respect to rotor resistance variations. Identifiability questions naturally arose and were answered using the concepts of persistency of excitation and nonlinear observability. Since 1991 the problem of designing a global output feedback tracking control which does not require rotor flux measurements, is adaptive with respect to load torque and rotor resistance variations, and has no singularities was posed and finally solved in 1999 following the indirect field-oriented approach.

More recently, an important line of research was focused on the design of feedback control algorithms based on stator current measurements only. The absence of rotor speed measurements, which improves the reliability of the motor and reduces its cost, forced the redesign of those control algorithms which make use of rotor speed measurements in many crucial steps. The question itself of speed and rotor flux observability from stator current measurements is rather delicate and leads to the discovery of operating conditions in which observability fails: of course, the concept of nonlinear observability has to be used since the motor model is nonlinear. The study of identifiability of rotor resistance and load torque from stator current measurements leads to the discovery of persistently exciting reference signals for the flux modulus, which is required to be time-varying. Several speed sensorless control algorithms were recently developed which show superior performance with respect to inverse system based controls but are of course inferior to controls which make use of speed sensors.

At the present stage of research on induction motor control a coherent collection of estimation and control algorithms is available, including the most recent speed sensorless controllers. This book collects and discusses, using a unified notation and a modern nonlinear control terminology, the most important steps and issues in the design of estimation and control algorithms for induction motors. Many estimation and control algorithms are reported: their stability is analyzed and their performance is illustrated by simulations and experiments on the same induction motor. An intense and challenging collective research effort (which also involved at various stages the authors of this book) is carefully documented and analyzed, with the aim of providing and clarifying the basic intuition and tools required in the analysis and design of nonlinear adaptive feedback control algorithms. This material should be of specific interest to engineers who are engaged in the design of control algorithms for electric motors and, more generally, to a broader audience interested in nonlinear control design. In fact, induction motor dynamics are surprisingly rich and their control is challenging even to engineers with a strong nonlinear control background. The induction motor is an excellent source of projects, examples, and exercises for courses in nonlinear control design since they can be physically and experimentally tested. The book can be used for graduate courses on the control of induction motors and for independent study.

This book is divided into six chapters and two appendices. Since the stability of controlled induction motors is carefully analyzed throughout the book, the basic definitions and tools from Lyapunov stability theory are recalled in Appendix A. Since in many instances the basic concepts and tools from nonlinear control theory (nonlinear change of coordinates, observability, feedback linearization, input–output decoupling) are used, they are recalled in Appendix B. In Chapter 1, the modeling issues and the basic assumptions are discussed; moreover the structural properties of the models such as observability, parameter identifiability, linearizability, inverse dynamics and steady-state behaviors, including power losses minimization, are analyzed. Chapter 2 is devoted to state feedback control to explore the performance which can be obtained using full state variables measurements, and to examine those controllers which could tolerate the replacements of sensors by asymptotic observers and those which can be made adaptive with respect to uncertain parameters. The estimation of state variables, in particular rotor fluxes and rotor currents, and the identification of physical parameters such as load torque and resistances are discussed in Chapter 3: adaptive rotor flux observers are also presented. In Chapter 4 output feedback controls based on rotor speed, stator current, and stator voltage measurements are presented: some algorithms incorporate flux observers to improve performance, while the most complex algorithm is adaptive with respect to both torque load and rotor resistance. In Chapter 5 speed-sensorless output feedback controls which are based only on stator current measurements are discussed along with their adaptive versions. Chapter 6 contains some concluding remarks. All the control schemes are numerically simulated for the same motor with similar references to illustrate their performance, so that they can be compared: advantages and drawbacks of each scheme are pointed out. Some estimation and control algorithms are validated by experiments. Experimental tests are also presented which validate the motor model and the parameters used. The bibliography collects more than 200 journal papers and books on induction motor control from 1971 to 2009; it is, however, far from being complete and only contains all the material which was actually used during the preparation of this book. Finally, frequent exchanges of ideas and fruitful collaborations with Professor Sergei Peresada on induction motor control are acknowledged with pleasure.

Rome,
December 2009

Riccardo Marino
Patrizio Tomei
Cristiano Maria Verrelli

Contents

Chapter 1
Dynamical Models and Structural Properties

Abstract Starting from the three physical stator and rotor windings, several state space dynamical models for the induction motor are introduced in this chapter. Each model clarifies specific dynamical properties. Their steady-state operating conditions are determined and analyzed: in particular the steady-state torque–speed characteristic curve is computed when sinusoidal voltages with constant amplitude and frequency are applied. This curve reveals many important nonlinear features: for instance, for a given load torque there may be two operating conditions, a stable one and an unstable one; they become closer and closer as the load torque increases up to a load torque bifurcation value. More generally, the dynamic inverse system is explicitly computed: it generates the voltage inputs which are required to track a desired time-varying rotor speed profile with the desired rotor flux modulus. The flux modulus parameterizes the control input: it may be chosen to minimize the power losses or to keep the voltage modulus constant or below a desired level (field weakening). The corresponding tracking dynamics are also computed: they determine limit cycles in the state space whose speed depends on the load torque and the desired rotor speed and flux. The structural properties of the motor from the control view point are studied: observability from stator currents and rotor speed measurements; observability from stator currents and rotor fluxes; observability from stator current measurements only; feedback linearizability, *i.e.* the possibility of transforming the motor model into a linear and controllable system by state feedback (either static or dynamic), which implies the controllability property; the identifiability, from different set of measurements, of critical parameters such as load torque and rotor resistance which may vary during operations. The induction motor turns out to be feedback linearizable by a dynamic state feedback; it is observable for any voltage input if stator currents and rotor speed are measured but it is not observable if only stator currents are measured and rotor speed and rotor fluxes are kept constant.

1.1 Modeling Assumptions

Consider a two-pole, three-phase symmetrical induction machine (see Figure 1.1). The stator windings are assumed to be identical with resistance R_s and equivalent

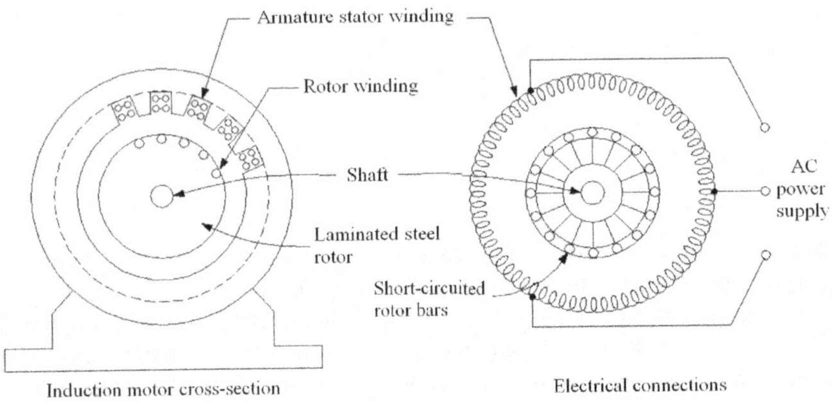

Fig. 1.1 Three-phase induction motor

turns N_s. The rotor windings are also assumed to be identical with resistance R_r and equivalent turns N_r. The air gap is assumed to be uniform. Stator and rotor windings are assumed to be approximated as sinusoidally distributed windings. The angle δ represents the rotor position with respect to the stator. We assume that the induction machine is operated as a motor, that is the rotor speed

$$\omega = \frac{\mathrm{d}\delta}{\mathrm{d}t}$$

and the load torque T_L have opposite signs. The rotor windings are short circuited while the stator windings are connected to a balanced three-phase source. When a balanced three-phase current is flowing in the stator windings an air gap magneto-motive force rotates about the air gap at a speed determined by the frequency of the stator currents and the number of poles. If the speed of the rotating magneto-motive force is different from the rotor speed, balanced three-phase currents will be induced in the short circuited rotor windings; the names of induction motor or asynchronous motor are due to this physical principle. The difference between the speed of the rotating magneto-motive force due to stator currents and the speed of the rotor determines the frequency of the induced rotor currents. If this speed difference is zero, that is the rotor rotates at the same speed as the magneto-motive force, no rotor currents are induced. Let

$$\Psi_s = [\psi_{s1}, \psi_{s2}, \psi_{s3}]^{\mathrm{T}}$$

$$\Psi_r = [\psi_{r1}, \psi_{r2}, \psi_{r3}]^{\mathrm{T}}$$

be the vectors whose components are the stator and rotor flux linkages, respectively, with 1, 2, 3 denoting the three phases. Similarly, let

$$\begin{aligned} I_s &= [i_{s1}, i_{s2}, i_{s3}]^{\mathrm{T}} \\ I_r &= [i_{r1}, i_{r2}, i_{r3}]^{\mathrm{T}} \end{aligned}$$

be the vectors whose components are the stator and rotor currents. Then, for an induction motor with one pole pair, we can write

$$\begin{aligned} \frac{\mathrm{d}\psi_{s1}}{\mathrm{d}t} + R_s i_{s1} &= u_{s1} \\ \frac{\mathrm{d}\psi_{s2}}{\mathrm{d}t} + R_s i_{s2} &= u_{s2} \\ \frac{\mathrm{d}\psi_{s3}}{\mathrm{d}t} + R_s i_{s3} &= u_{s3} \\ \frac{\mathrm{d}\psi_{r1}}{\mathrm{d}t} + R_r i_{r1} &= 0 \\ \frac{\mathrm{d}\psi_{r2}}{\mathrm{d}t} + R_r i_{r2} &= 0 \\ \frac{\mathrm{d}\psi_{r3}}{\mathrm{d}t} + R_r i_{r3} &= 0 \end{aligned} \tag{1.1}$$

where the stator and rotor fluxes, under the assumption of linear magnetic circuits, satisfy the linear relation

$$\begin{bmatrix} \Psi_s \\ \Psi_r \end{bmatrix} = \begin{bmatrix} l_s & l_{s,r} \\ l_{s,r}^{\mathrm{T}} & l_r \end{bmatrix} \begin{bmatrix} I_s \\ I_r \end{bmatrix} \tag{1.2}$$

with

$$\begin{aligned} l_s &= \begin{bmatrix} l_{sl} + l_{sm} & -\frac{l_{sm}}{2} & -\frac{l_{sm}}{2} \\ -\frac{l_{sm}}{2} & l_{sl} + l_{sm} & -\frac{l_{sm}}{2} \\ -\frac{l_{sm}}{2} & -\frac{l_{sm}}{2} & l_{sl} + l_{sm} \end{bmatrix} \\ l_{s,r} &= l_{sr} \begin{bmatrix} \cos(\delta) & \cos\left(\delta + \frac{2}{3}\pi\right) & \cos\left(\delta - \frac{2}{3}\pi\right) \\ \cos\left(\delta - \frac{2}{3}\pi\right) & \cos(\delta) & \cos\left(\delta + \frac{2}{3}\pi\right) \\ \cos\left(\delta + \frac{2}{3}\pi\right) & \cos\left(\delta - \frac{2}{3}\pi\right) & \cos(\delta) \end{bmatrix} \\ l_r &= \begin{bmatrix} l_{rl} + l_{rm} & -\frac{l_{rm}}{2} & -\frac{l_{rm}}{2} \\ -\frac{l_{rm}}{2} & l_{rl} + l_{rm} & -\frac{l_{rm}}{2} \\ -\frac{l_{rm}}{2} & -\frac{l_{rm}}{2} & l_{rl} + l_{rm} \end{bmatrix}. \end{aligned} \tag{1.3}$$

In (1.3) l_{sl} denotes the leakage inductance of the stator windings, l_{sm} denotes the magnetizing inductance of the stator windings, l_{rl} denotes the leakage inductance of the rotor windings, l_{rm} denotes the magnetizing inductance of the rotor windings, and l_{sr} denotes the amplitude of the mutual inductance between stator and rotor windings. Neglecting iron losses we set $l_{sl}=0$ and $l_{rl}=0$ so that (1.3) becomes

$$
\begin{aligned}
l_s &= l_{sm}\begin{bmatrix} 1 & -\frac{1}{2} & -\frac{1}{2} \\ -\frac{1}{2} & 1 & -\frac{1}{2} \\ -\frac{1}{2} & -\frac{1}{2} & 1 \end{bmatrix} \\
l_r &= l_{rm}\begin{bmatrix} 1 & -\frac{1}{2} & -\frac{1}{2} \\ -\frac{1}{2} & 1 & -\frac{1}{2} \\ -\frac{1}{2} & -\frac{1}{2} & 1 \end{bmatrix} .
\end{aligned}
$$

Denote by

$$
\begin{aligned}
L_s &= \frac{3}{2} l_{sm} \\
L_r &= \frac{3}{2} l_{rm} \\
M &= \frac{3}{2} l_{sr}
\end{aligned}
$$

the stator, rotor, and mutual inductances, respectively. Hence, from (1.3) we have

$$
\begin{aligned}
l_s &= \frac{2}{3} L_s \begin{bmatrix} 1 & -\frac{1}{2} & -\frac{1}{2} \\ -\frac{1}{2} & 1 & -\frac{1}{2} \\ -\frac{1}{2} & -\frac{1}{2} & 1 \end{bmatrix} \\
l_{s,r} &= \frac{2}{3} M \begin{bmatrix} \cos(\delta) & \cos\left(\delta+\frac{2}{3}\pi\right) & \cos\left(\delta-\frac{2}{3}\pi\right) \\ \cos\left(\delta-\frac{2}{3}\pi\right) & \cos(\delta) & \cos\left(\delta+\frac{2}{3}\pi\right) \\ \cos\left(\delta+\frac{2}{3}\pi\right) & \cos\left(\delta-\frac{2}{3}\pi\right) & \cos(\delta) \end{bmatrix} \\
l_r &= \frac{2}{3} L_r \begin{bmatrix} 1 & -\frac{1}{2} & -\frac{1}{2} \\ -\frac{1}{2} & 1 & -\frac{1}{2} \\ -\frac{1}{2} & -\frac{1}{2} & 1 \end{bmatrix} .
\end{aligned}
\tag{1.4}
$$

When the motor is operating in balanced conditions we have the constraints

$$
\begin{aligned}
i_{s1}+i_{s2}+i_{s3} &= 0 \\
i_{r1}+i_{r2}+i_{r3} &= 0 \\
u_{s1}+u_{s2}+u_{s3} &= 0
\end{aligned}
\tag{1.5}
$$

and, therefore, it is convenient to introduce the new variables

$$
\begin{bmatrix} i_{s0} \\ i_{sa} \\ i_{sb} \end{bmatrix} = \sqrt{\frac{2}{3}} \begin{bmatrix} \frac{\sqrt{2}}{2} & \frac{\sqrt{2}}{2} & \frac{\sqrt{2}}{2} \\ 1 & -\frac{1}{2} & -\frac{1}{2} \\ 0 & \frac{\sqrt{3}}{2} & -\frac{\sqrt{3}}{2} \end{bmatrix} \begin{bmatrix} i_{s1} \\ i_{s2} \\ i_{s3} \end{bmatrix} \triangleq U \begin{bmatrix} i_{s1} \\ i_{s2} \\ i_{s3} \end{bmatrix}
\tag{1.6}
$$

in which U is a unitary matrix ($U^{-1}=U^{\mathrm{T}}$ and $\det[U]=1$); similarly we define

$$
\begin{bmatrix} i_{r0} \\ i_{rd'} \\ i_{rq'} \end{bmatrix} = U \begin{bmatrix} i_{r1} \\ i_{r2} \\ i_{r3} \end{bmatrix}
$$
$$
\begin{bmatrix} \psi_{s0} \\ \psi_{sa} \\ \psi_{sb} \end{bmatrix} = U \begin{bmatrix} \psi_{s1} \\ \psi_{s2} \\ \psi_{s3} \end{bmatrix}
$$
$$
\begin{bmatrix} \psi_{r0} \\ \psi_{rd'} \\ \psi_{rq'} \end{bmatrix} = U \begin{bmatrix} \psi_{r1} \\ \psi_{r2} \\ \psi_{r3} \end{bmatrix}
$$
$$
\begin{bmatrix} u_{s0} \\ u_{sa} \\ u_{sb} \end{bmatrix} = U \begin{bmatrix} u_{s1} \\ u_{s2} \\ u_{s3} \end{bmatrix} \tag{1.7}
$$

where $(\psi_{rd'}, \psi_{rq'})$ and $(i_{rd'}, i_{rq'})$ denote the (d', q')-components of the rotor flux and current vectors in the (d', q') reference frame attached to the rotor, rotating at rotor speed $\omega = \dot{\delta}$ and identified by the rotor angle δ in the fixed (a, b) reference frame attached to the stator, while (ψ_{sa}, ψ_{sb}) and (u_{sa}, u_{sb}) denote the (a, b)-components of the stator flux and the stator voltage vectors in the fixed (a, b) frame.

1.2 State Space Models

Let T_e be the electromagnetic torque produced by the motor and J the motor moment of inertia. Recall that T_L is the load torque while R_s and R_r are the stator and rotor resistances, respectively. Since in balanced operating conditions (1.5) $i_{s0} = 0$, $i_{r0} = 0$ and $u_{s0} = 0$, on the basis of (1.1), (1.6), and (1.7), for an induction motor with one pole pair we can write (the damping friction torque is usually negligible in induction motors and it is therefore set equal to zero)

$$
\begin{aligned}
\dot{\psi}_{sa} + R_s i_{sa} &= u_{sa} \\
\dot{\psi}_{sb} + R_s i_{sb} &= u_{sb} \\
\dot{\psi}_{rd'} + R_r i_{rd'} &= 0 \\
\dot{\psi}_{rq'} + R_r i_{rq'} &= 0 \\
\dot{\delta} &= \omega \\
J\dot{\omega} &= T_e - T_L
\end{aligned} \tag{1.8}
$$

in which

$$
\begin{bmatrix} i_{rd'} \\ i_{rq'} \end{bmatrix} = \begin{bmatrix} \cos\delta & \sin\delta \\ -\sin\delta & \cos\delta \end{bmatrix} \begin{bmatrix} i_{ra} \\ i_{rb} \end{bmatrix}
$$
$$
\begin{bmatrix} \psi_{rd'} \\ \psi_{rq'} \end{bmatrix} = \begin{bmatrix} \cos\delta & \sin\delta \\ -\sin\delta & \cos\delta \end{bmatrix} \begin{bmatrix} \psi_{ra} \\ \psi_{rb} \end{bmatrix}
$$

with (ψ_{ra}, ψ_{rb}) and (i_{ra}, i_{rb}) denoting the (a,b)-components of the rotor flux and the rotor current in the fixed (a,b) frame. Since

$$\begin{aligned}\begin{bmatrix}\dot\psi_{ra}\\ \dot\psi_{rb}\end{bmatrix} &= \begin{bmatrix}\cos\delta & -\sin\delta\\ \sin\delta & \cos\delta\end{bmatrix}\begin{bmatrix}\dot\psi_{rd'}\\ \dot\psi_{rq'}\end{bmatrix} + \omega\begin{bmatrix}-\sin\delta & -\cos\delta\\ \cos\delta & -\sin\delta\end{bmatrix}\begin{bmatrix}\psi_{rd'}\\ \psi_{rq'}\end{bmatrix}\\ &= -\begin{bmatrix}R_r & 0\\ 0 & R_r\end{bmatrix}\begin{bmatrix}i_{ra}\\ i_{rb}\end{bmatrix} + \begin{bmatrix}-\omega\psi_{rb}\\ \omega\psi_{ra}\end{bmatrix}\end{aligned} \tag{1.9}$$

we can write in (a,b) coordinates

$$\begin{aligned}\begin{bmatrix}\dot\psi_{sa}\\ \dot\psi_{sb}\end{bmatrix} + \begin{bmatrix}R_s & 0\\ 0 & R_s\end{bmatrix}\begin{bmatrix}i_{sa}\\ i_{sb}\end{bmatrix} &= \begin{bmatrix}u_{sa}\\ u_{sb}\end{bmatrix}\\ \begin{bmatrix}\dot\psi_{ra}\\ \dot\psi_{rb}\end{bmatrix} + \begin{bmatrix}R_r & 0\\ 0 & R_r\end{bmatrix}\begin{bmatrix}i_{ra}\\ i_{rb}\end{bmatrix} + \begin{bmatrix}0 & \omega\\ -\omega & 0\end{bmatrix}\begin{bmatrix}\psi_{ra}\\ \psi_{rb}\end{bmatrix} &= \begin{bmatrix}0\\ 0\end{bmatrix}.\end{aligned} \tag{1.10}$$

From (1.2), (1.4), (1.6), and (1.7), the electromagnetic equations are

$$\begin{bmatrix}\psi_{sa}\\ \psi_{sb}\\ \psi_{ra}\\ \psi_{rb}\end{bmatrix} = \begin{bmatrix}L_s & 0 & M & 0\\ 0 & L_s & 0 & M\\ M & 0 & L_r & 0\\ 0 & M & 0 & L_r\end{bmatrix}\begin{bmatrix}i_{sa}\\ i_{sb}\\ i_{ra}\\ i_{rb}\end{bmatrix} \triangleq L\begin{bmatrix}i_{sa}\\ i_{sb}\\ i_{ra}\\ i_{rb}\end{bmatrix}. \tag{1.11}$$

The matrix L is positive definite, *i.e.* the quadratic form associated to L

$$\frac{1}{2}i^{\mathrm{T}}Li \tag{1.12}$$

is positive for any nonzero value of the current vector $i = [i_{sa}, i_{sb}, i_{ra}, i_{rb}]^{\mathrm{T}}$: (1.12) represents the magnetic energy. This implies that $L_sL_r > M^2$. Define

$$R \triangleq \begin{bmatrix}R_s & 0 & 0 & 0\\ 0 & R_s & 0 & 0\\ 0 & 0 & R_r & 0\\ 0 & 0 & 0 & R_r\end{bmatrix} \tag{1.13}$$

so that, if we eliminate $(\psi_{sa}, \psi_{sb}, \psi_{ra}, \psi_{rb})$ in (1.10) by using (1.11), we obtain the first state space model

$$\begin{aligned}L\begin{bmatrix}\frac{\mathrm{d}i_{sa}}{\mathrm{d}t}\\ \frac{\mathrm{d}i_{sb}}{\mathrm{d}t}\\ \frac{\mathrm{d}i_{ra}}{\mathrm{d}t}\\ \frac{\mathrm{d}i_{rb}}{\mathrm{d}t}\end{bmatrix} &= -R\begin{bmatrix}i_{sa}\\ i_{sb}\\ i_{ra}\\ i_{rb}\end{bmatrix} - \begin{bmatrix}0 & 0 & 0 & 0\\ 0 & 0 & 0 & 0\\ 0 & \omega M & 0 & \omega L_r\\ -\omega M & 0 & -\omega L_r & 0\end{bmatrix}\begin{bmatrix}i_{sa}\\ i_{sb}\\ i_{ra}\\ i_{rb}\end{bmatrix} + \begin{bmatrix}u_{sa}\\ u_{sb}\\ 0\\ 0\end{bmatrix}\\ J\frac{\mathrm{d}\omega}{\mathrm{d}t} &= T_e - T_L\end{aligned} \tag{1.14}$$

in which the state variables are $(i_{sa}, i_{sb}, i_{ra}, i_{rb}, \omega)$ and the electromagnetic torque T_e is still to be determined as a function of the state variables. This choice of state variables is naturally linked to the energy stored in the motor given by

$$E = \frac{1}{2} i^{\mathrm{T}} L i + \frac{1}{2} J \omega^2 \tag{1.15}$$

which is the sum of the magnetic energy (1.12) and of the kinetic energy

$$\frac{1}{2} J \omega^2 . \tag{1.16}$$

The expression of T_e can be obtained from the energy balance

$$\begin{aligned} \frac{\mathrm{d}E}{\mathrm{d}t} &= [i_{sa}, i_{sb}][u_{sa}, u_{sb}]^{\mathrm{T}} - T_L \omega - i^{\mathrm{T}} R i \\ &+ \omega [i_{ra}, i_{rb}] \begin{bmatrix} 0 & -M \\ M & 0 \end{bmatrix} [i_{sa}, i_{sb}]^{\mathrm{T}} + T_e \omega . \end{aligned} \tag{1.17}$$

Let P_{in}, P_{out}, and P_{loss} denote the input power, the output power, and the power losses, respectively; since

$$\begin{aligned} \frac{\mathrm{d}E}{\mathrm{d}t} &= P_{in} - P_{out} - P_{loss} \\ &= [i_{sa}, i_{sb}][u_{sa}, u_{sb}]^{\mathrm{T}} - T_L \omega - i^{\mathrm{T}} R i , \end{aligned} \tag{1.18}$$

comparing (1.17) with (1.18) it follows that

$$T_e = [i_{ra}, i_{rb}] \begin{bmatrix} 0 & -M \\ M & 0 \end{bmatrix} [i_{sa}, i_{sb}]^{\mathrm{T}} = M(i_{ra} i_{sb} - i_{rb} i_{sa}) . \tag{1.19}$$

Note that the electromagnetic torque T_e produced by the motor is a nonlinear function of the state variables and constitutes the main nonlinear term in the induction motor model. Since from (1.11)

$$\begin{aligned} i_{ra} &= -\frac{M}{L_r} i_{sa} + \frac{1}{L_r} \psi_{ra} \\ i_{rb} &= -\frac{M}{L_r} i_{sb} + \frac{1}{L_r} \psi_{rb} \end{aligned} \tag{1.20}$$

the electromagnetic torque T_e can also be expressed as

$$T_e = \frac{M}{L_r} (\psi_{ra} i_{sb} - \psi_{rb} i_{sa}) . \tag{1.21}$$

From (1.14) and (1.19) we obtain the overall state space model in terms of the state variables $(i_{sa}, i_{sb}, i_{ra}, i_{rb}, \omega)$ and the input variables (u_{sa}, u_{sb}, T_L)

$$B \begin{bmatrix} \frac{di_{sa}}{dt} \\ \frac{di_{sb}}{dt} \\ \frac{di_{ra}}{dt} \\ \frac{di_{rb}}{dt} \\ \frac{d\omega}{dt} \end{bmatrix} + (K+C) \begin{bmatrix} i_{sa} \\ i_{sb} \\ i_{ra} \\ i_{rb} \\ \omega \end{bmatrix} = \begin{bmatrix} u_{sa} \\ u_{sb} \\ 0 \\ 0 \\ -T_L \end{bmatrix} \tag{1.22}$$

with

$$B = \begin{bmatrix} L_s & 0 & M & 0 & 0 \\ 0 & L_s & 0 & M & 0 \\ M & 0 & L_r & 0 & 0 \\ 0 & M & 0 & L_r & 0 \\ 0 & 0 & 0 & 0 & J \end{bmatrix}, \qquad K = \begin{bmatrix} R_s & 0 & 0 & 0 & 0 \\ 0 & R_s & 0 & 0 & 0 \\ 0 & 0 & R_r & 0 & 0 \\ 0 & 0 & 0 & R_r & 0 \\ 0 & 0 & 0 & 0 & 0 \end{bmatrix},$$

$$C = \begin{bmatrix} 0 & 0 & 0 & 0 & 0 \\ 0 & 0 & 0 & 0 & 0 \\ 0 & 0 & 0 & \omega L_r & M i_{sb} \\ 0 & 0 & -\omega L_r & 0 & -M i_{sa} \\ 0 & 0 & -M i_{sb} & M i_{sa} & 0 \end{bmatrix}. \tag{1.23}$$

Note that the matrix C is skew-symmetric, *i.e.* $C = -C^{\mathrm{T}}$. If we differentiate with respect to time the total energy (1.15), which can also be expressed as

$$E = \frac{1}{2}[i^{\mathrm{T}}, \omega] B \begin{bmatrix} i \\ \omega \end{bmatrix}$$

we reobtain (1.18)

$$\begin{aligned} \frac{dE}{dt} &= -[i^{\mathrm{T}}, \omega] K \begin{bmatrix} i \\ \omega \end{bmatrix} + i_{sa} u_{sa} + i_{sb} u_{sb} - \omega T_L \\ &= -P_{loss} + P_{in} - P_{out} \end{aligned}$$

since $C = -C^{\mathrm{T}}$. The model (1.22) is very advantageous to analyze the energy balance: for this reason the model (1.22) will be referred to as the energy model. In fact, integrating with respect to time the power balance (1.18), we obtain the total energy balance from an initial time t_0 to the time t:

$$\begin{aligned} & E(t) - E(t_0) + \int_{t_0}^{t} i(\tau)^{\mathrm{T}} R i(\tau) d\tau \\ &= \frac{1}{2} i(t)^{\mathrm{T}} L i(t) - \frac{1}{2} i(t_0)^{\mathrm{T}} L i(t_0) + \frac{1}{2} J \omega^2(t) - \frac{1}{2} J \omega^2(t_0) + \int_{t_0}^{t} i(\tau)^{\mathrm{T}} R i(\tau) d\tau \\ &= \int_{t_0}^{t} [i_{sa}(\tau) u_{sa}(\tau) + i_{sb}(\tau) u_{sb}(\tau)] d\tau - \int_{t_0}^{t} T_L \omega(\tau) d\tau . \end{aligned}$$

Eliminating $(\psi_{sa}, \psi_{sb}, i_{ra}, i_{rb})$ in (1.10) by using (1.11), namely (recall also (1.20))

$$\begin{bmatrix} i_{ra} \\ i_{rb} \end{bmatrix} = \begin{bmatrix} -\frac{M}{L_r} & 0 \\ 0 & -\frac{M}{L_r} \end{bmatrix} \begin{bmatrix} i_{sa} \\ i_{sb} \end{bmatrix} + \begin{bmatrix} \frac{1}{L_r} & 0 \\ 0 & \frac{1}{L_r} \end{bmatrix} \begin{bmatrix} \psi_{ra} \\ \psi_{rb} \end{bmatrix} \tag{1.24}$$

$$\begin{bmatrix} \psi_{sa} \\ \psi_{sb} \end{bmatrix} = \begin{bmatrix} L_s - \frac{M^2}{L_r} & 0 \\ 0 & L_s - \frac{M^2}{L_r} \end{bmatrix} \begin{bmatrix} i_{sa} \\ i_{sb} \end{bmatrix} + \begin{bmatrix} \frac{M}{L_r} & 0 \\ 0 & \frac{M}{L_r} \end{bmatrix} \begin{bmatrix} \psi_{ra} \\ \psi_{rb} \end{bmatrix} \tag{1.25}$$

we obtain a state space model in terms of the state variables $(\omega, \psi_{ra}, \psi_{rb}, i_{sa}, i_{sb})$ and the input variables (u_{sa}, u_{sb}, T_L)

$$\begin{aligned} \frac{d\omega}{dt} &= \mu\left(\psi_{ra} i_{sb} - \psi_{rb} i_{sa}\right) - \frac{T_L}{J} \\ \frac{d\psi_{ra}}{dt} &= -\alpha\psi_{ra} - \omega\psi_{rb} + \alpha M i_{sa} \\ \frac{d\psi_{rb}}{dt} &= -\alpha\psi_{rb} + \omega\psi_{ra} + \alpha M i_{sb} \\ \frac{di_{sa}}{dt} &= -\gamma i_{sa} + \frac{u_{sa}}{\sigma} + \beta\alpha\psi_{ra} + \beta\omega\psi_{rb} \\ \frac{di_{sb}}{dt} &= -\gamma i_{sb} + \frac{u_{sb}}{\sigma} + \beta\alpha\psi_{rb} - \beta\omega\psi_{ra} \end{aligned} \tag{1.26}$$

in which the following reparameterization is used:

$$\begin{aligned} \mu &= \frac{M}{JL_r} \\ \alpha &= \frac{R_r}{L_r} \\ \sigma &= L_s\left(1 - \frac{M^2}{L_s Lr}\right) \\ \beta &= \frac{M}{\sigma L_r} \\ \gamma &= \frac{R_s}{\sigma} + \beta\alpha M\,. \end{aligned} \tag{1.27}$$

Note that since $\sigma > 0$, all the above parameters are greater than zero. From (1.26) it follows that

$$\begin{aligned} \frac{di_{sa}}{dt} &= -\frac{R_s}{\sigma} i_{sa} + \frac{u_{sa}}{\sigma} - \beta\frac{d\psi_{ra}}{dt} \\ \frac{di_{sb}}{dt} &= -\frac{R_s}{\sigma} i_{sb} + \frac{u_{sb}}{\sigma} - \beta\frac{d\psi_{rb}}{dt}\,. \end{aligned} \tag{1.28}$$

The state space model (1.26), which will be referred to as the fixed frame model, has some advantages from the control view point since it clarifies that the control inputs (u_{sa}, u_{sb}) directly affect the dynamics of the stator currents (i_{sa}, i_{sb}) which can be viewed as intermediate control variables since they control the rotor speed ω and the rotor flux modulus $\sqrt{\psi_{ra}^2 + \psi_{rb}^2}$ in the first three equations in (1.26). Note that the model (1.26) is highly nonlinear due to the expression of the produced electromagnetic torque $T_e = \mu(\psi_{ra}i_{sb} - \psi_{rb}i_{sa})$ and to the products $\omega\psi_{rb}$ and $\omega\psi_{ra}$ appearing in the last four equations in (1.26) which are originated by the rotation of the rotor at speed $\omega(t)$, according to (1.9).

Let us now introduce a time-varying (d,q) frame which rotates at an arbitrary speed $\omega_0(t)$ and is identified at each time t by the angle $\varepsilon_0(t)$ so that

$$\frac{\mathrm{d}\varepsilon_0}{\mathrm{d}t} = \omega_0 \tag{1.29}$$

with $\varepsilon_0(0)$ an arbitrary initial condition. Rotor fluxes (ψ_{ra}, ψ_{rb}), stator currents

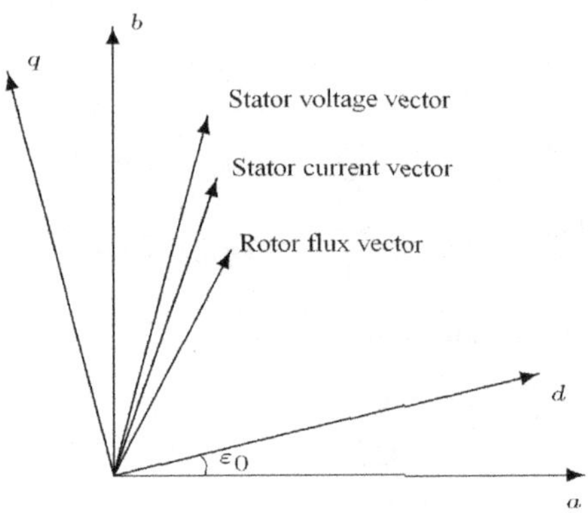

Fig. 1.2 (d,q) reference frame for the rotating frame model

(i_{sa}, i_{sb}), and stator voltages (u_{sa}, u_{sb}) are expressed with respect to the time-varying rotating (d,q) frame as (see Figure 1.2)

$$\begin{aligned}
\begin{bmatrix} \psi_{rd} \\ \psi_{rq} \end{bmatrix} &= \begin{bmatrix} \cos\varepsilon_0 & \sin\varepsilon_0 \\ -\sin\varepsilon_0 & \cos\varepsilon_0 \end{bmatrix} \begin{bmatrix} \psi_{ra} \\ \psi_{rb} \end{bmatrix} \\
\begin{bmatrix} i_{sd} \\ i_{sq} \end{bmatrix} &= \begin{bmatrix} \cos\varepsilon_0 & \sin\varepsilon_0 \\ -\sin\varepsilon_0 & \cos\varepsilon_0 \end{bmatrix} \begin{bmatrix} i_{sa} \\ i_{sb} \end{bmatrix} \\
\begin{bmatrix} u_{sd} \\ u_{sq} \end{bmatrix} &= \begin{bmatrix} \cos\varepsilon_0 & \sin\varepsilon_0 \\ -\sin\varepsilon_0 & \cos\varepsilon_0 \end{bmatrix} \begin{bmatrix} u_{sa} \\ u_{sb} \end{bmatrix} .
\end{aligned} \tag{1.30}$$

If the new state variables $(\omega, \psi_{rd}, \psi_{rq}, i_{sd}, i_{sq})$ and input variables (u_{sd}, u_{sq}, T_L) are used, in the new (d,q) rotating coordinates the induction motor model (1.26) becomes

$$\begin{aligned}
\frac{\mathrm{d}\omega}{\mathrm{d}t} &= \mu\left(\psi_{rd}i_{sq} - \psi_{rq}i_{sd}\right) - \frac{T_L}{J} \\
\frac{\mathrm{d}\psi_{rd}}{\mathrm{d}t} &= -\alpha\psi_{rd} + (\omega_0 - \omega)\,\psi_{rq} + \alpha M i_{sd}
\end{aligned}$$

$$\begin{aligned}
\frac{\mathrm{d}\psi_{rq}}{\mathrm{d}t} &= -\alpha\psi_{rq} - (\omega_0 - \omega)\,\psi_{rd} + \alpha M i_{sq} \\
\frac{\mathrm{d}i_{sd}}{\mathrm{d}t} &= -\gamma i_{sd} + \omega_0 i_{sq} + \beta\alpha\psi_{rd} + \beta\omega\psi_{rq} + \frac{u_{sd}}{\sigma} \\
\frac{\mathrm{d}i_{sq}}{\mathrm{d}t} &= -\gamma i_{sq} - \omega_0 i_{sd} + \beta\alpha\psi_{rq} - \beta\omega\psi_{rd} + \frac{u_{sq}}{\sigma}
\end{aligned} \tag{1.31}$$

which generalizes the fixed frame model (1.26) since (1.31) becomes (1.26) in the special case in which the speed ω_0 of the rotating coordinate frame is zero along with the initial angle $\varepsilon_0(0)$ (*i.e.* $\omega_0 = \varepsilon_0(0) = 0$ in (1.31)). The state space model (1.31) will be referred to as the rotating frame model.

In the third equation in (1.31) ω_0 can be freely chosen to our advantage. If we set, assuming $\psi_{rd} \neq 0$,

$$\omega_0 = \omega + \frac{\alpha M i_{sq}}{\psi_{rd}} \tag{1.32}$$

the third equation in (1.31) becomes

$$\frac{\mathrm{d}\psi_{rq}}{\mathrm{d}t} = -\alpha\psi_{rq} \tag{1.33}$$

which implies, since $\alpha > 0$, that $\psi_{rq}(t)$ tends exponentially to zero for any initial condition $\psi_{rq}(0)$, *i.e.*

$$\psi_{rq}(t) = \mathrm{e}^{-\alpha t}\psi_{rq}(0)\,. \tag{1.34}$$

If $\psi_{rq}(0) = 0$ then $\psi_{rq}(t) = 0$ for every $t \geq 0$. Equations (1.29) and (1.32) give

$$\frac{\mathrm{d}\varepsilon_0}{\mathrm{d}t} = \omega + \frac{\alpha M i_{sq}}{\psi_{rd}}\,. \tag{1.35}$$

Substituting (1.32) in (1.31) we obtain

$$\begin{aligned}
\frac{\mathrm{d}\omega}{\mathrm{d}t} &= \mu\,(\psi_{rd} i_{sq} - \psi_{rq} i_{sd}) - \frac{T_L}{J} \\
\frac{\mathrm{d}\psi_{rd}}{\mathrm{d}t} &= -\alpha\psi_{rd} + \frac{\alpha M i_{sq}}{\psi_{rd}}\psi_{rq} + \alpha M i_{sd} \\
\frac{\mathrm{d}\psi_{rq}}{\mathrm{d}t} &= -\alpha\psi_{rq} \\
\frac{\mathrm{d}i_{sd}}{\mathrm{d}t} &= -\gamma i_{sd} + \omega i_{sq} + \frac{\alpha M i_{sq}^2}{\psi_{rd}} + \beta\alpha\psi_{rd} + \beta\omega\psi_{rq} + \frac{u_{sd}}{\sigma} \\
\frac{\mathrm{d}i_{sq}}{\mathrm{d}t} &= -\gamma i_{sq} - \omega i_{sd} - \frac{\alpha M i_{sq} i_{sd}}{\psi_{rd}} + \beta\alpha\psi_{rq} - \beta\omega\psi_{rd} + \frac{u_{sq}}{\sigma}\,.
\end{aligned} \tag{1.36}$$

The model (1.36) is a special case of the rotating frame model (1.31) in which ω_0 is chosen according to (1.32). If $\psi_{rq}(0) = 0$ and consequently, according to (1.34),

$\psi_{rq}(t) = 0$ for every $t \geq 0$, then the (d,q) frame rotates so that the direct axis coincides with the rotor flux vector and ε_0 coincides with the angle ρ between the flux vector and the a-axis (see Figure 1.3), that is

$$\begin{aligned} \psi_{ra} &= \psi_{rd} \cos \rho \\ \psi_{rb} &= \psi_{rd} \sin \rho \end{aligned} \tag{1.37}$$

with

$$\begin{aligned} \psi_{rd} &= \sqrt{\psi_{ra}^2 + \psi_{rb}^2} \\ \rho &= \arctan\left(\frac{\psi_{rb}}{\psi_{ra}}\right) . \end{aligned} \tag{1.38}$$

In this case, *i.e.* when $\psi_{rq}(0) = 0$ or equivalently $\varepsilon_0 = \rho$, equations (1.36) become

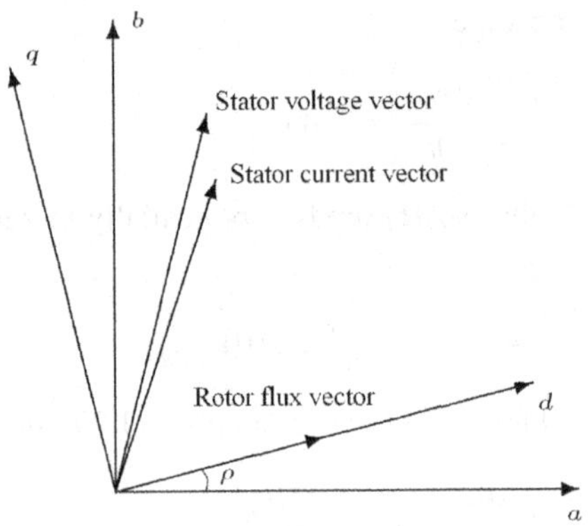

Fig. 1.3 (d,q) reference frame when $\varepsilon_0 = \rho$ for the field-oriented model

$$\begin{aligned} \frac{d\omega}{dt} &= \mu \psi_{rd} i_{sq} - \frac{T_L}{J} \\ \frac{d\psi_{rd}}{dt} &= -\alpha \psi_{rd} + \alpha M i_{sd} \\ \frac{d\rho}{dt} &= \omega + \frac{\alpha M i_{sq}}{\psi_{rd}} \\ \frac{di_{sd}}{dt} &= -\gamma i_{sd} + \omega i_{sq} + \frac{\alpha M i_{sq}^2}{\psi_{rd}} + \beta \alpha \psi_{rd} + \frac{u_{sd}}{\sigma} \\ \frac{di_{sq}}{dt} &= -\gamma i_{sq} - \omega i_{sd} - \frac{\alpha M i_{sq} i_{sd}}{\psi_{rd}} - \beta \omega \psi_{rd} + \frac{u_{sq}}{\sigma} \end{aligned} \tag{1.39}$$

which constitutes a motor state space model with state variables $(\omega, \psi_{rd}, \rho, i_{sd}, i_{sq})$ and input variables (u_{sd}, u_{sq}, T_L). The model (1.39) will be referred to as field-oriented model. Several important comments on the field-oriented model (1.39) are in order:

1. The difference between the rotor flux speed of rotation $\dot{\rho}$ and the rotor speed ω is equal to $\frac{\alpha M i_{sq}}{\psi_{rd}}$: it is usually called slip speed ω_s and is expressed as follows:

$$\omega_s = \dot{\rho} - \omega = \frac{\alpha M i_{sq}}{\psi_{rd}} = \frac{\alpha M T_e}{\mu \psi_{rd}^2}$$
$$= \frac{R_r M(\psi_{ra} i_{sb} - \psi_{rb} i_{sa})}{L_r(\psi_{ra}^2 + \psi_{rb}^2)}; \tag{1.40}$$

   it is proportional to the electromagnetic torque T_e and inversely proportional to the the flux modulus squared: the smaller the flux modulus, the larger the flux speed of rotation $\dot{\rho}$ while the larger the electromagnetic torque T_e, the larger the flux speed of rotation $\dot{\rho}$.
2. No matter how $\varepsilon_0(0)$, or equivalently $\psi_{rq}(0)$, is chosen, (1.39) describes the limiting behavior of (1.36) as t goes to infinity according to (1.34).
3. The field-oriented model (1.39) is an equivalent (except at $\psi_{ra} = \psi_{rb} = 0$) description of the fixed frame model (1.26) in the new state variables

$$\begin{aligned}
\omega &= \omega \\
\psi_{rd} &= \sqrt{\psi_{ra}^2 + \psi_{rb}^2} \\
\rho &= \arctan\left(\frac{\psi_{rb}}{\psi_{ra}}\right) \\
i_{sd} &= \frac{\psi_{ra} i_{sa} + \psi_{rb} i_{sb}}{\sqrt{\psi_{ra}^2 + \psi_{rb}^2}} = i_{sa}\cos\rho + i_{sb}\sin\rho \\
i_{sq} &= \frac{\psi_{ra} i_{sb} - \psi_{rb} i_{sa}}{\sqrt{\psi_{ra}^2 + \psi_{rb}^2}} = -i_{sa}\sin\rho + i_{sb}\cos\rho
\end{aligned} \tag{1.41}$$

   and new control input coordinates

$$\begin{aligned}
u_{sd} &= \frac{\psi_{ra} u_{sa} + \psi_{rb} u_{sb}}{\sqrt{\psi_{ra}^2 + \psi_{rb}^2}} = u_{sa}\cos\rho + u_{sb}\sin\rho \\
u_{sq} &= \frac{\psi_{ra} u_{sb} - \psi_{rb} u_{sa}}{\sqrt{\psi_{ra}^2 + \psi_{rb}^2}} = -u_{sa}\sin\rho + u_{sb}\cos\rho\,.
\end{aligned} \tag{1.42}$$

4. The field-oriented model (1.39) is the most advantageous from the control view point since the control inputs (u_{sd}, u_{sq}) directly affect the currents (i_{sd}, i_{sq}) dynamics only; the stator current vector (i_{sd}, i_{sq}) can be viewed as an intermediate control vector in the reduced order model

$$\frac{d\omega}{dt} = \mu \psi_{rd} i_{sq} - \frac{T_L}{J}$$
$$\frac{d\psi_{rd}}{dt} = -\alpha \psi_{rd} + \alpha M i_{sd}$$
$$\frac{d\rho}{dt} = \omega + \frac{\alpha M i_{sq}}{\psi_{rd}}$$

which is often called current-fed model. This model clarifies that the direct current component i_{sd} (which is sometimes called excitation current) is solely and directly responsible for the rotor flux modulus ψ_{rd} dynamics and indirectly responsible for the electromagnetic torque production through ψ_{rd} itself, while the quadrature current component i_{sq} is directly responsible for the electromagnetic torque production and therefore for the rotor speed ω dynamics.

5. As we shall see in the next section, the field-oriented model (1.39) allows us to determine immediately the sinusoidal steady-state operating condition corresponding to a constant speed ω^* and to a constant flux modulus $\psi^* = \sqrt{\psi_{ra}^2 + \psi_{rb}^2}$. Given ω^*, ψ^*, and T_L, if we set in (1.39) the time derivatives equal to zero and solve for $(i_{sd}^*, i_{sq}^*, u_{sd}^*, u_{sq}^*)$ the right-hand side in (1.39), we easily obtain (i_{sd}^*, i_{sq}^*) from the first two equations in (1.39) and (u_{sd}^*, u_{sq}^*) from the last two equations in (1.39), while the third equation gives the speed of rotation $\omega^* + \alpha M i_{sq}^* / \psi^*$ for the operating condition.

In conclusion, four state space models have been determined for the balanced, unsaturated induction motor:

1. The *energy model*

$$B \begin{bmatrix} \frac{di_{sa}}{dt} \\ \frac{di_{sb}}{dt} \\ \frac{di_{ra}}{dt} \\ \frac{di_{rb}}{dt} \\ \frac{d\omega}{dt} \end{bmatrix} + (K + C) \begin{bmatrix} i_{sa} \\ i_{sb} \\ i_{ra} \\ i_{rb} \\ \omega \end{bmatrix} = \begin{bmatrix} u_{sa} \\ u_{sb} \\ 0 \\ 0 \\ -T_L \end{bmatrix}$$

with

$$B = \begin{bmatrix} L_s & 0 & M & 0 & 0 \\ 0 & L_s & 0 & M & 0 \\ M & 0 & L_r & 0 & 0 \\ 0 & M & 0 & L_r & 0 \\ 0 & 0 & 0 & 0 & J \end{bmatrix}, \quad K = \begin{bmatrix} R_s & 0 & 0 & 0 & 0 \\ 0 & R_s & 0 & 0 & 0 \\ 0 & 0 & R_r & 0 & 0 \\ 0 & 0 & 0 & R_r & 0 \\ 0 & 0 & 0 & 0 & 0 \end{bmatrix},$$

$$C = \begin{bmatrix} 0 & 0 & 0 & 0 & 0 \\ 0 & 0 & 0 & 0 & 0 \\ 0 & 0 & 0 & \omega L_r & M i_{sb} \\ 0 & 0 & -\omega L_r & 0 & -M i_{sa} \\ 0 & 0 & -M i_{sb} & M i_{sa} & 0 \end{bmatrix}$$

in which $(i_{sa}, i_{sb}, i_{ra}, i_{rb}, \omega)$ are the state variables and (u_{sa}, u_{sb}) are the control inputs.

2. The *fixed frame model*

$$
\begin{aligned}
\frac{d\omega}{dt} &= \mu\left(\psi_{ra} i_{sb} - \psi_{rb} i_{sa}\right) - \frac{T_L}{J} \\
\frac{d\psi_{ra}}{dt} &= -\alpha\psi_{ra} - \omega\psi_{rb} + \alpha M i_{sa} \\
\frac{d\psi_{rb}}{dt} &= -\alpha\psi_{rb} + \omega\psi_{ra} + \alpha M i_{sb} \\
\frac{di_{sa}}{dt} &= -\gamma i_{sa} + \frac{u_{sa}}{\sigma} + \beta\alpha\psi_{ra} + \beta\omega\psi_{rb} \\
\frac{di_{sb}}{dt} &= -\gamma i_{sb} + \frac{u_{sb}}{\sigma} + \beta\alpha\psi_{rb} - \beta\omega\psi_{ra}
\end{aligned}
$$

in which $(\omega, \psi_{ra}, \psi_{rb}, i_{sa}, i_{sb})$ are the state variables and (u_{sa}, u_{sb}) are the control inputs.

3. The *rotating frame model*

$$
\begin{aligned}
\frac{d\omega}{dt} &= \mu\left(\psi_{rd} i_{sq} - \psi_{rq} i_{sd}\right) - \frac{T_L}{J} \\
\frac{d\psi_{rd}}{dt} &= -\alpha\psi_{rd} + (\omega_0 - \omega)\,\psi_{rq} + \alpha M i_{sd} \\
\frac{d\psi_{rq}}{dt} &= -\alpha\psi_{rq} - (\omega_0 - \omega)\,\psi_{rd} + \alpha M i_{sq} \\
\frac{di_{sd}}{dt} &= -\gamma i_{sd} + \omega_0 i_{sq} + \beta\alpha\psi_{rd} + \beta\omega\psi_{rq} + \frac{u_{sd}}{\sigma} \\
\frac{di_{sq}}{dt} &= -\gamma i_{sq} - \omega_0 i_{sd} + \beta\alpha\psi_{rq} - \beta\omega\psi_{rd} + \frac{u_{sq}}{\sigma}
\end{aligned}
$$

in which $(\omega, \psi_{rd}, \psi_{rq}, i_{sd}, i_{sq})$ are the state variables, (u_{sd}, u_{sq}) are the control inputs, and ω_0 is the speed of the rotating coordinate frame.

4. The *field-oriented model* ($\psi_{rd} > 0$)

$$
\begin{aligned}
\frac{d\omega}{dt} &= \mu\psi_{rd} i_{sq} - \frac{T_L}{J} \\
\frac{d\psi_{rd}}{dt} &= -\alpha\psi_{rd} + \alpha M i_{sd} \\
\frac{d\rho}{dt} &= \omega + \frac{\alpha M i_{sq}}{\psi_{rd}} \\
\frac{di_{sd}}{dt} &= -\gamma i_{sd} + \omega i_{sq} + \frac{\alpha M i_{sq}^2}{\psi_{rd}} + \beta\alpha\psi_{rd} + \frac{u_{sd}}{\sigma} \\
\frac{di_{sq}}{dt} &= -\gamma i_{sq} - \omega i_{sd} - \frac{\alpha M i_{sq} i_{sd}}{\psi_{rd}} - \beta\omega\psi_{rd} + \frac{u_{sq}}{\sigma}
\end{aligned}
$$

in which $(\omega, \psi_{rd}, \rho, i_{sd}, i_{sq})$ are the state variables, (u_{sd}, u_{sq}) are the control inputs and $\dot{\rho}$ is the speed of the rotating coordinate frame.

Throughout this book all numerical simulations are performed for a low power induction motor whose parameters are provided by the manufacturer and are given in Table 1.1. A detailed description of the experimental set-up, including the motor, is given in Section 1.8 where simulated and experimental data are compared.

Table 1.1 Nominal motor parameters

Motor-load inertia	$J = 0.0075\,\mathrm{Kgm}^2$
Stator resistance	$R_s = 5.3\,\Omega$
Rotor resistance	$R_r = 3.3\,\Omega$
Stator inductance	$L_s = 0.365\,\mathrm{H}$
Rotor inductance	$L_r = 0.375\,\mathrm{H}$
Mutual inductance	$M = 0.34\,\mathrm{H}$

1.3 Steady-state Operating Conditions with Sinusoidal Voltages

Let us first determine the constant vectors $(\psi^*, 0)$, (i_{sd}^*, i_{sq}^*), (u_{sd}^*, u_{sq}^*) which rotate at constant speed

$$\dot{\rho}^* = \omega^* + \frac{\alpha M i_{sq}^*}{\psi^*} = \omega^* + \omega_s^* \tag{1.43}$$

when the rotor speed ω^* is constant and constitute a steady-state solution for the field-oriented model (1.39) if $\rho^*(0) = \rho(0)$, *i.e.* they satisfy (1.43) and

$$\begin{aligned} \mu \psi^* i_{sq}^* - \frac{T_L}{J} &= 0 \\ -\alpha \psi^* + \alpha M i_{sd}^* &= 0 \\ -\gamma i_{sd}^* + \omega^* i_{sq}^* + \frac{\alpha M i_{sq}^{*2}}{\psi^*} + \beta \alpha \psi^* + \frac{u_{sd}^*}{\sigma} &= 0 \\ -\gamma i_{sq}^* - \omega^* i_{sd}^* - \frac{\alpha M i_{sq}^* i_{sd}^*}{\psi^*} - \beta \omega^* \psi^* + \frac{u_{sq}^*}{\sigma} &= 0\,. \end{aligned} \tag{1.44}$$

Solving for (i_{sd}^*, i_{sq}^*) the first two equations in (1.44) we obtain

$$i_{sd}^* = \frac{\psi^*}{M}$$

$$i_{sq}^* = \frac{T_L}{J\mu\psi^*} \tag{1.45}$$

so that the steady-state slip speed is from (1.43)

$$\omega_s^* = \frac{\alpha M T_L}{J\mu\psi^{*2}} = \frac{R_r T_L}{\psi^{*2}}.$$

Note that the steady-state slip speed ω_s^* depends on ψ^* and on two critical parameters R_r and T_L only, while the rotation speed of the steady-state operating condition is $\omega^* + R_r T_L/\psi^{*2}$ and depends on the references (ω^*, ψ^*) and on the parameters (R_r, T_L). Substituting (1.45) in the last two equations in (1.44) we obtain (recall that $\mu J = \frac{M}{L_r}$ and $\gamma = \frac{R_s}{\sigma} + \alpha\beta M$)

$$\begin{aligned} u_{sd}^* &= \frac{R_s}{M}\psi^* - \frac{\sigma T_L \omega^*}{J\mu\psi^*} - \frac{\sigma\alpha M T_L^2}{J^2\mu^2\psi^{*3}} \\ u_{sq}^* &= \frac{\sigma(\gamma+\alpha)T_L}{J\mu\psi^*} + \sigma\left(\frac{1}{M}+\beta\right)\omega^*\psi^*. \end{aligned} \tag{1.46}$$

Hence the steady-state, time-varying operating conditions in fixed coordinates for (1.26) are given by

$$\begin{aligned} \begin{bmatrix} \psi_{ra}^* \\ \psi_{rb}^* \end{bmatrix} &= \begin{bmatrix} \cos\rho^* & -\sin\rho^* \\ \sin\rho^* & \cos\rho^* \end{bmatrix} \begin{bmatrix} \psi^* \\ 0 \end{bmatrix} \\ \begin{bmatrix} i_{sa}^* \\ i_{sb}^* \end{bmatrix} &= \begin{bmatrix} \cos\rho^* & -\sin\rho^* \\ \sin\rho^* & \cos\rho^* \end{bmatrix} \begin{bmatrix} \frac{\psi^*}{M} \\ \frac{T_L}{J\mu\psi^*} \end{bmatrix} \\ \begin{bmatrix} u_{sa}^* \\ u_{sb}^* \end{bmatrix} &= \begin{bmatrix} \cos\rho^* & -\sin\rho^* \\ \sin\rho^* & \cos\rho^* \end{bmatrix} \begin{bmatrix} \frac{R_s}{M}\psi^* - \frac{\sigma T_L\omega^*}{J\mu\psi^*} - \frac{\sigma\alpha M T_L^2}{J^2\mu^2\psi^{*3}} \\ \frac{\sigma(\gamma+\alpha)T_L}{J\mu\psi^*} + \sigma\left(\frac{1}{M}+\beta\right)\omega^*\psi^* \end{bmatrix} \\ \dot\rho^* &= \omega^* + \frac{\alpha M T_L}{J\mu\psi^{*2}} = \omega^* + \frac{R_r T_L}{\psi^{*2}} = \omega^* + \omega_s^* \\ \rho^*(0) &= \arctan\left(\frac{\psi_{rb}(0)}{\psi_{ra}(0)}\right). \end{aligned} \tag{1.47}$$

The motor trajectories in the $(\psi_{ra}, \psi_{rb}, \omega)$ space are reported in Figure 1.4 for $\omega^* = 100\,\text{rad/s}$, $\psi^* = 1.16\,\text{Wb}$, and $T_L = 3\,\text{Nm}$. The steady-state solutions (1.47) are parametrized by the flux modulus ψ^*. In fact, given a desired rotor speed value ω^*, a load torque T_L and a fixed set of motor parameter values, there are infinitely many compatible operating conditions $(\psi^*, 0)$, $[i_{sd}^*(\psi^*), i_{sq}^*(\psi^*)]$, $[u_{sd}^*(\psi^*), u_{sq}^*(\psi^*)]$ which depend on the rotor flux modulus ψ^* and which rotate at speed $\dot\rho^* = \omega^* + \omega_s^*(\psi^*)$. Hence the rotor flux modulus ψ^* can be chosen for optimal steady-state performance and in particular to minimize the steady-state power losses given by $P_{loss} = i^{*\mathrm{T}} R i^*$ and to keep the voltage vector modulus constant or below a desired level (field weakening).

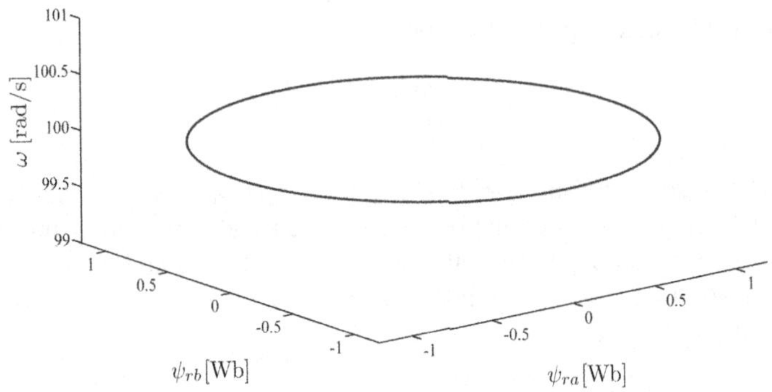

Fig. 1.4 Motor trajectories in the $(\psi_{ra}, \psi_{rb}, \omega)$ state space ($\omega^* = 100$ rad/s, $\psi^* = 1.16$ Wb, $T_L = 3$ Nm)

Remarks

1. If $T_L = 0$ then (1.46) and (1.47) become

$$\begin{bmatrix} u_{sa}^* \\ u_{sb}^* \end{bmatrix} = \begin{bmatrix} \cos\rho^* & -\sin\rho^* \\ \sin\rho^* & \cos\rho^* \end{bmatrix} \begin{bmatrix} \frac{R_s}{M}\psi^* \\ \sigma\left(\frac{1}{M}+\beta\right)\omega^*\psi^* \end{bmatrix}$$
$$\dot{\rho}^* = \omega^*$$

which are independent of R_r while

$$i_{sd}^* = \frac{\psi^*}{M}$$
$$i_{sq}^* = 0\,.$$

Hence, in the case of $T_L = 0$ the steady-state is independent of R_r: consequently R_r cannot be determined from steady-state measurements (ω^*, ψ_{ra}^*, ψ_{rb}^*, i_{sa}^*, i_{sb}^*, u_{sa}^*, u_{sb}^*), since the rotor currents are zero according to (1.24).

2. If rotor speed measurements are not available, the rotor speed ω^* and the parameter $\alpha = \frac{R_r}{L_r}$ cannot be uniquely determined from rotor fluxes $(\psi_{ra}^*, \psi_{rb}^*)$, stator currents (i_{sa}^*, i_{sb}^*), and stator voltages (u_{sa}^*, u_{sb}^*) since (1.47) can be rewritten as

$$\begin{bmatrix} u_{sa}^* \\ u_{sb}^* \end{bmatrix} = \begin{bmatrix} \cos\rho^* & -\sin\rho^* \\ \sin\rho^* & \cos\rho^* \end{bmatrix} \begin{bmatrix} \frac{R_s}{M}\psi^* - \frac{\sigma T_L \dot{\rho}^*}{J\mu\psi^*} \\ \frac{R_s T_L}{J\mu\psi^*} + \sigma\left(\frac{1}{M}+\beta\right)\psi^*\dot{\rho}^* \end{bmatrix}$$
$$\dot{\rho}^* = \omega^* + \frac{R_r T_L}{\psi^{*2}}\,.$$

Only the linear combination $\dot{\rho}^* = \omega^* + \frac{T_L}{\psi^{*2}} R_r$ can be determined from steady-state measurements $(\psi_{ra}^*, \psi_{rb}^*, i_{sa}^*, i_{sb}^*, u_{sa}^*, u_{sb}^*)$: hence ω^* can be uniquely determined only when $T_L = 0$.

1.3.1 Power Loss Minimization

The power which is required at the steady-state

$$P_{in}^* = u_{sa}^* i_{sa}^* + u_{sb}^* i_{sb}^*$$

to balance a given load torque T_L at a desired speed ω^* is given by

$$u_{sa}^* i_{sa}^* + u_{sb}^* i_{sb}^* = T_L \omega^* + P_{loss}^*$$

and is minimized if P_{loss}^* is minimized. Power losses are expressed as

$$P_{loss} = R_s \left(i_{sa}^2 + i_{sb}^2\right) + R_r \left(i_{ra}^2 + i_{rb}^2\right) .$$

Substituting (i_{ra}, i_{rb}) given in (1.20) we can also write

$$\begin{aligned} P_{loss} &= R_s \left(i_{sa}^2 + i_{sb}^2\right) + R_r \left[\left(-\frac{M}{L_r} i_{sa} + \frac{1}{L_r} \psi_{ra}\right)^2 + \left(-\frac{M}{L_r} i_{sb} + \frac{1}{L_r} \psi_{rb}\right)^2\right] \\ &= \left(R_s + \frac{R_r M^2}{L_r^2}\right)\left(i_{sa}^2 + i_{sb}^2\right) + \frac{R_r}{L_r^2}\left(\psi_{ra}^2 + \psi_{rb}^2\right) \\ &\quad - \frac{2 R_r M}{L_r^2}\left(i_{sa}\psi_{ra} + i_{sb}\psi_{rb}\right) . \end{aligned} \tag{1.48}$$

At steady-state, from (1.47) we compute

$$\begin{aligned} i_{sa}^{*2} + i_{sb}^{*2} &= \frac{\psi^{*2}}{M^2} + \frac{T_L^2}{J^2 \mu^2 \psi^{*2}} \\ i_{sa}^* \psi_{ra}^* + i_{sb}^* \psi_{rb}^* &= \frac{\psi^{*2}}{M} \end{aligned} \tag{1.49}$$

so that, from (1.48), we obtain the steady-state power losses (recall that $J\mu = \frac{M}{L_r}$)

$$P_{loss}^* = \frac{R_s}{M^2} \psi^{*2} + \left(R_s + \frac{R_r M^2}{L_r^2}\right) \frac{L_r^2 T_L^2}{M^2 \psi^{*2}} . \tag{1.50}$$

Differentiating with respect to ψ^{*2} and equating to zero we have

$$\frac{\mathrm{d}P_{loss}^*}{\mathrm{d}\psi^{*2}} = \frac{R_s}{M^2} - \left(R_s + \frac{R_r M^2}{L_r^2}\right) \frac{L_r^2 T_L^2}{M^2 \psi^{*4}} = 0$$

which gives the value ψ^* at which the minimum for P^*_{loss} is attained, *i.e.*

$$\psi^{*4} = \left(L_r^2 + \frac{R_r M^2}{R_s}\right) T_L^2$$

or

$$\psi^* = \sqrt[4]{\left(L_r^2 + \frac{R_r M^2}{R_s}\right) T_L^2} \,. \tag{1.51}$$

Note that the optimal flux modulus is independent of the desired speed (since no damping is assumed) and depends on four motor parameters (L_r, M, R_r, R_s) and on the load torque T_L. It does not depend on L_s. Hence, one of the motivations for online parameter estimation is the minimization of power losses by adjusting the desired flux modulus on the basis of the estimated parameter values. The most critical parameters which may vary online are the load torque, since it can abruptly change, and the rotor resistance, which changes due to rotor heating and cannot be measured.

1.3.2 Field Weakening

If we impose that the voltage vector modulus should not exceed (or should be equal to) a given desired value V^* for any desired speed ω^*, according to (1.47), from

$$\begin{aligned} u_{sa}^{*2} + u_{sb}^{*2} = &\left[\frac{R_s}{M}\psi^* - \frac{\sigma T_L \omega^*}{J\mu\psi^*} - \frac{\sigma\alpha M T_L^2}{J^2\mu^2\psi^{*3}}\right]^2 \\ &+ \left[\frac{\sigma(\gamma+\alpha)T_L}{J\mu\psi^*} + \sigma\left(\frac{1}{M}+\beta\right)\omega^*\psi^*\right]^2 \le V^{*2} \end{aligned} \tag{1.52}$$

or

$$\begin{aligned} u_{sa}^{*2} + u_{sb}^{*2} = &\left[\frac{R_s}{M}\psi^* - \frac{\sigma T_L \omega^*}{J\mu\psi^*} - \frac{\sigma\alpha M T_L^2}{J^2\mu^2\psi^{*3}}\right]^2 \\ &+ \left[\frac{\sigma(\gamma+\alpha)T_L}{J\mu\psi^*} + \sigma\left(\frac{1}{M}+\beta\right)\omega^*\psi^*\right]^2 = V^{*2} \end{aligned} \tag{1.53}$$

the value of ψ^* can be chosen in terms of ω^* and T_L. For instance, in the case in which the load torque T_L is zero the inequality (1.52) becomes

$$\frac{R_s^2}{M^2}\psi^{*2} + \sigma^2\left(\frac{1}{M}+\beta\right)^2 \omega^{*2}\psi^{*2} \le V^{*2} \,. \tag{1.54}$$

Solving (1.54) for ψ^{*2} we obtain

$$\psi^{*2} \leq \frac{V^{*2}}{\frac{R_s^2}{M^2} + \sigma^2\left(\frac{1}{M} + \beta\right)^2 \omega^{*2}}$$

which reveals that, given a desired value V^*, the desired flux must not exceed a critical value which decreases as the speed reference ω^* increases: this property is usually referred to as field weakening. On the other hand, the situation for nonzero load torque is not much different since, by using the approximation $L_s \simeq L_r \simeq M$, inequality (1.52) becomes (recall (1.27))

$$\frac{R_s^2}{M^2}\psi^{*2} + \omega^{*2}\psi^{*2} \leq V^{*2}$$

and therefore

$$\psi^{*2} \leq \frac{V^{*2}}{\frac{R_s^2}{M^2} + \omega^{*2}}$$

from which the field weakening property can still be seen.

1.3.3 Torque–Speed Characteristics

If, in addition to the requirement that the the voltage vector modulus should be equal to a given desired value V^*, the voltage vector is also required to rotate at a constant speed $\dot{\rho}^*$, there is an additional relationship for the triple (ω^*, ψ^*, T_L) to be satisfied together with (1.53):

$$\dot{\rho}^* = \omega^* + \frac{R_r T_L}{\psi^{*2}} \,. \tag{1.55}$$

On the basis of the two relationships (1.53) and (1.55) we can compute the so-called torque–speed motor characteristics. Assume that the load torque is nonzero; from (1.55) we obtain the rotor flux modulus

$$\psi^* = \sqrt{\frac{R_r T_L}{(\dot{\rho}^* - \omega^*)}} \,. \tag{1.56}$$

By substituting (1.56) in (1.53) and solving it for T_L we obtain

$$\begin{aligned} T_L(\omega^*) = V^{*2}M^2R_r(\dot{\rho}^* - \omega^*)\Big[&\dot{\rho}^{*4}L_s^2L_r^2 + 2R_sR_r\dot{\rho}^{*2}M^2 - 2\dot{\rho}^{*3}L_s^2L_r^2\omega^* \\ &-2\dot{\rho}^{*4}L_sL_rM^2 + \dot{\rho}^{*2}L_s^2L_r^2\omega^{*2} + \dot{\rho}^{*2}M^4\omega^{*2} - 2\dot{\rho}^{*3}M^4\omega^* + R_s^2R_r^2 \\ &+\dot{\rho}^{*4}M^4 - 2R_sR_r\dot{\rho}^*M^2\omega^* + 4\dot{\rho}^{*3}L_sL_rM^2\omega^* - 2\dot{\rho}^{*2}L_sL_r\omega^{*2}M^2 \\ &+R_s^2L_r^2\dot{\rho}^{*2} - 2R_s^2L_r^2\dot{\rho}^*\omega^* + R_S^2L_r^2\omega^{*2} + \dot{\rho}^{*2}R_r^2L_s^2\Big]^{-1} \end{aligned} \tag{1.57}$$

which is usually called torque–speed motor characteristics and describes, for given constant V^* and $\dot{\rho}^*$ which are the modulus and the frequency of the sinusoidal voltage input, the motor speed constant values corresponding to each admissible value of T_L. By substituting (1.57) into (1.56) we obtain the flux modulus

$$\psi^*(\omega^*) = \sqrt{\frac{R_r T_L(\omega^*)}{(\dot{\rho}^* - \omega^*)}} \tag{1.58}$$

while from (1.47) we obtain the stator current modulus

$$I^*(\omega^*) \stackrel{\triangle}{=} \sqrt{i_{sa}^{*2}(\omega^*) + i_{sb}^{*2}(\omega^*)} = \sqrt{\frac{\psi^{*2}(\omega^*)}{M^2} + \frac{T_L^2(\omega^*)}{J^2\mu^2\psi^{*2}(\omega^*)}} \,. \tag{1.59}$$

The functions (1.57), (1.58), and (1.59) are plotted in Figures 1.5, 1.6, and 1.7, respectively, for the nominal motor parameters given in Table 1.1. From (1.57) and Figure 1.5 we can first establish that there is one value for the load torque T_{Ls}, usually called load torque at stall or stalled torque,

$$\begin{aligned} T_{Ls} = V^{*2}M^2R_r\dot{\rho}^*\Big[&\dot{\rho}^{*4}L_s^2L_r^2 + 2R_sR_r\dot{\rho}^{*2}M^2 - 2\dot{\rho}^{*4}L_sL_rM^2 \\ &+R_s^2R_r^2 + \dot{\rho}^{*4}M^4 + R_s^2L_r^2\dot{\rho}^{*2} + \dot{\rho}^{*2}R_r^2L_s^2\Big]^{-1} \end{aligned} \tag{1.60}$$

which is compatible with $\omega^* = 0$ (motor at stall). Then, by differentiating $T_L(\omega^*)$ with respect to ω^* in (1.57) and equating to zero, we can obtain the value ω_p^* for ω^* which is usually called pull-out speed

$$\begin{aligned} \omega_p^* = \Bigg[&2\dot{\rho}^{*3}M^4 - 4\dot{\rho}^{*3}L_sL_rM^2 + 2R_s^2L_r^2\dot{\rho}^* + 2\dot{\rho}^{*3}L_s^2L_r^2 - 2\Big(R_s^4L_r^2R_r^2 \\ &+2\dot{\rho}^{*2}L_s^2L_r^2R_s^2R_r^2 + \dot{\rho}^{*4}L_s^4L_r^2R_r^2 + \dot{\rho}^{*4}M^4R_r^2L_s^2 + \dot{\rho}^{*2}M^4R_s^2R_r^2 \\ &-2\dot{\rho}^{*4}L_s^3L_rM^2R_r^2 - 2\dot{\rho}^{*2}L_sL_rM^2R_s^2R_r^2\Big)^{\frac{1}{2}}\Bigg]\Big[2\Big(\dot{\rho}^{*2}L_s^2L_r^2 \\ &+\dot{\rho}^{*2}M^4 - 2\dot{\rho}^{*2}L_sL_rM^2 + R_s^2L_r^2\Big)\Big]^{-1} \end{aligned} \tag{1.61}$$

corresponding to the maximum value T_{Lp} of T_L which is usually called pull-out torque, or peak torque, and is given by

$$\begin{aligned} T_{Lp} = V^{*2}M^2\Big(&\dot{\rho}^{*2}L_s^2 + R_s^2\Big)^{\frac{1}{2}}\Big(\dot{\rho}^{*2}L_s^2L_r^2 + \dot{\rho}^{*2}M^4 - 2\dot{\rho}^{*2}L_sL_rM^2 \\ &+R_s^2L_r^2\Big)^{\frac{1}{2}}\Bigg[2\Bigg(R_s^4L_r^2 + 2\dot{\rho}^{*2}L_s^2L_r^2R_s^2 + \dot{\rho}^{*4}L_s^4L_r^2 - 2\dot{\rho}^{*2}L_sL_rM^2R_s^2 \\ &-2\dot{\rho}^{*4}L_s^3L_rM^2 + \dot{\rho}^{*2}M^4R_s^2 + \dot{\rho}^*M^2\Big(\dot{\rho}^{*2}L_s^2 + R_s^2\Big)^{\frac{1}{2}}\Big(\dot{\rho}^{*2}L_s^2L_r^2 + \dot{\rho}^{*2}M^4 \end{aligned}$$

$$-2\dot{\rho}^{*2}L_sL_rM^2+R_s^2L_r^2\Big)^{\frac{1}{2}}R_s+\dot{\rho}^{*4}M^4L_s^2\Big)\Big]^{-1}. \qquad (1.62)$$

Note that the pull-out torque T_{Lp} does not depend on the rotor resistance R_r even though the pull-out speed ω_p^* does and increases when L_r decreases since $L_sL_r-M^2$ is positive in induction motors (the matrix L in (1.11) being positive definite). Finally recall that, in the case of zero load torque (unloaded motor), (1.56) does not hold since it has been obtained from (1.55) under the assumption that $T_L \neq 0$. When $T_L = 0$, from (1.55) we have $\dot{\rho}^* = \omega^*$ while the rotor flux modulus which is obtained by solving (1.53) with $T_L = 0$ and $\dot{\rho}^* = \omega^*$ is

$$\psi^* = \frac{MV^*}{\sqrt{R_s^2+\dot{\rho}^{*2}L_s^2}}$$

and does not depend on the rotor resistance R_r. Note that the functions $T_L(\omega^*)$ and $\psi^*(\omega^*)$ in (1.57) and (1.58) (and therefore $I^*(\omega^*)$ in (1.59)) are continuous on $[0,\dot{\rho}^*]$ since

$$\lim_{\omega^*\to\dot{\rho}^*} T_L(\omega^*) = 0$$

$$\lim_{\omega^*\to\dot{\rho}^*} \psi^*(\omega^*) = \frac{MV^*}{\sqrt{R_s^2+\dot{\rho}^{*2}L_s^2}}.$$

The torque–speed characteristics $T_L(\omega^*)$ obtained in (1.57) and the corresponding $\psi^*(\omega^*)$ and $I^*(\omega^*)$ given in (1.58) and (1.59) are plotted in Figures 1.8–1.16 for different values of V^*, $\dot{\rho}^*$ and motor parameters: recall that the nominal motor parameters are reported in Table 1.1.

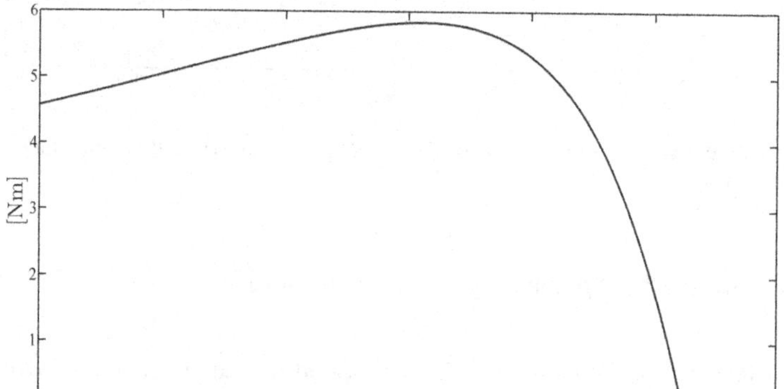

Fig. 1.5 Torque–speed characteristics $T_L(\omega^*)$ for $V^* = 110\,\text{V}$, $\dot{\rho}^* = 16.7\,\text{Hz}$ and nominal motor parameters given in (1.57)

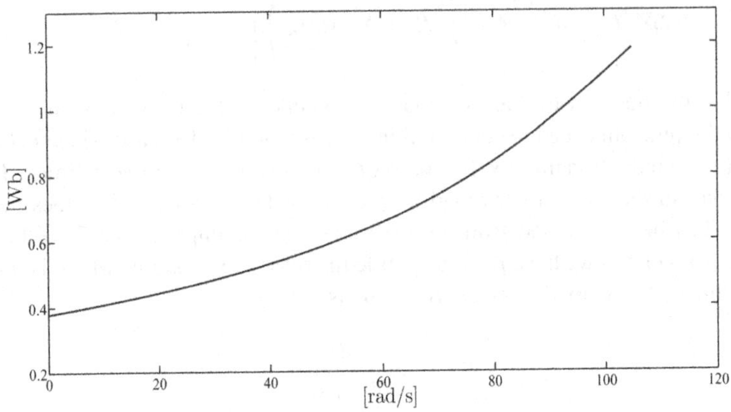

Fig. 1.6 Rotor flux modulus $\psi^*(\omega^*)$ for $V^* = 110\,\text{V}$, $\dot{\rho}^* = 16.7\,\text{Hz}$ and nominal motor parameters given in (1.58)

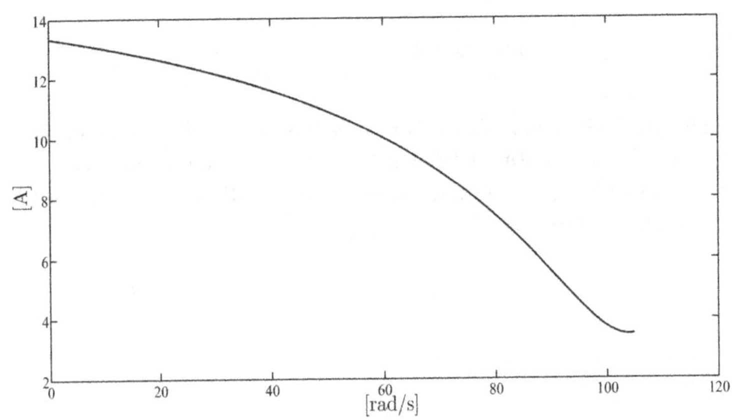

Fig. 1.7 Stator current modulus $I^*(\omega^*)$ for $V^* = 110\,\text{V}$, $\dot{\rho}^* = 16.7\,\text{Hz}$ and nominal motor parameters given in (1.59)

Hence, given V^* and $\dot{\rho}^*$, the key features of the torque–speed characteristic curve are:

1. There is only one T_L value, *i.e.* $T_L = T_{Ls}$ (load torque at stall), compatible with $\omega^* = 0$.
2. When $T_L = T_{Lp}$, there is only one compatible value $\omega^* = \omega_p^*$ for ω^*.
3. There is a unique value $\omega^* = \dot{\rho}^*$ for ω^* corresponding to zero load torque, *i.e.* to unloaded motors: the maximum speed is attained when the motor is unloaded.

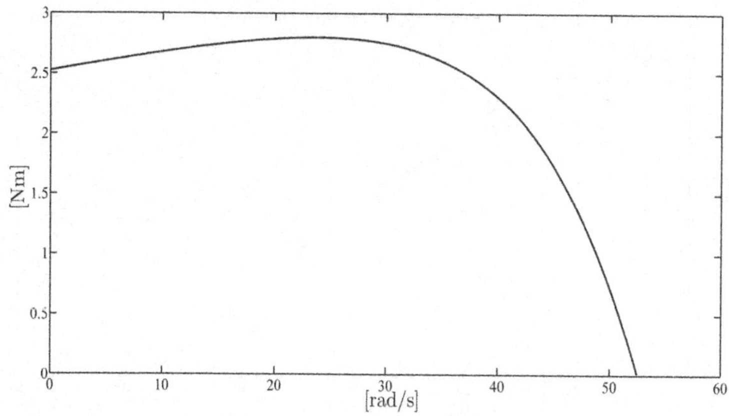

Fig. 1.8 Torque–speed characteristics $T_L(\omega^*)$ for $V^* = 50\,\text{V}$, $\dot{\rho}^* = 8.35\,\text{Hz}$ and nominal motor parameters

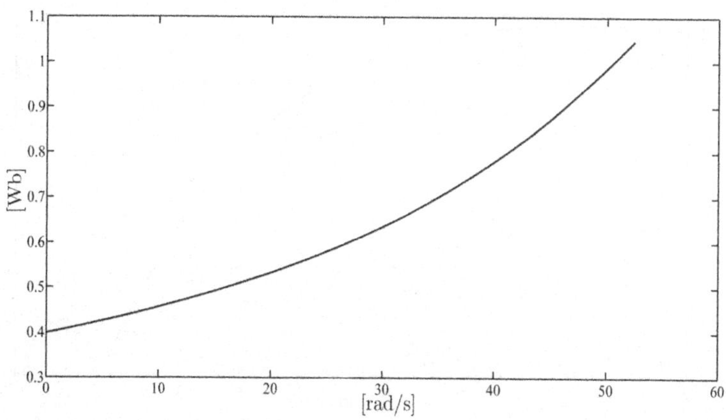

Fig. 1.9 Rotor flux modulus $\psi^*(\omega^*)$ for $V^* = 50\,\text{V}$, $\dot{\rho}^* = 8.35\,\text{Hz}$ and nominal motor parameters

4. When typically $T_L \in (T_{Ls}, T_{Lp})$ there are two compatible values for ω^* and ψ^* which correspond to two different steady-state operating conditions: as we shall see, for the specific motor considered in this book, the operating condition corresponding to the smaller ω^* is unstable while the operating condition corresponding to the larger ω^* is exponentially stable.
5. When $T_L < T_{Ls}$ there is only one operating condition.
6. If the difference $\dot{\rho}^* - \omega_p^*$ is small, then large variations of T_L in the range $[0, T_{Lp}]$ correspond to small variations for the steady-state rotor speed $\omega^* \in (\omega_p^*, \omega_0]$.
7. The comparison of the torque–speed characteristics at nominal rotor resistance value given in Figure 1.5 with the characteristics given in Figure 1.11 when the rotor resistance value is twice the nominal value shows that the rotor resistance

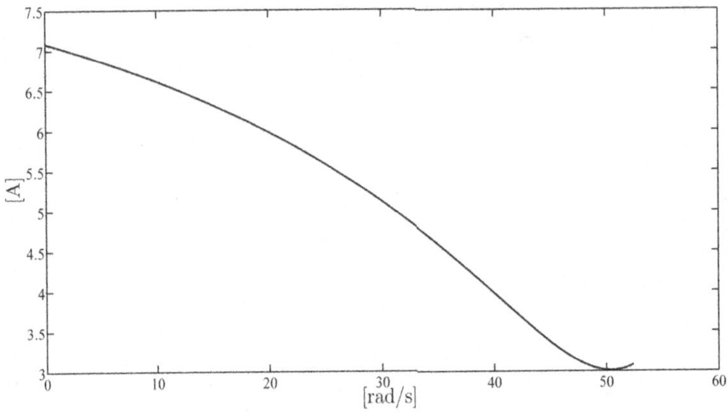

Fig. 1.10 Stator current modulus $I^*(\omega^*)$ for $V^* = 50\,\text{V}$, $\dot{\rho}^* = 8.35\,\text{Hz}$ and nominal motor parameters

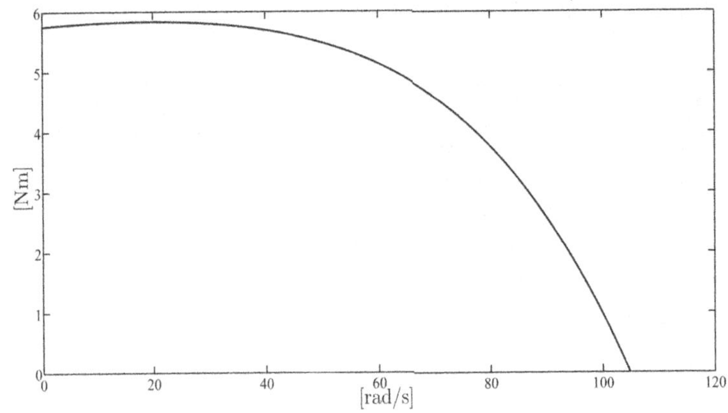

Fig. 1.11 Torque–speed characteristics $T_L(\omega^*)$ for $V^* = 110\,\text{V}$, $\dot{\rho}^* = 16.7\,\text{Hz}$ and $R_r = 6.6\,\Omega$ (twice the nominal value)

has a very important effect on the torque–speed characteristics: the peak torque T_{Lp} does not depend on R_r, as shown in (1.62), but it corresponds to a much smaller value for ω_p^* as R_r increases.

8. The comparison of Figure 1.5 corresponding to $L_r = 0.375\,\text{H}$ with Figure 1.14 corresponding to $L_r = 0.3375\,\text{H}$ confirms that the load torque increases if L_r decreases, according to (1.62).

Figure 1.6 shows that the rotor flux modulus increases as rotor speed increases: this also happens when smaller voltages and frequencies are applied as shown in Figure 1.9 and for parameters which are different from the nominal ones as shown in Figure 1.12 and in Figure 1.15. On the contrary, Figures 1.7, 1.10, 1.13, and 1.16

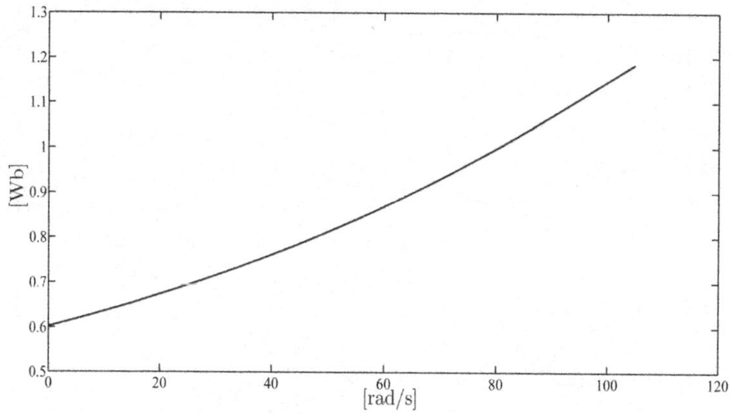

Fig. 1.12 Rotor flux modulus $\psi^*(\omega^*)$ for $V^* = 110\,\text{V}$, $\dot{\rho}^* = 16.7\,\text{Hz}$ and $R_r = 6.6\,\Omega$

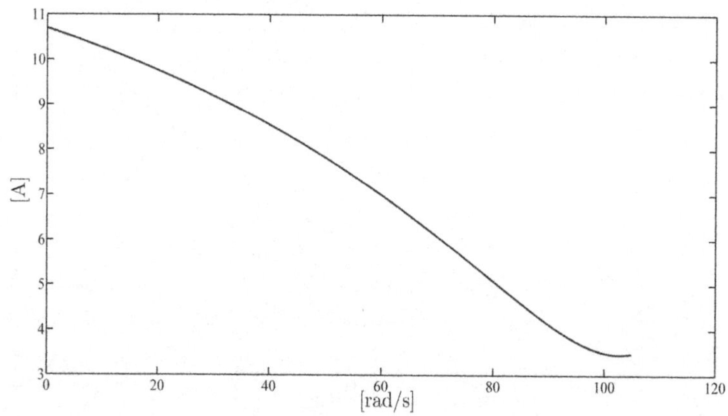

Fig. 1.13 Stator current modulus $I^*(\omega^*)$ for $V^* = 110\,\text{V}$, $\dot{\rho}^* = 16.7\,\text{Hz}$ and $R_r = 6.6\,\Omega$

show that the stator current modulus has its maximum at zero speed and then decreases as speed increases: this happens for different values of voltage modulus, voltage frequency, rotor resistance, and rotor inductances. It is apparent that the operating condition at zero speed is the most critical when sinusoidal inputs are applied, since it may be unstable and it corresponds to the minimum rotor flux modulus and to the maximum stator current modulus.

Let us now investigate in detail the stability and the attractivity properties of the operating conditions determined in (1.47). To this end, consider the rotating frame model (1.31) with $\omega_0 = \dot{\rho}^* = \omega^* + \omega_s^*$ as in (1.43); introduce the error variables

$$\begin{aligned}\tilde{\omega} &= \omega - \omega^* \\ \tilde{\psi}_{rd} &= \psi_{rd} - \psi^*\end{aligned}$$

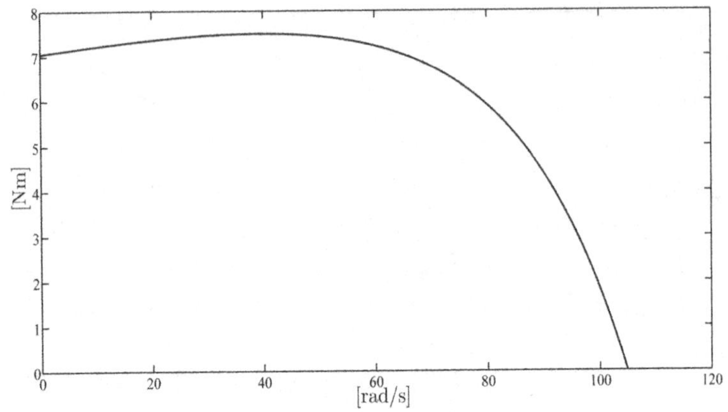

Fig. 1.14 Torque–speed characteristics $T_L(\omega^*)$ for $V^* = 110\,\text{V}$, $\dot{\rho}^* = 16.7\,\text{Hz}$ and $L_r = 0.3375\,\text{H}$

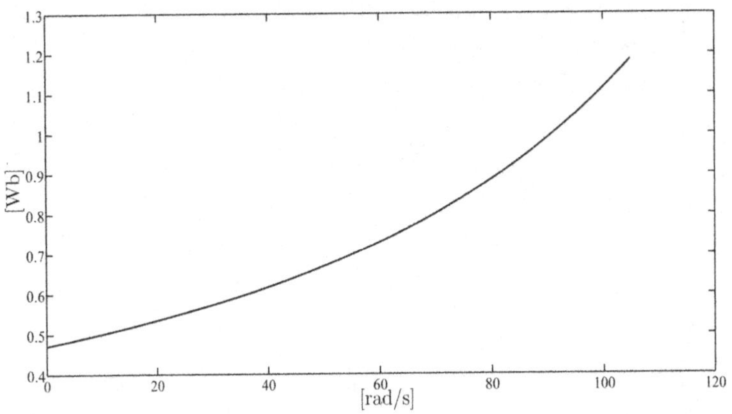

Fig. 1.15 Rotor flux modulus $\psi^*(\omega^*)$ for $V^* = 110\,\text{V}$, $\dot{\rho}^* = 16.7\,\text{Hz}$ and $L_r = 0.3375\,\text{H}$

$$\begin{aligned}
\tilde{\psi}_{rq} &= \psi_{rq} \\
\tilde{i}_{sd} &= i_{sd} - i^*_{sd} \\
\tilde{i}_{sq} &= i_{sq} - i^*_{sq}
\end{aligned}$$

with $(\omega^*, \psi^*, i^*_{sd}, i^*_{sq})$ satisfying (1.44). The error dynamics are given by

$$\begin{aligned}
\frac{\mathrm{d}\tilde{\omega}}{\mathrm{d}t} &= \mu\left[\tilde{\psi}_{rd} i^*_{sq} - \tilde{\psi}_{rq} i^*_{sd}\right] + \mu\psi^*\tilde{i}_{sq} + \mu\left[\tilde{\psi}_{rd}\tilde{i}_{sq} - \tilde{\psi}_{rq}\tilde{i}_{sd}\right] \\
\frac{\mathrm{d}\tilde{\psi}_{rd}}{\mathrm{d}t} &= -\alpha\tilde{\psi}_{rd} - \tilde{\omega}\tilde{\psi}_{rq} + \omega^*_s\tilde{\psi}_{rq} + \alpha M\tilde{i}_{sd} \\
\frac{\mathrm{d}\tilde{\psi}_{rq}}{\mathrm{d}t} &= -\alpha\tilde{\psi}_{rq} + \tilde{\omega}\tilde{\psi}_{rd} - \omega^*_s\tilde{\psi}_{rd} + \tilde{\omega}\psi^* + \alpha M\tilde{i}_{sq}
\end{aligned}$$

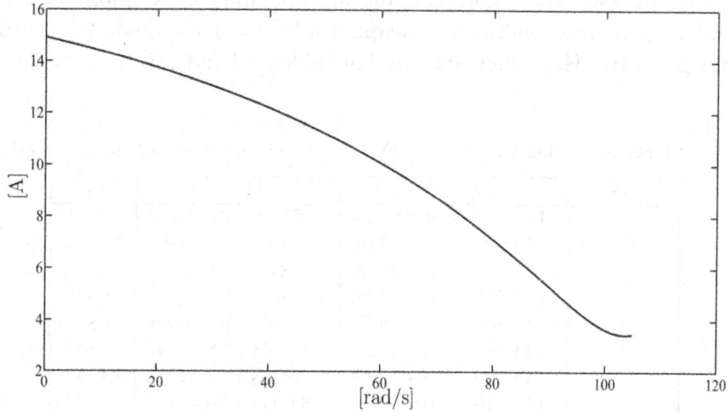

Fig. 1.16 Stator current modulus $I^*(\omega^*)$ for $V^* = 110\,\text{V}$, $\dot{\rho}^* = 16.7\,\text{Hz}$ and $L_r = 0.3375\,\text{H}$

$$\begin{aligned}
\frac{\mathrm{d}\tilde{i}_{sd}}{\mathrm{d}t} &= -\gamma\tilde{i}_{sd} + \omega_0^*\tilde{i}_{sq} + \beta\alpha\tilde{\psi}_{rd} + \beta\omega^*\tilde{\psi}_{rq} + \beta\tilde{\omega}\tilde{\psi}_{rq} \\
\frac{\mathrm{d}\tilde{i}_{sq}}{\mathrm{d}t} &= -\gamma\tilde{i}_{sq} - \omega_0^*\tilde{i}_{sd} + \beta\alpha\tilde{\psi}_{rq} - \beta\omega^*\tilde{\psi}_{rd} - \beta\tilde{\omega}\tilde{\psi}_{rd} - \beta\psi^*\tilde{\omega}\,.
\end{aligned} \tag{1.63}$$

The linear approximation about the origin $(\tilde{\omega}, \tilde{\psi}_{rd}, \tilde{\psi}_{rq}, \tilde{i}_{sd}, \tilde{i}_{sq}) = 0$ is given by

$$\begin{aligned}
\frac{\mathrm{d}\tilde{\omega}}{\mathrm{d}t} &= \mu i_{sq}^*\tilde{\psi}_{rd} - \mu i_{sd}^*\tilde{\psi}_{rq} + \mu\psi^*\tilde{i}_{sq} \\
\frac{\mathrm{d}\tilde{\psi}_{rd}}{\mathrm{d}t} &= -\alpha\tilde{\psi}_{rd} + \omega_s^*\tilde{\psi}_{rq} + \alpha M\tilde{i}_{sd} \\
\frac{\mathrm{d}\tilde{\psi}_{rq}}{\mathrm{d}t} &= \psi^*\tilde{\omega} - \omega_s^*\tilde{\psi}_{rd} - \alpha\tilde{\psi}_{rq} + \alpha M\tilde{i}_{sq} \\
\frac{\mathrm{d}\tilde{i}_{sd}}{\mathrm{d}t} &= \beta\alpha\tilde{\psi}_{rd} + \beta\omega^*\tilde{\psi}_{rq} - \gamma\tilde{i}_{sd} + (\omega^* + \omega_s^*)\tilde{i}_{sq} \\
\frac{\mathrm{d}\tilde{i}_{sq}}{\mathrm{d}t} &= -\beta\psi^*\tilde{\omega} - \beta\omega^*\tilde{\psi}_{rd} + \beta\alpha\tilde{\psi}_{rq} - (\omega^* + \omega_s^*)\tilde{i}_{sd} - \gamma\tilde{i}_{sq}\,.
\end{aligned} \tag{1.64}$$

Given the nominal motor parameters in Table 1.1, the five eigenvalues of the matrix in the linear approximation (1.64)

$$\begin{bmatrix}
0 & \mu i_{sq}^* & -\mu i_{sd}^* & 0 & \mu\psi^* \\
0 & -\alpha & \omega_s^* & \alpha M & 0 \\
\psi^* & -\omega_s^* & -\alpha & 0 & \alpha M \\
0 & \beta\alpha & \beta\omega^* & -\gamma & \omega^* + \omega_s^* \\
-\beta\psi^* & -\beta\omega^* & \beta\alpha & -\omega^* - \omega_s^* & -\gamma
\end{bmatrix}$$

are reported in Tables 1.2 and 1.3, for the operating conditions (ω^*, $T_L(\omega^*)$, $\psi^*(\omega^*), I(\omega^*)$) (corresponding to sinusoidal voltages of amplitude $V^* = 110\,\text{V}$ and frequency $\dot{\rho}^* = 16.7\,\text{Hz}$) which are listed in Tables 1.4 and 1.5.

Table 1.2 Linear approximation ($V^* = 110\,\text{V}$, $\dot{\rho}^* = 16.7\,\text{Hz}$): eigenvalues for $\omega^* < \omega_p^*$

ω^* (rad/s)	(λ_1, λ_2)	(λ_3, λ_4)	λ_5
0	$-144.0043 \pm 105.8864i$	$-7.5129 \pm 105.1025i$	2.9641
2	$-143.9905 \pm 105.1730i$	$-7.5434 \pm 103.8647i$	2.9974
4	$-143.9626 \pm 104.4608i$	$-7.5872 \pm 102.6282i$	3.02931
6	$-143.9206 \pm 103.7499i$	$-7.6444 \pm 101.3929i$	3.0596
8	$-143.8645 \pm 103.0407i$	$-7.7148 \pm 100.1586i$	3.0881
10	$-143.7942 \pm 102.3332i$	$-7.7983 \pm 98.9254i$	3.1145
12	$-143.7095 \pm 101.6279i$	$-7.8948 \pm 97.6931i$	3.1384
14	$-143.6105 \pm 100.9249i$	$-8.0043 \pm 96.4617i$	3.1593
16	$-143.4971 \pm 100.2245i$	$-8.1265 \pm 95.2310i$	3.1768
18	$-143.369 \pm 99.527i$	$-8.2614 \pm 94.001i$	3.1905
20	$-143.2263 \pm 98.8326i$	$-8.4088 \pm 92.7716i$	3.1998
22	$-143.0687 \pm 98.1416i$	$-8.5684 \pm 91.5427i$	3.2039
24	$-142.8962 \pm 97.4545i$	$-8.7401 \pm 90.3142i$	3.2023
26	$-142.7086 \pm 96.7714i$	$-8.9236 \pm 89.086i$	3.194
28	$-142.5057 \pm 96.0927i$	$-9.1186 \pm 87.858i$	3.1783
30	$-142.2873 \pm 95.4189i$	$-9.3249 \pm 86.63i$	3.154
32	$-142.0533 \pm 94.7503i$	$-9.542 \pm 85.4019i$	3.1201
34	$-141.8034 \pm 94.0873i$	$-9.7695 \pm 84.1736i$	3.0753
36	$-141.5374 \pm 93.4305i$	$-10.0069 \pm 82.9448i$	3.01825
38	$-141.255 \pm 92.7801i$	$-10.2537 \pm 81.7154i$	2.9471
40	$-140.9561 \pm 92.1369i$	$-10.5093 \pm 80.4851i$	2.8603
42	$-140.6402 \pm 91.5014i$	$-10.7728 \pm 79.2536i$	2.7557
44	$-140.3072 \pm 90.8741i$	$-11.0435 \pm 78.0206i$	2.6311
46	$-139.9566 \pm 90.2557i$	$-11.3205 \pm 76.7857i$	2.4838
48	$-139.5882 \pm 89.647i$	$-11.6025 \pm 75.5486i$	2.311
50	$-139.2015 \pm 89.0486i$	$-11.8884 \pm 74.3088i$	2.1095
52	$-138.7962 \pm 88.4615i$	$-12.1767 \pm 73.0656i$	1.8755
54	$-138.3719 \pm 87.8865i$	$-12.4657 \pm 71.8186i$	1.6048
56	$-137.9281 \pm 87.3247i$	$-12.7534 \pm 70.5669i$	1.2927
58	$-137.4643 \pm 86.777i$	$-13.0377 \pm 69.3099i$	0.9338
60	$-136.9802 \pm 86.2448i$	$-13.316 \pm 68.0467i$	0.5219
62	$-136.475 \pm 85.7292i$	$-13.5851 \pm 66.7763i$	0.05

The eigenvalues are also plotted in Figures 1.17–1.20 for different V^*, $\dot{\rho}^*$, and motor parameters, as rotor speed goes from zero to its maximum value. According to the linear approximation Theorem A.7 in Appendix A, we can conclude that the operating conditions $(T_L(\omega^*), \omega^*)$ are unstable for $\omega^* \in (0, \omega_p^*)$ since there is one positive real eigenvalue, while the steady-state conditions $(T_L(\omega^*), \omega^*)$ are locally exponentially stable for $\omega^* \in (\omega_p^*, \omega_0]$ since all eigenvalues have negative real parts. When $T_L = T_{Lp}$ the linear approximation (1.64) has one eigenvalue at zero and two pairs of complex conjugate eigenvalues with negative real parts: we are in a critical case in which the linear approximation Theorem A.7 in AppendixA does not apply. The motor transient behavior, when sinusoidal voltages of modulus $V^* = 110\,\text{V}$ and

Table 1.3 Linear approximation ($V^* = 110\,\text{V}$, $\dot{\rho}^* = 16.7\,\text{Hz}$): eigenvalues for $\omega^* > \omega_p^*$

ω^* (rad/s)	(λ_1, λ_2)	(λ_3, λ_4)	λ_5
64	$-135.9485 \pm 85.2317i$	$-13.8416 \pm 65.4978i$	-0.49
66	$-135.3999 \pm 84.7539i$	$-14.0814 \pm 64.2102i$	-1.1077
68	$-134.8287 \pm 84.2974i$	$-14.2996 \pm 62.9124i$	-1.8136
70	$-134.2345 \pm 83.8642i$	$-14.4904 \pm 61.6037i$	-2.6206
72	$-133.6166 \pm 83.4561i$	$-14.6470 \pm 60.2833i$	-3.5431
74	$-132.9746 \pm 83.0755i$	$-14.7614 \pm 58.9513i$	-4.5983
76	$-132.3079 \pm 82.7247i$	$-14.8241 \pm 57.6083i$	-5.8063
78	$-131.6162 \pm 82.4064i$	$-14.8237 \pm 56.2564i$	-7.1903
80	$130.8992 \pm 82.1234i$	$-14.7474 \pm 54.9001i$	-8.777
82	$-130.1568 \pm 81.8786i$	$-14.5803 \pm 53.547i$	-10.5961
84	$-129.3888 \pm 81.6752i$	$-14.3068 \pm 52.2102i$	-12.6789
86	$-128.5957 \pm 81.5166i$	$-13.9121 \pm 50.9092i$	-15.0546
88	$-127.7781 \pm 81.4062i$	$-13.3856 \pm 49.6722i$	-17.743
90	$-126.9368 \pm 81.3475i$	$-12.7265 \pm 48.5353i$	-20.7438
92	$-126.0733 \pm 81.3438i$	$-11.9497 \pm 47.5381i$	-24.0244
94	$-125.1896 \pm 81.3982i$	$-11.0887 \pm 46.7143i$	-27.5135
96	$-124.2884 \pm 81.5134i$	$-10.1916 \pm 46.0816i$	-31.1103
98	$-123.3729 \pm 81.6913i$	$-9.3096 \pm 45.6357i$	-34.7053
100	$-122.4472 \pm 81.9329i$	$-8.4867 \pm 45.3542i$	-38.2024
102	$-121.5160 \pm 82.2379i$	$-7.7543 \pm 45.2038i$	-41.5297
104	$-120.5847 \pm 82.6043i$	$-7.1314 \pm 45.1483i$	-44.6381

frequency $\dot{\rho}^* = 16.7$ Hz are imposed, is shown in Figures 1.21–1.24 for zero and nonzero applied load torques, respectively.

In conclusion:

1. Given constant references (ω^*, ψ^*), the steady-state operating conditions are

$$\begin{bmatrix} \psi_{ra}^* \\ \psi_{rb}^* \end{bmatrix} = \begin{bmatrix} \cos\rho^* & -\sin\rho^* \\ \sin\rho^* & \cos\rho^* \end{bmatrix} \begin{bmatrix} \psi^* \\ 0 \end{bmatrix}$$

$$\begin{bmatrix} i_{sa}^* \\ i_{sb}^* \end{bmatrix} = \begin{bmatrix} \cos\rho^* & -\sin\rho^* \\ \sin\rho^* & \cos\rho^* \end{bmatrix} \begin{bmatrix} \frac{\psi^*}{M} \\ \frac{T_L}{J\mu\psi^*} \end{bmatrix}$$

$$\begin{bmatrix} u_{sa}^* \\ u_{sb}^* \end{bmatrix} = \begin{bmatrix} \cos\rho^* & -\sin\rho^* \\ \sin\rho^* & \cos\rho^* \end{bmatrix} \begin{bmatrix} \frac{R_s}{M}\psi^* - \frac{\sigma T_L \omega^*}{J\mu\psi^*} - \frac{\sigma\alpha M T_L^2}{J^2\mu^2\psi^{*3}} \\ \frac{\sigma(\gamma+\alpha)T_L}{J\mu\psi^*} + \sigma\left(\frac{1}{M} + \beta\right)\omega^*\psi^* \end{bmatrix}$$

$$\dot{\rho}^* = \omega^* + \frac{\alpha M T_L}{J\mu\psi^{*2}} = \omega^* + \frac{R_r T_L}{\psi^{*2}} = \omega^* + \omega_s^*$$

$$\rho^*(0) = \arctan\left(\frac{\psi_{rb}(0)}{\psi_{ra}(0)}\right).$$

2. If ψ^* is chosen as

Table 1.4 $T_L(\omega^*)$, $\psi^*(\omega^*)$, $I^*(\omega^*)$ for $\omega^* < \omega_p^*$ ($V^* = 110$ V, $\dot{\rho}^* = 16.7$ Hz)

ω^* (rad/s)	T_L (Nm)	ψ^* (Wb)	I^* (A)
0	4.5685	0.3791	13.3366
2	4.6137	0.3847	13.2758
4	4.6594	0.3904	13.2130
6	4.7057	0.3963	13.1481
8	4.7524	0.4023	13.0811
10	4.7995	0.4086	13.0117
12	4.8471	0.4150	12.9399
14	4.8951	0.4216	12.8655
16	4.9435	0.4284	12.7886
18	4.9921	0.4355	12.7088
20	5.0410	0.4427	12.6260
22	5.0900	0.4502	12.5402
24	5.1391	0.4579	12.4510
26	5.1883	0.4659	12.3585
28	5.2373	0.4741	12.2623
30	5.2861	0.4827	12.1622
32	5.3345	0.4915	12.0581
34	5.3825	0.5006	11.9497
36	5.4297	0.5100	11.8368
38	5.4761	0.5198	11.7191
40	5.5213	0.5300	11.5963
42	5.5652	0.5404	11.4681
44	5.6074	0.5513	11.3341
46	5.6475	0.5626	11.1942
48	5.6853	0.5743	11.0477
50	5.7202	0.5865	10.8945
52	5.7518	0.5991	10.7340
54	5.7795	0.6123	10.5657
56	5.8026	0.6259	10.3892
58	5.8205	0.6401	10.2040
60	5.8322	0.6549	10.0095
62	5.8369	0.6703	9.8051

$$\psi^* = \sqrt[4]{\left(L_r^2 + \frac{R_r M^2}{R_s}\right) T_L^2}$$

then the *power losses* at steady-state

$$P_{loss}^* = \frac{R_s}{M^2}\psi^{*2} + \left(R_s + \frac{R_r M^2}{L_r^2}\right)\frac{L_r^2 T_L^2}{M^2 \psi^{*2}}$$

are minimized.

3. Given V^* and ω^*, the voltage constraint

$$u_{sa}^{*2} + u_{sb}^{*2} \le V^{*2}$$

Table 1.5 $T_L(\omega^*), \psi^*(\omega^*), I^*(\omega^*)$ for $\omega^* > \omega_p^*$ ($V^* = 110\,\text{V}, \dot{\rho}^* = 16.7\,\text{Hz}$)

ω^* (rad/s)	T_L (Nm)	ψ^* (Wb)	I^* (A)
64	5.8335	0.6863	9.5903
66	5.8206	0.7029	9.3642
68	5.7971	0.7203	9.1264
70	5.7612	0.7383	8.8760
72	5.7113	0.7572	8.6125
74	5.6452	0.7768	8.3350
76	5.5608	0.7972	8.0430
78	5.4554	0.8184	7.7358
80	5.3262	0.8406	7.4130
82	5.1699	0.8636	7.0744
84	4.9829	0.8875	6.7201
86	4.7612	0.9124	6.3506
88	4.5005	0.9381	5.9674
90	4.1958	0.9648	5.5732
92	3.8420	0.9923	5.1724
94	3.4335	1.0207	4.7726
96	2.9645	1.0498	4.3856
98	2.4288	1.0797	4.0299
100	1.8206	1.1100	3.7325
102	1.1342	1.1408	3.5299
104	0.3645	1.1717	3.4633

is satisfied if ψ^* is chosen so that

$$\left[\frac{R_s}{M}\psi^* - \frac{\sigma T_L \omega^*}{J\mu\psi^*} - \frac{\sigma\alpha M T_L^2}{J^2\mu^2\psi^{*3}}\right]^2 + \left[\frac{\sigma(\gamma+\alpha)T_L}{J\mu\psi^*} + \sigma\left(\frac{1}{M}+\beta\right)\omega^*\psi^*\right]^2 \leq V^{*2}.$$

4. Given V^* and $\dot{\rho}^*$, the load torque T_L and the rotor speed ω^* are related at steady-state by the motor *torque–speed characteristics*

$$\begin{aligned}T_L(\omega^*) = V^{*2}M^2R_r(\dot{\rho}^* - \omega^*)\Big[&\dot{\rho}^{*4}L_s^2L_r^2 + 2R_sR_r\dot{\rho}^{*2}M^2 - 2\dot{\rho}^{*3}L_s^2L_r^2\omega^* \\ &-2\dot{\rho}^{*4}L_sL_rM^2 + \dot{\rho}^{*2}L_s^2L_r^2\omega^{*2} + \dot{\rho}^{*2}M^4\omega^{*2} - 2\dot{\rho}^{*3}M^4\omega^* + R_s^2R_r^2 \\ &+\dot{\rho}^{*4}M^4 - 2R_sR_r\dot{\rho}^*M^2\omega^* + 4\dot{\rho}^{*3}L_sL_rM^2\omega^* - 2\dot{\rho}^{*2}L_sL_r\omega^{*2}M^2 \\ &+R_s^2L_r^2\dot{\rho}^{*2} - 2R_s^2L_r^2\dot{\rho}^*\omega^* + R_S^2L_r^2\omega^{*2} + \dot{\rho}^{*2}R_r^2L_s^2\Big]^{-1}.\end{aligned}$$

5. Given the nominal motor parameters in Table 1.1, for every ω^* such that $\omega^* > \omega_p^*$, with ω_p^* given in (1.61) corresponding to the maximum torque T_{Lp} given in (1.62), the operating condition is exponentially stable while

it is unstable with one positive real eigenvalue for every ω^* such that $0 \leq \omega^* < \omega_p^*$.

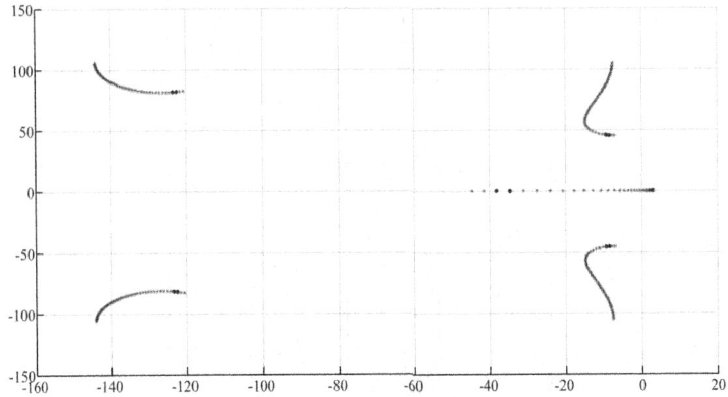

Fig. 1.17 Linear approximation: eigenvalues for $V^* = 110\,\text{V}$, $\dot{\rho}^* = 16.7\,\text{Hz}$ and nominal motor parameters

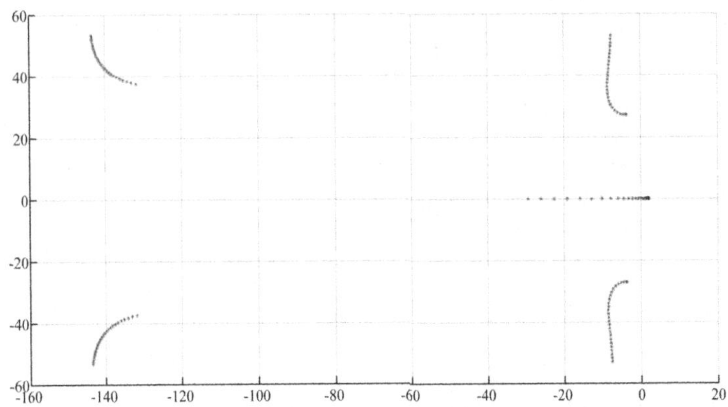

Fig. 1.18 Linear approximation: eigenvalues for $V^* = 50\,\text{V}$, $\dot{\rho}^* = 8.35\,\text{Hz}$ and nominal motor parameters

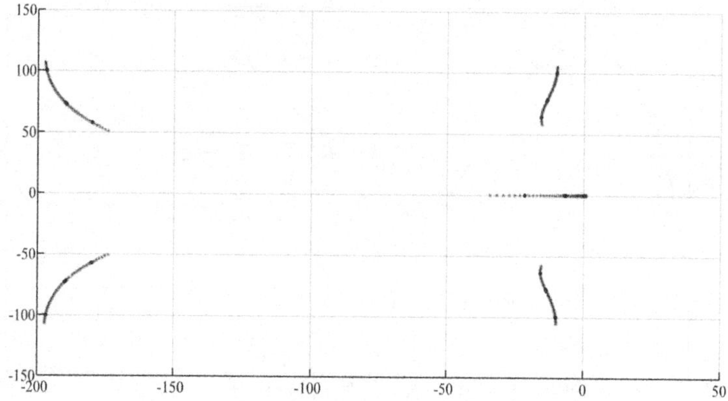

Fig. 1.19 Linear approximation: eigenvalues for $V^* = 110\,\text{V}$, $\dot{\rho}^* = 16.7\,\text{Hz}$ and $R_r = 6.6\,\Omega$

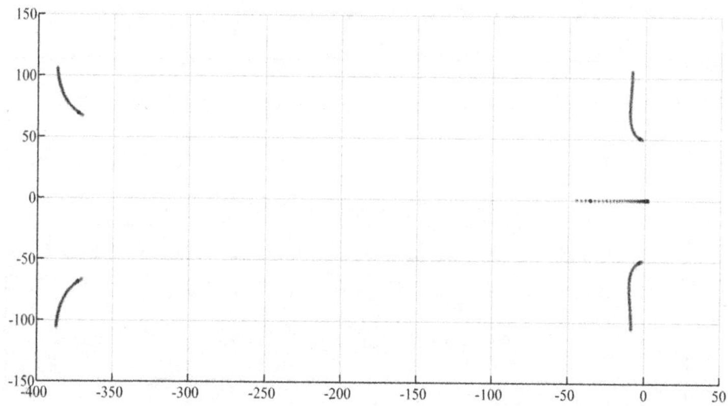

Fig. 1.20 Linear approximation: eigenvalues for $V^* = 110\,\text{V}$, $\dot{\rho}^* = 16.7\,\text{Hz}$ and $L_r = 0.3375\,\text{H}$

1.4 Inverse System and Tracking Dynamics

Consider the induction motor field-oriented model (1.39) expressed in state coordinates (1.41), *i.e.* in a (d,q) reference whose d-axis coincides with the flux vector and assume that the motor speed ω coincides, along with its time derivatives, with the desired rotor speed reference ω^* and that the flux modulus ψ_{rd} coincides, along with its time derivatives, with the desired reference ψ^*: the goal of this section is to determine the corresponding voltage control inputs. To this end, we compute the left inverse system, that is the dynamical system generating, on the basis of the reference signals (ω^*, ψ^*) and their time derivatives, the control inputs (u_{sd}^*, u_{sq}^*) which, applied to the field-oriented model (1.39) with compatible initial conditions, forces $\omega(t)$ to be equal to $\omega^*(t)$ and $\psi_{rd}(t)$ to be equal to $\psi^*(t)$ for all $t \geq 0$. In order to

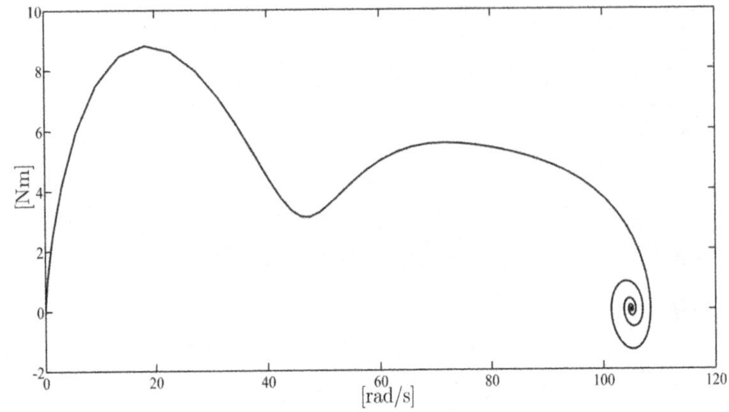

Fig. 1.21 Transient behavior for zero load torque in the $(\omega, J\dot{\omega})$ plane

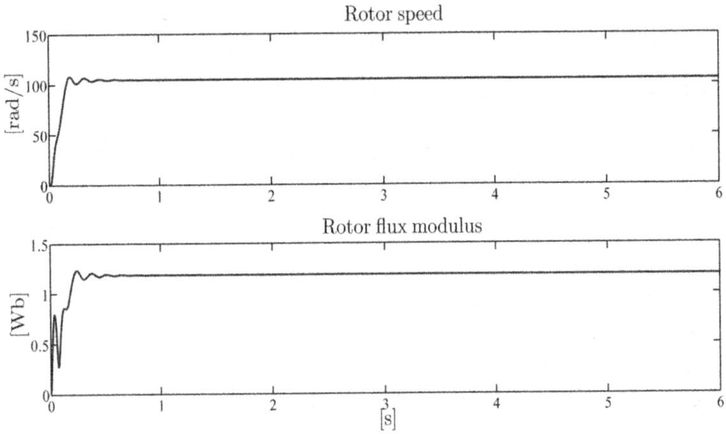

Fig. 1.22 Transient behavior for zero load torque: rotor speed and rotor flux modulus

compute the inverse system it is useful to compute first the tracking dynamics (see Appendix B), that is the motor dynamics constrained to the manifold in the state space which is characterized by

$$\begin{aligned} \omega(t) &= \omega^*(t), \quad \forall t \geq 0 \\ \psi_{rd}(t) &= \psi^*(t), \quad \forall t \geq 0. \end{aligned} \tag{1.65}$$

According to (1.65), from the first two equations in the field-oriented model (1.39) it follows that

$$\dot{\omega}^* = \mu \psi^* i_{sq}^* - \frac{T_L}{J}$$

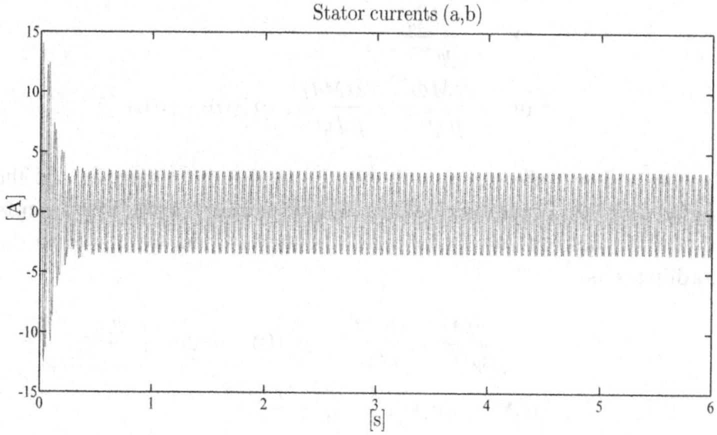

Fig. 1.23 Transient behavior for zero load torque: stator currents (i_{sa}, i_{sb})

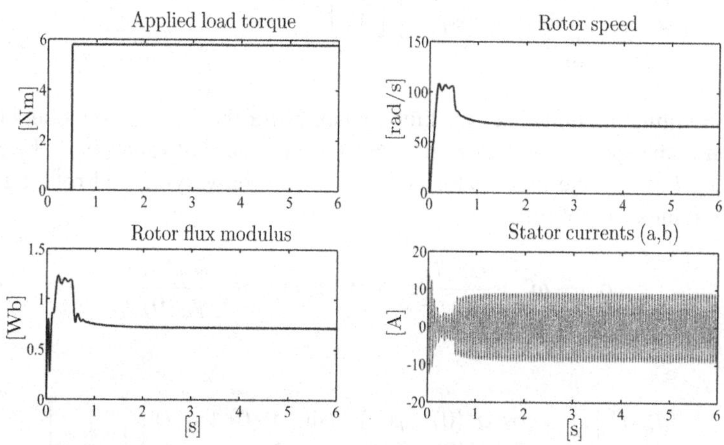

Fig. 1.24 Transient behavior for nonzero load torque: applied load torque, rotor speed, rotor flux modulus and stator currents (i_{sa}, i_{sb})

$$\dot{\psi}^* = -\alpha\psi^* + \alpha M i_{sd}^* \tag{1.66}$$

which constrains the currents (i_{sd}^*, i_{sq}^*) to be

$$\begin{aligned} i_{sd}^* &= \frac{\psi^*}{M} + \frac{\dot{\psi}^*}{\alpha M} \\ i_{sq}^* &= \frac{\dot{\omega}^*}{\mu\psi^*} + \frac{T_L}{J\mu\psi^*} \end{aligned} \tag{1.67}$$

and consequently the rotor flux angle ρ to satisfy the differential equation

$$
\begin{aligned}
\dot{\rho}^* &= \omega^* + \frac{\alpha M i_{sq}^*}{\psi^*} \\
&= \omega^* + \frac{\alpha M \dot{\omega}^*}{\mu \psi^{*2}} + \frac{\alpha M T_L}{\mu J \psi^{*2}}, \quad \rho^*(0) = \rho(0)
\end{aligned} \tag{1.68}
$$

which is obtained by substituting ω^*, ψ^*, i_{sq}^* in the third equation on the field-oriented model (1.39). Equations (1.67) and (1.68) constitute the first order tracking dynamics for the field-oriented model (1.39); they can be expressed in the (a,b) fixed coordinates as

$$
\begin{aligned}
\dot{\rho}^* &= \omega^* + \frac{\alpha M \dot{\omega}^*}{\mu \psi^{*2}} + \frac{\alpha M T_L}{\mu J \psi^{*2}}, \quad \rho^*(0) = \arctan\left(\frac{\psi_{rb}(0)}{\psi_{ra}(0)}\right) \\
\begin{bmatrix} i_{sa}^* \\ i_{sb}^* \end{bmatrix} &= \begin{bmatrix} \cos\rho^* & -\sin\rho^* \\ \sin\rho^* & \cos\rho^* \end{bmatrix} \begin{bmatrix} \frac{\psi^*}{M} + \frac{\dot{\psi}^*}{\alpha M} \\ \frac{\dot{\omega}^*}{\mu \psi^*} + \frac{T_L}{J \mu \psi^*} \end{bmatrix} \\
\begin{bmatrix} \psi_{ra}^* \\ \psi_{rb}^* \end{bmatrix} &= \begin{bmatrix} \cos\rho^* & -\sin\rho^* \\ \sin\rho^* & \cos\rho^* \end{bmatrix} \begin{bmatrix} \psi^* \\ 0 \end{bmatrix} \\
\omega^* &= \omega^*
\end{aligned} \tag{1.69}
$$

which constitute the first order tracking dynamics for the fixed frame model (1.26). Note that in the special cases in which (ω^*, ψ^*) are constant values both the current vector (i_{sa}^*, i_{sb}^*) and the flux vector $(\psi_{ra}^*, \psi_{rb}^*)$ are time-varying and rotating at the same constant speed of rotation

$$
\dot{\rho}^* = \omega^* + \frac{\alpha M T_L}{\mu J \psi^{*2}}, \quad \rho^*(0) = \arctan\left(\frac{\psi_{rb}(0)}{\psi_{ra}(0)}\right) \tag{1.70}
$$

that is

$$
\begin{aligned}
\begin{bmatrix} i_{sa}^*(t) \\ i_{sb}^*(t) \end{bmatrix} &= \begin{bmatrix} \cos(\rho^*(0) + \dot{\rho}^* t) & -\sin(\rho^*(0) + \dot{\rho}^* t) \\ \sin(\rho^*(0) + \dot{\rho}^* t) & \cos(\rho^*(0) + \dot{\rho}^* t) \end{bmatrix} \begin{bmatrix} \frac{\psi^*}{M} \\ \frac{T_L}{J \mu \psi^*} \end{bmatrix} \\
\begin{bmatrix} \psi_{ra}^*(t) \\ \psi_{rb}^*(t) \end{bmatrix} &= \begin{bmatrix} \cos(\rho^*(0) + \dot{\rho}^* t) & -\sin(\rho^*(0) + \dot{\rho}^* t) \\ \sin(\rho^*(0) + \dot{\rho}^* t) & \cos(\rho^*(0) + \dot{\rho}^* t) \end{bmatrix} \begin{bmatrix} \psi^* \\ 0 \end{bmatrix}.
\end{aligned} \tag{1.71}
$$

Hence, in this special case of constant references (ω^*, ψ^*) we reobtain the results in (1.47) of the previous Section 1.3. Equations (1.70) and (1.71) describe a hypersphere in the state space $(\omega, \psi_{ra}, \psi_{rb}, i_{sa}, i_{sb})$. The motor trajectories in the $(\psi_{ra}, \psi_{rb}, \omega)$ space are reported in Figure 1.25 for $\omega^* = 1$ rad/s, $\psi^*(t) = 1.16 + 0.2\sin(6\pi t)\,\text{Wb}$, and $T_L = 0.1\,\text{Nm}$. As we shall see, a time-varying reference is needed for the online estimation of the rotor resistance when the rotor speed is not measured. Figure 1.26 shows the motor trajectories in the $(\psi_{ra}, \psi_{rb}, \omega)$ space when $\psi^*(t) = 1.16 + 0.2\sin(6\pi t)\,\text{Wb}$; $\omega^*(t)$ is the output of the unitary gain third order linear filter with transfer function $F(s) = 1/(1+s/50)^3$ and input signal $\omega_f(t) = 1 + [45 + 12\sin(2\pi t)]H(t-5)$ with $H(t)$ the unit step function; $T_L = 0.1\,\text{Nm}$. If we substitute (1.65) and (1.67) in the last two equations in the field-

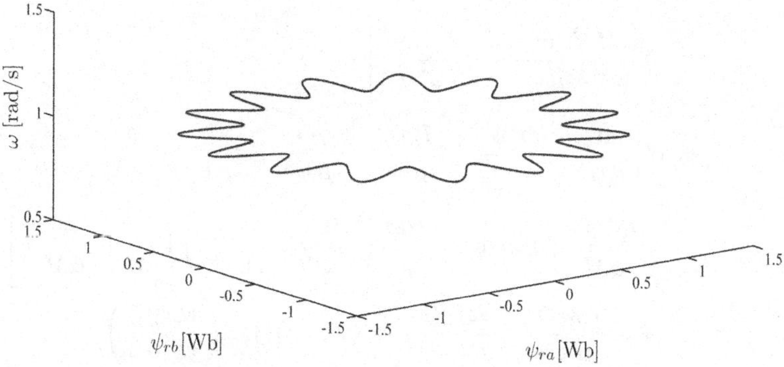

Fig. 1.25 Motor trajectories in the $(\psi_{ra}, \psi_{rb}, \omega)$ space ($\omega^* = 1\,\text{rad/s}$, $\psi^*(t) = 1.16 + 0.2\sin(6\pi t)\,\text{Wb}$, $T_L = 0.1\,\text{Nm}$)

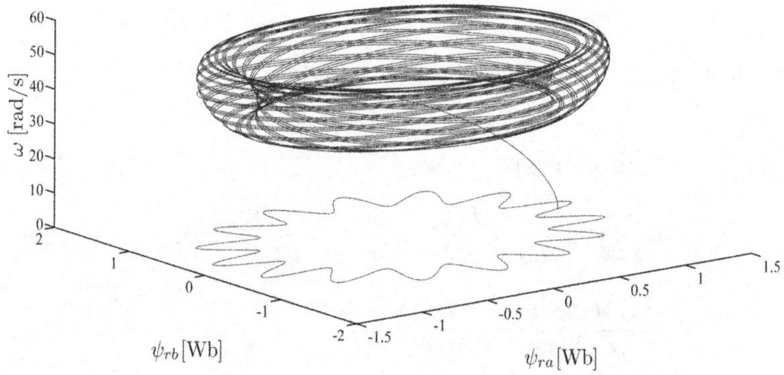

Fig. 1.26 Motor trajectories in the $(\psi_{ra}, \psi_{rb}, \omega)$ space with time-varying speed reference

oriented model (1.39) and solve for (u_{sd}^*, u_{sq}^*) we obtain the control inputs which are compatible with the tracking dynamics; equivalently, if we set $\omega = \omega^*$, $\psi_{rd} = \psi^*$, $\psi_{rq} = 0$, $i_{sd} = i_{sd}^*$, and $i_{sq} = i_{sq}^*$ in the last two equations in (1.36) we obtain

$$\begin{aligned}
\frac{\mathrm{d}i_{sd}^*}{\mathrm{d}t} &= -\gamma i_{sd}^* + \omega^* i_{sq}^* + \frac{\alpha M i_{sq}^{*2}}{\psi^*} + \beta\alpha\psi^* + \frac{u_{sd}^*}{\sigma} \\
\frac{\mathrm{d}i_{sq}^*}{\mathrm{d}t} &= -\gamma i_{sq}^* - \omega^* i_{sd}^* - \frac{\alpha M i_{sq}^* i_{sd}^*}{\psi^*} - \beta\omega^*\psi^* + \frac{u_{sq}^*}{\sigma} \, .
\end{aligned} \tag{1.72}$$

Substituting in (1.72) the currents (i_{sd}^*, i_{sq}^*) given in (1.67) and solving for (u_{sd}^*, u_{sq}^*) we obtain

$$
\begin{aligned}
u_{sd}^* &= \sigma\left[\frac{\dot\psi^*}{M} + \frac{\ddot\psi^*}{\alpha M} + \frac{R_s\psi^*}{\sigma M} + \frac{\gamma\dot\psi^*}{\alpha M} - \frac{\omega^*\dot\omega^*}{\mu\psi^*} - \frac{\omega^* T_L}{\mu J\psi^*}\right.\\
&\quad \left. -\frac{\alpha M}{\psi^*}\left(\frac{\dot\omega^*}{\mu\psi^*} + \frac{T_L}{J\mu\psi^*}\right)^2\right]\\
u_{sq}^* &= \sigma\left[\frac{\ddot\omega^*}{\mu\psi^*} - \frac{\dot\omega^*\dot\psi^*}{\mu\psi^{*2}} - \frac{T_L\dot\psi^*}{J\mu\psi^{*2}} + \frac{\gamma\dot\omega^*}{\mu\psi^*} + \frac{\gamma T_L}{J\mu\psi^*} + \frac{\omega^*\psi^*}{M}\right.\\
&\quad \left. +\frac{\omega^*\dot\psi^*}{\alpha M} + \beta\omega^*\psi^* + \frac{\alpha M}{\psi^*}\left(\frac{\dot\omega^*}{\mu\psi^*} + \frac{T_L}{J\mu\psi^*}\right)\left(\frac{\psi^*}{M} + \frac{\dot\psi^*}{\alpha M}\right)\right]\\
\dot\rho^* &= \omega^* + \frac{\alpha M\dot\omega^*}{\mu\psi^{*2}} + \frac{\alpha M T_L}{\mu J\psi^{*2}},\quad \rho^*(0) = \arctan\left(\frac{\psi_{rb}(0)}{\psi_{ra}(0)}\right)\\
\begin{bmatrix} u_{sa}^* \\ u_{sb}^* \end{bmatrix} &= \begin{bmatrix} \cos\rho^* & -\sin\rho^* \\ \sin\rho^* & \cos\rho^* \end{bmatrix}\begin{bmatrix} u_{sd}^* \\ u_{sq}^* \end{bmatrix}
\end{aligned}
\tag{1.73}
$$

which constitute a first order dynamic left inverse system for the induction motor model (1.26) with ω^*, $\dot\omega^*$, $\ddot\omega^*$, ψ^*, $\dot\psi^*$, $\dot\psi^*$, $\ddot\psi^*$ as inputs and u_{sa}^*, u_{sb}^* as outputs.

In conclusion, the first order *feedforward control*

$$
\begin{aligned}
\begin{bmatrix} u_{sa}^* \\ u_{sb}^* \end{bmatrix} &= \begin{bmatrix} \cos\rho^* & -\sin\rho^* \\ \sin\rho^* & \cos\rho^* \end{bmatrix}\begin{bmatrix} u_{sd}^* \\ u_{sq}^* \end{bmatrix}\\
u_{sd}^* &= \sigma\left[\frac{\dot\psi^*}{M} + \frac{\ddot\psi^*}{\alpha M} + \frac{R_s\psi^*}{\sigma M} + \frac{\gamma\dot\psi^*}{\alpha M} - \frac{\omega^*\dot\omega^*}{\mu\psi^*} - \frac{\omega^* T_L}{\mu J\psi^*}\right.\\
&\quad \left. -\frac{\alpha M}{\psi^*}\left(\frac{\dot\omega^*}{\mu\psi^*} + \frac{T_L}{J\mu\psi^*}\right)^2\right]\\
u_{sq}^* &= \sigma\left[\frac{\ddot\omega^*}{\mu\psi^*} - \frac{\dot\omega^*\dot\psi^*}{\mu\psi^{*2}} - \frac{T_L\dot\psi^*}{J\mu\psi^{*2}} + \frac{\gamma\dot\omega^*}{\mu\psi^*} + \frac{\gamma T_L}{J\mu\psi^*} + \frac{\omega^*\psi^*}{M}\right.\\
&\quad \left. +\frac{\omega^*\dot\psi^*}{\alpha M} + \beta\omega^*\psi^* + \frac{\alpha M}{\psi^*}\left(\frac{\dot\omega^*}{\mu\psi^*} + \frac{T_L}{J\mu\psi^*}\right)\left(\frac{\psi^*}{M} + \frac{\dot\psi^*}{\alpha M}\right)\right]\\
\dot\rho^* &= \omega^* + \frac{\alpha M\dot\omega^*}{\mu\psi^{*2}} + \frac{\alpha M T_L}{\mu J\psi^{*2}},\quad \rho^*(0) = \arctan\left(\frac{\psi_{rb}(0)}{\psi_{ra}(0)}\right)
\end{aligned}
\tag{1.74}
$$

is such that, when applied to the induction motor fixed frame model (1.26) with initial conditions

$$
\begin{aligned}
\omega(0) &= \omega^*(0)\\
\psi_{ra}(0) &= \psi^*(0)\cos\rho^*(0)
\end{aligned}
$$

$$\psi_{rb}(0) = \psi^*(0)\sin\rho^*(0)$$

$$\begin{bmatrix} i_{sa}(0) \\ i_{sb}(0) \end{bmatrix} = \begin{bmatrix} \cos\rho^*(0) & -\sin\rho^*(0) \\ \sin\rho^*(0) & \cos\rho^*(0) \end{bmatrix} \begin{bmatrix} \frac{\psi^*(0)}{M} + \frac{\dot{\psi}^*(0)}{\alpha M} \\ \frac{\dot{\omega}^*(0)}{\mu\psi^*(0)} + \frac{T_L}{J\mu\psi^*(0)} \end{bmatrix},$$

guarantees that

$$\omega(t) = \omega^*(t)$$
$$\psi_{ra}(t) = \psi^*(t)\cos\rho^*(t)$$
$$\psi_{rb}(t) = \psi^*(t)\sin\rho^*(t)$$
$$\begin{bmatrix} i_{sa}(t) \\ i_{sb}(t) \end{bmatrix} = \begin{bmatrix} \cos\rho^*(t) & -\sin\rho^*(t) \\ \sin\rho^*(t) & \cos\rho^*(t) \end{bmatrix} \begin{bmatrix} \frac{\psi^*(t)}{M} + \frac{\dot{\psi}^*(t)}{\alpha M} \\ \frac{\dot{\omega}^*(t)}{\mu\psi^*(t)} + \frac{T_L}{J\mu\psi^*(t)} \end{bmatrix}.$$

1.5 Observability

In this section Definition B.8 presented in Appendix B concerning local observability for nonlinear systems and the corresponding sufficient conditions given by Theorem B.11 will be used to investigate the observability properties of the induction motor. Since rotor flux measurements are not easily available, the first question to be addressed is the observability of rotor fluxes from stator currents, stator voltages, and rotor speed measurements. Since rotor speed measurements may also not be available due to sensor failures or on purpose to increase the reliability of the drive system, the second important question that will be discussed is the observability of rotor speed. Consider the induction motor fixed frame model (1.26) and assume that the rotor speed ω is constant so that the model becomes linear:

$$\begin{bmatrix} \frac{d\psi_{ra}}{dt} \\ \frac{d\psi_{rb}}{dt} \\ \frac{di_{sa}}{dt} \\ \frac{di_{sb}}{dt} \end{bmatrix} = \begin{bmatrix} -\alpha & -\omega & \alpha M & 0 \\ \omega & -\alpha & 0 & \alpha M \\ \beta\alpha & \beta\omega & -\gamma & 0 \\ -\beta\omega & \beta\alpha & 0 & -\gamma \end{bmatrix} \begin{bmatrix} \psi_{ra} \\ \psi_{rb} \\ i_{sa} \\ i_{sb} \end{bmatrix} + \begin{bmatrix} 0 & 0 \\ 0 & 0 \\ 1/\sigma & 0 \\ 0 & 1/\sigma \end{bmatrix} \begin{bmatrix} u_{sa} \\ u_{sb} \end{bmatrix}$$

$$\begin{bmatrix} i_{sa} \\ i_{sb} \end{bmatrix} = \begin{bmatrix} 0\,0\,1\,0 \\ 0\,0\,0\,1 \end{bmatrix} \begin{bmatrix} \psi_{ra} \\ \psi_{rb} \\ i_{sa} \\ i_{sb} \end{bmatrix}. \tag{1.75}$$

It is easy to check (see Problem 1.1) that the system (1.75) is observable for any constant value of rotor speed, including zero speed, and for any motor parameter $(\alpha, M, \beta, \gamma, \sigma)$: this implies that the rotor fluxes are observable from stator current measurements at constant rotor speed. If we also pose the problem of rotor speed observability from stator current measurements when the rotor speed ω is constant, the fixed frame model (1.26) should be considered

$$\begin{aligned}
\frac{d\omega}{dt} &= 0 \\
\frac{d\psi_{ra}}{dt} &= -\alpha\psi_{ra} - \omega\psi_{rb} + \alpha M i_{sa} \\
\frac{d\psi_{rb}}{dt} &= -\alpha\psi_{rb} + \omega\psi_{ra} + \alpha M i_{sb} \\
\frac{di_{sa}}{dt} &= -\gamma i_{sa} + \frac{u_{sa}}{\sigma} + \beta\alpha\psi_{ra} + \beta\omega\psi_{rb} \\
\frac{di_{sb}}{dt} &= -\gamma i_{sb} + \frac{u_{sb}}{\sigma} + \beta\alpha\psi_{rb} - \beta\omega\psi_{ra}
\end{aligned}$$

which can be rewritten as

$$\dot{x} = f(x) + g_a u_{sa} + g_b u_{sb}$$

with $x = [i_{sa}, i_{sb}, \psi_{ra}, \psi_{rb}, \omega]^{\mathrm{T}}$ and suitable vector fields f, g_a, g_b. Define the measured output functions

$$\begin{aligned}
h_1(i_{sa}) &= i_{sa} \\
h_2(i_{sb}) &= i_{sb} \; .
\end{aligned}$$

Compute the Lie derivatives

$$\begin{aligned}
L_f h_1 &= -\gamma i_{sa} + \beta\alpha\psi_{ra} + \beta\omega\psi_{rb} \\
L_f h_2 &= -\gamma i_{sb} + \beta\alpha\psi_{rb} - \beta\omega\psi_{ra} \\
L_f^2 h_1 &= -\gamma[-\gamma i_{sa} + \beta\alpha\psi_{ra} + \beta\omega\psi_{rb}] + \beta\alpha[-\alpha\psi_{ra} - \omega\psi_{rb} + \alpha M i_{sa}] \\
&\quad + \beta\omega[-\alpha\psi_{rb} + \omega\psi_{ra} + \alpha M i_{sb}] \\
L_f^2 h_2 &= -\gamma[-\gamma i_{sb} + \beta\alpha\psi_{rb} - \beta\omega\psi_{ra}] - \beta\omega[-\alpha\psi_{ra} - \omega\psi_{rb} + \alpha M i_{sa}] \\
&\quad + \beta\alpha[-\alpha\psi_{rb} + \omega\psi_{ra} + \alpha M i_{sb}] \; .
\end{aligned}$$

Consider the matrix

$$\begin{bmatrix} dh_1 \\ dh_2 \\ d(L_f h_1) \\ d(L_f h_2) \\ d(L_f^2 h_2) \end{bmatrix}$$

whose determinant is

$$-\beta^3(\alpha^2 + \omega^2)(-\alpha\psi_{ra} - \omega\psi_{rb} + \alpha M i_{sa}) \; .$$

Consider also the matrix

$$\begin{bmatrix} \mathrm{d}h_1 \\ \mathrm{d}h_2 \\ \mathrm{d}(L_f h_1) \\ \mathrm{d}(L_f h_2) \\ \mathrm{d}(L_f^2 h_1) \end{bmatrix}$$

whose determinant is

$$\beta^3(\alpha^2+\omega^2)(-\alpha\psi_{rb}+\omega\psi_{ra}+\alpha M i_{sb})\,.$$

Therefore, in the case of constant rotor speed, if

$$(-\alpha\psi_{ra}-\omega\psi_{rb}+\alpha M i_{sa})^2+(-\alpha\psi_{rb}+\omega\psi_{ra}+\alpha M i_{sb})^2>0$$

then, by Theorem B.11 in Appendix B, the induction motor is locally observable at any state in which the above inequality holds. In other words, when ω is constant, it is locally observable provided that the rotor flux vector rotates or its modulus is not constant.

In the general case in which the rotor speed ω is not constant and is available from measurements, consider the induction motor fixed frame model (1.26) which can be rewritten as

$$\dot{x} = f(x) + g_a u_{sa} + g_b u_{sb}$$

with $x = [i_{sa}, i_{sb}, \omega, \psi_{ra}, \psi_{rb}]^{\mathrm{T}}$ being the state variable vector and f, g_a, g_b being suitable vector fields. Define the measured output functions

$$\begin{aligned} h_1(i_{sa}) &= i_{sa} \\ h_2(i_{sb}) &= i_{sb} \\ h_3(\omega) &= \omega\,. \end{aligned}$$

Compute the Lie derivatives

$$\begin{aligned} L_f h_1 &= -\gamma i_{sa} + \beta\alpha\psi_{ra} + \beta\omega\psi_{rb} \\ L_f h_2 &= -\gamma i_{sb} + \beta\alpha\psi_{rb} - \beta\omega\psi_{ra} \\ L_f h_3 &= \mu\psi_{ra} i_{sb} - \mu\psi_{rb} i_{sa} - \frac{T_L}{J}\,. \end{aligned}$$

Since the determinant of the matrix

$$\begin{bmatrix} \mathrm{d}h_1 \\ \mathrm{d}h_2 \\ \mathrm{d}h_3 \\ \mathrm{d}(L_f h_1) \\ \mathrm{d}(L_f h_2) \end{bmatrix}$$

is $\beta[\alpha^2+\omega^2]$ and is always different from zero, it follows from Theorem B.11 in Appendix B that the induction motor is locally observable at any $x \in \mathbb{R}^5$. Moreover, the rotor fluxes (ψ_{ra}, ψ_{rb}) can be uniquely expressed in terms of the inputs (u_{sa}, u_{sb}), the outputs (ω, i_{sa}, i_{sb}), and their time derivatives. In fact, from the last two equations in the fixed frame model (1.26) we have

$$\begin{bmatrix} \alpha & \omega \\ -\omega & \alpha \end{bmatrix} \begin{bmatrix} \psi_{ra} \\ \psi_{rb} \end{bmatrix} = \begin{bmatrix} \frac{1}{\beta}\frac{di_{sa}}{dt} + \frac{\gamma i_{sa}}{\beta} - \frac{u_{sa}}{\sigma\beta} \\ \frac{1}{\beta}\frac{di_{sb}}{dt} + \frac{\gamma i_{sb}}{\beta} - \frac{u_{sb}}{\sigma\beta} \end{bmatrix}$$

from which it follows that

$$\begin{bmatrix} \psi_{ra} \\ \psi_{rb} \end{bmatrix} = \frac{1}{\alpha^2+\omega^2} \begin{bmatrix} \alpha & -\omega \\ \omega & \alpha \end{bmatrix} \begin{bmatrix} \frac{1}{\beta}\frac{di_{sa}}{dt} + \frac{\gamma i_{sa}}{\beta} - \frac{u_{sa}}{\sigma\beta} \\ \frac{1}{\beta}\frac{di_{sb}}{dt} + \frac{\gamma i_{sb}}{\beta} - \frac{u_{sb}}{\sigma\beta} \end{bmatrix}. \tag{1.76}$$

In other words, there cannot be two different solutions $(\psi_{ra1}(t),\ \psi_{rb1}(t))$ and $(\psi_{ra2}(t),\ \psi_{rb2}(t))$ giving rise to the same measurements $(i_{sa}(t),\ i_{sb}(t),\ \omega(t))$ when the same inputs $(u_{sa}(t),\ u_{sb}(t))$ are applied, so there cannot be any pair of indistinguishable initial states for any admissible control input. The induction motor is therefore observable from rotor speed and stator current measurements for any voltage input $(u_{sa}(t),\ u_{sb}(t))$.

The most difficult problem is the observability of the induction motor when the rotor speed ω is not measured and is not constant. To this end, let us consider the fixed frame model (1.26) with output functions

$$\begin{aligned} h_1(i_{sa}) &= i_{sa} \\ h_2(i_{sb}) &= i_{sb} \\ h_3(\psi_{ra}) &= \psi_{ra} \\ h_4(\psi_{rb}) &= \psi_{rb} \end{aligned}$$

which can be rewritten as

$$\begin{aligned} \dot{x} &= f(x) + g_a u_{sa} + g_b u_{sb} \\ y_1 &= h_1(x) \\ y_2 &= h_2(x) \\ y_3 &= h_3(x) \\ y_4 &= h_4(x) \end{aligned}$$

with $x = [\omega, \psi_{ra}, \psi_{rb}, i_{sa}, i_{sb}]^{\mathrm{T}}$ and f, g_a, g_b suitable vector fields. Compute the Lie derivatives

$$\begin{aligned} L_f h_1 &= -\gamma i_{sa} + \alpha\beta\psi_{ra} + \beta\omega\psi_{rb} \\ L_f h_2 &= -\gamma i_{sb} + \alpha\beta\psi_{rb} - \beta\omega\psi_{ra} \\ L_f h_3 &= -\alpha\psi_{ra} - \omega\psi_{rb} + \alpha M i_{sa} \end{aligned}$$

$$\begin{aligned}
L_f h_4 &= -\alpha\psi_{rb} + \omega\psi_{ra} + \alpha M i_{sb} \\
L_f^2 h_1 &= -\gamma(-\gamma i_{sa} + \alpha\beta\psi_{ra} + \beta\omega\psi_{rb}) \\
&\quad + \beta\alpha(-\alpha\psi_{ra} - \omega\psi_{rb} + \alpha M i_{sa}) \\
&\quad + \beta\omega(-\alpha\psi_{rb} + \omega\psi_{ra} + \alpha M i_{sb}) \\
&\quad + \beta\psi_{rb}\left[\mu(\psi_{ra} i_{sb} - \psi_{rb} i_{sa}) - \frac{T_L}{J}\right] \\
L_f^2 h_2 &= -\gamma(-\gamma i_{sb} + \alpha\beta\psi_{rb} - \beta\omega\psi_{ra}) \\
&\quad - \beta\omega(-\alpha\psi_{ra} - \omega\psi_{rb} + \alpha M i_{sa}) \\
&\quad + \beta\alpha(-\alpha\psi_{rb} + \omega\psi_{ra} + \alpha M i_{sb}) \\
&\quad - \beta\psi_{ra}\left[\mu(\psi_{ra} i_{sb} - \psi_{rb} i_{sa}) - \frac{T_L}{J}\right] .
\end{aligned}$$

In order to explore the observability from $(\psi_{ra}, \psi_{rb}, i_{sa}, i_{sb})$ measurements we consider the matrix

$$\begin{bmatrix} \mathrm{d}h_1 \\ \mathrm{d}h_2 \\ \mathrm{d}h_3 \\ \mathrm{d}h_4 \\ \mathrm{d}(L_f h_3) \\ \mathrm{d}(L_f h_4) \end{bmatrix} .$$

Since it has full rank provided that $\psi_{ra}^2 + \psi_{rb}^2 > 0$, it follows from Theorem B.11 in Appendix B that the induction motor is locally observable from stator current and rotor flux measurements at any state in which $\psi_{ra}^2 + \psi_{rb}^2 > 0$.

In order to explore the observability from stator current measurements only we consider the matrix

$$\begin{bmatrix} \mathrm{d}h_1 \\ \mathrm{d}h_2 \\ \mathrm{d}(L_f h_1) \\ \mathrm{d}(L_f h_2) \\ \mathrm{d}(L_f^2 h_1) \\ \mathrm{d}(L_f^2 h_2) \end{bmatrix} = \begin{bmatrix} 0 & I_{2\times 2} \\ J_3 & J_4 \end{bmatrix} .$$

We need to evaluate the rank of the matrix J_3 whose elements $J_{3,ij}$, $1 \le i \le 4$, $1 \le j \le 3$, are listed in the following:

$$\begin{aligned}
J_{3,11} &= \beta\psi_{rb} \\
J_{3,12} &= J_{3,23} = \alpha\beta \\
J_{3,13} &= -J_{3,22} = \beta\omega \\
J_{3,21} &= -\beta\psi_{ra} \\
J_{3,31} &= (-\gamma\beta - 2\beta\alpha)\psi_{rb} + 2\beta\omega\psi_{ra} + \alpha\beta M i_{sb}
\end{aligned}$$

$$\begin{aligned}
J_{3,32} &= -\alpha\beta\gamma - \alpha^2\beta + \beta\omega^2 + \beta\mu\psi_{rb}i_{sb} \\
J_{3,33} &= -\beta\gamma\omega - 2\beta\alpha\omega - 2\beta\mu\psi_{rb}i_{sa} \\
J_{3,41} &= (\gamma\beta + 2\beta\alpha)\psi_{ra} + 2\beta\omega\psi_{rb} - \alpha\beta M i_{sa} \\
J_{3,42} &= \beta\gamma\omega + 2\beta\alpha\omega - 2\beta\mu\psi_{ra}i_{sb} \\
J_{3,43} &= -\alpha\beta\gamma - \alpha^2\beta + \beta\omega^2 + \beta\mu\psi_{ra}i_{sa}\,.
\end{aligned}$$

We can conclude, by virtue of Theorem B.11 in Appendix B, that the induction motor model (1.26) is locally observable at any state $(\omega,\ \psi_{ra},\ \psi_{rb},\ i_{sa},\ i_{sb})$ in which rank $J_3 = 3$.

In conclusion:

1. If $\omega(t)$ is constant for every $t \geq 0$, the induction motor model (1.26) is locally observable from stator current measurements at any state $(\omega,\ \psi_{ra},\ \psi_{rb},\ i_{sa},\ i_{sb})$ such that
$$(-\alpha\psi_{ra} - \omega\psi_{rb} + \alpha M i_{sa})^2 + (-\alpha\psi_{rb} + \omega\psi_{ra} + \alpha M i_{sb})^2 > 0\,,$$
i.e. when $\dot{\psi}_{ra}^2 + \dot{\psi}_{rb}^2 > 0$.
2. The induction motor model (1.26) is observable from rotor speed and stator current measurements for any voltage inputs (u_{sa}, u_{sb}).
3. The induction motor model (1.26) is locally observable from rotor flux and stator current measurements at any state $(\omega,\ \psi_{ra},\ \psi_{rb},\ i_{sa},\ i_{sb})$ such that $\psi_{ra}^2 + \psi_{rb}^2 > 0$.

1.6 Parameter Identifiability

In this section the identifiability properties for the load torque and the rotor resistance will be investigated.

1.6.1 Load Torque Identifiability

Consider the induction motor fixed frame model (1.26) with the additional state variable T_L

$$\begin{aligned}
\frac{\mathrm{d}\omega}{\mathrm{d}t} &= \mu\left(\psi_{ra}i_{sb} - \psi_{rb}i_{sa}\right) - \frac{T_L}{J} \\
\frac{\mathrm{d}\psi_{ra}}{\mathrm{d}t} &= -\alpha\psi_{ra} - \omega\psi_{rb} + \alpha M i_{sa}
\end{aligned}$$

$$\begin{aligned}
\frac{\mathrm{d}\psi_{rb}}{\mathrm{d}t} &= -\alpha\psi_{rb} + \omega\psi_{ra} + \alpha M i_{sb} \\
\frac{\mathrm{d}i_{sa}}{\mathrm{d}t} &= -\gamma i_{sa} + \frac{u_{sa}}{\sigma} + \beta\alpha\psi_{ra} + \beta\omega\psi_{rb} \\
\frac{\mathrm{d}i_{sb}}{\mathrm{d}t} &= -\gamma i_{sb} + \frac{u_{sb}}{\sigma} + \beta\alpha\psi_{rb} - \beta\omega\psi_{ra} \\
\frac{\mathrm{d}T_L}{\mathrm{d}t} &= 0
\end{aligned} \tag{1.77}$$

which can be rewritten as

$$\dot{z} = f(z) + g_a u_{sa} + g_b u_{sb}$$

with $z = [i_{sa}, i_{sb}, \omega, \psi_{ra}, \psi_{rb}, T_L]^{\mathrm{T}}$ and suitable vector fields f, g_a, and g_b. Define the output functions

$$\begin{aligned}
h_1(i_{sa}) &= i_{sa} \\
h_2(i_{sb}) &= i_{sb} \\
h_3(\psi_{ra}) &= \psi_{ra} \\
h_4(\psi_{rb}) &= \psi_{rb} \\
h_5(\omega) &= \omega\,.
\end{aligned}$$

Compute the Lie derivatives

$$\begin{aligned}
L_f h_1 &= -\gamma i_{sa} + \alpha\beta\psi_{ra} + \beta\omega\psi_{rb} \\
L_f h_2 &= -\gamma i_{sb} + \alpha\beta\psi_{rb} - \beta\omega\psi_{ra} \\
L_f h_3 &= -\alpha\psi_{ra} - \omega\psi_{rb} + \alpha M i_{sa} \\
L_f h_4 &= -\alpha\psi_{rb} + \omega\psi_{ra} + \alpha M i_{sb} \\
L_f h_5 &= \mu(\psi_{ra} i_{sb} - \psi_{rb} i_{sa}) - \frac{T_L}{J}\,.
\end{aligned}$$

If the variables $(i_{sa}, i_{sb}, \omega, \psi_{ra}, \psi_{rb})$ are measured, then from the speed dynamics in (1.77) we obtain

$$T_L = J\mu(\psi_{ra} i_{sb} - \psi_{rb} i_{sa}) - J\dot{\omega}\,.$$

Hence, T_L can be obtained from the measured signals and their time derivatives and it is, therefore, identifiable.

If the only variables (i_{sa}, i_{sb}, ω) are measured, then consider the matrix

$$\begin{bmatrix} dh_1 \\ dh_2 \\ dh_5 \\ d(L_f h_1) \\ d(L_f h_2) \\ d(L_f h_5) \end{bmatrix} = \left[\begin{array}{c|ccc} I_{3\times 3} & & 0 & \\ \hline & \beta\alpha & \beta\omega & 0 \\ * & -\beta\omega & \beta\alpha & 0 \\ & \mu i_{sb} & -\mu i_{sa} & -1/J \end{array}\right]$$

whose determinant is

$$-\frac{\beta}{J}(\alpha^2+\omega^2)$$

which is different from zero for every extended state z: hence, according to Theorem B.11 in Appendix B, the system is locally observable at any extended state from (i_{sa}, i_{sb}, ω) measurements. Moreover, there cannot be two initial conditions z_1 and z_2 which are indistinguishable from (i_{sa}, i_{sb}, ω) measurements since the load torque T_L and the rotor fluxes (ψ_{ra}, ψ_{rb}) can be expressed in terms of the inputs (u_{sa}, u_{sb}), the outputs (i_{sa}, i_{sb}, ω), and their time derivatives as follows:

$$\begin{aligned} T_L &= J\mu(\psi_{ra} i_{sb} - \psi_{rb} i_{sa}) - J\dot{\omega} \\ \begin{bmatrix} \psi_{ra} \\ \psi_{rb} \end{bmatrix} &= \frac{1}{\beta(\alpha^2+\omega^2)} \begin{bmatrix} \alpha & -\omega \\ \omega & \alpha \end{bmatrix} \begin{bmatrix} \frac{di_{sa}}{dt} + \gamma i_{sa} - \frac{1}{\sigma} u_{sa} \\ \frac{di_{sb}}{dt} + \gamma i_{sb} - \frac{1}{\sigma} u_{sb} \end{bmatrix}. \end{aligned}$$

Hence, the induction motor is observable and identifiable with respect to T_L from (i_{sa}, i_{sb}, ω) measurements for any voltage inputs (u_{sa}, u_{sb}).

1.6.2 Rotor Resistance Identifiability

Consider the induction motor fixed frame model (1.26) with one additional state variable α

$$\begin{aligned} \frac{d\omega}{dt} &= \mu\left(\psi_{ra} i_{sb} - \psi_{rb} i_{sa}\right) - \frac{T_L}{J} \\ \frac{d\psi_{ra}}{dt} &= -\alpha\psi_{ra} - \omega\psi_{rb} + \alpha M i_{sa} \\ \frac{d\psi_{rb}}{dt} &= -\alpha\psi_{rb} + \omega\psi_{ra} + \alpha M i_{sb} \\ \frac{di_{sa}}{dt} &= -\left(\frac{R_s}{\sigma} + \alpha\beta M\right) i_{sa} + \beta\alpha\psi_{ra} + \beta\omega\psi_{rb} + \frac{u_{sa}}{\sigma} \\ \frac{di_{sb}}{dt} &= -\left(\frac{R_s}{\sigma} + \alpha\beta M\right) i_{sb} + \beta\alpha\psi_{rb} - \beta\omega\psi_{ra} + \frac{u_{sb}}{\sigma} \\ \dot{\alpha} &= 0 \end{aligned} \tag{1.78}$$

which can be rewritten as

$$\dot{x} = f(x) + g_a u_{sa} + g_b u_{sb}$$

with $x = [i_{sa},\ i_{sb},\ \omega,\ \psi_{ra},\ \psi_{rb},\ \alpha]^{\mathrm{T}}$ and suitable vector fields f, g_a, g_b. Define the output functions

$$\begin{aligned} h_1(i_{sa}) &= i_{sa} \\ h_2(i_{sb}) &= i_{sb} \\ h_3(\psi_{ra}) &= \psi_{ra} \\ h_4(\psi_{rb}) &= \psi_{rb} \\ h_5(\omega) &= \omega\,. \end{aligned} \tag{1.79}$$

Compute the Lie derivatives

$$\begin{aligned} L_f h_1 &= -\left(\frac{R_s}{\sigma} + \alpha\beta M\right) i_{sa} + \beta\alpha\psi_{ra} + \beta\omega\psi_{rb} \\ L_f h_2 &= -\left(\frac{R_s}{\sigma} + \alpha\beta M\right) i_{sb} + \beta\alpha\psi_{rb} - \beta\omega\psi_{ra} \\ L_f h_3 &= -\alpha\psi_{ra} - \omega\psi_{rb} + \alpha M i_{sa} \\ L_f h_4 &= -\alpha\psi_{rb} + \omega\psi_{ra} + \alpha M i_{sb} \\ L_f h_5 &= \mu(\psi_{ra} i_{sb} - \psi_{rb} i_{sa}) - \frac{T_L}{J}\,. \end{aligned}$$

If the variables $(i_{sa},\ i_{sb},\ \psi_{ra},\ \psi_{rb},\ \omega)$ are measured then from the flux dynamics in (1.26) we obtain

$$\begin{aligned} \alpha(M i_{sa} - \psi_{ra}) &= \dot{\psi}_{ra} + \omega\psi_{rb} \\ \alpha(M i_{sb} - \psi_{rb}) &= \dot{\psi}_{rb} - \omega\psi_{ra}\,. \end{aligned}$$

Hence, α can be obtained from the measured signals and their time derivatives and is therefore identifiable, provided that for all $t \geq 0$

$$\begin{bmatrix} \psi_{ra}(t) \\ \psi_{rb}(t) \end{bmatrix} \neq M \begin{bmatrix} i_{sa}(t) \\ i_{sb}(t) \end{bmatrix}. \tag{1.80}$$

If only the variables $(i_{sa},\ i_{sb},\ \omega)$ are measured then consider the matrix

$$\begin{bmatrix} \mathrm{d}h_1 \\ \mathrm{d}h_2 \\ \mathrm{d}h_5 \\ \mathrm{d}(L_f h_1) \\ \mathrm{d}(L_f h_2) \\ \mathrm{d}(L_f h_5) \end{bmatrix} = \left[\begin{array}{c|ccc} I_{3\times 3} & & 0 & \\ \hline & \beta\alpha & \beta\omega & \beta(\psi_{ra} - M i_{sa}) \\ * & -\beta\omega & \beta\alpha & \beta(\psi_{rb} - M i_{sb}) \\ & \mu i_{sb} & -\mu i_{sa} & 0 \end{array}\right]$$

whose determinant is

$$\mu\beta^2\left[-M\omega(i_{sa}^2+i_{sb}^2)+\omega(i_{sa}\psi_{ra}+i_{sb}\psi_{rb})-\alpha(\psi_{ra}i_{sb}-\psi_{rb}i_{sa})\right]\ . \quad (1.81)$$

If $\omega=0$ the determinant (1.81) is different from zero provided that the electromagnetic torque $J\mu(\psi_{ra}i_{sb}-\psi_{rb}i_{sa})$ is different from zero. If the electromagnetic torque is equal to zero the determinant (1.81) is different from zero when (1.80) is satisfied. If (1.81) is satisfied at an extended state $(i_{sa}, i_{sb}, \omega, \psi_{ra}, \psi_{rb}, \alpha)$ then, according to Theorem B.11 in Appendix B, the system is locally observable at such extended state from (i_{sa}, i_{sb}, ω) measurements.

If only the variables $(i_{sa}, i_{sb}, \psi_{ra}, \psi_{rb})$ are measured then consider the matrix

$$\begin{bmatrix} \mathrm{d}h_1 \\ \mathrm{d}h_2 \\ \mathrm{d}h_3 \\ \mathrm{d}h_4 \\ \mathrm{d}(L_f h_3) \\ \mathrm{d}(L_f h_4) \end{bmatrix}$$

which has full rank provided that

$$\psi_{ra}^2+\psi_{rb}^2-M(\psi_{ra}i_{sa}+\psi_{rb}i_{sb})\neq 0\ . \quad (1.82)$$

Note that

$$\frac{\mathrm{d}(\psi_{ra}^2+\psi_{rb}^2)}{\mathrm{d}t}=-2\alpha(\psi_{ra}^2+\psi_{rb}^2-M\psi_{ra}i_{sa}-M\psi_{rb}i_{sb})\ . \quad (1.83)$$

Hence, by virtue of Theorem B.11 in Appendix B the induction motor is locally observable and locally identifiable with respect to the parameter α at each $(i_{sa}, i_{sb}, \psi_{ra}, \psi_{rb}, \omega, \alpha)$ such that (1.82) holds, *i.e.* when (1.83) is different from zero.

In conclusion:

1. The load torque T_L is identifiable if the state variables $(i_{sa}, i_{sb}, \psi_{ra}, \psi_{rb}, \omega)$ are measured for any voltage inputs (u_{sa}, u_{sb}).
2. The induction motor is observable and identifiable with respect to T_L from (i_{sa}, i_{sb}, ω) measurements for any voltage inputs (u_{sa}, u_{sb}).
3. The rotor resistance R_r is identifiable if the state variables $(i_{sa}, i_{sb}, \psi_{ra}, \psi_{rb}, \omega)$ are measured along any trajectory such that

$$\begin{bmatrix} \psi_{ra}(t) \\ \psi_{rb}(t) \end{bmatrix} \neq M \begin{bmatrix} i_{sa}(t) \\ i_{sb}(t) \end{bmatrix}.$$

4. The induction motor model (1.26) is locally observable and the rotor resistance is locally identifiable from (i_{sa}, i_{sb}, ω) measurements at any $(i_{sa}, i_{sb}, \psi_{ra}, \psi_{rb}, \omega, R_r)$ such that

$$-M\omega(i_{sa}^2+i_{sb}^2)+\omega(i_{sa}\psi_{ra}+i_{sb}\psi_{rb})-\frac{R_r}{L_r}(\psi_{ra}i_{sb}-\psi_{rb}i_{sa})\neq 0\,.$$

5. The induction motor model (1.26) is locally observable and the rotor resistance is locally identifiable from $(i_{sa},\, i_{sb},\, \psi_{ra},\, \psi_{rb})$ measurements at any $(i_{sa},\, i_{sb},\, \psi_{ra},\, \psi_{rb},\, \omega,\, R_r)$ such that

$$\psi_{ra}^2+\psi_{rb}^2-M(\psi_{ra}i_{sa}+\psi_{rb}i_{sb})\neq 0$$

i.e. when

$$\frac{\mathrm{d}(\psi_{ra}^2+\psi_{rb}^2)}{\mathrm{d}t}\neq 0\,.$$

1.7 Feedback Linearizability

In this section we explore the controllability properties of the induction motor. Let us first reconsider the induction motor fixed frame model (1.26) and assume that the rotor speed ω is constant so that the model becomes (1.75), which is rewritten for convenience as

$$\begin{bmatrix}\frac{\mathrm{d}\psi_{ra}}{\mathrm{d}t}\\ \frac{\mathrm{d}\psi_{rb}}{\mathrm{d}t}\\ \frac{\mathrm{d}i_{sa}}{\mathrm{d}t}\\ \frac{\mathrm{d}i_{sb}}{\mathrm{d}t}\end{bmatrix}=\begin{bmatrix}-\alpha & -\omega & \alpha M & 0\\ \omega & -\alpha & 0 & \alpha M\\ \beta\alpha & \beta\omega & -\gamma & 0\\ -\beta\omega & \beta\alpha & 0 & -\gamma\end{bmatrix}\begin{bmatrix}\psi_{ra}\\ \psi_{rb}\\ i_{sa}\\ i_{sb}\end{bmatrix}+\begin{bmatrix}0 & 0\\ 0 & 0\\ 1/\sigma & 0\\ 0 & 1/\sigma\end{bmatrix}\begin{bmatrix}u_{sa}\\ u_{sb}\end{bmatrix}$$

$$\begin{bmatrix}i_{sa}\\ i_{sb}\end{bmatrix}=\begin{bmatrix}0\,0\,1\,0\\ 0\,0\,0\,1\end{bmatrix}\begin{bmatrix}\psi_{ra}\\ \psi_{rb}\\ i_{sa}\\ i_{sb}\end{bmatrix}\,. \tag{1.84}$$

It is easy to check (see Problem 1.2) that the linear system (1.84) is controllable with controllability indices $(2,2)$, for any constant value of rotor speed, including zero speed, and for any motor parameter $(\alpha,\, M,\, \beta,\, \gamma,\, \sigma)$: this implies that we can control the rotor fluxes from stator voltages when the rotor speed is constant. However, the main question is whether we can control the rotor speed. To answer this question positively we shall make use of the concept of feedback linearzation, *i.e.* the property for a nonlinear system to be transformed into a linear controllable system by a state feedback either static or dynamic. Let us introduce the following notations:

$$x=[\omega,\psi_{ra},\psi_{rb},i_{sa},i_{sb}]^{\mathrm{T}}$$

$$f(x) = \begin{bmatrix} \mu(\psi_{ra}i_{sb} - \psi_{rb}i_{sa}) - \frac{T_L}{J} \\ -\alpha\psi_{ra} - \omega\psi_{rb} + \alpha M i_{sa} \\ \omega\psi_{ra} - \alpha\psi_{rb} + \alpha M i_{sb} \\ \alpha\beta\psi_{ra} + \beta\omega\psi_{rb} - \gamma i_{sa} \\ -\beta\omega\psi_{ra} + \alpha\beta\psi_{rb} - \gamma i_{sb} \end{bmatrix}$$

$$g_a = \left[0,0,0,\frac{1}{\sigma},0\right]^{\mathrm{T}}$$

$$g_b = \left[0,0,0,0,\frac{1}{\sigma}\right]^{\mathrm{T}}.$$

The induction motor fixed frame model (1.26) can be rewritten as

$$\dot{x} = f(x) + g_a u_{sa} + g_b u_{sb}.$$

Let us compute

$$\mathrm{ad}_f g_a = \left[\frac{\mu\psi_{rb}}{\sigma}, -\frac{\alpha M}{\sigma}, 0, \frac{\gamma}{\sigma}, 0\right]^{\mathrm{T}}$$

$$\mathrm{ad}_f g_b = \left[-\frac{\mu\psi_{ra}}{\sigma}, 0, -\frac{\alpha M}{\sigma}, 0, \frac{\gamma}{\sigma}\right]^{\mathrm{T}}$$

$$\left[\mathrm{ad}_f g_a, \mathrm{ad}_f g_b\right] = \left[-\frac{2\alpha M\mu}{\sigma}, 0,0,0,0\right]^{\mathrm{T}}.$$

The distribution

$$\mathrm{span}\{g_a, g_b, \mathrm{ad}_f g_a, \mathrm{ad}_f g_b\}$$

is not involutive since

$$\left[\mathrm{ad}_f g_a, \mathrm{ad}_f g_b\right] \notin \mathrm{span}\{g_a, g_b, \mathrm{ad}_f g_a, \mathrm{ad}_f g_b\}.$$

In fact the vector fields $g_a, g_b, \mathrm{ad}_f g_a, \mathrm{ad}_f g_b, \left[\mathrm{ad}_f g_a, \mathrm{ad}_f g_b\right]$ are linearly independent for every $x \in \mathbb{R}^5$. Hence, according to Theorem B.8 in Appendix B, the system (1.26) is not feedback linearizable, *i.e.*, it cannot be transformed by a nonsingular state feedback into a linear controllable system in suitable state coordinates with two inputs. According to Theorem B.9 in Appendix B, the largest feedback linearizable subsystem has dimension 4. However there are many choices of four state variables whose dynamics can be made linear by state feedback. The most obvious choice is to make linear the dynamics of the state variables $(\psi_{ra}, \dot{\psi}_{ra}, \psi_{rb}, \dot{\psi}_{rb})$ by state feedback. Note in fact that the stator currents(i_{sa}, i_{sb}) can be replaced by $\dot{\psi}_{ra} = \Phi_a$ and $\dot{\psi}_{rb} = \Phi_b$ as new state variables since

$$\Phi_a = -\alpha\psi_{ra} - \omega\psi_{rb} + \alpha M i_{sa}$$

$$\Phi_b = -\alpha\psi_{rb} + \omega\psi_{ra} + \alpha M i_{sb}$$

and

$$i_{sa} = \frac{\Phi_a}{\alpha M} + \frac{\psi_{ra}}{M} + \frac{\omega\psi_{rb}}{\alpha M}$$
$$i_{sb} = \frac{\Phi_b}{\alpha M} + \frac{\psi_{rb}}{M} - \frac{\omega\psi_{ra}}{\alpha M}\,.$$

In the new state variables we have

$$\begin{aligned}
\dot{\psi}_{ra} &= \Phi_a \\
\dot{\Phi}_a &= \alpha^2\psi_{ra} + \alpha\omega\psi_{rb} - \alpha^2 M i_{sa} + \alpha\omega\psi_{rb} - \omega^2\psi_{ra} - \omega\alpha M i_{sb} \\
&\quad -\mu\psi_{ra}\psi_{rb}i_{sb} + \mu\psi_{rb}^2 i_{sa} + \frac{\psi_{rb}T_L}{J} - \alpha M\gamma i_{sa} + \alpha^2\beta M\psi_{ra} \\
&\quad +\alpha M\beta\omega\psi_{rb} + \frac{\alpha M}{\sigma}u_{sa} \\
&\triangleq -k_{a1}\psi_{ra} - k_{a2}\Phi_a + v_a \\
\dot{\psi}_{rb} &= \Phi_b \\
\dot{\Phi}_b &= \alpha^2\psi_{rb} - \alpha\omega\psi_{ra} - \alpha^2 M i_{sb} - \alpha\omega\psi_{ra} + \omega^2\psi_{rb} + \alpha\omega M i_{sa} \\
&\quad +\mu\psi_{ra}^2 i_{sb} - \mu\psi_{ra}\psi_{rb}i_{sa} - \frac{\psi_{ra}T_L}{J} - \alpha M\gamma i_{sb} + \alpha^2\beta M\psi_{rb} \\
&\quad -\alpha M\beta\omega\psi_{ra} + \frac{\alpha M}{\sigma}u_{sb} \\
&\triangleq -k_{a1}\psi_{rb} - k_{a2}\Phi_b + v_b \\
\dot{\omega} &= -\frac{\mu\left(\psi_{ra}^2 + \psi_{rb}^2\right)\omega}{\alpha M} + \frac{\mu}{\alpha M}\left(\psi_{ra}\Phi_b - \psi_{rb}\Phi_a\right) - \frac{T_L}{J}\,.
\end{aligned}$$

Hence the fluxes ψ_{ra} and ψ_{rb} can be controlled linearly from the new control inputs (v_a, v_b) and no singularities are introduced by the linearizing procedure. However, the speed dynamics remain nonlinear and it is difficult to control the speed ω through the rotor fluxes (ψ_{ra}, ψ_{rb}) and their time derivatives (Φ_a, Φ_b). If we define the dynamic extension

$$\frac{\mathrm{d}u_{sa}}{\mathrm{d}t} = \frac{v_{sa}}{\sigma} \tag{1.85}$$

the input u_{sa} becomes an additional state variable so that the extended state vector is

$$z = \left[\omega, \psi_{ra}, \psi_{rb}, i_{sa}, i_{sb}, u_{sa}\right]^{\mathrm{T}}$$

while v_{sa} becomes the new control input so that the new control input vector is

$$v = \left[v_{sa}, u_{sb}\right]^{\mathrm{T}}\,.$$

The extended system (1.26) and (1.85) becomes

$$\frac{\mathrm{d}z}{\mathrm{d}t} = \bar{f}(z) + \bar{g}_a v_{sa} + \bar{g}_b u_{sb} \tag{1.86}$$

with

$$\bar{f}(z) = \begin{bmatrix} \mu(\psi_{ra} i_{sb} - \psi_{rb} i_{sa}) - \frac{T_L}{J} \\ -\alpha\psi_{ra} - \omega\psi_{rb} + \alpha M i_{sa} \\ \omega\psi_{ra} - \alpha\psi_{rb} + \alpha M i_{sb} \\ \alpha\beta\psi_{ra} + \beta\omega\psi_{rb} - \gamma i_{sa} + \frac{u_{sa}}{\sigma} \\ -\beta\omega\psi_{ra} + \alpha\beta\psi_{rb} - \gamma i_{sb} \\ 0 \end{bmatrix}$$

$$\bar{g}_a = \left[0,0,0,0,0,0,\frac{1}{\sigma}\right]^{\mathrm{T}}$$

$$\bar{g}_b = \left[0,0,0,0,0,\frac{1}{\sigma},0\right]^{\mathrm{T}}.$$

Let us compute

$$\mathrm{ad}_{\bar{f}}\bar{g}_a = \left[0,0,0,-\frac{1}{\sigma^2},0,0\right]^{\mathrm{T}}$$

$$\mathrm{ad}_{\bar{f}}\bar{g}_b = \left[-\frac{\mu\psi_{ra}}{\sigma},0,-\frac{\alpha M}{\sigma},0,\frac{\gamma}{\sigma},0\right]^{\mathrm{T}}.$$

Since $\bar{g}_a$ and $\bar{g}_b$ are constant linearly independent vector fields, the distribution

$$\mathrm{span}\{\bar{g}_a, \bar{g}_b\}$$

is involutive of dimension 2 for every $z \in \mathbb{R}^6$. Since $\bar{g}_a$, $\bar{g}_b$, $\mathrm{ad}_{\bar{f}}\bar{g}_a$ are constant linearly independent vector fields and $\left[\mathrm{ad}_{\bar{f}}\bar{g}_b,\, \bar{g}_a\right] = 0$, $\left[\mathrm{ad}_{\bar{f}}\bar{g}_b,\, \bar{g}_b\right] = 0$, $\left[\mathrm{ad}_{\bar{f}}\bar{g}_a,\, \mathrm{ad}_{\bar{f}}\bar{g}_b\right] = 0$, the distribution

$$\mathrm{span}\{\bar{g}_a, \bar{g}_b, \mathrm{ad}_{\bar{f}}\bar{g}_a, \mathrm{ad}_{\bar{f}}\bar{g}_b\}$$

is involutive and its dimension is 4 for every $z \in \mathbb{R}^6$. Let us compute

$$\mathrm{ad}^2_{\bar{f}}\bar{g}_a = \left[-\frac{\mu\psi_{rb}}{\sigma^2},\frac{\alpha M}{\sigma^2},0,-\frac{\gamma}{\sigma^2},0,0\right]^{\mathrm{T}}$$

and

$$\mathrm{ad}_{\bar f}^2 \bar g_b = \begin{bmatrix} \frac{\mu}{\sigma}\alpha\psi_{ra} + \frac{\mu}{\sigma}\omega\psi_{rb} - 2\frac{\mu}{\sigma}\alpha M i_{sa} - \frac{\mu}{\sigma}\gamma\psi_{ra} \\ -\frac{\mu}{\sigma}\psi_{ra}\psi_{rb} - \alpha\frac{M}{\sigma}\omega \\ \frac{\mu}{\sigma}\psi_{ra}^2 - \frac{M}{\sigma}\alpha^2 - \frac{M}{\sigma}\alpha\gamma \\ \frac{\mu}{\sigma}\beta\psi_{ra}\psi_{rb} + \frac{M}{\sigma}\alpha\beta\omega \\ -\frac{\mu}{\sigma}\beta\psi_{ra}^2 + \frac{M}{\sigma}\alpha^2\beta + \frac{\gamma^2}{\sigma} \\ 0 \end{bmatrix}.$$

The distribution

$$\mathrm{span}\{\bar g_a, \bar g_b, \mathrm{ad}_{\bar f}\bar g_a, \mathrm{ad}_{\bar f}\bar g_b, \mathrm{ad}_{\bar f}^2\bar g_a, \mathrm{ad}_{\bar f}^2\bar g_b\}$$

has dimension 6 for every $z \in \mathbb{R}^6$ except singular points. Hence, the extended system (1.86) is feedback linearizable according to Theorem B.8 in Appendix B and, consequently, the system (1.26) is dynamically feedback linearizable.

In conclusion:

1. The induction motor model (1.26) is not linearizable by static state feedback.
2. The largest feedback linearizable subsystem by static state feedback of the induction motor model (1.26) has dimension 4.
3. The induction motor model (1.26) is *dynamically feedback linearizable* at nonsingular points.

1.8 Experimental Set-up

The experimental set-up which will be used throughout this book is illustrated in Figure 1.27. A current controlled direct current (DC) motor provides constant or time-varying load torque. A 32-bit digital signal processor (DSP) performs data acquisition, implements the control law using an improved Euler integration algorithm with a sampling time equal to 0.5 ms, and generates reference voltages for the power inverter with symmetrical pulse width modulation (PWM) and switching frequency of 15 kHz. A personal computer is used to program the DSP, to generate load torque or speed commands to the DC motor control system, and to display experimental data. The stator phase currents and voltages are measured by Hall-type sensors and three-phase isolated voltage sensors, respectively. The motor instantaneous speed is measured by an optical incremental encoder with 2,000 lines per revolution. The value of the stator resistance is updated using a temperature sensor mounted on the motor. All measured electrical signals are filtered by analog second order low-pass filters with cut off frequency equal to 2.6 kHz and converted by two parallel 12 bit analog/digital (A/D) converter channels with 25 μs conversion time. A 0.6-

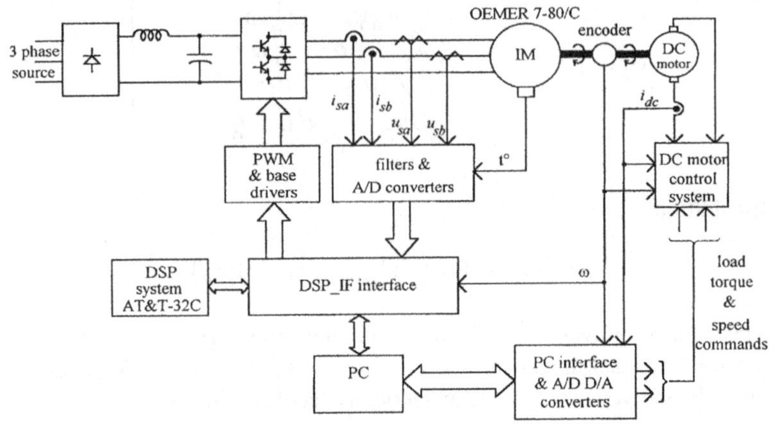

Fig. 1.27 Block diagram of the experimental set-up

kW induction motor (OEMER 7-80/C) has been used whose data, provided by the manufacturer, are given in Table 1.6. The accuracy of the model and its parame-

Table 1.6 Induction motor data

Rated power	600 W
Rated speed	1000 r/min
Rated torque	5.8 Nm
Maximum peak torque	28 Nm
Maximum mechanical speed	6000 r/min
Power supply three-phase	120 V
Rated current	4 A
Excitation current	2 A
Angular acceleration	17500 rad/s^2
Rotor inertia	0.0016 Kgm^2
Motor-Load inertia	0.0075 Kgm^2
Thermal time constant	30 min
Weight	12 Kg
Stator resistance	5.3 Ω
Rotor resistance	3.3 Ω
Stator inductance	0.365 H
Rotor inductance	0.375 H
Mutual inductance	0.34 H

ters was first tested. Figure 1.28 compares experimental (dashed line) and simulated (solid line) static torque–speed motor characteristics: a good matching is exhibited for small slip speed operating conditions, that is for load torque values smaller than the load torque at stall T_{Ls}. The transients corresponding to unloaded motor start up, when a phase input of 110 V/16.7 Hz is applied, are given in Figure 1.29: we note

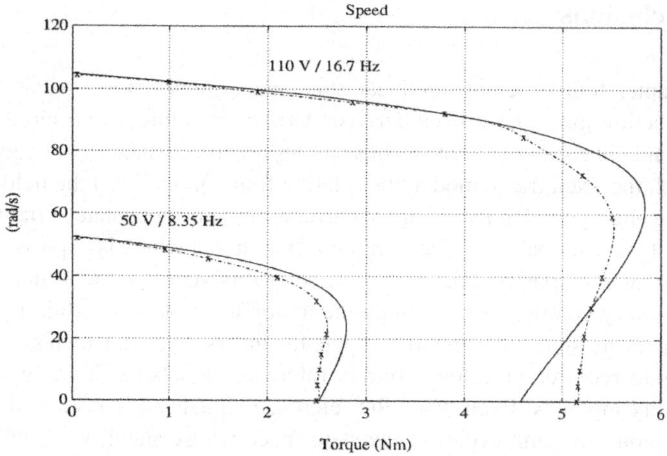

Fig. 1.28 Experimental (dashed line) and computed (solid line) static torque–speed characteristics

that the reference (dashed line) and the real (solid line) voltages are different due to the presence of high-order harmonics generated by the inverter. The discrepancies between simulated and experimental results are also due to parameter inaccuracies and unmodeled effects (such as magnetic saturation and skin effect).

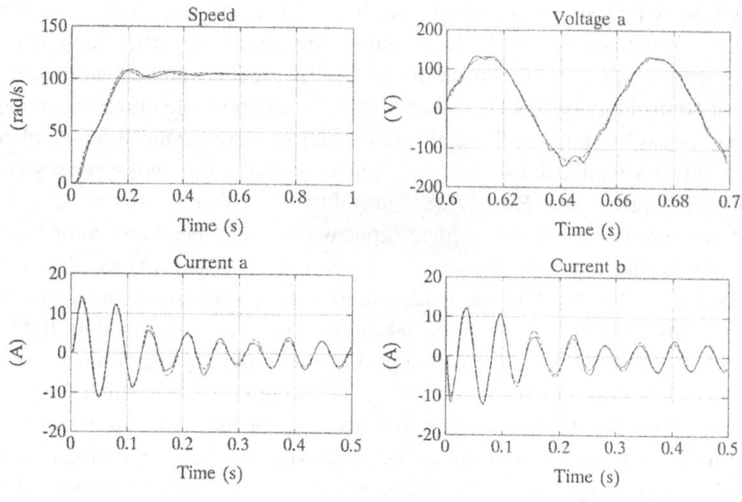

Fig. 1.29 Experimental (solid line) and simulated (dashed line) unloaded motor start-up

1.9 Conclusions

In this chapter four different nonlinear state space models for a balanced unsaturated induction motor have been derived starting from three short circuited rotor windings and three stator windings whose inputs are three balanced voltages: the energy model, the fixed frame model, the rotating frame model, and the field-oriented model. The energy model, in which the currents are chosen as state variables, clarifies the motor power balance. The rotating frame model which is expressed in a rotating frame whose axis coincides with the rotor flux vector is called field-oriented model: it is very helpful in determining the motor steady-state conditions when the voltage inputs are sinusoidal and to compute the inverse system which generates the control inputs required to follow arbitrary reference signals for the rotor speed and the rotor flux modulus. Even when the reference signals are constant, the steady-state conditions are limit cycles in the state space whose stability is studied using linear approximations and the corresponding eigenvalues. When the control inputs are sinusoidal voltages with constant modulus and frequency, the explicit formula for the torque–speed characteristic curve is determined which gives the steady-state rotor speed corresponding to a given load torque. As far as the specific motor considered in this book is concerned, the eigenvalues of the linear approximation at several operating conditions have been numerically determined: when the rotor speed is smaller than the critical value ω_p^*, the corresponding operating conditions are unstable since there is one eigenvalue with positive real part, while they are exponentially stable when the rotor speed is larger than ω_p^*. The explicit formula for the flux modulus which minimizes the power losses at steady-state is given: it depends on motor parameters and on the load torque. Structural properties such as observability, identifiability, and feedback linearizability are then studied using tools from nonlinear control theory. The induction motor is transformable into a linear and controllable system by state feedback if a first order dynamic extension is applied, *i.e.* it is dynamically feedback linearizable. Hence, the induction motor possesses good controllability properties. Rotor fluxes are observable from rotor speed and stator current measurements for any voltage input while rotor speed and rotor fluxes are locally observable from stator current measurements only, provided that a certain condition holds. The load torque is identifiable from rotor speed and stator current measurements for any voltage input, while the rotor resistance is identifiable even though the rotor speed is not measured provided that the time derivative of the rotor flux modulus is different from zero. The experimental set-up which will be used throughout this book to illustrate the performance of the control algorithms is described in Section 1.8. A good match between simulated and experimental data is obtained for operating conditions with small slip speed, while at higher slip speed some discrepancies arise due to unmodeled effects and measurement inaccuracies.

Problems

1.1. Consider the induction motor model (1.26) when the rotor speed ω is constant so that (1.26) becomes linear in the state variables $(\psi_{ra}, \psi_{rb}, i_{sa}, i_{sb})$. Show that, for any rotor speed ω and for any motor parameter value, it is observable from the measurements (i_{sa}, i_{sb}).

1.2. Consider the induction motor model (1.26) when the rotor speed ω is constant so that (1.26) becomes linear in the state variables $(\psi_{ra}, \psi_{rb}, i_{sa}, i_{sb})$. Show that, for any rotor speed ω and for any motor parameter value, it is controllable from the inputs (u_{sa}, u_{sb}) with controllability indices $(2,2)$.

1.3. Assume zero load torque and restrict the family of admissible inputs to the set of constant inputs (u_{sa}, u_{sb}). Show that the two equilibrium points for (1.26) with the additional state variable α

$$(\omega, \psi_{ra}, \psi_{rb}, i_{sa}, i_{sb}, \alpha) = \left(0, \frac{Mu_{sa}}{R_s}, \frac{Mu_{sb}}{R_s}, \frac{u_{sa}}{R_s}, \frac{u_{sb}}{R_s}, \alpha_1\right)$$

$$(\omega, \psi_{ra}, \psi_{rb}, i_{sa}, i_{sb}, \alpha) = \left(0, \frac{Mu_{sa}}{R_s}, \frac{Mu_{sb}}{R_s}, \frac{u_{sa}}{R_s}, \frac{u_{sb}}{R_s}, \alpha_2\right)$$

are indistinguishable from input–output measurements (u_{sa}, u_{sb}), $(\omega, \psi_{ra}, \psi_{rb}, i_{sa}, i_{sb})$.

1.4. Show that the induction motor is locally observable and locally identifiable from the outputs $(i_{sa}, i_{sb}, \psi_{ra}, \psi_{rb})$ with respect to both the load torque T_L and the parameter α provided that

$$-\alpha(\psi_{ra}^2 + \psi_{rb}^2) + \alpha M(\psi_{ra} i_{sa} + \psi_{rb} i_{sb}) \neq 0$$

i.e. when

$$\frac{\mathrm{d}\left(\psi_{ra}^2 + \psi_{rb}^2\right)}{\mathrm{d}t} \neq 0\,.$$

Suggestion: consider the model (1.26) with the two additional state variables T_L and α satisfying $\dot{T}_L = 0$ and $\dot{\alpha} = 0$, respectively.

1.5. Assume nonzero load torque and restrict the family of admissible inputs to the set of constant inputs (u_{sa}, u_{sb}) satisfying

$$\left(u_{sa}^2 + u_{sb}^2\right)^2 > \frac{4T_L^2 R_s^2}{J^2\mu^2\alpha^2 M^2}\,.$$

Show that the two equilibrium points for (1.26) ($\Delta = \frac{J^2\mu^2\alpha^2 M^2 \left(u_{sa}^2+u_{sb}^2\right)^2}{T_L^2 R_s^2} - 4$)

$$\omega_1 = -\frac{J\mu\alpha M}{2T_L R_s^2}\left(u_{sa}^2 + u_{sb}^2\right) + \frac{\sqrt{\Delta}}{2}$$

$$\begin{bmatrix} \psi_{ra1} \\ \psi_{rb1} \end{bmatrix} = \frac{\alpha M}{R_s\left(\alpha^2+\omega_1^2\right)} \begin{bmatrix} \alpha & -\omega_1 \\ \omega_1 & \alpha \end{bmatrix} \begin{bmatrix} u_{sa} \\ u_{sb} \end{bmatrix}$$
$$i_{sa} = \frac{u_{sa}}{R_s}$$
$$i_{sb} = \frac{u_{sb}}{R_s}$$

and

$$\omega_2 = -\frac{J\mu\alpha M}{2T_L R_s^2}\left(u_{sa}^2+u_{sb}^2\right) - \frac{\sqrt{\Delta}}{2}$$
$$\begin{bmatrix} \psi_{ra2} \\ \psi_{rb2} \end{bmatrix} = \frac{\alpha M}{R_s\left(\alpha^2+\omega_2^2\right)} \begin{bmatrix} \alpha & -\omega_2 \\ \omega_2 & \alpha \end{bmatrix} \begin{bmatrix} u_{sa} \\ u_{sb} \end{bmatrix}$$
$$i_{sa} = \frac{u_{sa}}{R_s}$$
$$i_{sb} = \frac{u_{sb}}{R_s}$$

are indistinguishable from input–output measurements (u_{sa}, u_{sb}), (i_{sa}, i_{sb}) for every input satisfying the above inequality.

1.6. Determine constant inputs $(u_{sa},\, u_{sb})$ and two equilibrium points for the model (1.26) with the additional state variable T_L which are indistinguishable from input–output measurements $(u_{sa},\, u_{sb})$, $(i_{sa},\, i_{sb})$. *Suggestion: see Problem 1.5.*

1.7. Assume nonzero load torque and restrict the family of admissible inputs to the set of inputs u_{sa} and u_{sb} guaranteeing, for suitable initial conditions,

$$\mu(\psi_{ra} i_{sb} - \psi_{rb} i_{sa}) = T_L$$
$$-2\alpha(\psi_{ra}^2 + \psi_{rb}^2) + 2\alpha M(\psi_{ra} i_{sa} + \psi_{rb} i_{sb}) = 0\,.$$

Show that there exist two state vectors for the model (1.26) with the additional state variable α which are indistinguishable from input–output measurements $(u_{sa},\, u_{sb})$, $(\psi_{ra},\, \psi_{rb},\, i_{sa},\, i_{sb})$.

1.8. Consider the induction motor model (1.39) and define the dynamic extension

$$\frac{du_{sq}}{dt} = v_q\,.$$

Show that the extended system is state feedback linearizable and determine the corresponding singularities.

1.9. Consider the induction motor model (1.39) and define

$$u_{sd} = \sigma\left[-\omega i_{sq} - \frac{\alpha M i_{sq}^2}{\psi_{rd}} - \beta\alpha\psi_{rd}\right] + v_d$$

$$u_{sq} = \sigma \left[\omega i_{sd} + \frac{\alpha M i_{sd} i_{sq}}{\psi_{rd}} + \beta \omega \psi_{rd} \right] + \xi$$

$$\frac{d\xi}{dt} = v_q \, .$$

Show that the extended system is state feedback linearizable and determine the corresponding singularities.

1.10. Consider the induction motor model (1.26) and define the dynamic extension

$$\frac{du_{sb}}{dt} = \frac{v_{sb}}{\sigma} \, .$$

Show that the extended system is state feedback linearizable.

Chapter 2
State Feedback Control

Abstract In this chapter we explore the advantages of feedback control assuming that all the state variables are measurable. This is not a realistic assumption since the rotor variables are usually not measured but it allows us to explore fully the potentiality of feedback control. In Section 2.1 we establish that the feedforward control does not guarantee the asymptotic stability of the desired operating point for every initial condition and for every parameter value: the load torque in particular is a critical parameter. Hence, feedback control is needed to achieve the asymptotic stability of the desired operating condition, for any load torque and for any initial condition. Six feedback control algorithms are then presented. The most complex is the dynamic feedback linearizing control presented in the last Section 2.6, which imposes an arbitrary linear dynamic behavior to the controlled motor. The input–output feedback linearizing control is presented in Section 2.4: it achieves arbitrary and decoupled linear dynamics for the two tracking errors of rotor speed and flux modulus; it is generalized in Section 2.5 by an adaptive input–output feedback linearizing control which identifies both the load torque and the rotor resistance in realistic operating conditions. The identification of these two parameters allows computation online of the optimal value of the rotor flux modulus which minimizes the power losses. All three feedback linearizing control schemes have excellent performances provided that the initial errors are sufficiently small: this is a significant limitation which is removed by the global control with arbitrary rate of convergence presented in Section 2.7. It is the evolution of the historically important direct field-oriented control which is presented in Section 2.2 and its variant, the indirect field-oriented control, which is discussed in Section 2.3 and can operate from any initial conditions. The field-oriented controls constitute a modification of the feedforward control discussed in Section 2.1 and contain the key steps to design the global control with arbitrary rate of convergence, which can operate from any motor initial conditions. The indirect field-oriented control is tested by experiments in Section 2.8 and its robustness with respect to rotor resistance variations is explored.

2.1 Stability Analysis of Feedforward Control

In this section we assume that the left inverse control (1.74) is applied as a feedforward control to the induction motor fixed frame model (1.26) and we study the case in which the initial conditions are not compatible in (1.74). We have seen in the previous chapter that if the initial conditions of the induction motor and of the inverse system are compatible and the inverse feedforward control (1.74) is applied, then the induction motor satisfies the equations

$$\begin{aligned}
\frac{\mathrm{d}\omega^*}{\mathrm{d}t} &= \mu\psi^* i_{sq}^* - \frac{T_L}{J} \\
\frac{\mathrm{d}\psi^*}{\mathrm{d}t} &= -\alpha\psi^* + \alpha M i_{sd}^* \\
0 &= -(\omega_0^* - \omega^*)\,\psi^* + \alpha M i_{sq}^* \\
\frac{\mathrm{d}i_{sd}^*}{\mathrm{d}t} &= -\gamma i_{sd}^* + \omega_0^* i_{sq}^* + \beta\alpha\psi^* + \frac{u_{sd}^*}{\sigma} \\
\frac{\mathrm{d}i_{sq}^*}{\mathrm{d}t} &= -\gamma i_{sq}^* - \omega_0^* i_{sd}^* - \beta\omega^*\psi^* + \frac{u_{sq}^*}{\sigma}
\end{aligned} \tag{2.1}$$

in which the (d,q) reference rotating frame rotates at speed

$$\frac{\mathrm{d}\varepsilon_0^*}{\mathrm{d}t} = \omega_0^* = \omega^* + \frac{\alpha M i_{sq}^*}{\psi^*} = \omega^* + \omega_s^* \tag{2.2}$$

and it is identified by the angle $\varepsilon_0^*(t) = \rho^*(t) = \rho(t)$ in the fixed (a,b) frame. The crucial question we are going to answer in this section is the following: what happens when the initial conditions of the inverse control (1.74) are not compatible and in particular when $\rho^*(0) \neq \arctan\frac{\psi_{rb}(0)}{\psi_{ra}(0)}$, which is very likely to happen since measurements of (ψ_{ra}, ψ_{rb}) are typically not available? In other words we are going to explore the stability and the attractivity (see Appendix A) of the steady-state solution $(\omega^*, \psi^*, 0, i_{sd}^*, i_{sq}^*)$. Since the rotor flux angle ρ no longer coincides with the angle ρ^*, consider the (d,q) frame identified by the angle ε_0^*. In the considered (d,q) frame the induction motor satisfies (1.31) with ω_0^* in place of ω_0. If we now subtract (2.1) from (1.31) we obtain the tracking error dynamics

$$\begin{aligned}
\frac{\mathrm{d}(\omega-\omega^*)}{\mathrm{d}t} &= \mu\left[(\psi_{rd}-\psi^*)\,i_{sq}^* - \psi_{rq} i_{sd}^*\right] + \mu\psi^*\left(i_{sq}-i_{sq}^*\right) \\
&\quad + \mu\left[(\psi_{rd}-\psi^*)\left(i_{sq}-i_{sq}^*\right) - \psi_{rq}\left(i_{sd}-i_{sd}^*\right)\right] \\
\frac{\mathrm{d}(\psi_{rd}-\psi^*)}{\mathrm{d}t} &= -\alpha(\psi_{rd}-\psi^*) - (\omega-\omega^*)\,\psi_{rq} \\
&\quad + \omega_s^*\psi_{rq} + \alpha M\left(i_{sd}-i_{sd}^*\right) \\
\frac{\mathrm{d}\psi_{rq}}{\mathrm{d}t} &= -\alpha\psi_{rq} + (\omega-\omega^*)(\psi_{rd}-\psi^*) \\
&\quad - \omega_s^*(\psi_{rd}-\psi^*) + (\omega-\omega^*)\,\psi^* + \alpha M\left(i_{sq}-i_{sq}^*\right)
\end{aligned}$$

$$\begin{aligned}
\frac{\mathrm{d}\left(i_{sd}-i_{sd}^*\right)}{\mathrm{d}t} &= -\gamma(i_{sd}-i_{sd}^*)+\omega_0^*\left(i_{sq}-i_{sq}^*\right)+\beta\alpha\left(\psi_{rd}-\psi^*\right)\\
&\quad +\beta\omega^*\psi_{rq}+\beta\left(\omega-\omega^*\right)\psi_{rq}\\
\frac{\mathrm{d}\left(i_{sq}-i_{sq}^*\right)}{\mathrm{d}t} &= -\gamma\left(i_{sq}-i_{sq}^*\right)-\omega_0^*\left(i_{sd}-i_{sd}^*\right)+\beta\alpha\psi_{rq}\\
&\quad -\beta\omega^*\left(\psi_{rd}-\psi^*\right)-\beta\left(\omega-\omega^*\right)\left(\psi_{rd}-\psi^*\right)\\
&\quad -\beta\left(\omega-\omega^*\right)\psi^*
\end{aligned} \tag{2.3}$$

where we recall that

$$\begin{aligned}
i_{sd}^* &= \frac{\psi^*}{M}+\frac{\dot\psi^*}{\alpha M}\\
i_{sq}^* &= \frac{\dot\omega^*}{\mu\psi^*}+\frac{T_L}{J\mu\psi^*}
\end{aligned} \tag{2.4}$$

and ω_s^* is defined in (2.2). The origin

$$\left[(\omega-\omega^*),(\psi_{rd}-\psi^*),\psi_{rq},(i_{sd}-i_{sd}^*),\left(i_{sq}-i_{sq}^*\right)\right]^{\mathrm{T}}=0\,,$$

which corresponds to zero tracking errors, is clearly an equilibrium point for the tracking error dynamics (2.3): this is the case of compatible initial conditions in which

$$\begin{aligned}
\omega(0) &= \omega^*(0)\\
\psi_{ra}(0) &= \psi^*(0)\cos\rho^*(0)\\
\psi_{rb}(0) &= \psi^*(0)\sin\rho^*(0)\\
\begin{bmatrix} i_{sa}(0)\\ i_{sb}(0)\end{bmatrix} &= \begin{bmatrix}\cos\rho^*(0) & -\sin\rho^*(0)\\ \sin\rho^*(0) & \cos\rho^*(0)\end{bmatrix}\begin{bmatrix}\frac{\psi^*(0)}{M}+\frac{\dot\psi^*(0)}{\alpha M}\\ \frac{\dot\omega^*(0)}{\mu\psi^*(0)}+\frac{T_L}{J\mu\psi^*(0)}\end{bmatrix}\\
\rho^*(0) &= \rho(0)\,.
\end{aligned}$$

In order to have a more compact notation, let us rewrite the tracking error dynamics in terms of the tracking errors

$$\begin{aligned}
\tilde\omega &= \omega-\omega^*\\
\tilde\psi_{rd} &= \psi_{rd}-\psi^*\\
\tilde\psi_{rq} &= \psi_{rq}\\
\tilde i_{sd} &= i_{sd}-i_{sd}^*\\
\tilde i_{sq} &= i_{sq}-i_{sq}^*
\end{aligned}$$

as

$$\frac{\mathrm{d}\tilde\omega}{\mathrm{d}t} = \mu i_{sq}^*\tilde\psi_{rd}-\mu i_{sd}^*\tilde\psi_{rq}+\mu\psi^*\tilde i_{sq}+\mu\tilde\psi_{rd}\tilde i_{sq}-\mu\tilde\psi_{rq}\tilde i_{sd}$$

$$
\begin{aligned}
\frac{\mathrm{d}\tilde{\psi}_{rd}}{\mathrm{d}t} &= -\alpha\tilde{\psi}_{rd} + \omega_s^*\tilde{\psi}_{rq} - \tilde{\omega}\tilde{\psi}_{rq} + \alpha M\tilde{i}_{sd} \\
\frac{\mathrm{d}\tilde{\psi}_{rq}}{\mathrm{d}t} &= -\alpha\tilde{\psi}_{rq} - \omega_s^*\tilde{\psi}_{rd} + \tilde{\omega}\tilde{\psi}_{rd} + \alpha M\tilde{i}_{sq} + \psi^*\tilde{\omega} \\
\frac{\mathrm{d}\tilde{i}_{sd}}{\mathrm{d}t} &= -\gamma\tilde{i}_{sd} + \omega_0^*\tilde{i}_{sq} + \beta\alpha\tilde{\psi}_{rd} + \beta\omega^*\tilde{\psi}_{rq} + \beta\tilde{\omega}\tilde{\psi}_{rq} \\
\frac{\mathrm{d}\tilde{i}_{sq}}{\mathrm{d}t} &= -\gamma\tilde{i}_{sq} - \omega_0^*\tilde{i}_{sd} + \beta\alpha\tilde{\psi}_{rq} - \beta\omega^*\tilde{\psi}_{rd} - \beta\tilde{\omega}\tilde{\psi}_{rd} - \beta\psi^*\tilde{\omega}\,. \qquad (2.5)
\end{aligned}
$$

Many questions are naturally posed on the tracking error dynamics (2.5): is the origin stable, asymptotically stable, exponentially stable, globally asymptotically stable, globally exponentially stable? How large is the region of attraction of the origin and what is the influence of critical parameters, such as rotor resistance R_r and load torque T_L, on the region of attraction and on the dynamic behavior? These questions have no simple answers since the system (2.5) is nonlinear and time-varying when the reference signals (ω^*, ψ^*) are time-varying. On the other hand, these questions are extremely important since, for instance, if the origin were globally exponentially stable with satisfactory transient properties and robustness with respect to parameter variations then no feedback control would be needed and the feedforward control (1.74) would achieve the tracking of (ω^*, ψ^*) from any initial condition. To examine one of those questions, consider the error system dynamics (2.5) and compute the equilibrium points for (2.5) in the case of constant references (ω^*, ψ^*) and $T_L \neq 0$. In order to use a more compact notation and simplify the analysis, define the tracking error variables

$$
\begin{aligned}
z_d &= \tilde{i}_{sd} + \beta\tilde{\psi}_{rd} \\
z_q &= \tilde{i}_{sq} + \beta\tilde{\psi}_{rq}
\end{aligned}
$$

so that the error system (2.5) in the case of constant references (ω^*, ψ^*) may be rewritten in the simpler form

$$
\begin{aligned}
\dot{\tilde{\omega}} &= A\tilde{\psi}_{rd} - B\tilde{\psi}_{rq} + Cz_q + \mu\tilde{\psi}_{rd}z_q - \mu\tilde{\psi}_{rq}z_d \\
\dot{\tilde{\psi}}_{rd} &= -D\tilde{\psi}_{rd} - \tilde{\omega}\tilde{\psi}_{rq} + E\tilde{\psi}_{rq} + \alpha M z_d \\
\dot{\tilde{\psi}}_{rq} &= -D\tilde{\psi}_{rq} + \tilde{\omega}\tilde{\psi}_{rd} - E\tilde{\psi}_{rd} + \psi^*\tilde{\omega} + \alpha M z_q \\
\dot{z}_d &= Fz_q - kz_d + k\beta\tilde{\psi}_{rd} \\
\dot{z}_q &= -Fz_d - kz_q + k\beta\tilde{\psi}_{rq} \qquad (2.6)
\end{aligned}
$$

in which the following reparameterization is used

$$
\begin{aligned}
A &= \frac{T_L}{J\psi^*}, \quad B = \mu\psi^*\left(\frac{1}{M} + \beta\right), \\
C &= \mu\psi^*, \quad D = \alpha + \alpha M\beta, \\
E &= \frac{\alpha M T_L}{\mu J\psi^{*2}}, \quad F = \omega^* + E, \quad k = \frac{R_s}{\sigma}\,.
\end{aligned}
$$

By direct computation it is possible to show that there exist constant references (ω^*, ψ^*) and $T_L \neq 0$, which are within physical bounds such that the tracking error dynamics (2.6) have an explicitly computable additional equilibrium point besides the origin whose first component

$$\begin{aligned}\tilde{\omega}_e = & \left(\beta kCF - \beta k\psi^*\mu F - k^2A - AF^2\right)^{-1}\Big[\alpha\beta^2MCk^2 - \alpha\beta MBk^2 \\ & -\alpha\beta MAFk - \beta CDk^2 + \beta kCFE - k^2AE + k^2BD + BF^2D - AF^2E\Big]\end{aligned}$$

is nonzero. Hence, for such references and load torque values, the origin is not a globally attractive equilibrium point for (2.6) and rotor speed tracking cannot be achieved for any motor initial condition. Furthermore, consider the linear approximation about the origin of system (2.5):

$$\begin{aligned}\frac{d\tilde{\omega}}{dt} &= \mu i_{sq}^*\tilde{\psi}_{rd} - \mu i_{sd}^*\tilde{\psi}_{rq} + \mu\psi^*\tilde{i}_{sq} \\ \frac{d\tilde{\psi}_{rd}}{dt} &= -\alpha\tilde{\psi}_{rd} + \omega_s^*\tilde{\psi}_{rq} + \alpha M\tilde{i}_{sd} \\ \frac{d\tilde{\psi}_{rq}}{dt} &= \psi^*\tilde{\omega} - \omega_s^*\tilde{\psi}_{rd} - \alpha\tilde{\psi}_{rq} + \alpha M\tilde{i}_{sq} \\ \frac{d\tilde{i}_{sd}}{dt} &= \beta\alpha\tilde{\psi}_{rd} + \beta\omega^*\tilde{\psi}_{rq} - \gamma\tilde{i}_{sd} + (\omega^* + \omega_s^*)\tilde{i}_{sq} \\ \frac{d\tilde{i}_{sq}}{dt} &= -\beta\psi^*\tilde{\omega} - \beta\omega^*\tilde{\psi}_{rd} + \beta\alpha\tilde{\psi}_{rq} - (\omega^* + \omega_s^*)\tilde{i}_{sd} - \gamma\tilde{i}_{sq}\,. \end{aligned} \tag{2.7}$$

According to the linear approximation Theorem A.7 in Appendix A, if the load torque satisfies the inequality

$$T_L^2 \geq \frac{(1+M\beta)^2\psi^{*4}}{L_r^2}$$

then the origin is an unstable equilibrium point for the error system (2.5). Recall that in the analysis of the torque–speed characteristic curve in Section 1.3 we observed in Figures 1.5 and 1.6 that unstable operating conditions at low rotor speed correspond to low rotor flux modulus and high load torques.

In conclusion, there exist constant references (ω^*, ψ^*) and $T_L \neq 0$ such that the feedforward control (1.74), specialized to the case of constant rotor speed and flux modulus references (see Figure 2.1 for current-fed motors),

$$\begin{bmatrix} u_{sa}^* \\ u_{sb}^* \end{bmatrix} = \begin{bmatrix} \cos\rho^* & -\sin\rho^* \\ \sin\rho^* & \cos\rho^* \end{bmatrix} \begin{bmatrix} u_{sd}^* \\ u_{sq}^* \end{bmatrix}$$

$$u_{sd}^* = \sigma\left[\frac{R_s\psi^*}{\sigma M} - \frac{\omega^*T_L}{\mu J\psi^*} - \frac{\alpha M}{\psi^*}\left(\frac{T_L}{J\mu\psi^*}\right)^2\right]$$

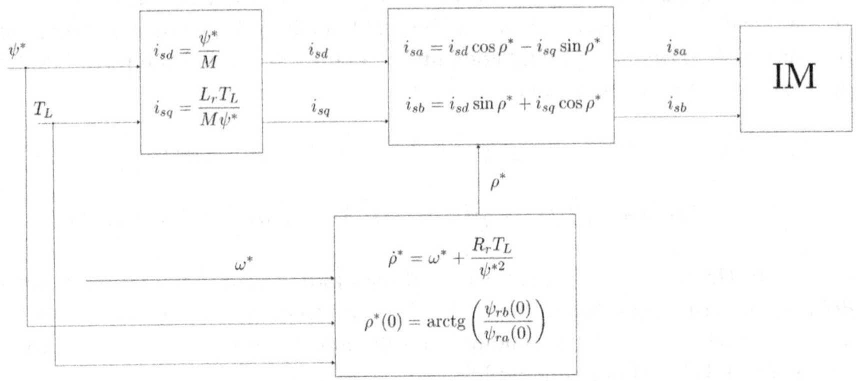

Fig. 2.1 Feedforward control for current-fed motors (constant references ω^*, ψ^*)

$$u_{sq}^* = \sigma\left[\frac{\gamma T_L}{J\mu\psi^*} + \frac{\omega^*\psi^*}{M} + \frac{\alpha T_L}{J\mu\psi^*} + \beta\omega^*\psi^*\right]$$
$$\dot{\rho}^* = \omega^* + \frac{\alpha M T_L}{\mu J \psi^{*2}} \tag{2.8}$$

depending on the reference signals (ω^*, ψ^*), on the initial condition $\rho^*(0)$, and on the machine parameters M, R_r, L_r, J, R_s, L_s, does not guarantee rotor speed tracking for any motor initial condition. Furthermore, the origin of the error system (2.5), which implies zero tracking error, is unstable if the inequality $|T_L| \geq \frac{(1+M\beta)\psi^{*2}}{L_r}$ is satisfied.

Illustrative Simulations

We tested the feedforward control (2.8) by simulations for the three-phase single pole pair 0.6-kW induction motor whose parameters have been reported in Chapter 1. All the motor initial conditions have been set equal to zero except for $\psi_{ra}(0) = \psi_{rb}(0) = 0.1\,\text{Wb}$. The references for the speed and flux modulus along with the applied load torque are reported in Figures 2.2–2.4. The rotor flux modulus reference signal starts from 0.001 Wb at $t = 0\,\text{s}$ and grows up to the constant value 1.16 Wb. The speed reference is zero until $t = 0.32\,\text{s}$ and grows up to the constant value 100 rad/s; at $t = 1.5\,\text{s}$ the reference for the flux modulus is reduced to 0.5 Wb. A 5.8-Nm load torque is applied to the motor and then is reduced to 4.8 Nm. Figures 2.3 and 2.4 show the time histories of rotor speed and flux modulus along with the corresponding tracking errors: the rotor speed and flux modulus are regulated to their corresponding references as long as the load torque satisfies the inequality $T_L < \frac{\psi^{*2}(1+M\beta)}{L_r}$, while rotor speed and rotor flux modulus regulation is not achieved when the load torque is greater than $\frac{\psi^{*2}(1+M\beta)}{L_r}$. In fact, for a load torque $T_L = 4.8\,\mathrm{Nm}$ and a constant rotor flux reference $\psi^* = 0.5\,\mathrm{Wb}$, the origin of the error system (2.5) is unstable while the computed additional equilibrium point for (2.5) is exponentially stable. Finally, the stator currents and voltages profiles, which are within the physical saturation limits, are reported in Figures 2.5 and 2.6.

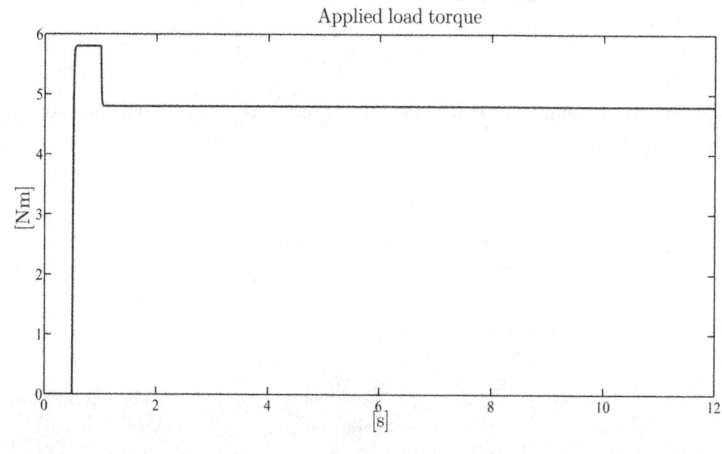

Fig. 2.2 Feedforward control: applied load torque T_L

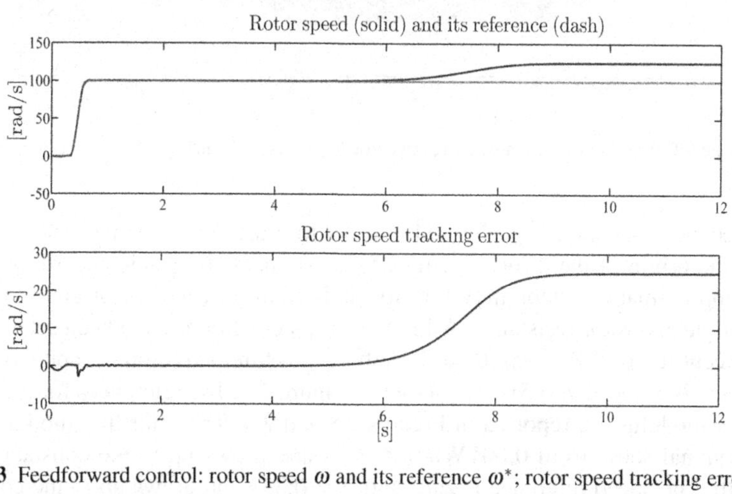

Fig. 2.3 Feedforward control: rotor speed ω and its reference ω^*; rotor speed tracking error

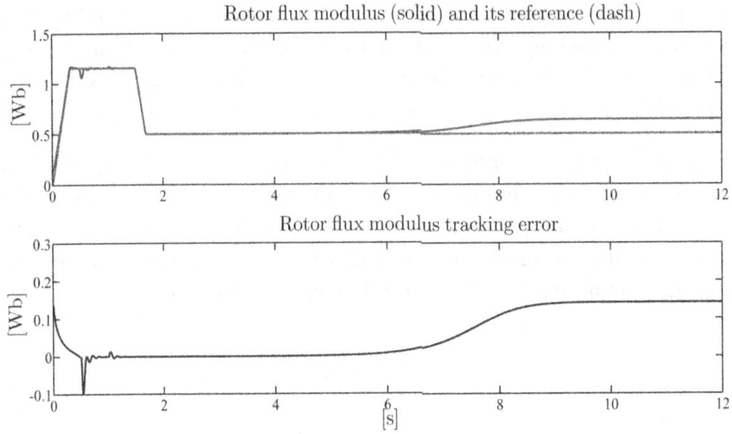

Fig. 2.4 Feedforward control: rotor flux modulus $\sqrt{\psi_{ra}^2 + \psi_{rb}^2}$ and its reference ψ^*; rotor flux modulus tracking error

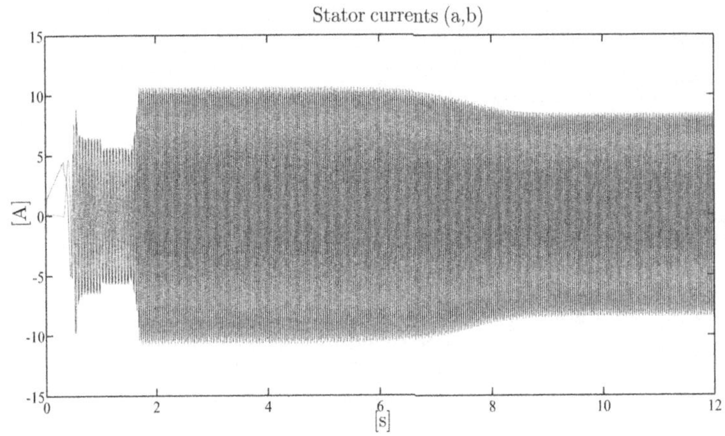

Fig. 2.5 Feedforward control: stator current vector (a,b)-components

A second simulation is performed in order to illustrate, in the particular case in which the origin of the error system (2.5) is exponentially stable according to the linear approximation Theorem A.7 in Appendix A, the effect of uncertainties in both load torque and rotor resistance. All motor initial conditions have been set equal to zero except for $\psi_{ra}(0) = \psi_{rb}(0) = 0.1\,\text{Wb}$. A 5.8-Nm load torque is applied to the motor and is reduced to 0.5 Nm as shown in Figure 2.7. The references for the speed and flux modulus are reported in Figures 2.8 and 2.9. The rotor flux modulus reference signal starts from 0.001 Wb at $t = 0\,\text{s}$ and grows up to the constant value 1.16 Wb. The speed reference is zero until $t = 0.32\,\text{s}$ and grows up to the constant

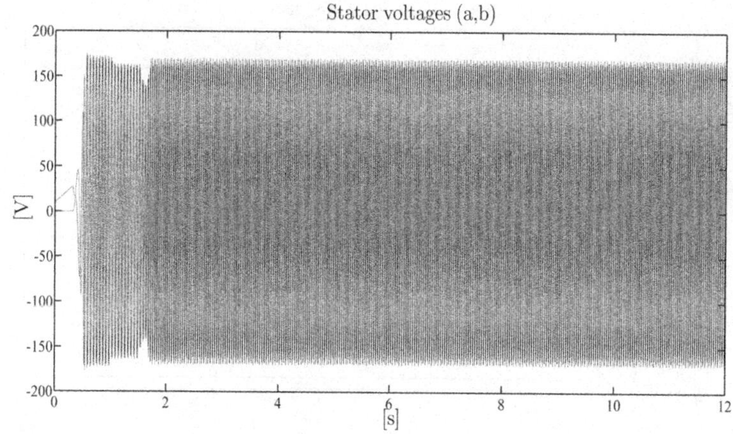

Fig. 2.6 Feedforward control: stator voltage vector (a,b)-components

value 100 rad/s; at $t = 1.5$ s the reference for the flux modulus is then reduced to 0.5 Wb. The value of the rotor resistance used by the feedforward control is 50% greater than the true value $R_r = 3.3\,\Omega$ while only a 0.2-Nm load torque is compensated by the controller for $t \geq 1$ s (see Figure 2.7). Figures 2.8 and 2.9 show the time histories of rotor speed and flux modulus along with the corresponding tracking errors: note that steady-state errors appear as expected. Finally, the stator currents and voltages profiles, which are within the physical saturation limits, are reported in Figures 2.10 and 2.11.

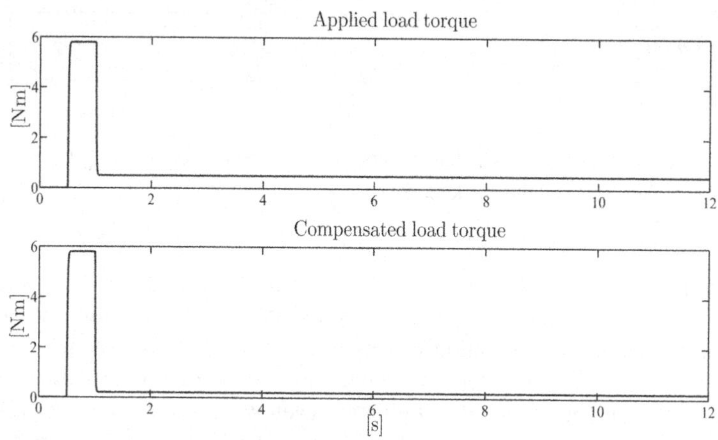

Fig. 2.7 Feedforward control with parameter uncertainties: applied load torque T_L and compensated load torque

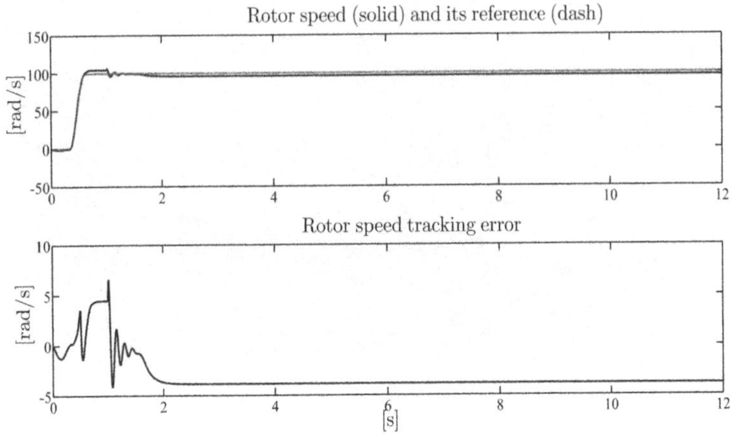

Fig. 2.8 Feedforward control with parameter uncertainties: rotor speed ω and its reference ω^*; rotor speed tracking error

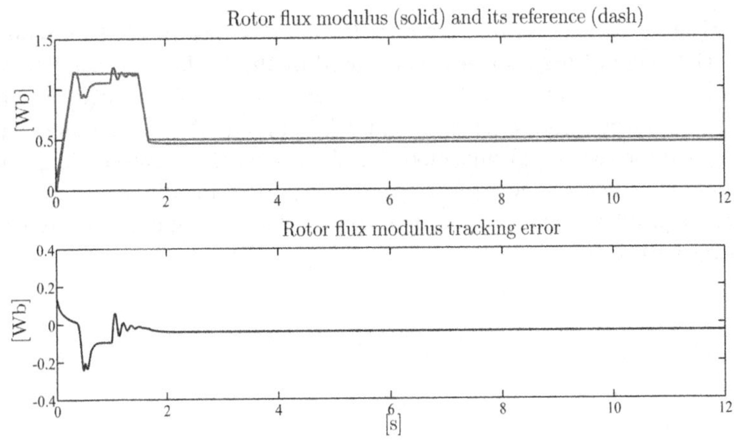

Fig. 2.9 Feedforward control with parameter uncertainties: rotor flux modulus $\sqrt{\psi_{ra}^2 + \psi_{rb}^2}$ and its reference ψ^*; rotor flux modulus tracking error

In summary, we have shown that the feedforward control (2.8), which is required to keep the motor at constant speed ω^* with a desired constant flux modulus ψ^*, does not guarantee the regulation to ω^* for any motor initial condition and may yield unstable steady-state operating conditions depending on load torque T_L, motor parameters, and desired flux modulus ψ^*, even when exact parameter values are used in (2.8). On the other hand, if the values of the critical parameters R_r and T_L used in (2.8) do not coincide with the corresponding actual values then the simulations show that steady-state tracking errors appear. Feedback is then required to guarantee asymptotically stable operating conditions for any initial condition and, at the same time, robustness with respect to uncertainties in critical motor parame-

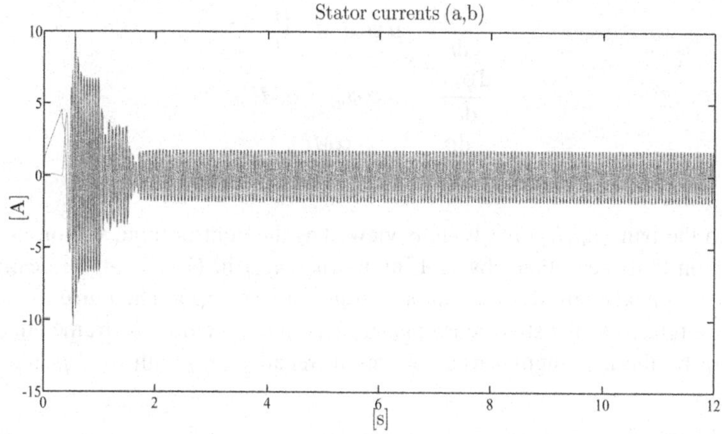

Fig. 2.10 Feedforward control with parameter uncertainties: stator current vector (a,b)-components

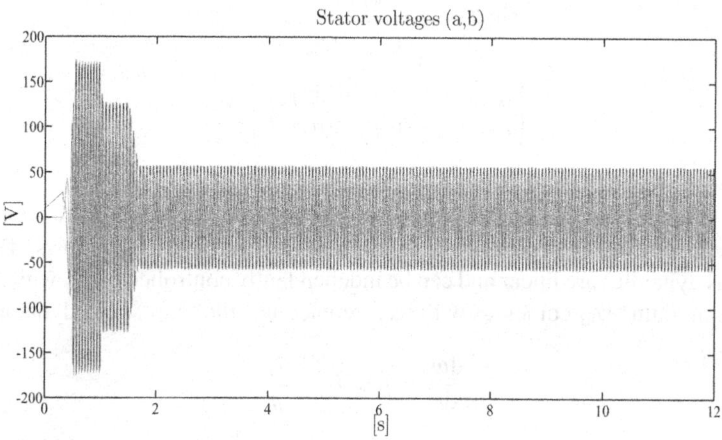

Fig. 2.11 Feedforward control with parameter uncertainties: stator voltage vector (a,b)-components

ters such as rotor resistance and load torque. This goal will be achieved in the next sections.

2.2 Direct Field-oriented Control

The purpose of this section is to illustrate the benefits of feedback control. Consider the reduced third-order model in the (d,q) frame rotating at rotor flux speed of rotation $\dot{\rho}$ and identified by the rotor flux angle ρ (see (1.39))

$$\begin{aligned}
\frac{d\omega}{dt} &= \mu \psi_{rd} i_{sq} - \frac{T_L}{J} \\
\frac{d\psi_{rd}}{dt} &= -\alpha \psi_{rd} + \alpha M i_{sd} \\
\frac{d\rho}{dt} &= \omega + \frac{\alpha M i_{sq}}{\psi_{rd}}
\end{aligned} \tag{2.9}$$

in which the pair (i_{sd}, i_{sq}) in (1.41) is viewed as the control input vector, under the assumption that the actual physical inputs (u_{sd}, u_{sq}) in (1.42) can be designed to track very quickly any desired stator current pair (i_{sd}, i_{sq}). The currents (i_{sd}, i_{sq}), which are related to the stator currents (i_{sa}, i_{sb}) in the fixed (a, b) frame attached to the stator by the nonsingular transformation (recall (1.30) with $\varepsilon_0 = \rho$ and (1.37), (1.38))

$$\begin{aligned}
\cos\rho &= \frac{\psi_{ra}}{\sqrt{\psi_{ra}^2 + \psi_{rb}^2}} \\
\sin\rho &= \frac{\psi_{rb}}{\sqrt{\psi_{ra}^2 + \psi_{rb}^2}} \\
\begin{bmatrix} i_{sa} \\ i_{sb} \end{bmatrix} &= \begin{bmatrix} \cos\rho & -\sin\rho \\ \sin\rho & \cos\rho \end{bmatrix} \begin{bmatrix} i_{sd} \\ i_{sq} \end{bmatrix}
\end{aligned} \tag{2.10}$$

are to be designed to track the desired signals (ω^*, ψ^*).

The structure of system (2.9) in the rotating frame is very suitable for multivariable feedback control design as the field-oriented control strategy shows. The flux modulus dynamics are linear and can be independently controlled by i_{sd} while, when the flux modulus ψ_{rd} coincides with its reference ψ^*, the rotor speed dynamics

$$\frac{d\omega}{dt} = \mu \psi^* i_{sq} - \frac{T_L}{J} \tag{2.11}$$

are also linear with respect to i_{sq} and can be independently controlled by i_{sq}. Since the unforced direct flux ψ_{rd} dynamics

$$\frac{d\psi_{rd}}{dt} = -\alpha \psi_{rd} \tag{2.12}$$

are asymptotically stable ($\alpha > 0$) when $i_{sd} = 0$, the control input i_{sd} can be designed as the following feedforward signal (recall (1.67))

$$i_{sd} = \frac{\psi^*}{M} + \frac{\dot{\psi}^*}{\alpha M} \tag{2.13}$$

which, substituted in (2.9), gives (recall that $\tilde{\psi}_{rd} = \psi_{rd} - \psi^*$)

$$\frac{d\tilde{\psi}_{rd}}{dt} = -\alpha \tilde{\psi}_{rd}\,. \tag{2.14}$$

On the other hand, from the first equation in (2.9), the speed error dynamics are (recall that $\tilde{\omega} = \omega - \omega^*$)

$$\begin{aligned}\frac{d\tilde{\omega}}{dt} &= \mu \psi_{rd} i_{sq} - \frac{T_L}{J} - \dot{\omega}^* \\ &= \mu \psi^* i_{sq} - \frac{T_L}{J} + \mu \tilde{\psi}_{rd} i_{sq} - \dot{\omega}^* .\end{aligned} \tag{2.15}$$

Define the control for i_{sq} as (k_ω is a positive control parameter)

$$i_{sq} = \frac{1}{\mu \psi^*}\left[-k_\omega \tilde{\omega} + \dot{\omega}^* + \frac{T_L}{J}\right] \tag{2.16}$$

and substitute it into the first equation in (2.9) or, equivalently, in (2.15) so that

$$\frac{d\tilde{\omega}}{dt} = -k_\omega \tilde{\omega} + \mu \tilde{\psi}_{rd} i_{sq} \tag{2.17}$$

in which $\tilde{\psi}_{rd}(t)$, according to (2.14), is an exponentially decaying signal for any $\psi_{rd}(0)$. According to (2.16), (2.17) may be rewritten as

$$\frac{d\tilde{\omega}}{dt} = -k_\omega \left[1 + \frac{\tilde{\psi}_{rd}}{\psi^*}\right]\tilde{\omega} + \frac{\tilde{\psi}_{rd}}{\psi^*}\left(\dot{\omega}^* + \frac{T_L}{J}\right) . \tag{2.18}$$

On the basis of (2.14) and (2.18), we conclude that $\tilde{\omega}(t)$ is bounded in the time interval $[0, t_*]$, with t_* any positive real. On the other hand, according to (2.14), for any initial condition $\psi_{rd}(0)$ and any positive real $\eta < 1$ there exists $\tilde{t}_* \geq 0$ such that, for all $t \geq \tilde{t}_*$,

$$\left|\frac{\tilde{\psi}_{rd}(t)}{\psi^*(t)}\right| \leq 1 - \eta .$$

Therefore, according to (2.18), $\tilde{\omega}(t)$ is an exponentially decaying signal for any initial condition $\omega(0)$. Note that in order to avoid the singularities in the controller at $\psi_{rd} = \sqrt{\psi_{ra}^2 + \psi_{rb}^2} = 0$ which appear in (2.10), according to (2.14) $\psi_{rd}(0)$ must be greater than zero so that $\psi_{rd}(t) > 0$ for all $t \geq 0$.

In conclusion: the *direct field-oriented control* (see Figure 2.12) is defined as

$$\begin{aligned}\begin{bmatrix} i_{sa} \\ i_{sb} \end{bmatrix} &= \begin{bmatrix} \cos\rho & -\sin\rho \\ \sin\rho & \cos\rho \end{bmatrix}\begin{bmatrix} i_{sd} \\ i_{sq} \end{bmatrix} \\ i_{sd} &= \frac{\psi^*}{M} + \frac{\dot{\psi}^*}{\alpha M} \\ i_{sq} &= \frac{1}{\mu \psi^*}\left[-k_\omega(\omega - \omega^*) + \dot{\omega}^* + \frac{T_L}{J}\right]\end{aligned}$$

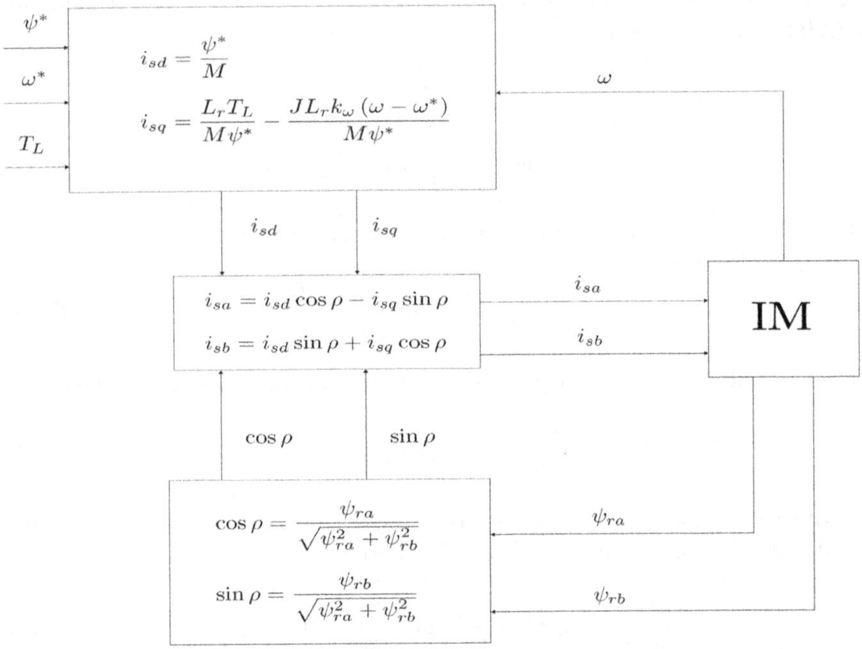

Fig. 2.12 Direct field-oriented control for current-fed motors (constant references ω^*, ψ^*)

$$\cos\rho = \frac{\psi_{ra}}{\sqrt{\psi_{ra}^2 + \psi_{rb}^2}}$$
$$\sin\rho = \frac{\psi_{rb}}{\sqrt{\psi_{ra}^2 + \psi_{rb}^2}}; \tag{2.19}$$

it is a static nonlinear feedback control algorithm which depends on the measurements of the state variables $(\omega, \psi_{ra}, \psi_{rb})$, on the reference signals (ω^*, ψ^*), on the positive control parameter k_ω, on the load torque T_L, and on the machine parameters M, R_r, L_r, J, since $\mu = \frac{M}{JL_r}$ and $\alpha = \frac{R_r}{L_r}$; it guarantees that, for any initial condition of the current-fed reduced order motor model (2.9) such that $\psi_{rd}(0) > 0$, the rotor speed and rotor flux modulus tracking errors decay exponentially to zero.

Remarks

1. The name direct field-oriented control is due to the fact that the stator current vector (i_{sa}, i_{sb}) rotates at the same speed $\dot{\rho}$ at which the rotor flux vector (ψ_{ra}, ψ_{rb}) rotates, *i.e.* it follows the orientation of the flux vector. The quadrature axis component i_{sq} of the stator current vector is responsible, according to (2.16), for the rotor speed tracking and depends on the load torque T_L, while the direct axis component i_{sd} of the stator current vector is responsible, according to (2.13), for the tracking of the rotor flux modulus.
2. The direct field-oriented control (2.19) is a static state feedback control which generates bounded currents (i_{sa}, i_{sb}) for every state value $(\omega, \psi_{rd}, \rho)$; it is not well defined at $\psi_{rd} = 0$ where (2.9) is no longer an equivalent description of the fixed frame model (1.26).
3. The measurements of (ψ_{ra}, ψ_{rb}) are required; the critical parameter R_r, which appears in i_{sd} through $\alpha = \frac{R_r}{L_r}$ when the reference ψ^* is not constant, is required only when the reference ψ^* is time-varying; if ψ^* is constant the direct field-oriented control does not depend on R_r.
4. If the flux modulus and the rotor speed are constant and equal to the desired values (ω^*, ψ^*) then the rotor flux rotates at constant speed $w = (\omega^* + \omega_s)$, with $\omega_s = \frac{\alpha M T_L}{\mu \psi^{*2}}$, and the induction motor is driven by the sinusoidal currents obtained from (2.19):

$$\begin{bmatrix} i_{sa} \\ i_{sb} \end{bmatrix} = \begin{bmatrix} \cos(\rho(0)+wt) & -\sin(\rho(0)+wt) \\ \sin(\rho(0)+wt) & \cos(\rho(0)+wt) \end{bmatrix} \begin{bmatrix} \frac{\psi^*}{M} \\ \frac{T_L}{J\mu\psi^*} \end{bmatrix}.$$

5. The direct field-oriented control (2.19) achieves asymptotic input–output feedback linearization: according to (2.14), the closed-loop dynamics for $(\psi_{rd} - \psi^*)$ are linear with a time constant equal to $\alpha^{-1} = L_r R_r^{-1}$ depending on the machine parameters; according to (2.17), once ψ_{rd} tends to its reference ψ^*, the closed-loop dynamics for $\tilde{\omega}$ tend to be linear with arbitrary time constant k_ω^{-1}. During the transient, the nonlinear term $J\mu\psi_{rd} i_{sq}$, which represents the electromagnetic torque T_e in the first equation in (2.9), makes the first two equations in (2.9) still nonlinear and coupled: for this reason the speed transients may be unsatisfactory.
6. The direct field-oriented control (2.19) can be simply modified by replacing ψ_{rd} with its reference ψ^* in the dynamic equation which generates the angle of rotation of the (d,q) reference frame, so that the indirect field-oriented control which will be discussed in the next section ($\varepsilon_0(0)$ is an arbitrary initial condition)

$$\begin{bmatrix} i_{sa} \\ i_{sb} \end{bmatrix} = \begin{bmatrix} \cos\varepsilon_0 & -\sin\varepsilon_0 \\ \sin\varepsilon_0 & \cos\varepsilon_0 \end{bmatrix} \begin{bmatrix} i_{sd} \\ i_{sq} \end{bmatrix}$$

$$\frac{d\varepsilon_0}{dt} = \omega_0 = \omega + \frac{\alpha M i_{sq}}{\psi^*}$$

$$i_{sd} = \frac{\psi^*}{M} + \frac{\dot{\psi}^*}{\alpha M}$$

$$i_{sq} = \frac{1}{\mu \psi^*}\left[-k_\omega(\omega - \omega^*) + \dot{\omega}^* + \frac{T_L}{J}\right] \tag{2.20}$$

is obtained. Note that (2.20) is always well defined even at $\psi_{rd} = 0$ where (2.19) is, on the contrary, not well defined. Furthermore, (2.20) does not require the measurement of rotor fluxes (ψ_{ra}, ψ_{rb}) but only the measurement of ω while ε_0 no longer coincides with the rotor flux angle ρ.

7. The direct field-oriented control (2.19) can be modified to obtain input–output feedback linearization (and not only asymptotic input–output feedback linearization) by using ψ_{rd} in place of ψ^* in the i_{sq} espression and by adding a feedback term in the i_{sd} espression in (2.19) as follows:

$$\begin{aligned}
\begin{bmatrix} i_{sa} \\ i_{sb} \end{bmatrix} &= \begin{bmatrix} \cos\rho & -\sin\rho \\ \sin\rho & \cos\rho \end{bmatrix} \begin{bmatrix} i_{sd} \\ i_{sq} \end{bmatrix} \\
i_{sd} &= \frac{\psi^*}{M} + \frac{\dot{\psi}^*}{\alpha M} - \frac{k_\psi(\psi_{rd} - \psi^*)}{\alpha M} \\
i_{sq} &= \frac{1}{\mu \psi_{rd}}\left[-k_\omega(\omega - \omega^*) + \dot{\omega}^* + \frac{T_L}{J}\right] \\
\psi_{rd} &= \psi_{ra}\cos\rho + \psi_{rb}\sin\rho \\
\cos\rho &= \frac{\psi_{ra}}{\sqrt{\psi_{ra}^2 + \psi_{rb}^2}} \\
\sin\rho &= \frac{\psi_{rb}}{\sqrt{\psi_{ra}^2 + \psi_{rb}^2}};
\end{aligned} \tag{2.21}$$

substituting (2.21) in (2.9) we have for the rotor speed and flux modulus tracking errors

$$\begin{aligned}
\frac{d\tilde{\omega}}{dt} &= -k_\omega \tilde{\omega} \\
\frac{d\tilde{\psi}_{rd}}{dt} &= -(\alpha + k_\psi)\tilde{\psi}_{rd}
\end{aligned} \tag{2.22}$$

which clarifies that the dynamics for the speed and flux modulus tracking errors are decoupled and linear with arbitrary time constants k_ω^{-1} and $(\alpha + k_\psi)^{-1}$. Note, however, that exact input–output decoupling and linearization have been achieved by the controller (2.21) at the expense of a singularity at $\psi_{rd} = 0$ which, in contrast to the indirect field-oriented control (2.20), may imply very large currents (i_{sa}, i_{sb}) when ψ_{rd} is close to zero.

It is instructive to compare the structures of the following four control algorithms: the feedforward control (1.74), the indirect field-oriented control (2.20), the direct field-oriented control (2.19) and the input–output feedback linearizing control (2.21). They are listed in terms of increasing complexity. In fact, the feedforward control (1.74) requires no state variable measurements but precise initialization for $\rho^*(0)$. The indirect field-oriented control (2.20) requires only rotor speed measurements which are used to introduce the feedback term $-k_\omega(\omega - \omega^*)$ in the expression of i_{sq} and to replace ω^* by ω in the dynamic equation which produces the rotation angle ε_0 and can be arbitrarily initialized: the first action is the classical feedback proportional to the speed tracking error, while the second action is unconventional and is aimed to achieve the field orientation without flux measurements, as we shall see in the next section. Both the direct field-oriented control and the input–output feedback linearizing control generate the rotation matrix on the basis of rotor flux measurements (ψ_{ra}, ψ_{rb}) without computing the angles ρ^* or ε_0 by integration: they differ since the input–output linearizing control contains an additional feedback term $-k_\psi(\psi_{rd} - \psi^*)/(\alpha M)$ in the expression of i_{sd} and makes use of ψ_{rd} instead of its reference ψ^* in the expression of i_{sq}.

Illustrative Simulations

We tested the direct field-oriented control by simulations for the current-fed three-phase single pole pair 0.6-kW induction motor whose parameters have been reported in Chapter 1: stator currents dynamics have been neglected so that the stator currents (i_{sa}, i_{sb}) constitute the motor control inputs. The rotor speed initial condition has been set equal to zero while the rotor fluxes initial conditions have been set equal to $\psi_{ra}(0) = \psi_{rb}(0) = 0.1\,\text{Wb}$. The control algorithm has been tested using the control parameter (the value is in SI units) $k_\omega = 12$, which directly affects the speed tracking error dynamics. The references for the speed and flux modulus along with the applied load torque are reported in Figures 2.13–2.15. The rotor flux modulus reference signal starts from 0.001 Wb at $t = 0\,\text{s}$ and grows up to the constant value 1.16 Wb. The speed reference is zero until $t = 0.32\,\text{s}$ and grows up to the constant value 100 rad/s; at $t = 1.5\,\text{s}$ the speed is required to go up to the value 200 rad/s, while the reference for the flux modulus is reduced to 0.5 Wb. A 5.8-Nm load torque is first applied to the motor and then is reduced to 1.8 Nm. Figures 2.14 and 2.15 show the time histories of rotor speed and flux modulus along with the corresponding tracking errors: the rotor speed and flux modulus track their references tightly. As illustrated by Figure 2.16, the motor trajectories in the state space tend to two attractive limit cycles corresponding to the two constant operating conditions imposed by the reference signals. Finally, the stator currents profiles, which are within the physical saturation limits, are reported in Figure 2.17. It is very interesting to compare Figures 2.2–2.5 which illustrate the feedforward control performance with Figures 2.13–2.17 which illustrate the performance of the feedback direct field-oriented control for the same parameters and similar reference signals. The rotor speed tracking errors (see Figures 2.3 and 2.14) are two orders of magnitude smaller in the feedback case (maximum error of 5 rad/s and 0.09 rad/s, respectively). Moreover, the feedback control also achieves very precise tracking when the load torque reduces its value to 1.8 Nm. Similar improvements are obtained in rotor flux tracking (compare Figures 2.4 and 2.15).

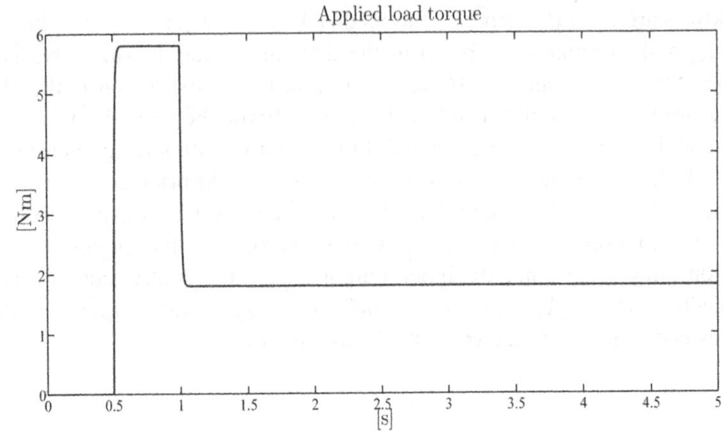

Fig. 2.13 Direct field-oriented control: applied load torque T_L

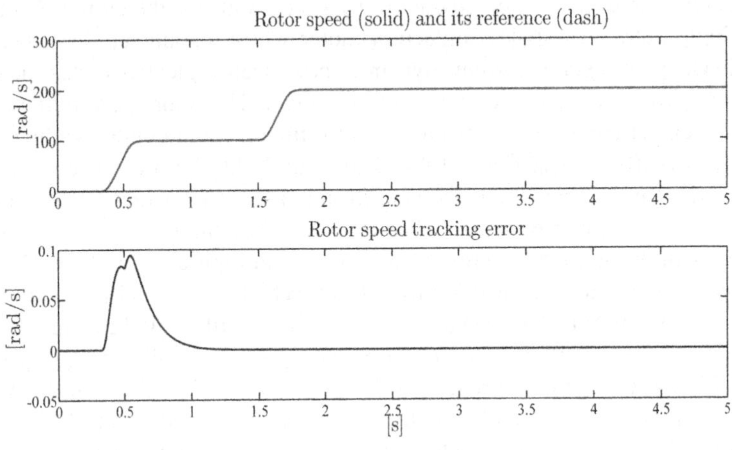

Fig. 2.14 Direct field-oriented control: rotor speed ω and its reference ω^*; rotor speed tracking error

So far we have designed the direct field-oriented control (2.19) and its variants (2.20) and (2.21) on the basis of the reduced order model (2.9) in which the stator current dynamics have been neglected, which is clearly an approximation. Reconsider now the full order model (1.39) expressed in the state coordinates (1.41) and in the control coordinates (u_{sd}, u_{sq}) which are related to the original control inputs (u_{sa}, u_{sb}) by

$$\begin{bmatrix} u_{sa} \\ u_{sb} \end{bmatrix} = \begin{bmatrix} \cos\rho & -\sin\rho \\ \sin\rho & \cos\rho \end{bmatrix} \begin{bmatrix} u_{sd} \\ u_{sq} \end{bmatrix}. \tag{2.23}$$

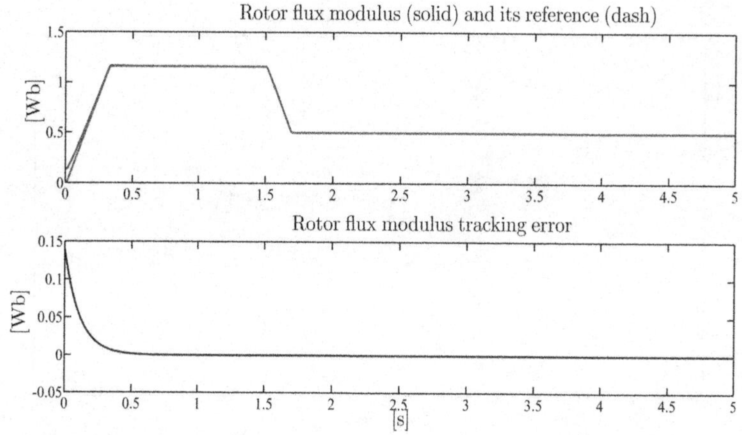

Fig. 2.15 Direct field-oriented control: rotor flux modulus $\sqrt{\psi_{ra}^2 + \psi_{rb}^2}$ and its reference ψ^*; rotor flux modulus tracking error

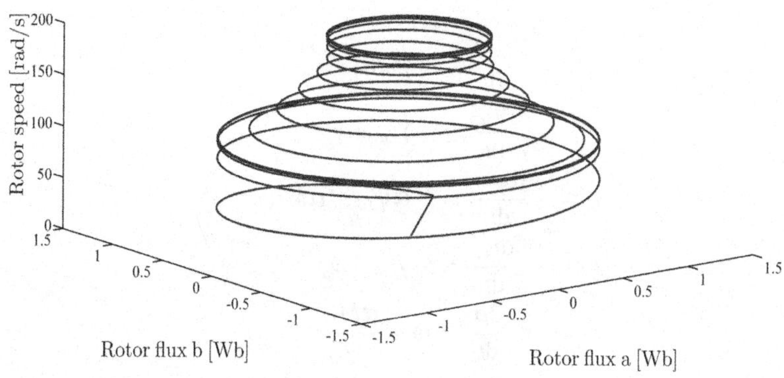

Fig. 2.16 Direct field-oriented control: $(\psi_{ra}, \psi_{rb}, \omega)$-trajectories

Let us now design (u_{sd}, u_{sq}) as state feedback controls so that the desired references (ω^*, ψ^*) are asymptotically tracked. From the last two equations in (1.39) define the state feedback control

$$
\begin{aligned}
u_{sd} &= \sigma\left[-\omega i_{sq} - \frac{\alpha M i_{sq}^2}{\psi_{rd}} - \beta\alpha\psi_{rd} + v_d\right] \\
u_{sq} &= \sigma\left[\omega i_{sd} + \frac{\alpha M i_{sq} i_{sd}}{\psi_{rd}} + \beta\omega\psi_{rd} + v_q\right]
\end{aligned}
\tag{2.24}
$$

in which (v_d, v_q) are new control inputs yet to be designed. Note that the state feedback control (2.23), (2.24) introduces a singularity at $\psi_{rd} = 0$ so that very large input

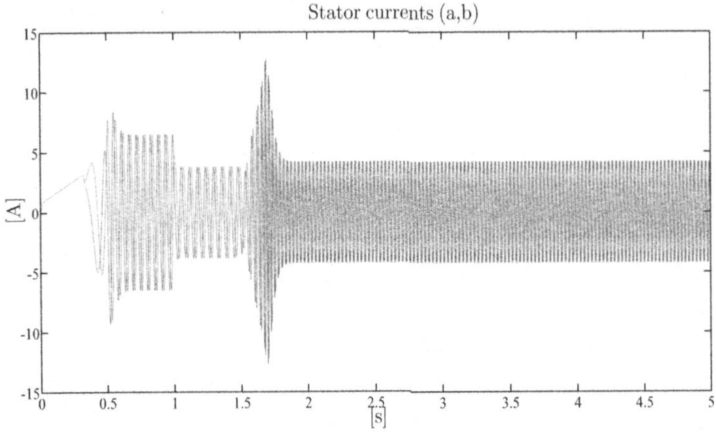

Fig. 2.17 Direct field-oriented control: stator current vector (a,b)-components (i_{sa}, i_{sb})

voltages are to be expected when the rotor flux modulus is close to zero. Substituting (2.24) in (1.39) we obtain the closed-loop system

$$\begin{aligned}\frac{\mathrm{d}\omega}{\mathrm{d}t} &= \mu\psi_{rd}i_{sq} - \frac{T_L}{J}\\ \frac{\mathrm{d}i_{sq}}{\mathrm{d}t} &= -\gamma i_{sq} + v_q\\ \frac{\mathrm{d}\psi_{rd}}{\mathrm{d}t} &= -\alpha\psi_{rd} + \alpha M i_{sd}\\ \frac{\mathrm{d}i_{sd}}{\mathrm{d}t} &= -\gamma i_{sd} + v_d\\ \frac{\mathrm{d}\rho}{\mathrm{d}t} &= \omega + \frac{\alpha M i_{sq}}{\psi_{rd}}\,. \end{aligned}\tag{2.25}$$

In other words, the system (1.26) is transformed into (2.25) by the state space change of coordinates (1.41) and the state feedback control (2.23), (2.24) provided that $\psi_{rd} \neq 0$, since in $\psi_{rd} = 0$ the field-oriented model (1.39) is no longer an equivalent description of the fixed frame model (1.26). The closed-loop system (2.25) has a much simpler structure than system (1.26): the flux amplitude dynamics are linear

$$\begin{aligned}\frac{\mathrm{d}\psi_{rd}}{\mathrm{d}t} &= -\alpha\psi_{rd} + \alpha M i_{sd}\\ \frac{\mathrm{d}i_{sd}}{\mathrm{d}t} &= -\gamma i_{sd} + v_d \end{aligned}\tag{2.26}$$

and can be independently controlled by v_d for instance via a proportional-integral (PI) controller (k_{dp} and k_{di} are positive control parameters)

$$v_d(t) = -k_{dp}(\psi_{rd}(t) - \psi^*(t)) - k_{di}\int_0^t (\psi_{rd}(\tau) - \psi^*(\tau))\mathrm{d}\tau\,. \tag{2.27}$$

Once the flux amplitude ψ_{rd} tracks its reference ψ^*, the rotor speed dynamics are also linear

$$\begin{aligned}\frac{\mathrm{d}\omega}{\mathrm{d}t} &= \mu\psi^* i_{sq} - \frac{T_L}{J}\\ \frac{\mathrm{d}i_{sq}}{\mathrm{d}t} &= -\gamma i_{sq} + v_q\end{aligned} \tag{2.28}$$

and can be independently controlled by v_q for instance via two nested loops of PI controllers (k_{qp} and k_{qi} are positive control parameters)

$$v_q(t) = -k_{qp}(T_e(t) - T_e^*(t)) - k_{qi}\int_0^t (T_e(\tau) - T_e^*(\tau))\mathrm{d}\tau \tag{2.29}$$

with (k_{Tp} and k_{Ti} are positive control parameters)

$$\begin{aligned}T_e(t) &= \mu\psi_{rd}(t) i_{sq}(t)\\ T_e^*(t) &= -k_{Tp}(\omega(t) - \omega^*(t)) - k_{Ti}\int_0^t (\omega(\tau) - \omega^*(\tau))\mathrm{d}\tau\,.\end{aligned} \tag{2.30}$$

We can then say that the state feedback control (2.23), (2.24) achieves asymptotic input–output linearization and decoupling since the first four equations in (2.25) tend to the linear ones (2.26) and (2.28) as ψ_{rd} tends to its reference ψ^*; moreover the outputs ω and ψ_{rd} can be independently controlled by the new inputs v_d and v_q, respectively. Note, however, that during flux transients the nonlinear term $J\mu\psi_{rd}i_{sq}$, which represents the electromagnetic torque T_e in the first equation in (2.25), makes the first four equations in (2.25) still nonlinear and coupled; consequently, the speed transients are difficult to evaluate and may result unacceptable when the flux undergoes a transient to improve efficiency.

2.3 Indirect Field-oriented Control

The indirect field-oriented control (2.20) is a modification of the direct field-oriented control (2.19) which uses, instead of the rotor flux angle ρ, an arbitrary rotating angle ε_0 which is generated replacing ψ_{rd} by its reference ψ^* in the differential equation defining the ε_0 dynamics, *i.e.*

$$\begin{aligned}\begin{bmatrix} i_{sa}\\ i_{sb}\end{bmatrix} &= \begin{bmatrix}\cos\varepsilon_0 & -\sin\varepsilon_0\\ \sin\varepsilon_0 & \cos\varepsilon_0\end{bmatrix}\begin{bmatrix}\frac{\psi^*}{M} + \frac{\dot\psi^*}{\alpha M}\\ \frac{1}{\mu\psi^*}\left[-k_\omega(\omega-\omega^*) + \dot\omega^* + \frac{T_L}{J}\right]\end{bmatrix}\\ \frac{\mathrm{d}\varepsilon_0}{\mathrm{d}t} &= \omega_0 = \omega + \frac{\alpha M i_{sq}}{\psi^*}\end{aligned}$$

$$= \omega + \frac{\alpha M}{\mu \psi^{*2}} \left[-k_\omega (\omega - \omega^*) + \dot{\omega}^* + \frac{T_L}{J} \right] \tag{2.31}$$

with arbitrary $\varepsilon_0(0)$.

The indirect field-oriented control (2.31) may also be designed directly on the basis of the first three equations in the rotating frame model (1.31):

$$\begin{aligned}
\frac{d\omega}{dt} &= \mu\left(\psi_{rd} i_{sq} - \psi_{rq} i_{sd}\right) - \frac{T_L}{J} \\
\frac{d\psi_{rd}}{dt} &= -\alpha \psi_{rd} + (\omega_0 - \omega)\, \psi_{rq} + \alpha M i_{sd} \\
\frac{d\psi_{rq}}{dt} &= -\alpha \psi_{rq} - (\omega_0 - \omega)\, \psi_{rd} + \alpha M i_{sq}\,.
\end{aligned} \tag{2.32}$$

Let $(\omega^*(t), \psi^*(t), 0)$ be the reference signals for $(\omega(t), \psi_{rd}(t), \psi_{rq}(t))$: (2.32) can be rewritten as ($\tilde{\omega} = \omega - \omega^*$, $\tilde{\psi}_{rd} = \psi_{rd} - \psi^*$, $\tilde{\psi}_{rq} = \psi_{rq}$)

$$\begin{aligned}
\frac{d\tilde{\omega}}{dt} &= \mu\left(\tilde{\psi}_{rd} i_{sq} - \tilde{\psi}_{rq} i_{sd}\right) + \mu \psi^* i_{sq} - \dot{\omega}^* - \frac{T_L}{J} \\
\frac{d\tilde{\psi}_{rd}}{dt} &= -\alpha \tilde{\psi}_{rd} + (\omega_0 - \omega)\, \tilde{\psi}_{rq} - \alpha \psi^* - \dot{\psi}^* + \alpha M i_{sd} \\
\frac{d\tilde{\psi}_{rq}}{dt} &= -\alpha \tilde{\psi}_{rq} - (\omega_0 - \omega)\, \tilde{\psi}_{rd} - (\omega_0 - \omega)\psi^* + \alpha M i_{sq}\,.
\end{aligned} \tag{2.33}$$

Our goal is to design a dynamic feedback control

$$\begin{aligned}
\begin{bmatrix} i_{sa} \\ i_{sb} \end{bmatrix} &= \begin{bmatrix} \cos\varepsilon_0 & -\sin\varepsilon_0 \\ \sin\varepsilon_0 & \cos\varepsilon_0 \end{bmatrix} \begin{bmatrix} i_{sd} \\ i_{sq} \end{bmatrix} \\
\dot{\varepsilon}_0 &= \omega_0
\end{aligned} \tag{2.34}$$

so that the tracking errors $(\tilde{\omega}, \tilde{\psi}_{rd}, \tilde{\psi}_{rq})$ tend exponentially to zero from any initial condition. Note that by introducing a dynamic feedback we add an extra degree of freedom in the design since we can now freely choose in (2.33) three independent feedback terms $(i_{sd}, i_{sq}, \omega_0)$ in place of (i_{sd}, i_{sq}) which are at our disposal if a static feedback is designed. We design $(i_{sd}, i_{sq}, \omega_0)$ as feedback terms so that

$$\begin{aligned}
\mu \psi^* i_{sq} - \dot{\omega}^* - \frac{T_L}{J} &= -k_\omega(\omega - \omega^*) \\
-\alpha \psi^* - \dot{\psi}^* + \alpha M i_{sd} &= 0 \\
-(\omega_0 - \omega)\psi^* + \alpha M i_{sq} &= 0
\end{aligned} \tag{2.35}$$

since, substituting (2.35) in (2.33), we obtain

$$\begin{aligned}
\frac{d\tilde{\omega}}{dt} &= -k_\omega \tilde{\omega} + \mu\left(\tilde{\psi}_{rd} i_{sq} - \tilde{\psi}_{rq} i_{sd}\right) \\
\frac{d\tilde{\psi}_{rd}}{dt} &= -\alpha \tilde{\psi}_{rd} + (\omega_0 - \omega)\, \tilde{\psi}_{rq}
\end{aligned}$$

$$\frac{\mathrm{d}\tilde{\psi}_{rq}}{\mathrm{d}t} = -\alpha\tilde{\psi}_{rq} - (\omega_0 - \omega)\,\tilde{\psi}_{rd}\,.$$

Solving (2.35) for $(i_{sd}, i_{sq}, \omega_0)$, we explicitly obtain the feedback terms to be used in (2.34)

$$\begin{aligned} i_{sd} &= \frac{\psi^*}{M} + \frac{\dot{\psi}^*}{\alpha M} \\ i_{sq} &= \frac{1}{\mu\psi^*}\left[-k_\omega(\omega - \omega^*) + \dot{\omega}^* + \frac{T_L}{J}\right] \\ \omega_0 &= \omega + \frac{\alpha M}{\mu\psi^{*2}}\left[-k_\omega(\omega - \omega^*) + \dot{\omega}^* + \frac{T_L}{J}\right] \end{aligned}$$

which, substituted in (2.34), gives exactly the indirect field-oriented control (2.31) in (a,b) coordinates.

The advantages of (2.31) with respect to (2.19) are:

1. while (2.19) is not well defined at $\psi_{ra}^2 + \psi_{rb}^2 = 0$, the indirect field-oriented control (2.31) is always well defined since $\psi^*(t) \geq c_\psi > 0$ for all $t \geq 0$;
2. the control (2.31) requires only the measurement of rotor speed ω while (2.19) also requires the measurement of (ψ_{ra}, ψ_{rb}) to compute $\cos\rho$ and $\sin\rho$;
3. the control (2.31) is a dynamic first order output feedback controller for the current-fed model (2.9) from rotor speed measurements while (2.19) is a static state feedback controller for the same reduced order model (2.9) from rotor speed and flux measurements.

We now substitute (2.31) in the first three equations of the fixed frame model (1.26): to evaluate the closed-loop stability it is more convenient to use directly the rotating frame model (1.31) in the (d,q) reference frame rotating at the speed

$$\frac{\mathrm{d}\varepsilon_0}{\mathrm{d}t} = \omega_0 \tag{2.36}$$

defined in (2.31). Substituting

$$\begin{aligned} i_{sd} &= \frac{\psi^*}{M} + \frac{\dot{\psi}^*}{\alpha M} \\ i_{sq} &= \frac{1}{\mu\psi^*}\left[-k_\omega\,(\omega - \omega^*) + \dot{\omega}^* + \frac{T_L}{J}\right] \end{aligned} \tag{2.37}$$

in the first three equations of model (1.31) we obtain for the second equation in (1.31)

$$\frac{\mathrm{d}\tilde{\psi}_{rd}}{\mathrm{d}t} = -\alpha\tilde{\psi}_{rd} + (\omega_0 - \omega)\,\tilde{\psi}_{rq} \tag{2.38}$$

while for the third equation in (1.31), since

$$(\omega_0 - \omega)\,\psi^* = \alpha M i_{sq}\,, \tag{2.39}$$

we have

$$\frac{\mathrm{d}\tilde{\psi}_{rq}}{\mathrm{d}t} = -\alpha\tilde{\psi}_{rq} - (\omega_0 - \omega)\,\tilde{\psi}_{rd}\,. \tag{2.40}$$

The first equation in (1.31) can be rewritten as

$$\frac{\mathrm{d}\omega}{\mathrm{d}t} = \mu\psi^* i_{sq} - \frac{T_L}{J} + \mu\left(\tilde{\psi}_{rd} i_{sq} - \psi_{rq} i_{sd}\right) \tag{2.41}$$

so that, substituting (2.37) in (2.41), we obtain

$$\frac{\mathrm{d}\tilde{\omega}}{\mathrm{d}t} = -k_\omega\tilde{\omega} + \mu\left(\tilde{\psi}_{rd} i_{sq} - \tilde{\psi}_{rq} i_{sd}\right) \tag{2.42}$$

which along with (2.38) and (2.40), namely

$$\begin{bmatrix} \dot{\tilde{\psi}}_{rd} \\ \dot{\tilde{\psi}}_{rq} \end{bmatrix} = \begin{bmatrix} -\alpha & (\omega_0 - \omega) \\ -(\omega_0 - \omega) & -\alpha \end{bmatrix} \begin{bmatrix} \tilde{\psi}_{rd} \\ \tilde{\psi}_{rq} \end{bmatrix}, \tag{2.43}$$

constitute the closed-loop linear, time-varying tracking dynamics in the (d,q) frame defined by (2.36). Consider the positive definite function

$$V = \frac{1}{2}\tilde{\psi}_{rd}^2 + \frac{1}{2}\tilde{\psi}_{rq}^2 \tag{2.44}$$

whose time derivative along the trajectories of the closed-loop system (2.43) is

$$\dot{V} = -\alpha\tilde{\psi}_{rd}^2 - \alpha\tilde{\psi}_{rq}^2 = -2\alpha V\,. \tag{2.45}$$

Since, according to (2.45),

$$V(t) = \mathrm{e}^{-2\alpha t} V(0) \tag{2.46}$$

we can establish that $\tilde{\psi}_{rd}(t)$ and $\tilde{\psi}_{rq}(t)$ exponentially tend to zero for any initial condition $(\psi_{rd}(0), \psi_{rq}(0))$. In particular, since $\psi_{rq}(t)$ tends to zero, $(\varepsilon_0(t) - \rho(t))$ tends to zero, *i.e.* the (d,q) frame rotating at speed $\omega_0(t)$ given by (2.36) will tend to have its d-axis coincident with the rotating rotor flux vector: field orientation is asymptotically achieved. According to (2.37), (2.42) may be rewritten as

$$\begin{aligned} \frac{\mathrm{d}\tilde{\omega}}{\mathrm{d}t} &= -k_\omega\left[1 + \frac{\tilde{\psi}_{rd}}{\psi^*}\right]\tilde{\omega} \\ &\quad + \frac{\tilde{\psi}_{rd}}{\psi^*}\left(\dot{\omega}^* + \frac{T_L}{J}\right) - \mu\tilde{\psi}_{rq}\left[\frac{\psi^*}{M} + \frac{\dot{\psi}^*}{\alpha M}\right]. \end{aligned} \tag{2.47}$$

According to (2.46) and (2.47), $\tilde{\omega}(t)$ is bounded on $[0, t_*]$ with t_* any positive real. On the other hand, according to (2.46), for any initial condition $(\psi_{rd}(0), \psi_{rq}(0))$ and any positive real $\eta < 1$ there exists $\tilde{t}_* \geq 0$ such that for all $t \geq \tilde{t}_*$

$$\left|\frac{\tilde{\psi}_{rd}(t)}{\psi^*(t)}\right| \leq 1-\eta\,.$$

Therefore, according to (2.47), $\tilde{\omega}(t)$ is an exponentially decaying signal for any initial condition $\omega(0)$.

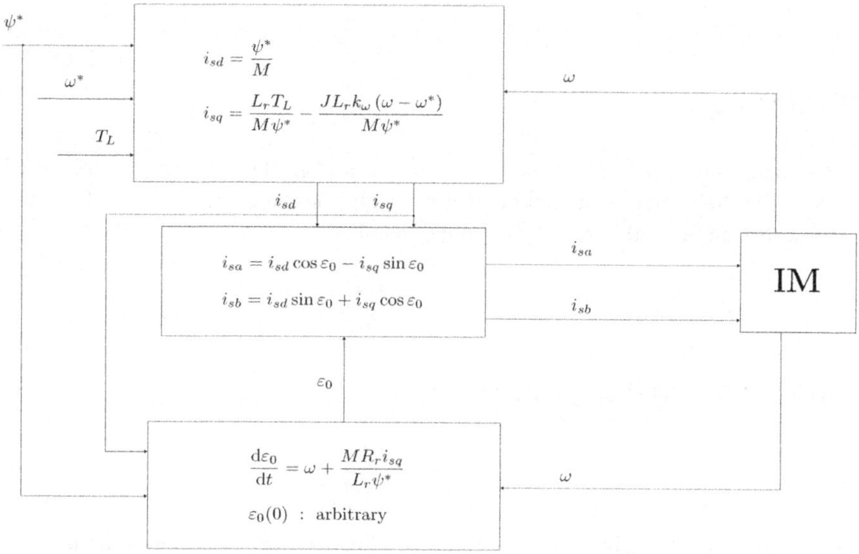

Fig. 2.18 Indirect field-oriented control for current-fed motors (constant references ω^*, ψ^*)

In conclusion: the *indirect field-oriented control* (see Figure 2.18) is defined as

$$\begin{aligned}
\begin{bmatrix} i_{sa} \\ i_{sb} \end{bmatrix} &= \begin{bmatrix} \cos\varepsilon_0 & -\sin\varepsilon_0 \\ \sin\varepsilon_0 & \cos\varepsilon_0 \end{bmatrix} \begin{bmatrix} i_{sd} \\ i_{sq} \end{bmatrix} \\
i_{sd} &= \frac{\psi^*}{M} + \frac{\dot{\psi}^*}{\alpha M} \\
i_{sq} &= \frac{1}{\mu\psi^*}\left[-k_\omega(\omega-\omega^*) + \dot{\omega}^* + \frac{T_L}{J}\right] \\
\frac{d\varepsilon_0}{dt} &= \omega + \frac{\alpha M i_{sq}}{\psi^*}\,;
\end{aligned} \tag{2.48}$$

it is a first order dynamic control algorithm which depends on the measurements of the rotor speed ω, on the reference signals (ω^*, ψ^*), on the arbitrary initial condition $\varepsilon_0(0)$, on the positive control parameter k_ω, on the load torque

T_L, and on the machine parameters M, R_r, L_r, J, since $\mu = \frac{M}{JL_r}$ and $\alpha = \frac{R_r}{L_r}$; it guarantees that, for any $\varepsilon_0(0)$ and for any initial condition of the current-fed reduced order motor model (2.32), the rotor speed and flux modulus tracking errors decay exponentially to zero.

Remarks

1. The name indirect field-oriented control arises from the fact that it is obtained from the direct field-oriented control by using the angle ε_0 in place of the the rotor flux angle ρ: the rotor flux angle dynamics
$$\frac{d\rho}{dt} = \omega + \frac{\alpha M i_{sq}}{\psi_{rd}}$$
with $\rho(0) = \arctan\left(\frac{\psi_{rb}(0)}{\psi_{ra}(0)}\right)$ are replaced by
$$\frac{d\varepsilon_0}{dt} = \omega + \frac{\alpha M i_{sq}}{\psi^*}$$
with arbitrary $\varepsilon_0(0)$ in which the reference flux ψ^* simply replaces the unmeasured ψ_{rd}. Note that, even though ε_0 does not coincide with the rotor flux angle ρ, $[\varepsilon_0(t) - \rho(t)]$ exponentially tends to zero for every initial condition $\varepsilon_0(0)$: the remarkable fact is that field orientation is achieved for any $\varepsilon_0(0)$.
2. As in direct field-oriented control, according to (2.37), the quadrature axis component i_{sq} of the stator current vector is responsible for the rotor speed tracking and depends on T_L, while the direct axis component i_{sd} of the stator current vector is responsible for the tracking of the rotor flux modulus.
3. The critical parameter R_r is required to generate the angle ε_0 in (2.48), since $\alpha = R_r/L_r$. This feature makes the indirect field-oriented control more sensitive with respect to α than the direct field-oriented control, as the experiments reported in Section 2.8 confirm. As we shall see in Chapter 3, the online identification of α is related to the estimation of the rotor flux, so that the need for rotor flux estimation, which is seemingly eliminated by the indirect field-oriented control, reappears through the critical parameter α, whose identification is very closely related to the flux estimation.
4. The measurement of ω is always required in (2.48) even when stringent specifications on speed dynamics are not required so that the gain k_ω in i_{sq} is set equal to zero. In fact, ω is still needed to compute the angle ε_0.
5. As in direct field-oriented control, if the flux modulus and the rotor speed are constant and equal to the desired values (ω^*, ψ^*), then the rotor flux rotates at

constant speed $w = (\omega^* + \omega_s)$, with $\omega_s = \frac{\alpha M T_L}{\mu \psi^{*2}}$, and the induction motor is driven by the sinusoidal currents obtained from (2.48):

$$\begin{bmatrix} i_{sa} \\ i_{sb} \end{bmatrix} = \begin{bmatrix} \cos(\varepsilon_0(0) + wt) & -\sin(\varepsilon_0(0) + wt) \\ \sin(\varepsilon_0(0) + wt) & \cos(\varepsilon_0(0) + wt) \end{bmatrix} \begin{bmatrix} \frac{\psi^*}{M} \\ \frac{T_L}{J\mu\psi^*} \end{bmatrix}.$$

6. According to (2.47), once ψ_{rd} tends to its reference ψ^* and ψ_{rq} tends to zero, the closed-loop dynamics for $\tilde{\omega}$ tend to be linear with arbitrary time constant k_ω^{-1}. During the transient, the nonlinear term $J\mu(\psi_{rd} i_{sq} - \psi_{rq} i_{sd})$, which represents the electromagnetic torque T_e in the first equation in (1.31), makes the first two equations in (1.31) still nonlinear and coupled: for this reason the speed transients may be unsatisfactory, for instance when the flux modulus is adjusted to improve power efficiency.

Illustrative Simulations

We tested the indirect field-oriented control by simulations for the current-fed, three-phase, single pole pair 0.6-kW induction motor whose parameters have been reported in Chapter 1: stator currents dynamics have been neglected so that the stator currents (i_{sa}, i_{sb}) constitute the motor control inputs. The rotor speed initial condition has been set equal to zero while the rotor fluxes initial conditions have been set equal to $\psi_{ra}(0) = \psi_{rb}(0) = 0.1$ Wb. The control algorithm has been tested with the control parameter (the value is in SI units) $k_\omega = 12$, which directly affects the speed tracking error dynamics, and the initial condition $\varepsilon_0(0) = 0$. The references for the speed and flux modulus along with the applied load torque are reported in Figures 2.19–2.21. The rotor flux modulus reference signal starts from 0.001 Wb at $t = 0$ s and grows up to the constant value 1.16 Wb. The speed reference is zero until $t = 0.32$ s and grows up to the constant value 100 rad/s; at $t = 1.5$ s the speed is required to go up to the value 200 rad/s, while the reference for the flux modulus is reduced to 0.5 Wb. A 5.8-Nm load torque is applied to the motor at $t = 0.5$ s and is reduced to 1.8 Nm at $t = 1$ s. Figures 2.20 and 2.21 show the time histories of rotor speed and flux modulus along with the corresponding tracking errors: both the rotor speed and the flux modulus track their references tightly. As illustrated by Figure 2.22 the motor trajectories in the state space tend to two attractive limit cycles corresponding to the two constant operating conditions imposed by the reference signals. Finally, the stator currents profiles, which are within physical saturation limits, are reported in Figure 2.23. We now compare the performance of the direct field-oriented control illustrated in Figures 2.13–2.17 with the performance obtained by the indirect field-oriented control illustrated in Figures 2.19–2.23 for the same motor, same initial conditions, same motor parameters, same control parameters ($k_\omega = 12$), and same reference signals. While the stator current inputs can be hardly distinguished (compare Figures 2.17 and 2.23) and the state space trajectories are very similar (compare Figures 2.16 and 2.22), the only difference is in the transient behavior of the speed errors (compare Figures 2.14 and 2.20), which shows a worse speed transient during the time interval $[0, 1]$ s in which the rotor flux error is different from zero (see Figure 2.21) when the indirect field-oriented control is used.

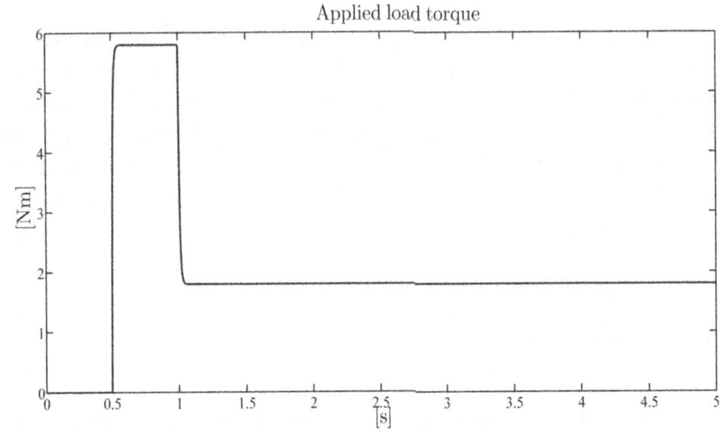

Fig. 2.19 Indirect field-oriented control: applied load torque T_L

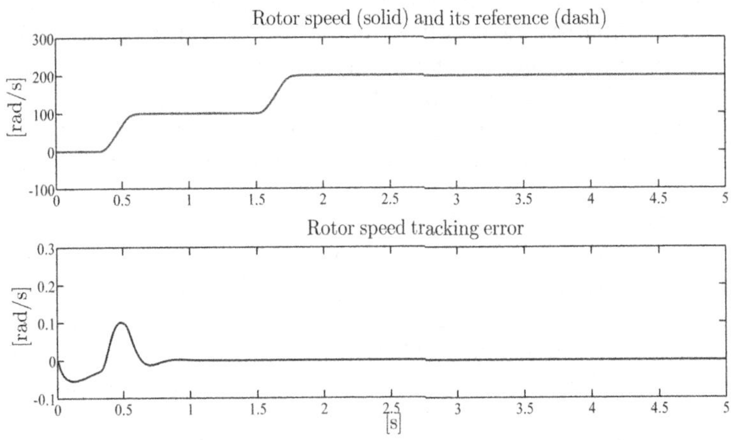

Fig. 2.20 Indirect field-oriented control: rotor speed ω and its reference ω^*; rotor speed tracking error

So far we have designed the indirect field-oriented control (2.48) on the basis of the reduced order model (2.9). If the full order model (1.31), expressed in the

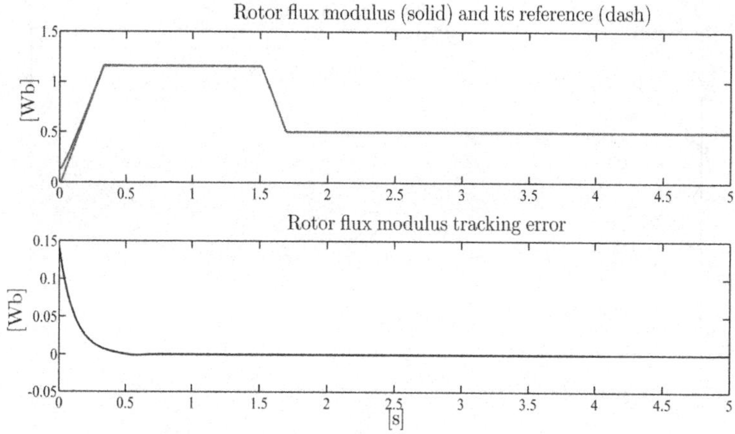

Fig. 2.21 Indirect field-oriented control: rotor flux modulus $\sqrt{\psi_{ra}^2+\psi_{rb}^2}$ and its reference ψ^*; rotor flux modulus tracking error

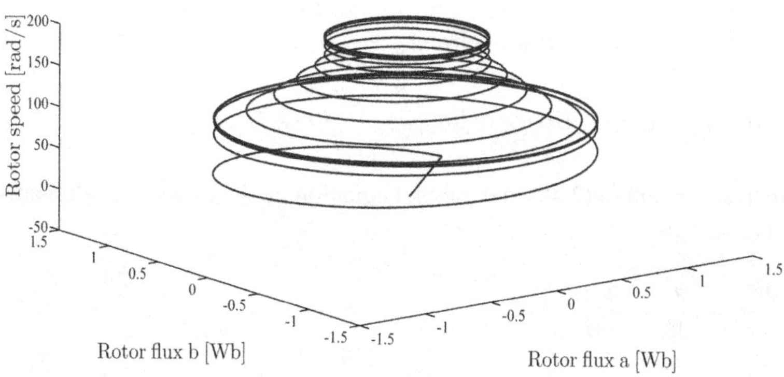

Fig. 2.22 Indirect field-oriented control: $(\psi_{ra}, \psi_{rb}, \omega)$-trajectories

state and control coordinates (1.30), is considered, then (u_{sd}, u_{sq}) are to be designed as state feedback controls so that the reference signals (ω^*, ψ^*) are asymptotically tracked. This can actually be achieved by defining the speed of the rotating (d,q) frame and the reference signals for the stator current vector (d,q) components as

$$\begin{aligned}
\frac{\mathrm{d}\varepsilon_0}{\mathrm{d}t} &= \omega_0 = \omega + \frac{\alpha M i_{sq}}{\psi^*} \\
i_{sd}^* &= \frac{\psi^*}{M} + \frac{\dot{\psi}^*}{\alpha M} \\
i_{sq}^* &= \frac{1}{\mu\psi^*}\left[-k_\omega(\omega-\omega^*)+\dot{\omega}^*+\frac{T_L}{J}\right]
\end{aligned} \tag{2.49}$$

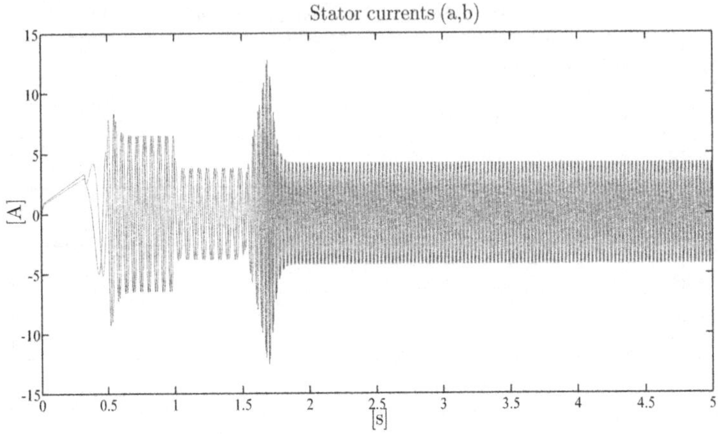

Fig. 2.23 Indirect field-oriented control: stator current vector (a,b)-components (i_{sa}, i_{sb})

and the stator voltages (u_{sd}, u_{sq}) in the (d,q) frame rotating at speed ω_0 as (k_i is a positive control parameter)

$$
\begin{aligned}
u_{sd} &= \sigma\left[\gamma i^*_{sd} - \omega_0 i_{sq} - \beta\alpha\psi_{rd} - \beta\omega\psi_{rq} + \frac{\mathrm{d}i^*_{sd}}{\mathrm{d}t} - k_i\left(i_{sd} - i^*_{sd}\right) - \alpha M\left(\psi_{rd} - \psi^*\right)\right] \\
u_{sq} &= \sigma\left[\gamma i^*_{sq} + \omega_0 i_{sd} - \beta\alpha\psi_{rq} + \beta\omega\psi_{rd} + \frac{\mathrm{d}i^*_{sq}}{\mathrm{d}t} - k_i\left(i_{sq} - i^*_{sq}\right)\right]
\end{aligned} \tag{2.50}
$$

in which, according to (2.49) and the first equation in (1.31), the time derivatives of i^*_{sd} and i^*_{sq} are given by

$$
\begin{aligned}
\frac{\mathrm{d}i^*_{sd}}{\mathrm{d}t} &= \frac{\dot{\psi}^*}{M} + \frac{\ddot{\psi}^*}{\alpha M} \\
\frac{\mathrm{d}i^*_{sq}}{\mathrm{d}t} &= \frac{1}{\mu\psi^*}\left[-k_\omega\left(\dot{\omega} - \dot{\omega}^*\right) + \ddot{\omega}^*\right] - \frac{\dot{\psi}^*}{\mu\psi^{*2}}\left[-k_\omega\left(\omega - \omega^*\right) + \dot{\omega}^* + \frac{T_L}{J}\right] \\
&= \frac{1}{\mu\psi^*}\left[-k_\omega\left[\mu\left(\psi_{rd} i_{sq} - \psi_{rq} i_{sd}\right) - \frac{T_L}{J} - \dot{\omega}^*\right] + \ddot{\omega}^*\right] \\
&\quad - \frac{\dot{\psi}^*}{\mu\psi^{*2}}\left[-k_\omega\left(\omega - \omega^*\right) + \dot{\omega}^* + \frac{T_L}{J}\right].
\end{aligned} \tag{2.51}
$$

Exponential rotor speed and flux modulus tracking can be proved for any motor initial condition (see Problem 2.6). However, as in the current-fed case, the rate of convergence depends on α, according to (2.46).

2.4 Input–Output Feedback Linearizing Control

The input–output feedback linearizing control (2.21) for current-fed induction motors has the structure given in Figure 2.24 (compare with Figures 2.1 and 2.12). In this section we design an input–output feedback linearizing control both for the field-oriented model (1.39) and for the fixed frame model (1.26). The direct field-

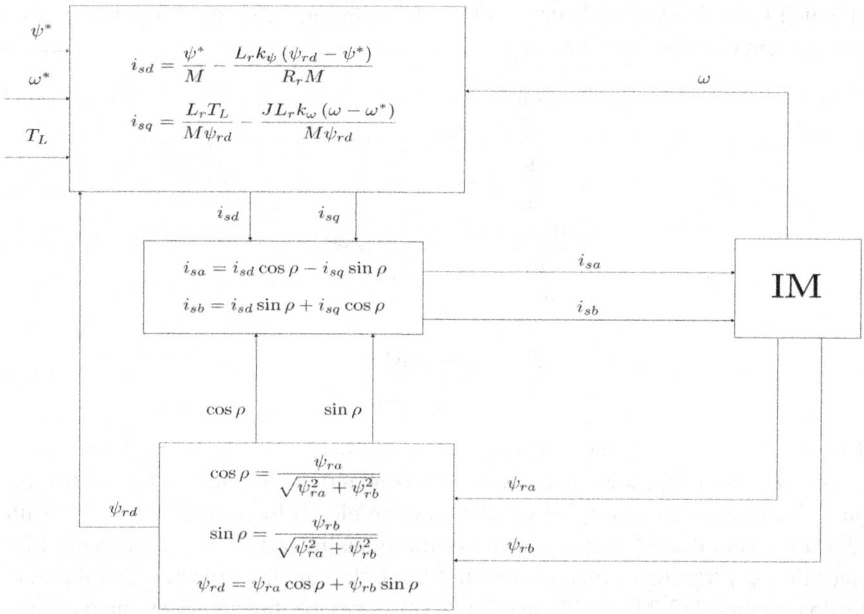

Fig. 2.24 Input–output feedback linearizing control for current-fed motors (constant references ω^*, ψ^*)

oriented control design which led us to the state feedback control (2.23), (2.24) can be improved by designing the additional control input v_q so that the first two equations in (2.25) are made linear by state feedback and input–output feedback linearization and decoupling are achieved: to this end, introduce the rotor angular acceleration

$$a = \mu \psi_{rd} i_{sq} - \frac{T_L}{J} \tag{2.52}$$

in place of i_{sq} as a new state variable so that the first two equations in (2.25) become (recall that T_L is assumed to be constant)

$$\frac{d\omega}{dt} = a$$

$$\begin{aligned}\frac{\mathrm{d}a}{\mathrm{d}t} &= \mu\frac{\mathrm{d}\psi_{rd}}{\mathrm{d}t}i_{sq} + \mu\psi_{rd}\frac{\mathrm{d}i_{sq}}{\mathrm{d}t} \\ &= -\mu\alpha\psi_{rd}i_{sq} + \mu\alpha M i_{sd}i_{sq} - \gamma\mu\psi_{rd}i_{sq} + \mu\psi_{rd}v_q\,. \end{aligned} \tag{2.53}$$

Define the additional state feedback term

$$v_q = \alpha i_{sq} + \gamma i_{sq} - \frac{\alpha M i_{sd} i_{sq}}{\psi_{rd}} + \frac{1}{\mu\psi_{rd}}v'_q \tag{2.54}$$

in which v'_q is an additional control input. Substituting (2.54) in (2.53), the closed-loop system (2.25), (2.54) becomes

$$\begin{aligned}\frac{\mathrm{d}\omega}{\mathrm{d}t} &= a \\ \frac{\mathrm{d}a}{\mathrm{d}t} &= v'_q \\ \frac{\mathrm{d}\psi_{rd}}{\mathrm{d}t} &= -\alpha\psi_{rd} + \alpha M i_{sd} \\ \frac{\mathrm{d}i_{sd}}{\mathrm{d}t} &= -\gamma i_{sd} + v_d \\ \frac{\mathrm{d}\rho}{\mathrm{d}t} &= \omega + \frac{\alpha M i_{sq}}{\psi_{rd}}\,. \end{aligned} \tag{2.55}$$

The first four equations in (2.55) are linear and decoupled since the dynamics of ω and ψ_{rd} are independent and can be independently controlled by the control inputs v'_q and v_d, respectively. If we compare the closed-loop dynamics (2.25) with (2.55) we note that while rotor speed ω transients in (2.55) are not influenced by rotor flux ψ_{rd} transients, this is not so in (2.25). We can then conclude that the state feedback control (2.23), (2.24), and (2.54) improves the direct field-oriented control since it achieves input–output feedback linearization and decoupling as shown by the closed-loop dynamics (2.55). To design v_d in (2.26), introduce the time derivative of ψ_{rd}

$$v_{\psi d} = -\alpha\psi_{rd} + \alpha M i_{sd}$$

so that (2.26) is transformed into

$$\begin{aligned}\frac{\mathrm{d}\psi_{rd}}{\mathrm{d}t} &= v_{\psi d} \\ \frac{\mathrm{d}v_{\psi d}}{\mathrm{d}t} &= -\alpha(\alpha M i_{sd} - \alpha\psi_{rd}) + \alpha M(-\gamma i_{sd} + v_d)\,. \end{aligned}$$

Define

$$v_d = \gamma i_{sd} - \frac{\alpha\psi_{rd}}{M} + \alpha i_{sd} + v'_d \tag{2.56}$$

so that we can write

$$\frac{d\omega}{dt} = a$$
$$\frac{da}{dt} = v'_q$$
$$\frac{d\psi_{rd}}{dt} = v_{\psi d}$$
$$\frac{dv_{\psi d}}{dt} = v'_d \, .$$

In order to track the desired references ω^* and ψ^* for ω and ψ_{rd}, the input signals (v'_d, v'_q) are designed as

$$\begin{aligned} v'_d(t) &= -k_{\psi p}\left(\psi_{rd}(t) - \psi(t)\right) - k_{\psi d}\left(v_{\psi d}(t) - \dot{\psi}^*(t)\right) + \ddot{\psi}^*(t) \\ v'_q(t) &= -k_{\omega p}\left(\omega(t) - \omega^*(t)\right) - k_{\omega d}\left(a(t) - \dot{\omega}^*(t)\right) + \ddot{\omega}^*(t) \end{aligned} \tag{2.57}$$

where $k_{\psi p}$, $k_{\psi d}$, $k_{\omega p}$ and $k_{\omega d}$ are positive constant design parameters to be determined in order to make the decoupled, linear, time-invariant, second order systems ($\tilde{\omega} = \omega - \omega^*$, $\tilde{\psi}_{rd} = \psi_{rd} - \psi^*$)

$$\frac{d^2\tilde{\omega}}{dt^2} = -k_{\omega p}\tilde{\omega} - k_{\omega d}\frac{d\tilde{\omega}}{dt}$$
$$\frac{d^2\tilde{\psi}_{rd}}{dt^2} = -k_{\psi p}\tilde{\psi}_{rd} - k_{\psi d}\frac{d\tilde{\psi}_{rd}}{dt}$$

exponentially stable and to shape their dynamics.

Note that the same result can be achieved directly in fixed (a, b) coordinates without introducing any rotating frame at all. It is enough to perform the following state space change of coordinates $z = \varphi(\omega, \psi_{ra}, \psi_{rb}, i_{sa}, i_{sb})$ with $z = [z_1, \ldots, z_5]$ and

$$\begin{aligned} z_1 &= \omega \\ z_2 &= a = \mu\left(\psi_{ra} i_{sb} - \psi_{rb} i_{sa}\right) - \frac{T_L}{J} \\ z_3 &= \psi_{ra}^2 + \psi_{rb}^2 \\ z_4 &= \frac{d\left(\psi_{ra}^2 + \psi_{rb}^2\right)}{dt} = -2\alpha\left(\psi_{ra}^2 + \psi_{rb}^2\right) + 2\alpha M\left(\psi_{ra} i_{sa} + \psi_{rb} i_{sb}\right) \\ z_5 &= \rho = \arctan\left(\frac{\psi_{rb}}{\psi_{ra}}\right) . \end{aligned} \tag{2.58}$$

According to the Inverse Function Theorem B.1 in Appendix B, for any point $p = (\omega, \psi_{ra}, \psi_{rb}, i_{sa}, i_{sb})$ satisfying $\psi_{ra}^2 + \psi_{rb}^2 \neq 0$ there exists an open neighborhood U of p such that $\varphi(p) = (z_1(p), \ldots, z_n(p)) : U \to \varphi(U)$ is a diffeomorphism, that is a bijection with $\varphi(\cdot)$ and $\varphi^{-1}(\cdot)$ smooth maps. In the new z-coordinates, system (1.26) becomes

$$\begin{bmatrix} \dot{z}_1 \\ \dot{z}_2 \\ \dot{z}_3 \\ \dot{z}_4 \\ \dot{z}_5 \end{bmatrix} = \begin{bmatrix} z_2 \\ \Gamma_1 \\ z_4 \\ \Gamma_2 \\ \omega + \dfrac{\alpha M (\psi_{ra} i_{sb} - \psi_{rb} i_{sa})}{(\psi_{ra}^2 + \psi_{rb}^2)} \end{bmatrix} + \begin{bmatrix} 0\,0 \\ 1\,0 \\ 0\,0 \\ 0\,1 \\ 0\,0 \end{bmatrix} \begin{bmatrix} -\dfrac{\mu \psi_{rb}}{\sigma} & \dfrac{\mu \psi_{ra}}{\sigma} \\ \dfrac{2\alpha M \psi_{ra}}{\sigma} & \dfrac{2\alpha M \psi_{rb}}{\sigma} \end{bmatrix} \begin{bmatrix} u_{sa} \\ u_{sb} \end{bmatrix} \tag{2.59}$$

in which the nonlinear terms

$$\begin{aligned} \Gamma_1(\omega, \psi_{ra}, \psi_{rb}, i_{sa}, i_{sb}) &= -\mu\beta\omega\left(\psi_{ra}^2 + \psi_{rb}^2\right) - \mu(\alpha+\gamma)(\psi_{ra} i_{sb} - \psi_{rb} i_{sa}) \\ &\quad -\mu\omega(\psi_{ra} i_{sa} + \psi_{rb} i_{sb}) \\ \Gamma_2(\omega, \psi_{ra}, \psi_{rb}, i_{sa}, i_{sb}) &= \left(4\alpha^2 + 2\alpha^2\beta M\right)\left(\psi_{ra}^2 + \psi_{rb}^2\right) + 2\alpha M\omega(\psi_{ra} i_{sb} - \psi_{rb} i_{sa}) \\ &\quad - \left(6\alpha^2 M + 2\alpha\gamma M\right)(\psi_{ra} i_{sa} + \psi_{rb} i_{sb}) + 2\alpha^2 M^2\left(i_{sa}^2 + i_{sb}^2\right) \end{aligned}$$

appear and the decoupling matrix

$$D(\psi_{ra}, \psi_{rb}) = \begin{bmatrix} -\dfrac{\mu \psi_{rb}}{\sigma} & \dfrac{\mu \psi_{ra}}{\sigma} \\ \dfrac{2\alpha M \psi_{ra}}{\sigma} & \dfrac{2\alpha M \psi_{rb}}{\sigma} \end{bmatrix}$$

whose determinant is

$$\det[D](\psi_{ra}, \psi_{rb}) = -\frac{2\mu\alpha M}{\sigma^2}\left(\psi_{ra}^2 + \psi_{rb}^2\right)$$

is nonsingular provided that $(\psi_{ra}^2 + \psi_{rb}^2) \neq 0$: hence, Theorem B.10 in Appendix B applies with control characteristic indices $\rho_1 = 2$, $\rho_2 = 2$. As a matter of fact, Theorem B.10 applies directly to the induction motor fixed frame model (1.26) with outputs ω and $\psi_{ra}^2 + \psi_{rb}^2$ and decoupling indices $\rho_1 = 2$, $\rho_2 = 2$. The input–output state feedback linearizing control is

$$\begin{bmatrix} u_{sa} \\ u_{sb} \end{bmatrix} = D(\psi_{ra}, \psi_{rb})^{-1} \begin{bmatrix} -\Gamma_1 + v_a \\ -\Gamma_2 + v_b \end{bmatrix} \tag{2.60}$$

which substituted in (2.59) gives

$$\begin{bmatrix} \dot{z}_1 \\ \dot{z}_2 \\ \dot{z}_3 \\ \dot{z}_4 \\ \dot{z}_5 \end{bmatrix} = \begin{bmatrix} z_2 \\ v_a \\ z_4 \\ v_b \\ \omega + \dfrac{\alpha M(\psi_{ra} i_{sb} - \psi_{rb} i_{sa})}{(\psi_{ra}^2 + \psi_{rb}^2)} \end{bmatrix}$$

$$= \begin{bmatrix} z_2 \\ v_a \\ z_4 \\ v_b \\ z_1 + \dfrac{\alpha M\left(z_2 + \frac{T_L}{J}\right)}{\mu z_3} \end{bmatrix} = \begin{bmatrix} z_2 \\ v_a \\ z_4 \\ v_b \\ z_1 + \dfrac{R_r(Jz_2 + T_L)}{z_3} \end{bmatrix}. \tag{2.61}$$

The last equation in (2.61) represents the dynamics which have been made unobservable from the outputs ω and $(\psi_{ra}^2 + \psi_{rb}^2)$ by the state feedback control (2.60). The same interpretation can be given to the last equation in (2.55).

Remarks

1. The closed-loop system (2.61) is input–output decoupled and linear: the input–output map consists of a pair of second order linear time-invariant systems. This allows for an independent tracking of the outputs so that transient responses can be decoupled: this is an improvement with respect to both the direct and the indirect field-oriented controls. Note however that the exact input–output decoupling and linearization have been achieved by the controller (2.60) at the expense of a singularity at $(\psi_{ra}^2 + \psi_{rb}^2) = 0$ introduced by the inversion of the decoupling matrix

$$D(\psi_{ra}, \psi_{rb})^{-1} = [\det[D](\psi_{ra}, \psi_{rb})]^{-1} \begin{bmatrix} \frac{2\alpha M \psi_{rb}}{\sigma} & -\frac{\mu \psi_{ra}}{\sigma} \\ -\frac{2\alpha M \psi_{ra}}{\sigma} & -\frac{\mu \psi_{rb}}{\sigma} \end{bmatrix}$$

$$= \left[-\frac{2\mu\alpha M}{\sigma^2}(\psi_{ra}^2 + \psi_{rb}^2)\right]^{-1} \begin{bmatrix} \frac{2\alpha M \psi_{rb}}{\sigma} & -\frac{\mu \psi_{ra}}{\sigma} \\ -\frac{2\alpha M \psi_{ra}}{\sigma} & -\frac{\mu \psi_{rb}}{\sigma} \end{bmatrix} \tag{2.62}$$

in (2.60) which, in contrast to the indirect field-oriented control, may imply very large voltages (u_{sa}, u_{sb}) when $(\psi_{ra}^2 + \psi_{rb}^2)$ is close to zero. Recall that the direct field-oriented control also cannot operate when the rotor flux modulus is zero. Note that (2.62) can be rewritten as

$$D(\psi_{ra}, \psi_{rb})^{-1} = \begin{bmatrix} -\sin\rho & \cos\rho \\ \cos\rho & \sin\rho \end{bmatrix} \begin{bmatrix} \frac{\sigma}{\mu\sqrt{\psi_{ra}^2 + \psi_{rb}^2}} & 0 \\ 0 & \frac{\sigma}{2\alpha M\sqrt{\psi_{ra}^2 + \psi_{rb}^2}} \end{bmatrix}$$

with $\sin\rho = \psi_{rb}/\sqrt{\psi_{ra}^2+\psi_{rb}^2}$ and $\cos\rho = \psi_{ra}/\sqrt{\psi_{ra}^2+\psi_{rb}^2}$. The rotation matrix has been reobtained without introducing any rotating frame from the beginning.
2. Measurements of the state variables are required both by the control (2.60) and by the direct field-oriented control.
3. While the indirect field-oriented control is a first order output feedback (from ω) dynamic control, the control (2.60) is a static state feedback control law.
4. Since it has been previously shown that the induction motor model (1.26) is not feedback linearizable and that the largest feedback linearizable subsystem has dimension 4, the control (2.60) provides the largest linearizable subsystem in the closed-loop.

We now design the input signals v_a and v_b in (2.60) in order to track the desired references ω^* and ψ^{*2} for the rotor speed $z_1 = \omega$ and $z_3 = (\psi_{ra}^2+\psi_{rb}^2)$. Similarly to the design of v'_d and v'_q, we choose

$$\begin{aligned}
v_a &= -k_{\omega p}(z_1-\omega^*) - k_{\omega d}(\dot z_1 - \dot\omega^*) + \ddot\omega^* \\
&= -k_{\omega p}(\omega-\omega^*) - k_{\omega d}\left(\mu(\psi_{ra}i_{sb}-\psi_{rb}i_{sa}) - \frac{T_L}{J} - \dot\omega^*\right) + \ddot\omega^* \\
v_b &= -k_{\psi p}\left(z_3-\psi^{*2}\right) - k_{\psi d}(\dot z_3 - 2\psi^*\dot\psi^*) + 2\dot\psi^{*2} + 2\psi^*\ddot\psi^* \\
&= -k_{\psi p}\left[(\psi_{ra}^2+\psi_{rb}^2)-\psi^{*2}\right] \\
&\quad -k_{\psi d}\left[-2\alpha(\psi_{ra}^2+\psi_{rb}^2) + 2\alpha M(\psi_{ra}i_{sa}+\psi_{rb}i_{sb}) - 2\psi^*\dot\psi^*\right] \\
&\quad +2\dot\psi^{*2} + 2\psi^*\ddot\psi^*
\end{aligned}$$

where $k_{\omega p}, k_{\omega d}, k_{\psi p}, k_{\psi d}$ are constant design parameters to be determined in order to make the decoupled, linear, time-invariant, second order systems ($\tilde\Psi = \psi_{ra}^2+\psi_{rb}^2 - \psi^{*2}$)

$$\begin{aligned}
\frac{d^2\tilde\omega}{dt^2} &= -k_{\omega p}\tilde\omega - k_{\omega d}\frac{d\tilde\omega}{dt} \\
\frac{d^2\tilde\Psi}{dt^2} &= -k_{\psi p}\tilde\Psi - k_{\psi d}\frac{d\tilde\Psi}{dt}
\end{aligned} \tag{2.63}$$

exponentially stable and to shape their dynamics.

In conclusion: the *input–output feedback linearizing control* is defined as

$$\begin{bmatrix} u_{sa} \\ u_{sb}\end{bmatrix} = \left[-\frac{2\mu\alpha M}{\sigma^2}(\psi_{ra}^2+\psi_{rb}^2)\right]^{-1}\begin{bmatrix} \frac{2\alpha M\psi_{rb}}{\sigma} & -\frac{\mu\psi_{ra}}{\sigma} \\ -\frac{2\alpha M\psi_{ra}}{\sigma} & -\frac{\mu\psi_{rb}}{\sigma}\end{bmatrix}\begin{bmatrix} -\Gamma_1 + v_a \\ -\Gamma_2 + v_b\end{bmatrix}$$

$$\begin{aligned}
\Gamma_1 &= -\mu\beta\omega(\psi_{ra}^2+\psi_{rb}^2) - \mu(\alpha+\gamma)(\psi_{ra}i_{sb}-\psi_{rb}i_{sa}) \\
&\quad -\mu\omega(\psi_{ra}i_{sa}+\psi_{rb}i_{sb})
\end{aligned}$$

$$
\begin{aligned}
\Gamma_2 &= \left(4\alpha^2+2\alpha^2\beta M\right)\left(\psi_{ra}^2+\psi_{rb}^2\right)+2\alpha M\omega\left(\psi_{ra}i_{sb}-\psi_{rb}i_{sa}\right)\\
&\quad -\left(6\alpha^2 M+2\alpha\gamma M\right)\left(\psi_{ra}i_{sa}+\psi_{rb}i_{sb}\right)+2\alpha^2 M^2\left(i_{sa}^2+i_{sb}^2\right)\\
v_a &= -k_{\omega p}\left(\omega-\omega^*\right)-k_{\omega d}\left(\mu\left(\psi_{ra}i_{sb}-\psi_{rb}i_{sa}\right)-\frac{T_L}{J}-\dot{\omega}^*\right)+\ddot{\omega}^*\\
v_b &= -k_{\psi p}\left[\left(\psi_{ra}^2+\psi_{rb}^2\right)-\psi^{*2}\right]\\
&\quad -k_{\psi d}\left[-2\alpha\left(\psi_{ra}^2+\psi_{rb}^2\right)+2\alpha M\left(\psi_{ra}i_{sa}+\psi_{rb}i_{sb}\right)-2\psi^*\dot{\psi}^*\right]\\
&\quad +2\dot{\psi}^{*2}+2\psi^*\ddot{\psi}^*;
\end{aligned}
\tag{2.64}
$$

it is a static nonlinear feedback control algorithm which depends on the measurements of the state variables $(\omega, \psi_{ra}, \psi_{rb}, i_{sa}, i_{sb})$, on the reference signals (ω^*, ψ^*), on the positive control parameters $k_{\omega p}$, $k_{\omega d}$, $k_{\psi p}$, $k_{\psi d}$, on the load torque T_L and on the machine parameters M, R_r, L_r, J, R_s, L_s, since $\mu = \dfrac{M}{JL_r}$, $\alpha = \dfrac{R_r}{L_r}$, $\sigma = L_s\left(1-\dfrac{M^2}{L_sLr}\right)$, $\beta = \dfrac{M}{\sigma L_r}$, $\gamma = \dfrac{R_s}{\sigma}+\beta\alpha M$; it guarantees that, for suitable motor initial conditions such that $\psi_{ra}^2(t)+\psi_{rb}^2(t) \geq c_\psi > 0$ for all $t \geq 0$, the rotor speed and flux modulus tracking errors have decoupled dynamics and decay exponentially to zero according to (2.63).

Illustrative Simulations

We tested the input–output feedback linearizing control by simulations for the three-phase single pole pair 0.6-kW induction motor whose parameters have been reported in Chapter 1. All the motor initial conditions have been set equal to zero except for $\psi_{ra}(0) = \psi_{rb}(0) = 0.1$ Wb. The control algorithm has been tested with the control parameters (all the values are in SI units) $k_{\omega p} = 8100$, $k_{\omega d} = 180$, $k_{\psi p} = 8100$, $k_{\psi d} = 180$; real coincident eigenvalues are assigned to the matrices associated with the decoupled, linear time-invariant second order systems (2.63). The references for the speed and flux modulus along with the applied load torque are reported in Figures 2.25–2.27. The rotor flux modulus reference signal starts from 0.001 Wb at $t = 0$ s and grows up to the constant value 1.16 Wb. The speed reference is zero until $t = 0.32$ s and grows up to the constant value 100 rad/s; at $t = 1.5$ s the speed is required to go up to the value 200 rad/s, while the reference for the flux modulus is reduced to 0.5 Wb. A 5.8-Nm load torque is applied to the motor and is reduced to 1.8 Nm. Figures 2.26 and 2.27 show the time histories of rotor speed and flux modulus along with the corresponding tracking errors: the rotor speed and flux modulus track tightly their references. Note that there is no coupling between rotor speed tracking and rotor flux modulus tracking at $t = 0.5$ s and $t = 1$ s when rotor speed is perturbed by the uncompensated load torque time derivative. Finally, the stator current and voltages profiles, which are within the physical saturation limits, are reported in Figures 2.28 and 2.29.

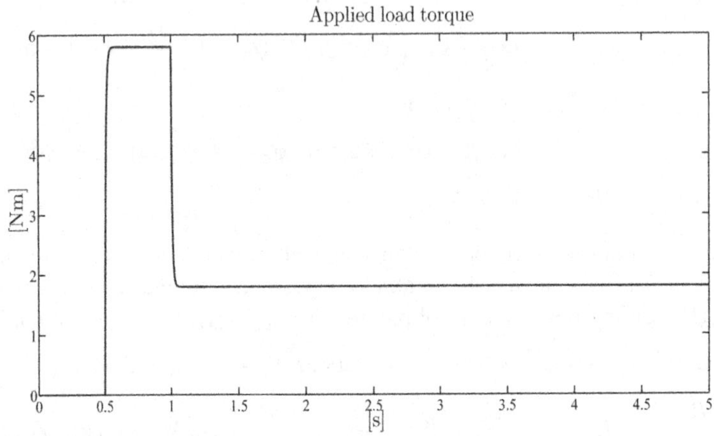

Fig. 2.25 Input–output feedback linearizing control: applied load torque T_L

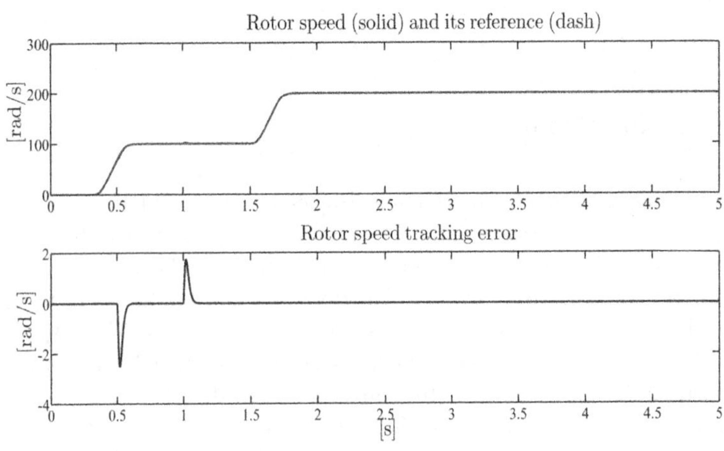

Fig. 2.26 Input–output feedback linearizing control: rotor speed ω and its reference ω^*; rotor speed tracking error

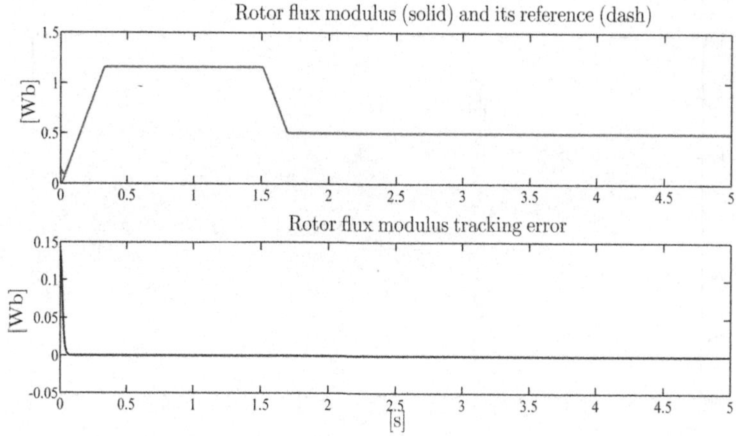

Fig. 2.27 Input–output feedback linearizing control: rotor flux modulus $\sqrt{\psi_{ra}^2 + \psi_{rb}^2}$ and its reference ψ^*; rotor flux modulus tracking error

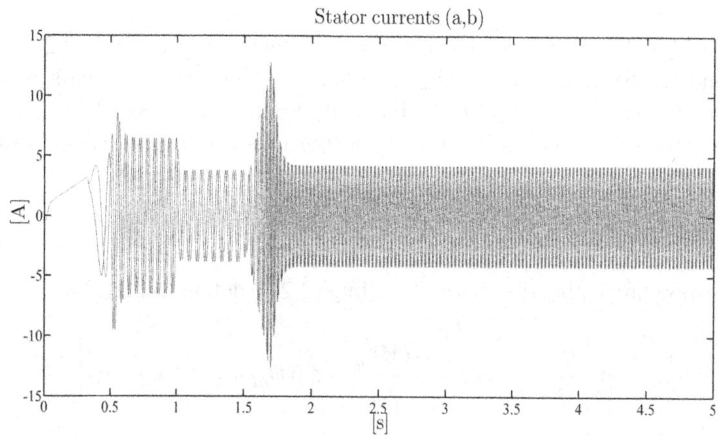

Fig. 2.28 Input–output feedback linearizing control: stator current vector (a,b)-components (i_{sa}, i_{sb})

2.5 Adaptive Input–Output Feedback Linearizing Control

So far we have designed the state feedback control algorithms by assuming the knowledge of the rotor resistance R_r and of the load torque T_L. While the load torque depends on applications, the rotor resistance may vary up to 100% during operations due to rotor heating: thus they constitute typically uncertain parameters. Experiments reported in Section 2.8 will show how critical the rotor resistance parameter is for the control design. In this section we shall estimate its value online along with the load torque.

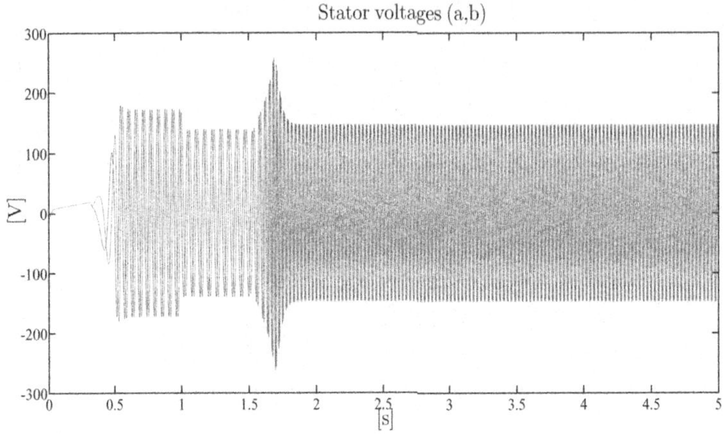

Fig. 2.29 Input–output feedback linearizing control: stator voltage vector (a,b)-components (u_{sa}, u_{sb})

Reconsider the state feedback input–output linearizing control (2.23), (2.24), (2.54) and (2.56); note that (u_{sd}, u_{sq}) defined in (2.24) are linear with respect to the unknown parameter $\alpha = R_r L_r^{-1}$. Hence, the controller (2.23), (2.24), (2.54) and (2.56) constitutes a good starting point to design an adaptation with respect to α. To this end, let us denote by $\hat{\alpha}$ the estimate of the parameter α and by

$$\tilde{\alpha} = \alpha - \hat{\alpha}$$

the corresponding estimation error. Recalling (2.24), define (u_{sd}, u_{sq}) as

$$\begin{aligned} u_{sd} &= \sigma \left[\frac{R_s}{\sigma} i_{sd} - \omega i_{sq} - \frac{\hat{\alpha} M i_{sq}^2}{\psi_{rd}} - \hat{\alpha} \beta \psi_{rd} + \hat{\alpha} M \beta i_{sd} + v_d \right] \\ u_{sq} &= \sigma \left[\frac{R_s}{\sigma} i_{sq} + \omega i_{sd} + \frac{\hat{\alpha} M i_{sq} i_{sd}}{\psi_{rd}} + \beta \omega \psi_{rd} + \hat{\alpha} M \beta i_{sq} + v_q \right] \end{aligned} \tag{2.65}$$

in which (v_d, v_q) are additional control inputs to be designed. Substituting (2.65) in (1.39) we obtain

$$\begin{aligned} \frac{\mathrm{d}\omega}{\mathrm{d}t} &= \mu \psi_{rd} i_{sq} - \frac{T_L}{J} \\ \frac{\mathrm{d}i_{sq}}{\mathrm{d}t} &= -\left(\frac{M i_{sq} i_{sd}}{\psi_{rd}} + M \beta i_{sq} \right) \tilde{\alpha} + v_q \\ \frac{\mathrm{d}\psi_{rd}}{\mathrm{d}t} &= -\alpha \psi_{rd} + \alpha M i_{sd} \\ \frac{\mathrm{d}i_{sd}}{\mathrm{d}t} &= \left(\frac{M i_{sq}^2}{\psi_{rd}} + \beta \psi_{rd} - M \beta i_{sd} \right) \tilde{\alpha} + v_d \end{aligned}$$

$$\frac{\mathrm{d}\rho}{\mathrm{d}t} = \omega + \frac{\alpha M i_{sq}}{\psi_{rd}} \,. \tag{2.66}$$

Let us denote by $\hat{T}_L$ the estimate of the load torque T_L and by

$$\tilde{T}_L = T_L - \hat{T}_L$$

the corresponding estimation error. Since the load torque appears additively in the rotor speed dynamics given by the first equation in (1.31), we design the load torque estimate $\hat{T}_L$ as the output of a linear time-invariant one-dimensional system

$$\begin{aligned} \dot{\xi} &= -\lambda\xi + \lambda J\mu\psi_{rd} i_{sq} + \lambda^2 J\omega - v_T \\ \hat{T}_L &= \xi - \lambda J\omega \end{aligned} \tag{2.67}$$

in which the term v_T is yet to be defined. The above choice is justified by the fact that, according to (2.67), $\dot{\hat{T}}_L$ satisfies

$$\begin{aligned} \dot{\hat{T}}_L &= -\lambda J\mu\psi_{rd} i_{sq} + \lambda T_L + \dot{\xi} \\ &= \lambda T_L - \lambda(\xi - \lambda J\omega) - v_T \\ &= \lambda(T_L - \hat{T}_L) - v_T \end{aligned} \tag{2.68}$$

and therefore

$$\dot{\tilde{T}}_L = -\lambda\tilde{T}_L + v_T \tag{2.69}$$

which will be useful in the overall stability analysis. Introduce two new state variables: the estimated rotor angular acceleration

$$\hat{a} = \mu\psi_{rd} i_{sq} - \frac{\hat{T}_L}{J} \tag{2.70}$$

and the estimated time derivative of the rotor flux direct component ψ_{rd}

$$\hat{v}_{\psi d} = -\hat{\alpha}\psi_{rd} + \hat{\alpha} M i_{sd} \,. \tag{2.71}$$

In new state coordinates $(\omega, \hat{a}, \psi_{rd}, \hat{v}_{\psi d}, \rho)$, (2.66) are rewritten as (recall (2.53))

$$\begin{aligned} \frac{\mathrm{d}\omega}{\mathrm{d}t} &= \hat{a} - \frac{\tilde{T}_L}{J} \\ \frac{\mathrm{d}\hat{a}}{\mathrm{d}t} &= \mu(-\alpha\psi_{rd} + \alpha M i_{sd}) i_{sq} - \mu\tilde{\alpha} M i_{sq} i_{sd} + \mu\psi_{rd} v_q - \mu\tilde{\alpha}\psi_{rd}\beta M i_{sq} \\ &\quad -\frac{\lambda}{J}\tilde{T}_L + \frac{v_T}{J} \\ \frac{\mathrm{d}\psi_{rd}}{\mathrm{d}t} &= \hat{v}_{\psi d} + (M i_{sd} - \psi_{rd})\tilde{\alpha} \\ \frac{\mathrm{d}\hat{v}_{\psi d}}{\mathrm{d}t} &= \dot{\hat{\alpha}}(M i_{sd} - \psi_{rd}) - \hat{\alpha}(\alpha M i_{sd} - \alpha\psi_{rd}) \end{aligned}$$

$$+\hat{\alpha}M\left[\left(\frac{Mi_{sq}^2}{\psi_{rd}}+\beta\psi_{rd}-M\beta i_{sd}\right)\tilde{\alpha}+v_d\right]$$

$$\frac{\mathrm{d}\rho}{\mathrm{d}t}=\omega+\alpha M\left(\frac{\hat{a}}{\mu\psi_{rd}^2}+\frac{\hat{T}_L}{J\mu\psi_{rd}^2}\right). \tag{2.72}$$

Define

$$\begin{aligned} v_q &= \hat{\alpha}i_{sq}-\frac{\hat{\alpha}Mi_{sd}i_{sq}}{\psi_{rd}}-\frac{v_T}{J\mu\psi_{rd}} \\ &\quad +\frac{1}{\mu\psi_{rd}}\left[-k_{\omega p}(\omega-\omega^*)-k_{\omega d}(\hat{a}-\dot{\omega}^*)+\ddot{\omega}^*\right] \\ v_d &= -\frac{\hat{\alpha}\psi_{rd}}{M}+\hat{\alpha}i_{sd}+\frac{1}{\hat{\alpha}M}\Bigg[-\dot{\hat{\alpha}}(Mi_{sd}-\psi_{rd}) \\ &\quad -k_{\psi p}(\psi_{rd}-\psi^*)-k_{\psi d}(\hat{v}_{\psi d}-\dot{\psi}^*)+\ddot{\psi}^*\Bigg] \end{aligned} \tag{2.73}$$

which substituted in (2.72) give

$$\begin{aligned} \frac{\mathrm{d}\omega}{\mathrm{d}t} &= \hat{a}-\frac{\tilde{T}_L}{J} \\ \frac{\mathrm{d}\hat{a}}{\mathrm{d}t} &= -k_{\omega p}(\omega-\omega^*)-k_{\omega d}(\hat{a}-\dot{\omega}^*)+\ddot{\omega}^*-\mu\tilde{\alpha}(1+\beta M)\psi_{rd}i_{sq}-\frac{\lambda}{J}\tilde{T}_L \\ \frac{\mathrm{d}\psi_{rd}}{\mathrm{d}t} &= \hat{v}_{\psi d}+(Mi_{sd}-\psi_{rd})\tilde{\alpha} \\ \frac{\mathrm{d}\hat{v}_{\psi d}}{\mathrm{d}t} &= -k_{\psi p}(\psi_{rd}-\psi^*)-k_{\psi d}(\hat{v}_{\psi d}-\dot{\psi}^*)+\ddot{\psi}^* \\ &\quad +\tilde{\alpha}\left[-\hat{\alpha}M(1+\beta M)i_{sd}+\hat{\alpha}(1+\beta M)\psi_{rd}+\frac{\hat{\alpha}M^2i_{sq}^2}{\psi_{rd}}\right] \\ \frac{\mathrm{d}\rho}{\mathrm{d}t} &= \omega+\alpha M\left(\frac{\hat{a}}{\mu\psi_{rd}^2}+\frac{\hat{T}_L}{J\mu\psi_{rd}^2}\right). \end{aligned} \tag{2.74}$$

Let P_ω and P_ψ be the positive definite solutions of the Lyapunov matrix equations (see Theorem A.6 in Appendix A)

$$\begin{bmatrix} 0 & -k_{\omega p} \\ 1 & -k_{\omega d} \end{bmatrix} P_\omega + P_\omega \begin{bmatrix} 0 & 1 \\ -k_{\omega p} & -k_{\omega d} \end{bmatrix} = -I_2 \tag{2.75}$$

$$\begin{bmatrix} 0 & -k_{\psi p} \\ 1 & -k_{\psi d} \end{bmatrix} P_\psi + P_\psi \begin{bmatrix} 0 & 1 \\ -k_{\psi p} & -k_{\psi d} \end{bmatrix} = -I_2 \tag{2.76}$$

in which I_2 is the 2×2 identity matrix. Define the tracking errors

$$\tilde{\omega}=\omega-\omega^*$$

$$\begin{aligned}\tilde{a} &= \hat{a} - \dot{\omega}^* \\ \tilde{\psi}_{rd} &= \psi_{rd} - \psi^* \\ \tilde{v}_{\psi d} &= \hat{v}_{\psi d} - \dot{\psi}^*\end{aligned}$$

and introduce the error vectors

$$\begin{aligned}w_1 &= [\tilde{\omega}, \tilde{a}]^{\mathrm{T}} \\ w_2 &= [\tilde{\psi}_{rd}, \tilde{v}_{\psi d}]^{\mathrm{T}} .\end{aligned}$$

Consider the Lyapunov function

$$V = w_1^{\mathrm{T}} P_\omega w_1 + w_2^{\mathrm{T}} P_\psi w_2 + \tilde{T}_L^2 + \frac{\tilde{\alpha}^2}{\lambda_\alpha} \tag{2.77}$$

in which λ_α is a positive control parameter. The time derivative of function V along the trajectories of the closed-loop system (2.74) is

$$\begin{aligned}\dot{V} =& -\|w_1\|^2 - \|w_2\|^2 - 2\lambda \tilde{T}_L^2 + \tilde{T}_L \left(2v_T + 2w_1^{\mathrm{T}} P_\omega \left[-\frac{1}{J}, -\frac{\lambda}{J} \right]^{\mathrm{T}} \right) \\ &+ \tilde{\alpha} \left(2\lambda_\alpha^{-1} \dot{\hat{\alpha}} + 2w_1^{\mathrm{T}} P_\omega \left[0, -(1+M\beta)\mu \psi_{rd} i_{sq} \right]^{\mathrm{T}} \right. \\ &\left. + 2w_2^{\mathrm{T}} P_\psi \left[M i_{sd} - \psi_{rd}, -(1+M\beta)\hat{\alpha} M i_{sd} + (1+M\beta)\hat{\alpha}\psi_{rd} + \frac{\hat{\alpha} M^2 i_{sq}^2}{\psi_{rd}} \right]^{\mathrm{T}} \right) .\end{aligned} \tag{2.78}$$

If we design the yet undefined term v_T and the estimation law for $\hat{\alpha}$ as

$$\begin{aligned}v_T =& -w_1^{\mathrm{T}} P_\omega \left[-\frac{1}{J}, -\frac{\lambda}{J} \right]^{\mathrm{T}} \\ \dot{\hat{\alpha}} =& \lambda_\alpha \left(w_1^{\mathrm{T}} P_\omega \left[0, -(1+M\beta)\mu \psi_{rd} i_{sq} \right]^{\mathrm{T}} \right. \\ &\left. + w_2^{\mathrm{T}} P_\psi \left[M i_{sd} - \psi_{rd}, -(1+M\beta)\hat{\alpha} M i_{sd} + (1+M\beta)\hat{\alpha}\psi_{rd} + \frac{\hat{\alpha} M^2 i_{sq}^2}{\psi_{rd}} \right]^{\mathrm{T}} \right)\end{aligned}$$

then from (2.78) we obtain

$$\dot{V} = -\|w_1\|^2 - \|w_2\|^2 - 2\lambda \tilde{T}_L^2 . \tag{2.79}$$

Since the previous equation implies that for all $t \geq 0$

$$V(t) \leq V(0)$$

then we can restrict the initial conditions for the tracking and the estimation errors such that $V(0)$ is sufficiently small to guarantee $\psi_{rd}(t) \geq c_1 > 0$ and $\hat{\alpha}(t) \geq c_2 > 0$ for all $t \geq 0$. On the other hand, from (2.77) and (2.79) we can establish that $w_1(t)$, $w_2(t)$, $\tilde{T}_L(t)$, and $\tilde{\alpha}(t)$ are bounded functions on $[0,\infty)$ and therefore, according to (2.70) and (2.71), $\mu\psi_{rd}(t)i_{sq}(t)$ and $(Mi_{sd}(t) - \psi_{rd}(t))$ are bounded functions on $[0,\infty)$. Since $\psi_{rd}(t)$ is a bounded function on $[0,\infty)$, it follows that $i_{sd}(t)$ and $i_{sq}(t)$ are bounded functions on $[0,\infty)$. Therefore $\dot{\omega}(t)$, $\dot{\psi}_{rd}(t)$, $\dot{\hat{T}}_L(t)$, and $\dot{\hat{a}}(t)$ are bounded functions on $[0,\infty)$ so that $\tilde{\omega}(t)$, $\tilde{\psi}_{rd}(t)$, and $\tilde{T}_L(t)$ are uniformly continuous functions on $[0,\infty)$. Since for any $t \geq 0$

$$\int_0^t (\tilde{\omega}^2(\tau) + \tilde{\psi}_{rd}^2(\tau) + \tilde{a}^2(\tau) + 2\lambda \tilde{T}_L^2(\tau))\mathrm{d}\tau \leq V(0) - V(t) \leq V(0) \quad (2.80)$$

which implies

$$\lim_{t\to\infty} \int_0^t [\tilde{\omega}^2(\tau) + \tilde{\psi}_{rd}^2(\tau) + \tilde{a}^2(\tau) + 2\lambda \tilde{T}_L^2(\tau)]\mathrm{d}\tau \leq V(0) \quad (2.81)$$

by applying Barbalat's Lemma A.2 in Appendix A, we have

$$\begin{aligned} \lim_{t\to\infty} \tilde{\omega}(t) &= 0 \\ \lim_{t\to\infty} \tilde{\psi}_{rd}(t) &= 0 \\ \lim_{t\to\infty} \tilde{a}(t) &= 0 \\ \lim_{t\to\infty} \tilde{T}_L(t) &= 0\,. \end{aligned} \quad (2.82)$$

Hence, asymptotic rotor speed and flux modulus tracking of the reference signals ω^* and ψ^* is achieved along with estimation of the unknown load torque.

On the other hand, since $\ddot{\tilde{a}}(t)$ is a bounded function on $[0,\infty)$, $\dot{\tilde{a}}(t)$ is a uniformly continuous function on $[0,\infty)$ so that, by Barbalat's Lemma A.2,

$$\lim_{t\to\infty} \dot{\tilde{a}}(t) = 0\,.$$

Since from (2.74) we have

$$\begin{aligned} \dot{\tilde{a}} &= -k_{\omega p}\tilde{\omega} - k_{\omega d}\tilde{a} - \tilde{\alpha}(1+\beta M)\left(\hat{a} + \frac{\hat{T}_L}{J}\right) - \frac{\lambda}{J}\tilde{T}_L \\ &= -k_{\omega p}\tilde{\omega} - k_{\omega d}\tilde{a} - \tilde{\alpha}(1+\beta M)\left(\tilde{a} - \frac{\tilde{T}_L}{J}\right) - \frac{\lambda}{J}\tilde{T}_L - \tilde{\alpha}(1+\beta M)\left(\dot{\omega}^* + \frac{T_L}{J}\right), \end{aligned}$$

according to (2.82) we can finally establish that

$$\lim_{t\to\infty} \left[\left(\dot{\omega}^*(t) + \frac{T_L}{J}\right)\tilde{\alpha}(t)\right] = 0 \quad (2.83)$$

which implies that $\tilde{\alpha}(t)$ asymptotically converges to zero provided that $\dot{\omega}^*(t) + T_L/J$ is different from zero for all $t \geq 0$. A stronger result is obtained since, if there exist two positive reals T_p and c_p such that

$$\int_t^{t+T_p} \left[\left(\dot{\omega}^*(\tau) + \frac{T_L}{J} \right)^2 \right] \mathrm{d}\tau \geq c_p, \quad \forall t \geq 0\,, \tag{2.84}$$

then exponential convergence to zero of the rotor speed and rotor flux modulus tracking errors is guaranteed, for sufficiently small initial conditions, along with exponential estimation of both the uncertain parameters α and T_L.

In conclusion: the *adaptive input–output feedback linearizing control* is defined as

$$\begin{aligned}
\begin{bmatrix} u_{sa} \\ u_{sb} \end{bmatrix} &= \begin{bmatrix} \cos\rho & -\sin\rho \\ \sin\rho & \cos\rho \end{bmatrix} \begin{bmatrix} u_{sd} \\ u_{sq} \end{bmatrix} \\
u_{sd} &= \sigma \left[\frac{R_s}{\sigma} i_{sd} - \omega i_{sq} - \frac{\hat{\alpha} M i_{sq}^2}{\psi_{rd}} - \hat{\alpha}\beta\psi_{rd} + \hat{\alpha} M \beta i_{sd} + v_d \right] \\
u_{sq} &= \sigma \left[\frac{R_s}{\sigma} i_{sq} + \omega i_{sd} + \frac{\hat{\alpha} M i_{sq} i_{sd}}{\psi_{rd}} + \beta\omega\psi_{rd} + \hat{\alpha} M \beta i_{sq} + v_q \right] \\
v_d &= -\frac{\hat{\alpha}\psi_{rd}}{M} + \hat{\alpha} i_{sd} + \frac{1}{\hat{\alpha} M} \Big[-\dot{\hat{\alpha}}(M i_{sd} - \psi_{rd}) \\
&\quad -k_{\psi p}(\psi_{rd} - \psi^*) - k_{\psi d}(\hat{v}_{\psi d} - \dot{\psi}^*) + \ddot{\psi}^* \Big] \\
v_q &= \hat{\alpha} i_{sq} - \frac{\hat{\alpha} M i_{sd} i_{sq}}{\psi_{rd}} - \frac{v_T}{J\mu\psi_{rd}} \\
&\quad + \frac{1}{\mu\psi_{rd}} [-k_{\omega p}(\omega - \omega^*) - k_{\omega d}(\hat{a} - \dot{\omega}^*) + \ddot{\omega}^*] \\
\psi_{rd} &= \psi_{ra}\cos\rho + \psi_{rb}\sin\rho \\
\begin{bmatrix} i_{sd} \\ i_{sq} \end{bmatrix} &= \begin{bmatrix} \cos\rho & \sin\rho \\ -\sin\rho & \cos\rho \end{bmatrix} \begin{bmatrix} i_{sa} \\ i_{sb} \end{bmatrix} \\
-I_2 &= \begin{bmatrix} 0 & -k_{\omega p} \\ 1 & -k_{\omega d} \end{bmatrix} P_\omega + P_\omega \begin{bmatrix} 0 & 1 \\ -k_{\omega p} & -k_{\omega d} \end{bmatrix} \\
-I_2 &= \begin{bmatrix} 0 & -k_{\psi p} \\ 1 & -k_{\psi d} \end{bmatrix} P_\psi + P_\psi \begin{bmatrix} 0 & 1 \\ -k_{\psi p} & -k_{\psi d} \end{bmatrix} \\
\dot{\xi} &= -\lambda\xi + \lambda J\mu\psi_{rd} i_{sq} + \lambda^2 J\omega - v_T \\
\hat{T}_L &= \xi - \lambda J\omega
\end{aligned}$$

$$v_T = w_1^T P_\omega \left[\frac{1}{J}, \frac{\lambda}{J}\right]^T$$

$$\dot{\hat{\alpha}} = \lambda_\alpha \left(w_1^T P_\omega [0, -(1+M\beta)\mu\psi_{rd} i_{sq}]^T + w_2^T P_\psi \left[M i_{sd} - \psi_{rd}, \right.\right.$$

$$\left.\left. -(1+M\beta)\hat{\alpha} M i_{sd} + (1+M\beta)\hat{\alpha}\psi_{rd} + \frac{\hat{\alpha} M^2 i_{sq}^2}{\psi_{rd}} \right]^T \right); \quad (2.85)$$

it is a second order dynamic nonlinear feedback control algorithm which depends on the measurements of the state variables $(\omega, \psi_{ra}, \psi_{rb}, i_{sa}, i_{sb})$, on the reference signals (ω^*, ψ^*), on the positive control parameters $k_{\omega p}$, $k_{\omega d}$, $k_{\psi p}$, $k_{\psi d}$, λ, λ_α, and on the machine parameters M, L_r, J, R_s, L_s since $\mu = \frac{M}{JL_r}$, $\alpha = \frac{R_r}{L_r}$, $\sigma = L_s\left(1 - \frac{M^2}{L_s Lr}\right)$, $\beta = \frac{M}{\sigma L_r}$; it guarantees that, for any initial condition such that $\psi_{rd}(t) \geq c_1 > 0$ and $\hat{\alpha}(t) \geq c_2 > 0$ for all $t \geq 0$, the rotor speed and flux modulus tracking errors tend asymptotically to zero; moreover, the rotor speed and flux modulus tracking errors along with the rotor resistance and load torque estimation errors decay exponentially to zero provided that there exist two positive reals T_p and c_p such that the persistency of excitation condition (2.84) is satisfied.

Remarks

1. Note that even for constant speed and zero load torque, that is when the rotor resistance estimate is not guaranteed to converge to the true value, both the rotor speed and the rotor flux modulus tracking errors asymptotically converge to zero.
2. When both the critical parameters T_L and α are known, so that we can set $\hat{T}_L \equiv T_L$ and $\hat{\alpha} \equiv \alpha$, the controller (2.85) reduces to the input–output feedback linearizing controller designed in the first part of Section 2.4.
3. When the critical parameter α is known, so that we can set $\hat{\alpha} = \alpha$, the controller (2.85) guarantees exponential rotor speed and flux modulus tracking along with exponential load torque estimation.
4. The resulting controller (2.85) shows singularities at $\psi_{rd} = 0$ and $\hat{\alpha} = 0$ which can imply very large voltages when ψ_{rd} and $\hat{\alpha}$ are close to zero.

Illustrative Simulations

We tested the adaptive input–output feedback linearizing control by simulations for the three-phase single pole pair 0.6-kW induction motor whose parameters have been reported in Chapter 1. All the motor initial conditions have been set equal to zero except for $\psi_{ra}(0) = \psi_{rb}(0) = 0.1\,\text{Wb}$. The control algorithm has been tested with the control parameters (all the values are in SI units): $k_{\omega p} = 8100$, $k_{\omega d} = 180$, $k_{\psi p} = 8100$, $k_{\psi d} = 180$. Real coincident eigenvalues are assigned to the matrices associated with the decoupled, linear, time-invariant, second order systems describing the dynamics of the rotor speed and tracking errors when the estimates of the unknown parameters are equal to the corresponding true parameter values. Those values are the same as those used for the nonadaptive version of the previous section. The parameters λ and λ_α, which are the adaptation gains for the estimates of the load torque T_L and of the parameter α, have been set equal to $\lambda = 120$ and $\lambda_\alpha = 0.09$. The initial condition $\hat{T}_L(0)$ has been set equal to zero while the initial condition for the estimate of α has been set equal to $\hat{\alpha}(0) = 13.2\,\text{s}^{-1}$ which is 50% greater than the true parameter value $\alpha = 8.8\ \text{s}^{-1}$. The references for the speed and flux modulus along with the applied load torque are reported in Figures 2.30–2.32. The rotor flux modulus reference signal starts from 0.001 Wb at $t = 0\,\text{s}$ and grows up to the constant value 1.16 Wb. The speed reference is zero until $t = 0.32\,\text{s}$ and grows up to the constant value 100 rad/s; at $t = 1.5\,\text{s}$ the speed is required to go up to the value 200 rad/s, while the reference for the flux modulus is reduced to 0.5 Wb. A 5.8-Nm load torque is applied to the motor and is reduced to 1.8 Nm. Figures 2.30 and 2.31 show the time histories of rotor speed and flux modulus along with the corresponding tracking errors: the rotor speed tracks its reference tightly even though load torque sharply changes since, according to Figure 2.32, the load torque estimate quickly recovers the applied unknown load torque. The rotor flux modulus tracks its reference: there is, however, a coupling with rotor speed tracking at $t = 0.5\,\text{s}$ and $t = 1\,\text{s}$ when the rotor speed tracking error is perturbed by the unknown load torque. Also the estimate of α quickly converges, according to Figure 2.33, to the true value (unknown to the controller). Stator currents and voltages are within the saturation limits, as illustrated by Figures 2.34 and 2.35.

2.6 Dynamic Feedback Linearizing Control

The key idea of field-oriented control is to use the direct current component i_{sd} to control the flux modulus ψ_{rd} and the quadrature current component i_{sq} to control the speed ω in (1.39): there is no concern on controlling the dynamics of the remaining state variable in (2.9). However, while the dynamics of ψ_{rd} are linear and the dynamics of ω tend to be linear if ψ_{rd} tends to be constant, the dynamics of ρ in (2.9) remain nonlinear in terms of the states (ω, ψ_{rd}) and of the control input i_{sq} in (2.9). Similarly, the input–output feedback linearizing control (2.23), (2.24), (2.54), (2.56) aims at controlling linearly and independently ω and ψ_{rd} while leaving the

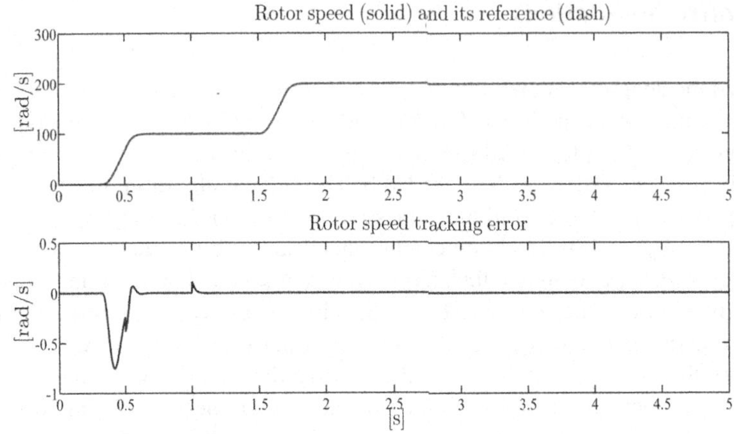

Fig. 2.30 Adaptive input–output feedback linearizing control: rotor speed ω and its reference ω^*; rotor speed tracking error

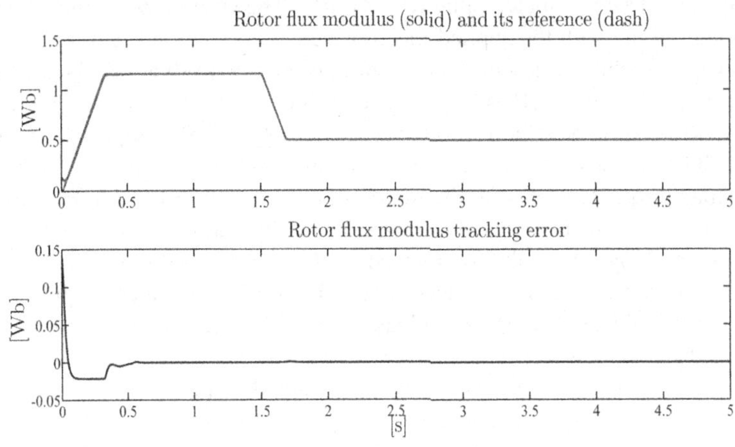

Fig. 2.31 Adaptive input–output feedback linearizing control: rotor flux modulus $\sqrt{\psi_{ra}^2 + \psi_{rb}^2}$ and its reference ψ^*; rotor flux modulus tracking error

dynamics of ρ nonlinearly dependent on the state variables $(\omega, \psi_{rd}, i_{sq})$ in (2.55). For this reason the state feedback control (2.23), (2.24), (2.54), (2.56) linearizes the input–output behavior from (v_d, v_q') to (ω, ψ_{rd}) but fails to make the controlled system dynamics (2.55) linear since the dynamics of ρ remain nonlinear.

On the other hand, if the goal of the control is to make the system, in suitable state coordinates, linear by state feedback, then it is convenient to consider the four equations in (1.39)

$$\frac{d\omega}{dt} = \mu \psi_{rd} i_{sq} - \frac{T_L}{J}$$

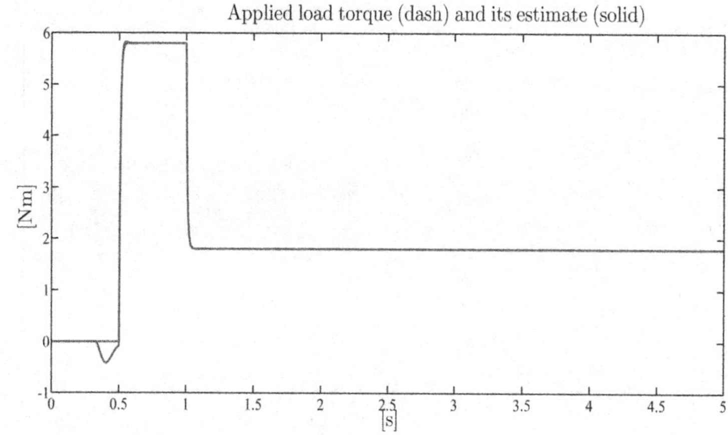

Fig. 2.32 Adaptive input–output feedback linearizing control: applied load torque T_L and its estimate $\hat{T}_L$

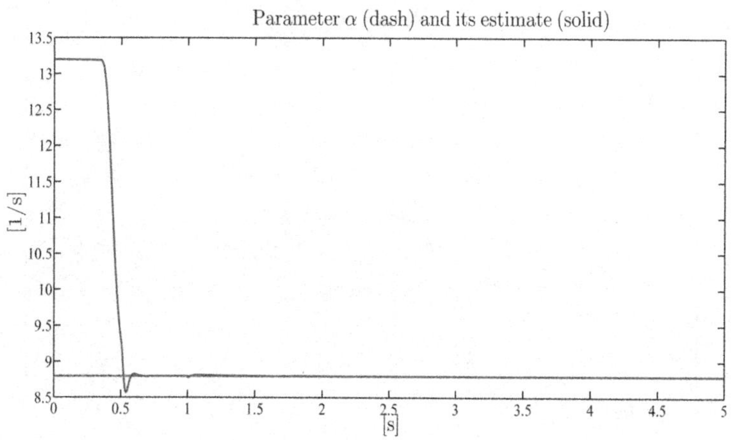

Fig. 2.33 Adaptive input–output feedback linearizing control: parameter α and its estimate $\hat{\alpha}$

$$\begin{aligned}
\frac{\mathrm{d}i_{sq}}{\mathrm{d}t} &= -\gamma i_{sq} - \omega i_{sd} - \frac{\alpha M i_{sq} i_{sd}}{\psi_{rd}} - \beta \omega \psi_{rd} + \frac{u_{sq}}{\sigma} \\
\frac{\mathrm{d}\rho}{\mathrm{d}t} &= \omega + \frac{\alpha M i_{sq}}{\psi_{rd}} \\
\frac{\mathrm{d}\psi_{rd}}{\mathrm{d}t} &= -\alpha \psi_{rd} + \alpha M i_{sd}
\end{aligned} \tag{2.86}$$

in which u_{sq} and i_{sd} are viewed as the control inputs.

Consider the state space change of coordinates from $(\omega, i_{sq}, \rho, \psi_{rd})$ to $(\omega, \dot{\omega}, \rho, \dot{\rho})$, which is nonsingular provided that $\psi_{rd} \neq 0$ and $i_{sq} \neq 0$ since from

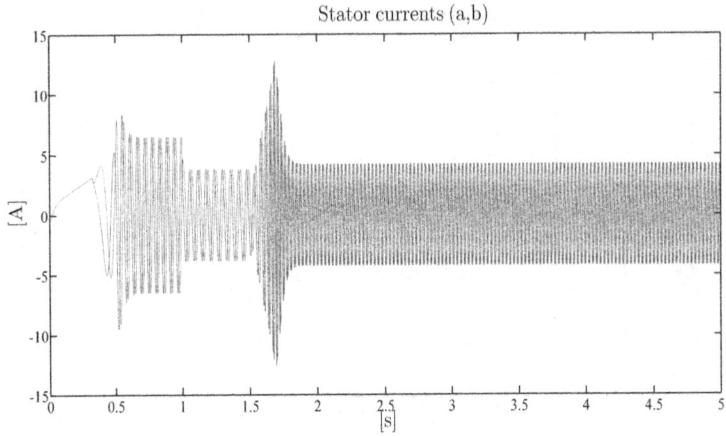

Fig. 2.34 Adaptive input–output feedback linearizing control: stator current vector (a,b)-components (i_{sa}, i_{sb})

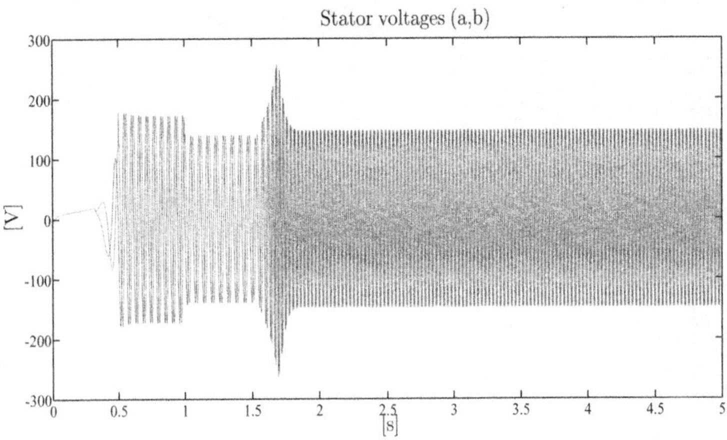

Fig. 2.35 Adaptive input–output feedback linearizing control: stator voltage vector (a,b)-components (u_{sa}, u_{sb})

$$\dot{\rho} - \omega = \frac{\alpha M i_{sq}}{\psi_{rd}} = \frac{\alpha M}{\mu \psi_{rd}^2} \left(\dot{\omega} + \frac{T_L}{J} \right)$$

we can solve for ψ_{rd} obtaining (if $i_{sq} \neq 0$)

$$\psi_{rd} = \sqrt{\frac{\alpha M \left(\dot{\omega} + \dfrac{T_L}{J} \right)}{\mu (\dot{\rho} - \omega)}} \tag{2.87}$$

while

$$i_{sq} = \frac{\left(\dot{\omega} + \frac{T_L}{J}\right)}{\mu \psi_{rd}} = \sqrt{\frac{\left(\dot{\omega} + \frac{T_L}{J}\right)(\dot{\rho} - \omega)}{\mu \alpha M}}. \tag{2.88}$$

In new coordinates $(\omega, \dot{\omega}, \rho, \dot{\rho})$ the dynamics become

$$\begin{aligned}
\begin{bmatrix} \ddot{\omega} \\ \ddot{\rho} \end{bmatrix} &= \begin{bmatrix} -\mu(\gamma+\alpha)\psi_{rd} i_{sq} - \mu\beta\omega\psi_{rd}^2 \\ \mu\psi_{rd} i_{sq} - \frac{T_L}{J} + \frac{\alpha M(\alpha-\gamma) i_{sq}}{\psi_{rd}} - \alpha M\beta\omega \end{bmatrix} \\
&+ \begin{bmatrix} \frac{\mu\psi_{rd}}{\sigma} & -\mu\omega\psi_{rd} \\ \frac{\alpha M}{\sigma\psi_{rd}} & -\frac{2\alpha^2 M^2 i_{sq}}{\psi_{rd}^2} - \frac{\alpha M\omega}{\psi_{rd}} \end{bmatrix} \begin{bmatrix} u_{sq} \\ i_{sd} \end{bmatrix} \\
&\triangleq \begin{bmatrix} -\mu(\gamma+\alpha)\psi_{rd} i_{sq} - \mu\beta\omega\psi_{rd}^2 \\ \mu\psi_{rd} i_{sq} - \frac{T_L}{J} + \frac{\alpha M(\alpha-\gamma) i_{sq}}{\psi_{rd}} - \alpha M\beta\omega \end{bmatrix} + D_d \begin{bmatrix} u_{sq} \\ i_{sd} \end{bmatrix}.
\end{aligned} \tag{2.89}$$

Since

$$\det[D_d] = -\frac{2\mu\alpha^2 M^2 i_{sq}}{\sigma\psi_{rd}},$$

provided that $\psi_{rd} \neq 0$ and $i_{sq} \neq 0$, the signals u_{sq} and i_{sd} can be uniquely expressed in terms of $(\omega, \dot{\omega}, \ddot{\omega}, \rho, \dot{\rho}, \ddot{\rho})$ from (2.87), (2.88) and (2.89) as

$$\begin{bmatrix} u_{sq} \\ i_{sd} \end{bmatrix} = D_d^{-1} \begin{bmatrix} \ddot{\omega} + \mu(\gamma+\alpha)\psi_{rd} i_{sq} + \mu\beta\omega\psi_{rd}^2 \\ \ddot{\rho} - \mu\psi_{rd} i_{sq} + \frac{T_L}{J} - \frac{\alpha M(\alpha-\gamma) i_{sq}}{\psi_{rd}} + \alpha M\beta\omega \end{bmatrix} \tag{2.90}$$

with ψ_{rd} and i_{sq} given by (2.87) and (2.88). Hence, if a new input v_{sq} is defined as

$$\frac{du_{sq}}{dt} = \frac{v_{sq}}{\sigma} \tag{2.91}$$

so that the variable u_{sq} becomes an additional state, (1.39) together with (2.91) become

$$\begin{aligned}
\frac{d\omega}{dt} &= \mu\psi_{rd} i_{sq} - \frac{T_L}{J} \\
\frac{di_{sq}}{dt} &= -\gamma i_{sq} - \omega i_{sd} - \frac{\alpha M i_{sd} i_{sq}}{\psi_{rd}} - \beta\omega\psi_{rd} + \frac{u_{sq}}{\sigma} \\
\frac{du_{sq}}{dt} &= \frac{v_{sq}}{\sigma} \\
\frac{d\rho}{dt} &= \omega + \frac{\alpha M i_{sq}}{\psi_{rd}}
\end{aligned}$$

$$
\begin{aligned}
\frac{d\psi_{rd}}{dt} &= -\alpha\psi_{rd} + \alpha M i_{sd} \\
\frac{di_{sd}}{dt} &= -\gamma i_{sd} + \omega i_{sq} + \frac{\alpha M i_{sq}^2}{\psi_{rd}} + \beta\alpha\psi_{rd} + \frac{u_{sd}}{\sigma}
\end{aligned} \tag{2.92}
$$

in which $(\omega, i_{sq}, u_{sq}, \rho, \psi_{rd}, i_{sd})$ are the state variables and (v_{sq}, u_{sd}) are the control input variables. From (2.87), (2.88), and (2.90) it follows that $(\omega, \dot{\omega}, \ddot{\omega}, \rho, \dot{\rho}, \ddot{\rho})$ defined in (2.86) and (2.89) constitute an equivalent set of state variables if $\psi_{rd} \neq 0$ and $i_{sq} \neq 0$ since $(\omega, i_{sq}, u_{sq}, \rho, \psi_{rd}, i_{sd})$ can be uniquely expressed in terms of $(\omega, \dot{\omega}, \ddot{\omega}, \rho, \dot{\rho}, \ddot{\rho})$. Hence, differentiating (2.89) with respect to time we can express the dynamics (2.92) in new coordinates $(\omega, \dot{\omega}, \ddot{\omega}, \rho, \dot{\rho}, \ddot{\rho})$ as follows

$$
\begin{bmatrix} \dddot{\omega} \\ \dddot{\rho} \end{bmatrix} = \begin{bmatrix} \Phi_\omega \\ \Phi_\rho \end{bmatrix} + \frac{D_d}{\sigma} \begin{bmatrix} v_{sq} \\ u_{sd} \end{bmatrix}
$$

with

$$
\begin{aligned}
\Phi_\omega &= -\mu(\gamma+\alpha)[\Delta_1 i_{sq} + \Delta_2\psi_{rd}] - \mu\beta[\Delta_3\psi_{rd}^2 + 2\omega\psi_{rd}\Delta_1] \\
&\quad + \frac{\mu\Delta_1 u_{sq}}{\sigma} - \mu\Delta_1\omega i_{sd} - \mu\omega\psi_{rd}\Delta_4 - \mu\Delta_3\psi_{rd} i_{sd} \\
\Phi_\rho &= \mu\Delta_1 i_{sq} + \mu\psi_{rd}\Delta_2 + \frac{\alpha M(\alpha-\gamma)\Delta_2}{\psi_{rd}} - \frac{\alpha M(\alpha-\gamma) i_{sq}\Delta_1}{\psi_{rd}^2} - \alpha M\beta\Delta_3 \\
&\quad - \frac{\alpha M\Delta_1 u_{sq}}{\sigma\psi_{rd}^2} - \frac{\alpha M\Delta_3 i_{sd}}{\psi_{rd}} + \frac{\alpha M\omega\Delta_1 i_{sd}}{\psi_{rd}^2} - \frac{2\alpha^2 M^2\Delta_2 i_{sd}}{\psi_{rd}^2} \\
&\quad + \frac{4\alpha^2 M^2 i_{sq} i_{sd}\Delta_1}{\psi_{rd}^3} - \left(\frac{2\alpha^2 M^2 i_{sq}}{\psi_{rd}^2} + \frac{\alpha M\omega}{\psi_{rd}}\right)\Delta_4
\end{aligned}
$$

and

$$
\begin{aligned}
\Delta_1 &= -\alpha\psi_{rd} + \alpha M i_{sd} \\
\Delta_2 &= -\gamma i_{sq} - \omega i_{sd} - \frac{\alpha M i_{sd} i_{sq}}{\psi_{rd}} - \beta\omega\psi_{rd} + \frac{u_{sq}}{\sigma} \\
\Delta_3 &= \mu\psi_{rd} i_{sq} - \frac{T_L}{J} \\
\Delta_4 &= -\gamma i_{sd} + \omega i_{sq} + \frac{\alpha M i_{sq}^2}{\psi_{rd}} + \beta\alpha\psi_{rd} \,.
\end{aligned}
$$

By defining the state feedback control law

$$
\begin{bmatrix} v_{sq} \\ u_{sd} \end{bmatrix} = \sigma D_d^{-1} \begin{bmatrix} -\Phi_\omega - k_{\omega 1}(\omega-\omega^*) - k_{\omega 2}(\dot{\omega}-\dot{\omega}^*) - k_{\omega 3}(\ddot{\omega}-\ddot{\omega}^*) + \dddot{\omega}^* \\ -\Phi_\rho - k_{\rho 1}(\rho-\rho^*) - k_{\rho 2}(\dot{\rho}-\dot{\rho}^*) - k_{\rho 3}(\ddot{\rho}-\ddot{\rho}^*) + \dddot{\rho}^* \end{bmatrix}
$$

the closed-loop linear dynamics are obtained

$$\begin{aligned}\frac{d^3\tilde{\omega}}{dt^3} &= -k_{\omega 1}\tilde{\omega} - k_{\omega 2}\frac{d\tilde{\omega}}{dt} - k_{\omega 3}\frac{d^2\tilde{\omega}}{dt^2} \\ \frac{d^3\tilde{\rho}}{dt^3} &= -k_{\rho 1}\tilde{\rho} - k_{\rho 2}\frac{d\tilde{\rho}}{dt} - k_{\rho 3}\frac{d^2\tilde{\rho}}{dt^2}\end{aligned} \tag{2.93}$$

with $\tilde{\omega} = \omega - \omega^*$, $\tilde{\rho} = \rho - \rho^*$, ρ^* defined in (1.70) and $k_{\omega 1}, k_{\omega 2}, k_{\omega 3}, k_{\rho 1}, k_{\rho 2}, k_{\rho 3}$ positive control parameters such that all the roots of the polynomials

$$\begin{aligned}\pi_\omega(s) &= s^3 + k_{\omega 3}s^2 + k_{\omega 2}s + k_{\omega 1} \\ \pi_\rho(s) &= s^3 + k_{\rho 3}s^2 + k_{\rho 2}s + k_{\rho 1}\end{aligned}$$

have negative real parts.

In conclusion: the *dynamic feedback linearizing control* is defined as

$$\begin{aligned}
\begin{bmatrix} u_{sa} \\ u_{sb} \end{bmatrix} &= \begin{bmatrix} \cos\rho & -\sin\rho \\ \sin\rho & \cos\rho \end{bmatrix} \begin{bmatrix} u_{sd} \\ u_{sq} \end{bmatrix} \\
\frac{du_{sq}}{dt} &= \frac{v_{sq}}{\sigma} \\
\begin{bmatrix} v_{sq} \\ u_{sd} \end{bmatrix} &= \sigma D_d^{-1} \begin{bmatrix} -\Phi_\omega - k_{\omega 1}(\omega - \omega^*) - k_{\omega 2}(\dot{\omega} - \dot{\omega}^*) - k_{\omega 3}(\ddot{\omega} - \ddot{\omega}^*) + \dddot{\omega}^* \\ -\Phi_\rho - k_{\rho 1}(\rho - \rho^*) - k_{\rho 2}(\dot{\rho} - \dot{\rho}^*) - k_{\rho 3}(\ddot{\rho} - \ddot{\rho}^*) + \dddot{\rho}^* \end{bmatrix} \\
\dot{\rho} &= \omega + \frac{\alpha M i_{sq}}{\psi_{rd}} \\
\dot{\rho}^* &= \omega^* + \frac{\alpha M T_L}{\mu J \psi^{*2}}, \quad \rho^*(0) = \arctan\left(\frac{\psi_{rb}(0)}{\psi_{ra}(0)}\right) \\
\psi_{rd} &= \psi_{ra}\cos\rho + \psi_{rb}\sin\rho \\
\begin{bmatrix} i_{sd} \\ i_{sq} \end{bmatrix} &= \begin{bmatrix} \cos\rho & \sin\rho \\ -\sin\rho & \cos\rho \end{bmatrix} \begin{bmatrix} i_{sa} \\ i_{sb} \end{bmatrix} \\
\Phi_\omega &= -\mu(\gamma + \alpha)[\Delta_1 i_{sq} + \Delta_2 \psi_{rd}] - \mu\beta[\Delta_3 \psi_{rd}^2 + 2\omega\psi_{rd}\Delta_1] \\
&\quad + \frac{\mu\Delta_1 u_{sq}}{\sigma} - \mu\Delta_1\omega i_{sd} - \mu\omega\psi_{rd}\Delta_4 - \mu\Delta_3\psi_{rd} i_{sd} \\
\Phi_\rho &= \mu\Delta_1 i_{sq} + \mu\psi_{rd}\Delta_2 + \frac{\alpha M(\alpha - \gamma)\Delta_2}{\psi_{rd}} - \frac{\alpha M(\alpha - \gamma) i_{sq}\Delta_1}{\psi_{rd}^2} - \alpha M\beta\Delta_3 \\
&\quad - \frac{\alpha M\Delta_1 u_{sq}}{\sigma\psi_{rd}^2} - \frac{\alpha M\Delta_3 i_{sd}}{\psi_{rd}} + \frac{\alpha M\omega\Delta_1 i_{sd}}{\psi_{rd}^2} - \frac{2\alpha^2 M^2 \Delta_2 i_{sd}}{\psi_{rd}^2} \\
&\quad + \frac{4\alpha^2 M^2 i_{sq} i_{sd}\Delta_1}{\psi_{rd}^3} - \left(\frac{2\alpha^2 M^2 i_{sq}}{\psi_{rd}^2} + \frac{\alpha M\omega}{\psi_{rd}}\right)\Delta_4 \\
\Delta_1 &= -\alpha\psi_{rd} + \alpha M i_{sd} \\
\Delta_2 &= -\gamma i_{sq} - \omega i_{sd} - \frac{\alpha M i_{sd} i_{sq}}{\psi_{rd}} - \beta\omega\psi_{rd} + \frac{u_{sq}}{\sigma}
\end{aligned}$$

$$\Delta_3 = \mu \psi_{rd} i_{sq} - \frac{T_L}{J}$$

$$\Delta_4 = -\gamma i_{sd} + \omega i_{sq} + \frac{\alpha M i_{sq}^2}{\psi_{rd}} + \beta \alpha \psi_{rd}\,; \tag{2.94}$$

it is a third order dynamic nonlinear feedback control algorithm which depends on the measurements of the state variables $(\omega, \psi_{ra}, \psi_{rb}, i_{sa}, i_{sb})$, on the reference signals (ω^*, ψ^*), on the positive control parameters $k_{\omega 1}$, $k_{\omega 2}$, $k_{\omega 3}$, $k_{\rho 1}$, $k_{\rho 2}$, $k_{\rho 3}$, on the load torque T_L, and on the machine parameters M, R_r, L_r, J, R_s, L_s, since $\mu = \dfrac{M}{JL_r}$, $\alpha = \dfrac{R_r}{L_r}$, $\sigma = L_s\left(1 - \dfrac{M^2}{L_s Lr}\right)$, $\beta = \dfrac{M}{\sigma L_r}$, $\gamma = \dfrac{R_s}{\sigma} + \beta \alpha M$; it guarantees that, for suitable initial and operating conditions such that $\psi_{rd}(t) \geq c_1 > 0$ and $i_{sq}(t) \geq c_2 > 0$ for all $t \geq 0$, the rotor speed and flux angle tracking errors have decoupled dynamics and decay exponentially to zero according to (2.93), with c_1 and c_2 depending on the initial tracking errors.

Illustrative Simulations

We tested the dynamic feedback linearizing control by simulations for the three-phase single pole pair 0.6-kW induction motor whose parameters have been reported in Chapter 1. All the motor initial conditions have been set equal to zero except for $\psi_{ra}(0) = \psi_{rb}(0) = 0.1\,\text{Wb}$. The motor is driven by the feedforward control until $t = 0.7\,\text{s}$ in order to avoid the singularities appearing in the dynamic feedback linearizing control; at $t = 0.7\,\text{s}$ the dynamic feedback linearizing control is applied. The control parameters used in the simulation are (all the values are in SI units): $k_{\omega 1} = 10^6$, $k_{\omega 2} = 3 \times 10^4$, $k_{\omega 3} = 3 \times 10^2$, $k_{\rho 1} = 10^6$, $k_{\rho 2} = 3 \times 10^4$, $k_{\rho 3} = 3 \times 10^2$; real coincident eigenvalues are assigned to the matrices associated to the decoupled, linear time-invariant third order systems (2.93). The controller initial condition has been set equal to $u_{sq}(0) = 171.46922\,\text{V}$ in order to avoid discontinuities at the switching time $t = 0.7\,\text{s}$. The references for the speed and flux modulus along with the applied load torque are reported in Figures 2.36–2.38. The rotor flux modulus reference signal starts from 0.001 Wb at $t = 0\,\text{s}$ and grows up to the constant value 1.16 Wb. The speed reference is zero until $t = 0.32\,\text{s}$ and grows up to the constant value 100 rad/s. A 5.8-Nm load torque is applied to the motor and is reduced to 4.8 Nm. Figures 2.37 and 2.38 show the time histories of the rotor speed and rotor flux modulus along with the corresponding tracking errors: the rotor speed and rotor flux modulus track their references. Note that at $t = 1.5\,\text{s}$ the speed and flux modulus tracking errors are perturbed by the uncompensated load torque time derivatives.

Finally, the stator current and voltages profiles, which are within the physical saturation limits, are reported in Figures 2.39 and 2.40.

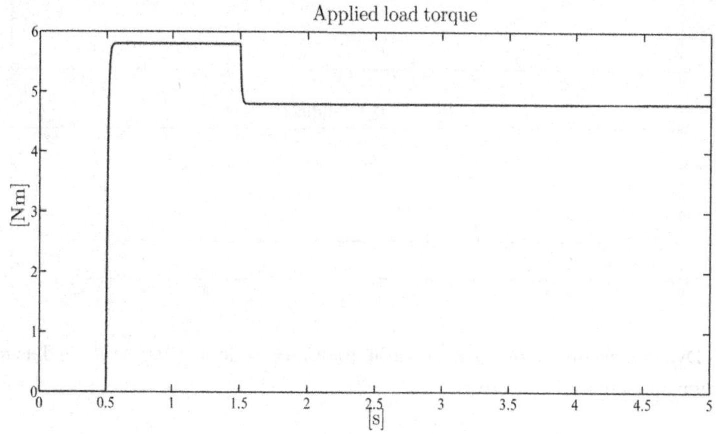

Fig. 2.36 Dynamic feedback linearizing control: applied load torque T_L

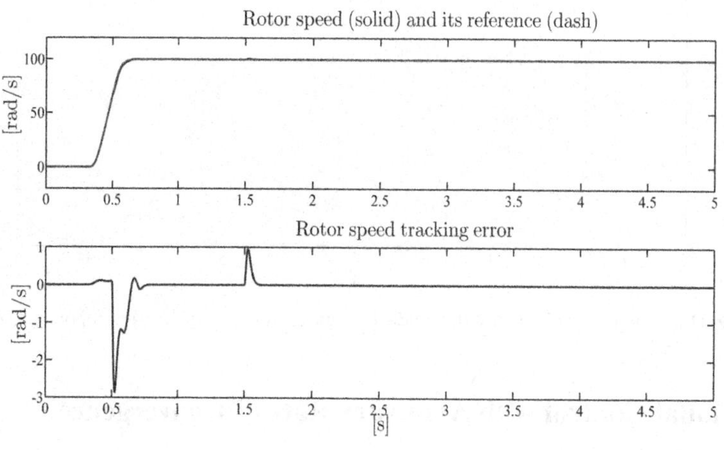

Fig. 2.37 Dynamic feedback linearizing control: rotor speed ω and its reference ω^*; rotor speed tracking error

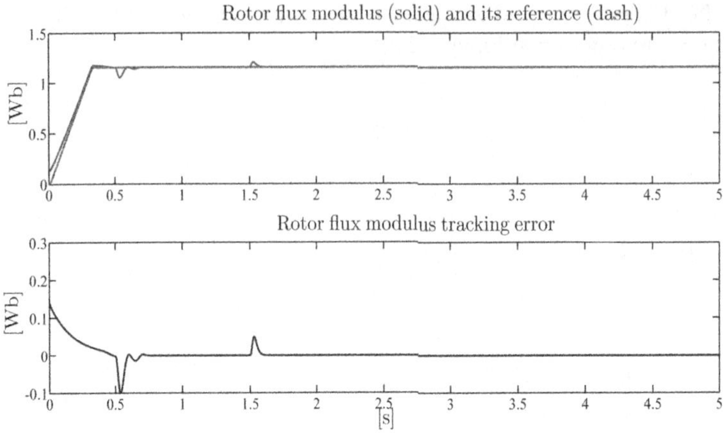

Fig. 2.38 Dynamic feedback linearizing control: rotor flux modulus $\sqrt{\psi_{ra}^2 + \psi_{rb}^2}$ and its reference ψ^*; rotor flux modulus tracking error

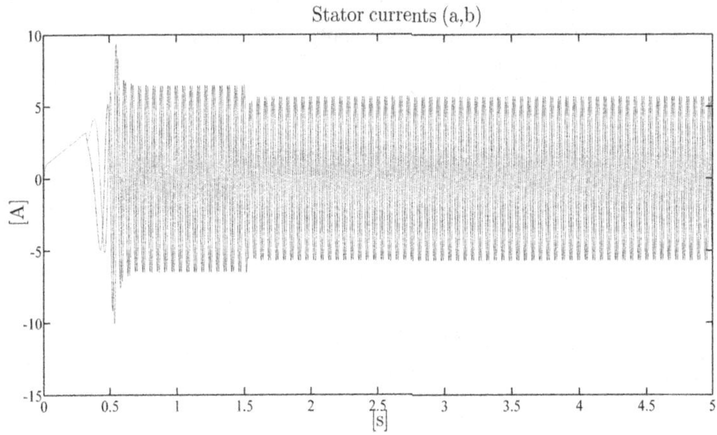

Fig. 2.39 Dynamic feedback linearizing control: stator current vector (a,b)-components (i_{sa}, i_{sb})

2.7 Global Control with Arbitrary Rate of Convergence

The goal of this section is to improve the indirect field-oriented control presented in Section 2.3, which has the advantage of allowing for any motor initial condition but has the drawback of not guaranteeing exponential tracking of reference signals $(\omega^*(t), \psi^*(t))$ with arbitrary rate of convergence. On the other hand, we have seen in Section 2.4 that it is possible to design a state feedback control which achieves exponential tracking with arbitrary rate from sufficiently small initial errors. We would like to bridge the gap between these two control schemes. To this end, we reconsider the control algorithm (2.48) and modify the reference for the stator current

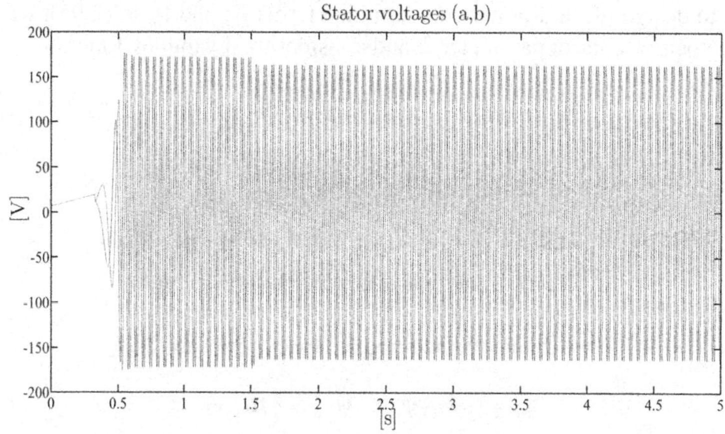

Fig. 2.40 Dynamic feedback linearizing control: stator voltage vector (a,b)-components (u_{sa}, u_{sb})

vector d-component and the speed of the rotating (d,q) frame as follows:

$$\begin{aligned} i_{sd}^* &= \frac{\psi^*}{M} + \frac{\dot{\psi}^*}{\alpha M} + \frac{\eta_d}{\alpha M} \\ \omega_0 &= \omega + \frac{\alpha M i_{sq}}{\psi^*} - \frac{\eta_q}{\psi^*} , \end{aligned} \tag{2.95}$$

i.e. by adding two feedback terms η_d and η_q which will be designed in the following. The reference for the stator current vector q-component remains the same as in (2.48), *i.e.*

$$i_{sq}^* = \frac{1}{\mu \psi^*} \left[-k_\omega(\omega - \omega^*) + \dot{\omega}^* + \frac{T_L}{J} \right] . \tag{2.96}$$

Introduce the tracking errors

$$\begin{aligned} \tilde{\omega} &= \omega - \omega^* \\ \tilde{\psi}_{rd} &= \psi_{rd} - \psi^* \\ \tilde{\psi}_{rq} &= \psi_{rq} \\ \tilde{i}_{sd} &= i_{sd} - i_{sd}^* \\ \tilde{i}_{sq} &= i_{sq} - i_{sq}^* \end{aligned}$$

for the rotor speed, rotor flux vector (d,q)-components and stator current vector (d,q)-components, respectively. The dynamics for $\tilde{\omega}$, $\tilde{\psi}_{rd}$, and $\tilde{\psi}_{rq}$ are given by

$$\begin{aligned} \dot{\tilde{\omega}} &= -k_\omega \tilde{\omega} + \mu(\tilde{\psi}_{rd} i_{sq} - \tilde{\psi}_{rq} i_{sd}) + \mu \psi^* \tilde{i}_{sq} \\ \dot{\tilde{\psi}}_{rd} &= -\alpha \tilde{\psi}_{rd} + (\omega_0 - \omega)\tilde{\psi}_{rq} + \alpha M \tilde{i}_{sd} + \eta_d \\ \dot{\tilde{\psi}}_{rq} &= -\alpha \tilde{\psi}_{rq} - (\omega_0 - \omega)\tilde{\psi}_{rd} + \eta_q . \end{aligned} \tag{2.97}$$

In order to design the additional undetermined terms η_d and η_q in (2.95), we introduce the positive control parameter λ and consider the Lyapunov function

$$W = \frac{1}{2}\left(\lambda\tilde{\omega}^2 + \tilde{\psi}_{rd}^2 + \tilde{\psi}_{rq}^2\right) \tag{2.98}$$

whose time derivative along the trajectories of (2.97) is

$$\begin{aligned} \dot{W} &= -\lambda k_\omega \tilde{\omega}^2 + \lambda\mu(\tilde{\psi}_{rd} i_{sq} - \tilde{\psi}_{rq} i_{sd})\tilde{\omega} + \lambda\mu\psi^* \tilde{i}_{sq}\tilde{\omega} \\ &\quad -\alpha(\tilde{\psi}_{rd}^2 + \tilde{\psi}_{rq}^2) + \alpha M \tilde{i}_{sd}\tilde{\psi}_{rd} + \eta_d\tilde{\psi}_{rd} + \eta_q\tilde{\psi}_{rq}\,. \end{aligned} \tag{2.99}$$

Since $i_{sq} = \tilde{i}_{sq} + i_{sq}^*$, we define the additive feedback terms in (2.95) as

$$\begin{aligned} \eta_d &= -k_\psi(\psi_{rd} - \psi^*) - \lambda\mu i_{sq}^*\tilde{\omega} \\ \eta_q &= -k_\psi\psi_{rq} + \lambda\mu i_{sd}\tilde{\omega} \end{aligned} \tag{2.100}$$

in which k_ψ is a positive control parameter and (ψ_{rd}, ψ_{rq}) are the measured rotor flux vector components in the (d,q) frame which is identified by the rotating angle ε_0, whose dynamics are

$$\frac{\mathrm{d}\varepsilon_0}{\mathrm{d}t} = \omega_0 = \omega + \frac{\alpha M i_{sq}}{\psi^*} + \frac{k_\psi\psi_{rq}}{\psi^*} - \frac{\lambda\mu i_{sd}\tilde{\omega}}{\psi^*}$$

with arbitrary initial condition $\varepsilon_0(0)$. Recall that

$$\begin{aligned} \begin{bmatrix} \psi_{rd} \\ \psi_{rq} \end{bmatrix} &= \begin{bmatrix} \cos\varepsilon_0 & \sin\varepsilon_0 \\ -\sin\varepsilon_0 & \cos\varepsilon_0 \end{bmatrix} \begin{bmatrix} \psi_{ra} \\ \psi_{rb} \end{bmatrix} \\ \begin{bmatrix} i_{sd} \\ i_{sq} \end{bmatrix} &= \begin{bmatrix} \cos\varepsilon_0 & \sin\varepsilon_0 \\ -\sin\varepsilon_0 & \cos\varepsilon_0 \end{bmatrix} \begin{bmatrix} i_{sa} \\ i_{sb} \end{bmatrix}. \end{aligned}$$

By substituting (2.100) in (2.99) we obtain

$$\dot{W} = -\lambda k_\omega\tilde{\omega}^2 - (\alpha + k_\psi)(\tilde{\psi}_{rd}^2 + \tilde{\psi}_{rq}^2) + \lambda\mu\psi_{rd}\tilde{i}_{sq}\tilde{\omega} + \alpha M\tilde{i}_{sd}\tilde{\psi}_{rd}\,. \tag{2.101}$$

Note that the feedback terms depending on the arbitrary control parameter k_ψ introduced in (2.100) are beneficial since k_ψ is added to the given motor parameter α in (2.101). The influence of the last two terms in (2.101) will then be compensated by a suitable choice of the stator voltages (u_{sd}, u_{sq}). To this end, let us compute

$$\frac{\mathrm{d}i_{sq}^*}{\mathrm{d}t} = \Gamma_q \tag{2.102}$$

in which the term Γ_q depending on known signals is

$$\Gamma_q = \frac{1}{\mu\psi^*}\left[k_\omega^2\tilde{\omega} - k_\omega\mu\psi^*\tilde{i}_{sq} + \ddot{\omega}^*\right] - \frac{\dot{\psi}^*}{\mu\psi^{*2}}\left[-k_\omega\tilde{\omega} + \frac{T_L}{J} + \dot{\omega}^*\right]$$

$$-\frac{k_\omega i_{sq}}{\psi^*}\tilde{\psi}_{rd} + \frac{k_\omega i_{sd}}{\psi^*}\tilde{\psi}_{rq} \,. \tag{2.103}$$

Similarly, let us compute

$$\frac{\mathrm{d}i^*_{sd}}{\mathrm{d}t} = \Gamma_d \tag{2.104}$$

in which the term Γ_d depending on known signals is

$$\begin{aligned}\Gamma_d = {} & \frac{\dot{\psi}^*}{M} + \frac{\ddot{\psi}^*}{\alpha M} - \frac{\lambda\mu^2\psi^*\tilde{i}_{sq}i^*_{sq}}{\alpha M} - \frac{\lambda\mu\Gamma_q\tilde{\omega}}{\alpha M} + \frac{\lambda\mu k_\omega\tilde{\omega}i^*_{sq}}{\alpha M} + \frac{k_\psi\dot{\psi}^*}{\alpha M} \\ & - \frac{k_\psi}{\alpha M}\left[-\alpha\psi_{rd} + (\omega_0 - \omega)\psi_{rq} + \alpha M i_{sd}\right] \\ & - \frac{\lambda\mu^2\tilde{\psi}_{rd}i_{sq}i^*_{sq}}{\alpha M} + \frac{\lambda\mu^2\tilde{\psi}_{rq}i_{sd}i^*_{sq}}{\alpha M} \,.\end{aligned} \tag{2.105}$$

On the basis of (2.102) and (2.104) the dynamics for the stator currents tracking errors $\tilde{i}_{sd}$ and $\tilde{i}_{sq}$ can be computed as follows

$$\begin{aligned}\frac{\mathrm{d}\tilde{i}_{sd}}{\mathrm{d}t} &= -\gamma i_{sd} + \omega_0 i_{sq} + \alpha\beta\psi_{rd} + \beta\omega\psi_{rq} + \frac{u_{sd}}{\sigma} - \Gamma_d \\ \frac{\mathrm{d}\tilde{i}_{sq}}{\mathrm{d}t} &= -\gamma i_{sq} - \omega_0 i_{sd} + \alpha\beta\psi_{rq} - \beta\omega\psi_{rd} + \frac{u_{sq}}{\sigma} - \Gamma_q \,.\end{aligned} \tag{2.106}$$

Design the control inputs (u_{sd}, u_{sq}) as

$$\begin{aligned}u_{sd} &= \sigma\left[\gamma i^*_{sd} - \omega_0 i_{sq} - \alpha\beta\psi_{rd} - \beta\omega\psi_{rq} + \Gamma_d - k_i\tilde{i}_{sd} + v_d\right] \\ u_{sq} &= \sigma\left[\gamma i^*_{sq} + \omega_0 i_{sd} - \alpha\beta\psi_{rq} + \beta\omega\psi_{rd} + \Gamma_q - k_i\tilde{i}_{sq} + v_q\right]\end{aligned} \tag{2.107}$$

where v_d and v_q are yet to be designed and k_i is a positive control parameter, so that (2.106) becomes

$$\begin{aligned}\frac{\mathrm{d}\tilde{i}_{sd}}{\mathrm{d}t} &= -(\gamma + k_i)\tilde{i}_{sd} + v_d \\ \frac{\mathrm{d}\tilde{i}_{sq}}{\mathrm{d}t} &= -(\gamma + k_i)\tilde{i}_{sq} + v_q \,.\end{aligned} \tag{2.108}$$

In order to choose the undefined terms v_d and v_q, consider the Lyapunov function for the overall tracking error dynamics

$$V = W + \frac{1}{2}\left(\tilde{i}^2_{sd} + \tilde{i}^2_{sq}\right) \tag{2.109}$$

whose time derivative along the trajectories of the closed-loop system (2.97), (2.106) and (2.107) satisfies

$$\dot{V} = -\lambda k_\omega\tilde{\omega}^2 - (\alpha + k_\psi)\left(\tilde{\psi}^2_{rd} + \tilde{\psi}^2_{rq}\right) + \lambda\mu\psi_{rd}\tilde{i}_{sq}\tilde{\omega} + \alpha M\tilde{i}_{sd}\tilde{\psi}_{rd}$$

$$-(\gamma+k_i)(\tilde{i}_{sd}^2+\tilde{i}_{sq}^2)+v_d\tilde{i}_{sd}+v_q\tilde{i}_{sq}\,. \tag{2.110}$$

If we design the yet undefined terms v_d and v_q as

$$\begin{aligned} v_d &= -\alpha M\tilde{\psi}_{rd} \\ v_q &= -\lambda\mu\psi_{rd}\tilde{\omega} \end{aligned} \tag{2.111}$$

then from (2.110) we obtain

$$\begin{aligned} \dot{V} &\leq -\lambda k_\omega\tilde{\omega}^2-(\alpha+k_\psi)\left(\tilde{\psi}_{rd}^2+\tilde{\psi}_{rq}^2\right)-(\gamma+k_i)(\tilde{i}_{sd}^2+\tilde{i}_{sq}^2) \\ &\leq -2\min\{k_\omega,\alpha+k_\psi,\gamma+k_i\}V\,. \end{aligned} \tag{2.112}$$

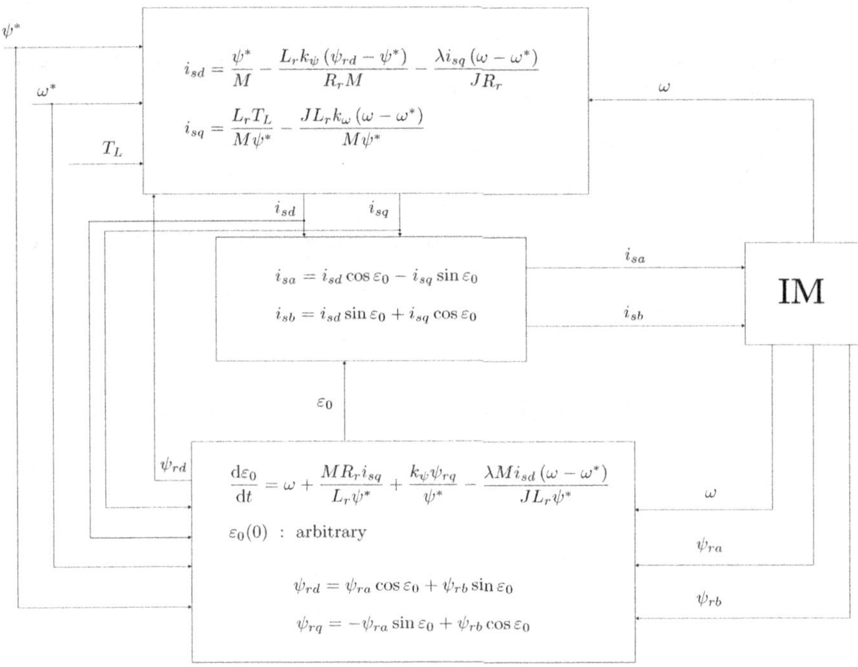

Fig. 2.41 Global control with arbitrary rate of convergence for current-fed motors (constant references ω^*,ψ^*)

In conclusion, the first order nonlinear state feedback *global control with arbitrary rate of convergence* (see Figure 2.41)

$$
\begin{aligned}
u_{sd} &= \sigma\left[\gamma i_{sd}^* - \omega_0 i_{sq} - \alpha\beta\psi_{rd} - \beta\omega\psi_{rq} + \Gamma_d - k_i(i_{sd} - i_{sd}^*) + v_d\right]\\
u_{sq} &= \sigma\left[\gamma i_{sq}^* + \omega_0 i_{sd} - \alpha\beta\psi_{rq} + \beta\omega\psi_{rd} + \Gamma_q - k_i(i_{sq} - i_{sq}^*) + v_q\right]\\
v_d &= -\alpha M(\psi_{rd} - \psi^*)\\
v_q &= -\lambda\mu\psi_{rd}(\omega - \omega^*)
\end{aligned}
\tag{2.113}
$$

with Γ_d and Γ_q given in (2.105) and (2.103) and

$$
\begin{aligned}
\begin{bmatrix} u_{sa} \\ u_{sb} \end{bmatrix} &= \begin{bmatrix} \cos\varepsilon_0 & -\sin\varepsilon_0 \\ \sin\varepsilon_0 & \cos\varepsilon_0 \end{bmatrix} \begin{bmatrix} u_{sd} \\ u_{sq} \end{bmatrix}\\
i_{sd}^* &= \frac{\psi^*}{M} + \frac{\dot\psi^*}{\alpha M} + \frac{\eta_d}{\alpha M}\\
i_{sq}^* &= \frac{1}{\mu\psi^*}\left[-k_\omega(\omega - \omega^*) + \dot\omega^* + \frac{T_L}{J}\right]\\
\frac{\mathrm{d}\varepsilon_0}{\mathrm{d}t} &= \omega_0 = \omega + \frac{\alpha M i_{sq}}{\psi^*} - \frac{\eta_q}{\psi^*}\\
\eta_d &= -k_\psi(\psi_{rd} - \psi^*) - \lambda\mu i_{sq}^*(\omega - \omega^*)\\
\eta_q &= -k_\psi\psi_{rq} + \lambda\mu i_{sd}(\omega - \omega^*)\\
\begin{bmatrix} i_{sd} \\ i_{sq} \end{bmatrix} &= \begin{bmatrix} \cos\varepsilon_0 & \sin\varepsilon_0 \\ -\sin\varepsilon_0 & \cos\varepsilon_0 \end{bmatrix} \begin{bmatrix} i_{sa} \\ i_{sb} \end{bmatrix}\\
\begin{bmatrix} \psi_{rd} \\ \psi_{rq} \end{bmatrix} &= \begin{bmatrix} \cos\varepsilon_0 & \sin\varepsilon_0 \\ -\sin\varepsilon_0 & \cos\varepsilon_0 \end{bmatrix} \begin{bmatrix} \psi_{ra} \\ \psi_{rb} \end{bmatrix}
\end{aligned}
\tag{2.114}
$$

with (k_ω, k_ψ, k_i) arbitrary positive design parameters, is such that $(\omega - \omega^*)$, $(\psi_{rd} - \psi^*)$, ψ_{rq}, $(i_{sd} - i_{sd}^*)$, $(i_{sq} - i_{sq}^*)$ tend exponentially to zero from any initial condition with arbitrary rate of convergence $\min\{k_\omega, \alpha + k_\psi, \gamma + k_i\}$.

As we shall see, in Chapter 4 a global adaptive version of the controller (2.114) will be presented which does not rely on rotor flux measurements and is adaptive with respect to both critical parameters T_L and α: no arbitrary exponential rate of convergence will, however, be obtaïned when the parameters T_L and α are uncertain.

Illustrative Simulations

We tested the global control with arbitrary rate of convergence by simulations for the three-phase single pole pair 0.6-kW induction motor whose parameters have been reported in Chapter 1. All the motor and controller initial conditions have been set to zero except for $\psi_{ra}(0) = \psi_{rb}(0) = 0.1$ Wb. The control parameters are (all values are in SI units) $\lambda = 0.005$, $k_\omega = 450$, $k_i = 800$, $k_\psi = 12$. The references for the speed and flux modulus along with the applied load torque are reported in

Figures 2.42–2.44. The rotor flux modulus reference signal starts from 0.01 Wb at $t = 0$ s and grows up to the constant value 1.16 Wb. The speed reference is zero until $t = 0.32$ s and grows up to the constant value 100 rad/s; at $t = 1.5$ s the speed is required to go up to the value 200 rad/s, while the reference for the flux modulus is reduced to 0.5 Wb. A 5.8-Nm load torque is applied to the motor and is reduced to 1.8 Nm. Figures 2.43 and 2.44 show the time histories of the rotor speed and the flux modulus along with the corresponding tracking errors: the rotor speed and the flux modulus track their references tightly. Finally, the stator current and voltages profiles (which are within physical saturation limits) are reported in Figures 2.45 and 2.46.

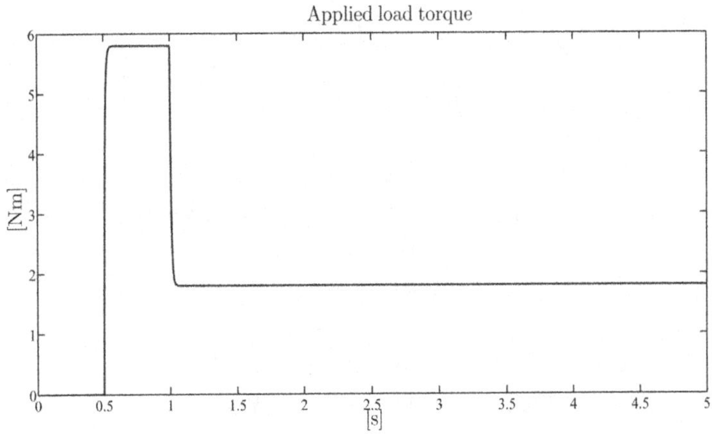

Fig. 2.42 Global control with arbitrary rate of convergence: applied load torque T_L

2.8 Experimental Results

Two experiments have been performed in order to test how critical the parameter α is in practice. The indirect field-oriented control (2.49) has been tested with u_{sd} and u_{sq} in (2.50) simplified by PI controls on the current errors $i_{sd} - i_{sd}^*$ and $i_{sq} - i_{sq}^*$ and T_L replaced by its estimate (2.115) (see Problem 2.5), in which in place of the true value of α a constant estimate $\hat{\alpha}$ has been used. The reference signals for speed and rotor flux modulus in the experiments are reported in Figure 2.47: the flux modulus is first required to reach its desired value of 1.16 Wb before 0.5 s when the rotor speed is then required to reach its desired value of 100 rad/s. After start-up, a constant load torque of 5.8 Nm is applied. In the first experiment $\hat{\alpha}$ underestimates the correct value of α, *i.e.* $\hat{\alpha}/\alpha = 0.7$, while in the second one α is overestimated, *i.e.* $\hat{\alpha}/\alpha = 1.5$. The following control parameters and initial conditions have been used: $k_\omega = 300$, $k_T = 187$, $\hat{T}_L(0) = 0$, $\varepsilon_0(0) = 0$. The gains k_P and k_I of the PI controllers

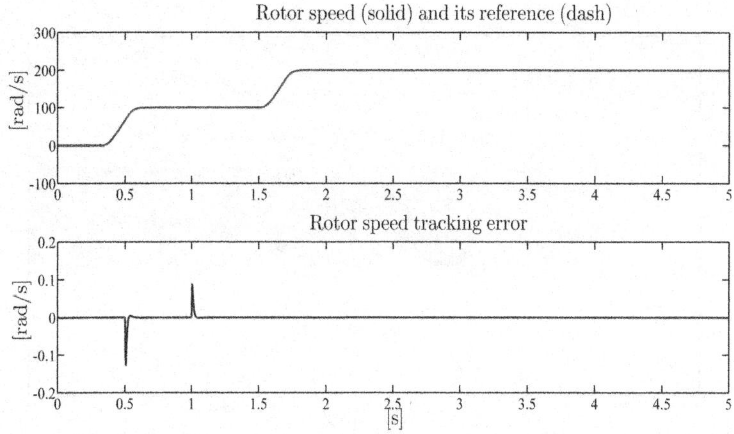

Fig. 2.43 Global control with arbitrary rate of convergence: rotor speed ω and its reference ω^*; rotor speed tracking error

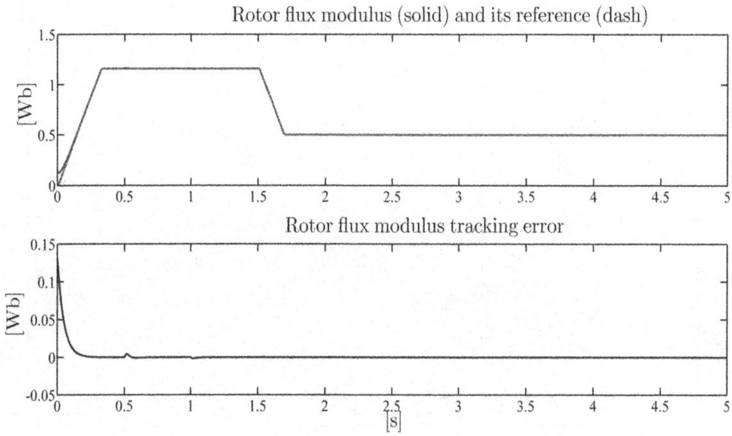

Fig. 2.44 Global control with arbitrary rate of convergence: rotor flux modulus $\sqrt{\psi_{ra}^2 + \psi_{rb}^2}$ and its reference ψ^*; rotor flux modulus tracking error

for the voltages (u_{sd}, u_{sq}) are chosen so that a unit step reference is tracked with a settling time of about 2.5 ms. The flux is estimated by the open-loop rotor flux observer (3.8) which will be given in Section 3.1.1 (converging outside the magnetic saturation region), which makes use of the true value of α. The performance achieved by the controller in the two cases are reported in Figures 2.48 and 2.49: while the speed error is still satisfactory, the flux modulus is above the reference rated value when $\hat{\alpha}/\alpha = 0.7$ and below when $\hat{\alpha}/\alpha = 1.5$ and, therefore, the power efficiency degrades. In both cases higher i_q currents (when compared with the corresponding i_q obtained in simulations) are required to produce the rated torque: this is

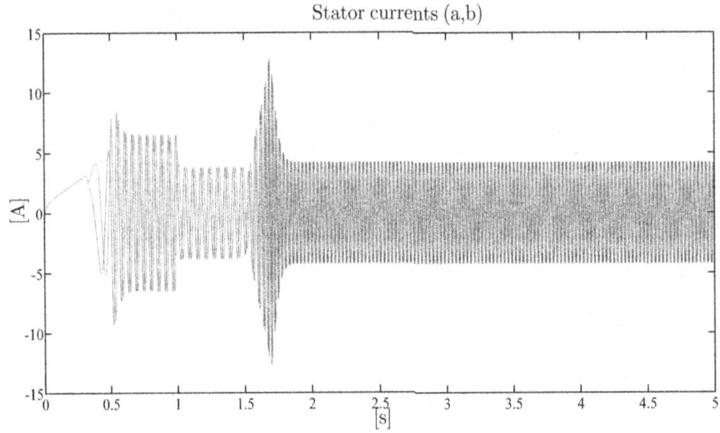

Fig. 2.45 Global control with arbitrary rate of convergence: stator current vector (a,b)-components (i_{sa}, i_{sb})

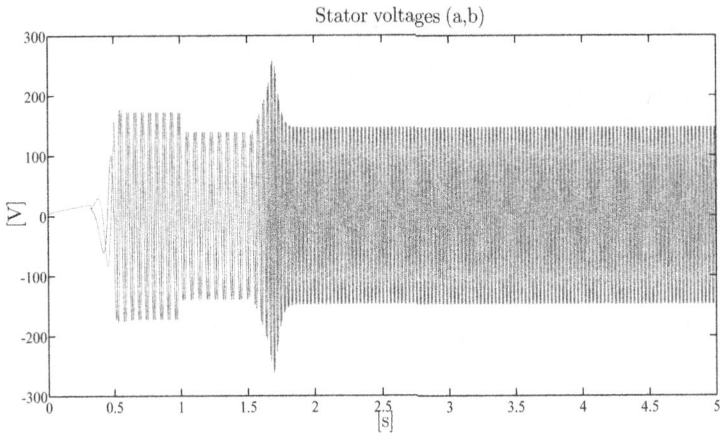

Fig. 2.46 Global control with arbitrary rate of convergence: stator voltage vector (a,b)-components (u_{sa}, u_{sb})

due to the magnetic saturation when $\hat{\alpha}/\alpha = 0.7$ and to the low flux modulus when $\hat{\alpha}/\alpha = 1.5$.

2.9 Conclusions

In this chapter the potentiality of feedback control for induction motors has been fully explored under the assumption that all the state variables are available for feedback. The motivation for introducing feedback actions in the controller comes

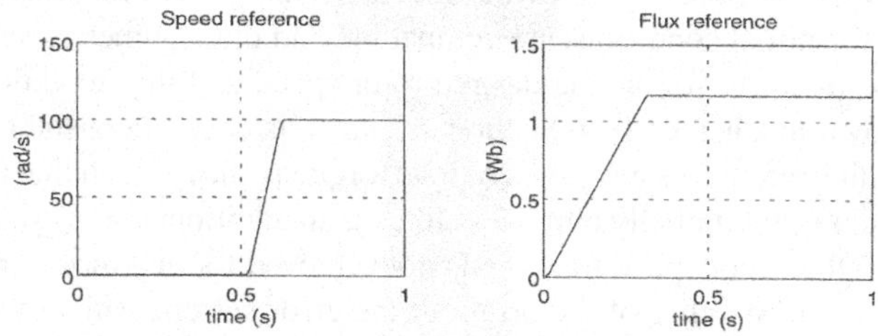

Fig. 2.47 Reference signals in the experiments

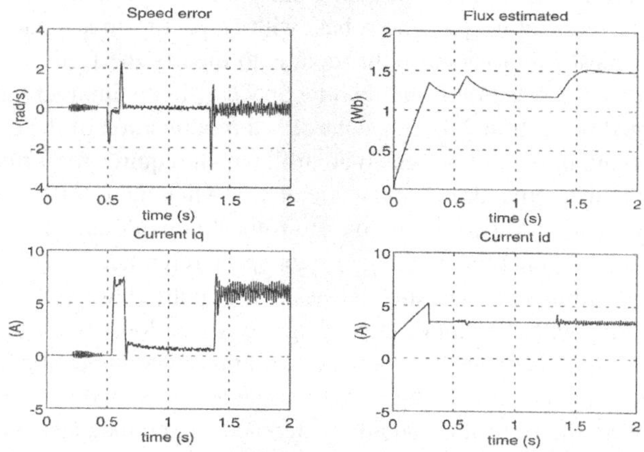

Fig. 2.48 Experimental results with the indirect field-oriented control and underestimated rotor resistance

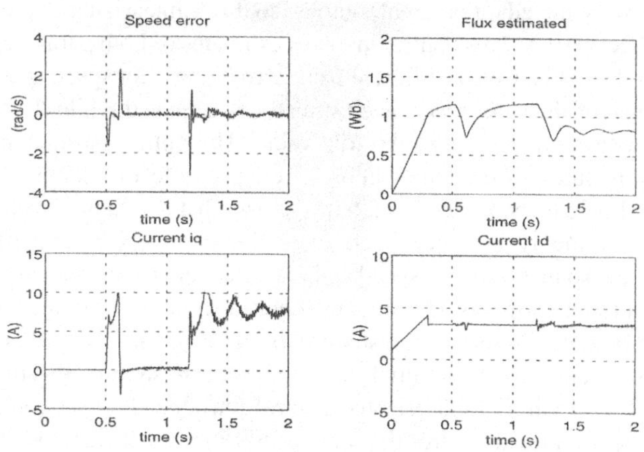

Fig. 2.49 Experimental results with the indirect field-oriented control and overestimated rotor resistance

from the analysis in Section 2.1 of the motor driven by the feedforward control which, when the initial conditions are compatible and the parameters values are exact, forces the motor to follow the desired rotor speed and flux modulus reference signals as shown in Chapter 1. In Section 2.1 the existence of constant rotor speed and flux modulus references and nonzero load torque is shown such that the origin of the error system is not globally attractive. In fact, an additional nonzero equilibrium point exists so that rotor speed tracking is not achieved for any motor initial condition. Moreover, the stability of the origin of the error system critically depends on the load torque which turns an asymptotically stable operating condition into an unstable one by crossing a critical value. Hence, the main goal of the feedback control is to make the desired operating condition attractive for any motor initial condition, for any value of the load torque, and robust with respect to motor parameter variations. The simplest way to achieve this goal is to modify the feedforward control by including a feedback action from the rotor speed as in the indirect field-oriented control presented in Section 2.3. This control is a modification of the original, historically important, direct field-oriented control, which requires rotor flux measurements and has a distinctive drawback: a singularity when the rotor flux modulus is zero requires the motor to start from nonzero rotor flux modulus initial conditions, in order to achieve exponential tracking of rotor speed reference signals for any load torque. If one is willing to accept singularities in the control law and to rely on rotor flux measurements with the aim of improving the closed-loop motor performance, then the input–output feedback linearizing control presented in Section 2.4 achieves an independent exponential tracking of rotor speed and flux modulus reference signals. For instance, the rotor flux modulus reference can be independently adjusted to minimize the power losses, as discussed in Chapter 1, without affecting the speed tracking: this feature is very appealing in electric traction applications. This control strategy can be rendered adaptive with respect to uncertainties in the load torque and rotor resistance by the adaptive input–output feedback linearizing control which is presented in Section 2.5. This control involves estimates of both parameters: assuming that either the load torque is different from zero or the rotor speed is not constant, the convergence of the rotor resistance estimator is achieved while the load torque estimator is always convergent to the true value. The online estimation of critical parameters constitutes a strong motivation for the use of feedback control and leads to an improved efficiency. In Section 2.6 it is shown that a dynamic third order state feedback control, the dynamic feedback linearizing control, can render the closed-loop motor linear so that the rotor speed and rotor flux angle can independently track their desired reference signals. However the control shows singularities when either the rotor flux modulus is zero or the stator current vector quadrature component is zero and it works only when the initial tracking errors are sufficiently small. Finally, in Section 2.7 the indirect field-oriented control introduced in Section 2.4 is generalized to obtain a global control that has several important features: it works for any motor initial condition and for any load torque; it achieves exponential tracking of desired reference signals $(\omega^*(t), \psi^*(t))$ with arbitrary rate of convergence; it is linear with respect to the rotor fluxes, the load torque, and the rotor resistance, so that, as we shall see in Chapter 4, rotor flux observers and uncertain parameter estimators can be incorporated. An indirect field-oriented control, which is the only implementable since it does not require flux measurements, has been experimentally tested in Section 2.8 to evaluate how critical the exact knowledge of the rotor resistance is. It turns out that an uncertain rotor resistance causes errors in the flux regulation so that power efficiency degrades. Many control algorithms presented in this chapter require flux measurements and exact parameter values. It will be discussed in Chapter 3 how to design rotor flux observers, adaptive observers, and parameter estimators for load torque and rotor resistance. The global control with arbitrary rate of convergence presented in Section 2.7 will be the starting point in Chapter 4 to design control algorithms which achieve rotor speed and flux modulus tracking for any initial condition, without the need of rotor flux measurements and of load torque and rotor resistance exact values.

Problems

2.1. Given any positive constant rotor speed and flux modulus reference values (ω^*, ψ^*), show that, in the case of zero load torque, exponential rotor speed and flux modulus tracking are guaranteed for any motor initial condition by the feedforward control (2.8). *Suggestion: use the positive definite function* $V = \gamma_1(\omega - \omega^*)^2 + \gamma_2(\psi_{rd} - \psi^*)^2 + \gamma_2\psi_{rq}^2 + \gamma_3(i_{sd} - i_{sd}^*)^2 + \gamma_3(i_{sq} - i_{sq}^*)^2$ *with* $\gamma_1\mu = \gamma_2 M = \gamma_3\beta$ *and* γ_1, γ_2, γ_3 *positive reals.*

2.2. Show that the constant rotor flux modulus reference minimizing the power losses given by (1.51) in Chapter 1 makes the origin of the error system exponentially stable.

2.3. Show that exponential rotor speed and flux modulus tracking along with exponential load torque estimation can be achieved for current-fed motors by both the direct and the indirect field-oriented controls (2.19) and (2.48) with the load torque estimate (k_T is a positive control parameter)

$$\hat{T}_L(t) = \hat{T}_L(0) - k_T \int_0^t (\omega(\tau) - \omega^*(\tau))d\tau \tag{2.115}$$

in place of T_L. *Suggestion: follow the analysis performed for the nonadaptive case.*

2.4. Design a modified version of the indirect field-oriented control (2.48) for current-fed motors which guarantees, for any motor initial condition using rotor speed measurements only, exponential rotor speed and flux modulus tracking with a rate of decay which depends on α. *Suggestion: use the positive definite function* $V = \lambda(\omega - \omega^*)^2 + (\psi_{rd} - \psi^*)^2 + \psi_{rq}^2$ *with* λ *a positive real.*

2.5. Consider the indirect field-oriented control algorithm (2.48) with $\hat{T}_L$ given in Problem 2.3 in place of T_L

$$
\begin{aligned}
\begin{bmatrix} i_{sa}^* \\ i_{sb}^* \end{bmatrix} &= \begin{bmatrix} \cos\varepsilon_0 & -\sin\varepsilon_0 \\ \sin\varepsilon_0 & \cos\varepsilon_0 \end{bmatrix} \begin{bmatrix} i_{sd}^* \\ i_{sq}^* \end{bmatrix} \\
i_{sd}^* &= \frac{\psi^*}{M} + \frac{\dot{\psi}^*}{\alpha M} \\
i_{sq}^* &= \frac{1}{\mu \psi^*} \left[-k_\omega(\omega - \omega^*) + \dot{\omega}^* + \frac{\hat{T}_L}{J} \right] \\
\dot{\varepsilon}_0 &= \omega_0 = \omega + \frac{\alpha M i_{sq}}{\psi^*} \\
\hat{T}_L(t) &= \hat{T}_L(0) - k_T \int_0^t [\omega(\tau) - \omega^*(\tau)] \mathrm{d}\tau
\end{aligned}
$$

and the PI control inputs

$$
\begin{aligned}
u_{sa}(t) &= -k_P[i_{sa}(t) - i_{sa}^*(t)] - k_I \int_0^t [i_{sa}(\tau) - i_{sa}^*(\tau)] \mathrm{d}\tau \\
u_{sb}(t) &= -k_P[i_{sb}(t) - i_{sb}^*(t)] - k_I \int_0^t [i_{sb}(\tau) - i_{sb}^*(\tau)] \mathrm{d}\tau
\end{aligned}
$$

to drive the stator current tracking errors $(i_{sa} - i_{sa}^*)$ and $(i_{sb} - i_{sb}^*)$ quickly to zero. Simulate the closed-loop performance and compare with the experimental results given in Section 2.8.

2.6. Show that exponential rotor speed and flux modulus tracking is guaranteed for any initial condition of the full order model by the indirect field-oriented control (2.49)–(2.51). *Suggestion: follow the analysis performed for the third-order model.*

2.7. Design a modified version of the indirect field-oriented control (2.49)–(2.51) which is global (*i.e.* it works for any motor initial condition) and adaptive with respect to the critical parameters T_L (load torque) and R_r (rotor resistance). *Suggestion: use the positive definite function $V = \gamma_1(\omega - \omega^*)^2 + \gamma_2(\psi_{rd} - \psi^*)^2 + \gamma_2\psi_{rq}^2 + (i_{sd} - i_{sd}^*)^2 + (i_{sq} - i_{sq}^*)^2 + \gamma_3(T_L - \hat{T}_L)^2 + \gamma_4(\alpha - \hat{\alpha})^2$ with γ_1, γ_2, γ_3, and γ_4 positive reals.*

2.8. Design a modified version of the input–output feedback linearizing control (2.85) which is adaptive with respect to the load torque T_L and both rotor and stator resistances R_r, R_s. *Suggestion: follow, in the control design, steps similar to those presented in Section 2.5.*

2.9. By following the ideas presented in Section 2.6, design a dynamic feedback linearizing control by choosing $\left(\psi_{rd}, \omega - \frac{\mu \psi_{rd}^2 \rho}{\alpha M}\right)$ in place of (ω, ρ).

2.10. Analyze the closed-loop behavior of the full order motor model (1.39) controlled by the algorithm (2.24), (2.27), (2.29), and (2.30) when ω^* and ψ^* are constant while T_L and α are constant uncertain parameters. What happens if the feedback terms $-\alpha M i_{sq}^2/\psi_{rd}$ and $\alpha M i_{sq} i_{sd}/\psi_{rd}$ are dropped in (2.24)?

2.11. Rewrite the input–output feedback linearizing control (2.64) as

$$\begin{bmatrix} u_{sa} \\ u_{sb} \end{bmatrix} = \begin{bmatrix} \cos\rho & -\sin\rho \\ \sin\rho & \cos\rho \end{bmatrix} \begin{bmatrix} v_{sd} \\ v_{sq} \end{bmatrix}$$

with $\cos\rho = \psi_{ra}/\sqrt{\psi_{ra}^2 + \psi_{rb}^2}$, $\sin\rho = \psi_{rb}/\sqrt{\psi_{ra}^2 + \psi_{rb}^2}$ and compare (v_{sd}, v_{sq}) with (u_{sd}, u_{sq}) in (2.24), (2.54), (2.56), and (2.57).

2.12. Show that if the condition (2.84) holds, then the control (2.85) guarantees for suitable initial conditions exponential convergence to zero of the rotor speed and the rotor flux modulus tracking errors and of the estimation errors as well. *Suggestion: use the Persistency of Excitation Lemma A.3 in Appendix A.*

2.13. Design an adaptive version of the global control with arbitrary rate of convergence (2.113) when the load torque T_L is unknown: replace T_L with its estimate $\hat{T}_L$ and then design an adaptation law for $\hat{T}_L$ using the function [recall (2.98) and (2.109)]

$$V_T = V + \frac{1}{2\lambda_T}(T_L - \hat{T}_L)^2$$

with λ_T a positive real design parameter. Compare the resulting controller with that obtained by setting $\hat{\alpha} = \alpha$ in (2.85).

2.14. Simulate the adaptive control

$$\begin{aligned}
\begin{bmatrix} u_{sa} \\ u_{sb} \end{bmatrix} &= \begin{bmatrix} \cos\rho & -\sin\rho \\ \sin\rho & \cos\rho \end{bmatrix} \begin{bmatrix} u_{sd} \\ u_{sq} \end{bmatrix} \\
u_{sd} &= \sigma\left[\gamma i_{sd}^* - \left(\omega + \frac{\alpha M i_{sq}}{\psi_{rd}}\right) i_{sq} - \alpha\beta\psi_{rd} + \Gamma_d - k_i \tilde{i}_{sd} - \alpha M \tilde{\psi}_{rd}\right] \\
u_{sq} &= \sigma\left[\gamma i_{sq}^* + \left(\omega + \frac{\alpha M i_{sq}}{\psi_{rd}}\right) i_{sd} + \beta\omega\psi_{rd} + \Gamma_q - k_i \tilde{i}_{sq} - \mu\tilde{\omega}\tilde{\psi}_{rd}\right] \\
i_{sd}^* &= \frac{\psi^*}{M} + \frac{\dot{\psi}^*}{\alpha M} - \frac{k_\psi \tilde{\psi}_{rd}}{\alpha M} \\
i_{sq}^* &= \frac{1}{\mu\psi_{rd}}\left(-k_\omega\tilde{\omega} + \frac{\hat{T}_L}{J} + \dot{\omega}^*\right) \\
\hat{T}_L &= \xi - \lambda J\omega \\
\dot{\xi} &= -\lambda\xi + \lambda J\mu\psi_{rd} i_{sq} + \lambda^2 J\omega - v_T \\
v_T &= \frac{(k_\omega + \lambda)}{J\mu\psi_{rd}}\tilde{i}_{sq} + \frac{\tilde{\omega}}{J} \\
\Gamma_d &= \frac{\dot{\psi}^*}{M} + \frac{\ddot{\psi}^*}{\alpha M} - \frac{k_\psi}{\alpha M}(-\alpha\psi_{rd} + \alpha M i_{sd} - \dot{\psi}^*) \\
\Gamma_q &= \frac{1}{\mu\psi_{rd}}\left[-k_\omega\left(\mu\psi_{rd} i_{sq} - \frac{\hat{T}_L}{J} - \dot{\omega}^*\right) - \frac{v_T}{J} + \ddot{\omega}^*\right]
\end{aligned}$$

$$-\frac{1}{\mu\psi_{rd}^2}\left(-k_\omega\tilde{\omega}+\frac{\hat{T}_L}{J}+\dot{\omega}^*\right)(-\alpha\psi_{rd}+\alpha M i_{sd})$$
$$\psi_{rd}=\psi_{ra}\cos\rho+\psi_{rb}\sin\rho$$
$$\begin{bmatrix} i_{sd} \\ i_{sq} \end{bmatrix}=\begin{bmatrix} \cos\rho & \sin\rho \\ -\sin\rho & \cos\rho \end{bmatrix}\begin{bmatrix} i_{sa} \\ i_{sb} \end{bmatrix}$$

by choosing the control parameters $(k_i,\, k_\psi,\, k_\omega,\, \lambda)$; compare the results with those obtained by (2.85) with $\hat{\alpha}=\alpha$ and $k_{\omega p}=8,100$, $k_{\omega d}=180$, $k_{\psi p}=8,100$, $k_{\psi d}=180$, $\lambda=120$.

2.15. Consider the current-fed model (2.32) in an arbitrary rotating frame and the dynamic feedback control (2.34): design $(i_{sd},\, i_{sq},\, \omega_0)$ assuming that $(\omega,\, \psi_{rd},\, \psi_{rq})$ are measured so that the closed-loop system (2.32), (2.34) becomes linear and decoupled (k_ω, k_d, and k_q are positive reals)

$$\begin{aligned}\dot{\tilde{\omega}} &= -k_\omega\tilde{\omega}\\ \dot{\tilde{\psi}}_{rd} &= -k_d\tilde{\psi}_{rd}\\ \dot{\tilde{\psi}}_{rq} &= -k_q\tilde{\psi}_{rq}\end{aligned}$$

provided that $\psi_{rd}^2(t)+\psi_{rq}^2(t)\geq c>0$ for all $t\geq 0$.

2.16. Design an adaptive version of the global control with arbitrary rate of convergence (2.113) which is adaptive with respect to all model parameters and guarantees the asymptotic tracking of both rotor speed and flux modulus references from any motor initial conditions. *Suggestion: reparameterize and use projection algorithms.*

2.17. Consider the indirect field-oriented control (2.49) and replace the parameter α by a constant estimate $\hat{\alpha}$. Compute the error dynamics and their equilibrium points in terms of the estimation error $\tilde{\alpha}=\alpha-\hat{\alpha}$ and evaluate their stability.

Chapter 3
Flux Observers and Parameter Estimation

Abstract Under the assumption that the rotor speed, the stator currents, and the stator voltages are available from measurements, this chapter is devoted to the design of rotor flux (or rotor current) asymptotic observers and their adaptive versions when the rotor resistance is uncertain. Load torque estimators are also designed which can be used in conjunction with the flux observers. The estimation algorithms which are presented and analyzed in this chapter are intended to complement the control algorithms which were obtained in Chapter 2 under the assumption that the rotor fluxes are available for feedback and that the load torque and the rotor/stator resistances are known. The rotor fluxes have been shown to be observable in Chapter 1 so that global rotor flux observers with arbitrary exponential rate of convergence are designed in this chapter. Adaptive observers show that the rotor resistance can be estimated online along with rotor fluxes, provided that persistency of excitation conditions are satisfied.

3.1 Nonadaptive Observers

In this section we show how different observers can be designed assuming that the rotor speed ω and the stator currents i_{sa} and i_{sb} are available from measurements and that the motor parameters are known exactly. First a rotor flux observer is designed, then a rotor current observer is analyzed, and, finally, a rotor flux observer with arbitrary rate of convergence is discussed.

3.1.1 Open-loop Rotor Flux Observer

Recall that it was established in Section 1.5 that the rotor fluxes (ψ_{ra}, ψ_{rb}) are observable from the output measurements (ω, i_{sa}, i_{sb}) since, according to (1.76), the rotor fluxes can be uniquely expressed in terms of rotor speed, stator currents and their time derivatives, and stator voltage inputs. Consider the rotor flux dynamic equations in (1.26):

$$\begin{aligned}\frac{\mathrm{d}\psi_{ra}}{\mathrm{d}t} &= -\alpha\psi_{ra} - \omega\psi_{rb} + \alpha M i_{sa}\\ \frac{\mathrm{d}\psi_{rb}}{\mathrm{d}t} &= -\alpha\psi_{rb} + \omega\psi_{ra} + \alpha M i_{sb}\,.\end{aligned} \tag{3.1}$$

Note that if $i_{sa} = i_{sb} = 0$ then the rotor fluxes decay exponentially to zero since the function

$$V = \frac{1}{2}\left(\psi_{ra}^2 + \psi_{rb}^2\right)$$

has time derivative along the trajectories of (3.1)

$$\dot{V} = -\alpha\left(\psi_{ra}^2 + \psi_{rb}^2\right) = -2\alpha V$$

which implies, integrating with respect to time on the time interval $[0,t]$,

$$V(t) = V(0)\mathrm{e}^{-2\alpha t}\,.$$

We take advantage of this property in the design of the following open-loop rotor flux observer

$$\begin{aligned}\frac{\mathrm{d}\hat{\psi}_{ra}}{\mathrm{d}t} &= -\alpha\hat{\psi}_{ra} - \omega\hat{\psi}_{rb} + \alpha M i_{sa}\\ \frac{\mathrm{d}\hat{\psi}_{rb}}{\mathrm{d}t} &= -\alpha\hat{\psi}_{rb} + \omega\hat{\psi}_{ra} + \alpha M i_{sb}\end{aligned} \tag{3.2}$$

which is a copy of the rotor flux dynamic equations (3.1) in which the flux estimate replaces the true flux. Note that (3.2) requires the measurements of (ω, i_{sa}, i_{sb}) and the knowledge of the motor parameters $\alpha = R_r/L_r$ and M. Defining the flux estimation errors as

$$\begin{aligned}\tilde{\psi}_{ra} &= \psi_{ra} - \hat{\psi}_{ra}\\ \tilde{\psi}_{rb} &= \psi_{rb} - \hat{\psi}_{rb}\end{aligned} \tag{3.3}$$

the resulting estimation error dynamics are given by

$$\begin{aligned}\frac{\mathrm{d}\tilde{\psi}_{ra}}{\mathrm{d}t} &= -\alpha\tilde{\psi}_{ra} - \omega\tilde{\psi}_{rb}\\ \frac{\mathrm{d}\tilde{\psi}_{rb}}{\mathrm{d}t} &= -\alpha\tilde{\psi}_{rb} + \omega\tilde{\psi}_{ra}\,.\end{aligned} \tag{3.4}$$

Now, we can repeat for (3.4) the analysis developed for (3.1) with $i_{sa} = i_{sb} = 0$ and consider the Lyapunov function

$$V = \frac{1}{2}\left(\tilde{\psi}_{ra}^2 + \tilde{\psi}_{rb}^2\right) \tag{3.5}$$

whose time derivative along the trajectories of (3.4) is given by

$$\dot{V} = -\alpha\left(\tilde{\psi}_{ra}^2 + \tilde{\psi}_{rb}^2\right) = -2\alpha V\ ; \tag{3.6}$$

integrating with respect to time on the time interval $[0,t]$, we obtain

$$V(t) = V(0)\mathrm{e}^{-2\alpha t} \tag{3.7}$$

which implies that the norm of the rotor flux error vector $\tilde{\psi}_r(t) = [\tilde{\psi}_{ra}(t),\ \tilde{\psi}_{rb}(t)]^{\mathrm{T}}$ decays exponentially with rate of convergence α, for any initial estimation error $V(0) = \frac{1}{2}(\tilde{\psi}_{ra}^2(0) + \tilde{\psi}_{rb}^2(0))$. Note that the exponential rate of convergence is $\alpha = R_r/L_r$: hence, the rotor flux observer (3.2) does not have an arbitrary rate of convergence, since it depends on the motor parameters R_r and L_r.

In conclusion: the *open-loop rotor flux observer*

$$\begin{aligned} \dot{\hat{\psi}}_{ra} &= -\alpha\hat{\psi}_{ra} - \omega\hat{\psi}_{rb} + \alpha M i_{sa} \\ \dot{\hat{\psi}}_{rb} &= -\alpha\hat{\psi}_{rb} + \omega\hat{\psi}_{ra} + \alpha M i_{sb} \end{aligned} \tag{3.8}$$

guarantees that for any initial condition $(\hat{\psi}_{ra}(0), \hat{\psi}_{rb}(0))$ the estimation errors $\tilde{\psi}_{ra}(t) = \psi_{ra}(t) - \hat{\psi}_{ra}(t)$ and $\tilde{\psi}_{rb}(t) = \psi_{rb}(t) - \hat{\psi}_{rb}(t)$ converge exponentially to zero with rate of convergence $\alpha = R_r/L_r$; it requires the measurements of $\omega(t)$, $i_{sa}(t)$, and $i_{sb}(t)$ and depends on the motor parameters R_r, L_r, and M.

Illustrative Simulations

We tested the open-loop rotor flux observer (3.8) by simulations for the three-phase single pole pair 0.6-kW induction motor whose parameters have been reported in Chapter 1. The motor (with initial conditions $\psi_{ra}(0) = \psi_{rb}(0) = 0.1\,\mathrm{Wb}$) is controlled by the input–output feedback linearizing control (with control parameters, rotor speed and flux modulus references, and applied load torque as reported in Section 2.4). The observer initial conditions are set to zero, *i.e.* $\hat{\psi}_{ra}(0) = \hat{\psi}_{rb}(0) = 0$. The rotor flux vector (a,b) components along with the corresponding estimation errors are reported in Figures 3.1 and 3.2: they show that exponentially converging estimation is achieved. The influence of the rotor resistance uncertainty, when the rotor resistance value used by the observer is 50% greater than the motor rotor resistance value, is illustrated by Figures 3.3 and 3.4 which show that steady-state estimation errors appear.

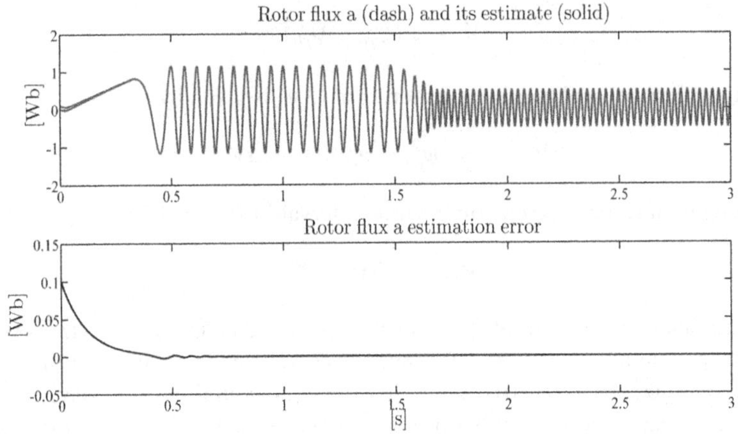

Fig. 3.1 Open-loop rotor flux observer: rotor flux ψ_{ra} and its estimate; rotor flux ψ_{ra} estimation error

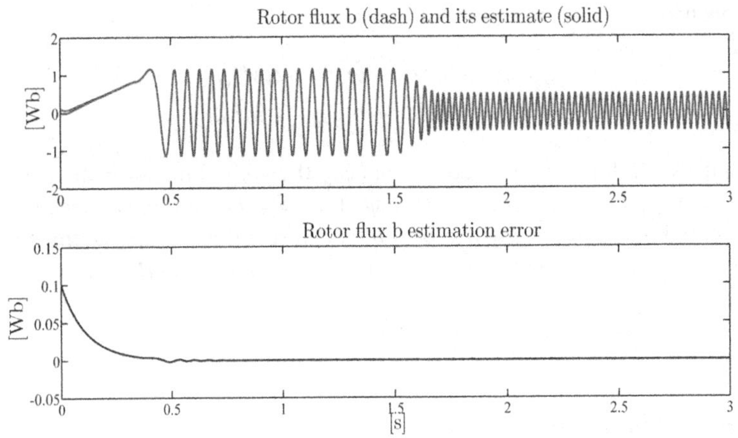

Fig. 3.2 Open-loop rotor flux observer: rotor flux ψ_{rb} and its estimate; rotor flux ψ_{rb} estimation error

3.1.2 Open-loop Rotor Current Observer

Since the rotor fluxes are assumed to be related to the rotor and stator currents through the linear equations

$$\begin{aligned}\psi_{ra} &= Mi_{sa} + L_r i_{ra}\\ \psi_{rb} &= Mi_{sb} + L_r i_{rb}\,,\end{aligned} \tag{3.9}$$

an estimate of the rotor fluxes can be obtained from the measurements of the stator currents and a rotor current observer. From (1.14), we obtain the rotor current

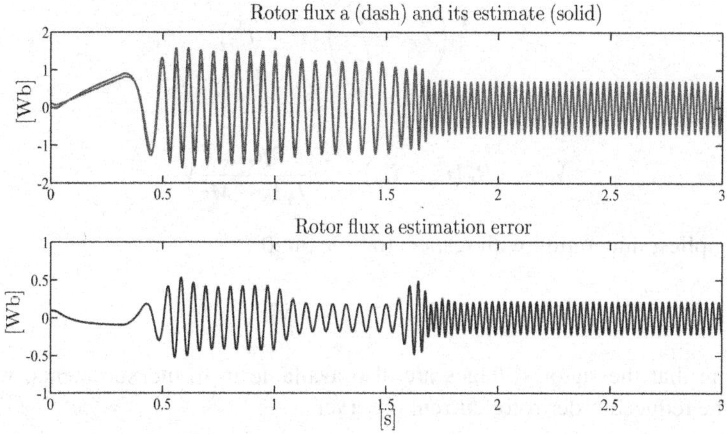

Fig. 3.3 Open-loop rotor flux observer in the case of rotor resistance uncertainty: rotor flux ψ_{ra} and its estimate; rotor flux ψ_{ra} estimation error

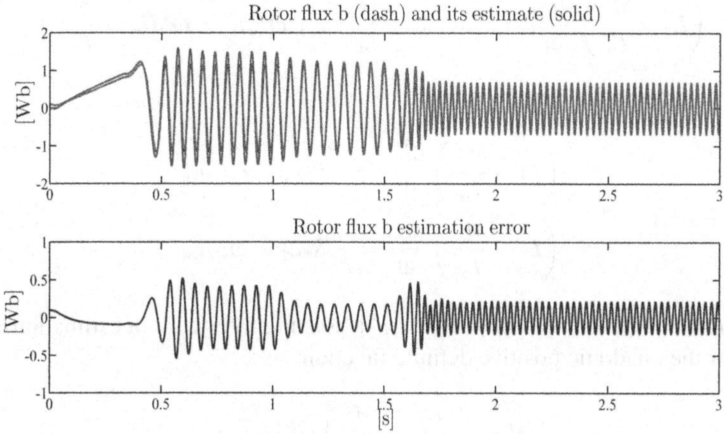

Fig. 3.4 Open-loop rotor flux observer in the case of rotor resistance uncertainty: rotor flux ψ_{rb} and its estimate; rotor flux ψ_{rb} estimation error

dynamics:

$$\begin{aligned}\left(L_r - \frac{M^2}{L_s}\right)\frac{\mathrm{d}i_{ra}}{\mathrm{d}t} &= -R_r i_{ra} + \frac{MR_s}{L_s} i_{sa} - \omega L_r i_{rb} - \omega M i_{sb} + \frac{M}{L_s} u_{sa} \\ \left(L_r - \frac{M^2}{L_s}\right)\frac{\mathrm{d}i_{rb}}{\mathrm{d}t} &= -R_r i_{rb} + \frac{MR_s}{L_s} i_{sb} + \omega L_r i_{ra} + \omega M i_{sa} + \frac{M}{L_s} u_{sb}\,. \end{aligned} \quad (3.10)$$

Note that if $u_{sa} = u_{sb} = 0$ and $i_{sa} = i_{sb} = 0$ then the rotor currents $i_{ra}(t)$ and $i_{rb}(t)$ decay exponentially to zero since the function

$$V = \frac{1}{2}\left(L_r - \frac{M^2}{L_s}\right)\left(i_{ra}^2 + i_{rb}^2\right)$$

has time derivative

$$\dot{V} = -R_r\left(i_{ra}^2 + i_{rb}^2\right) = -\frac{2R_rL_s}{L_rL_s - M^2}V$$

which implies, integrating with respect to time on $[0,t]$,

$$V(t) = V(0)\mathrm{e}^{-\frac{2R_rL_s}{L_rL_s-M^2}t}.$$

Assuming that the stator voltages are also available from measurements, we can design the reduced order rotor current observer

$$\begin{aligned}\left(L_r - \frac{M^2}{L_s}\right)\frac{\mathrm{d}\hat{i}_{ra}}{\mathrm{d}t} &= -R_r\hat{i}_{ra} + \frac{MR_s}{L_s}i_{sa} - \omega L_r\hat{i}_{rb} - \omega M i_{sb} + \frac{M}{L_s}u_{sa}\\ \left(L_r - \frac{M^2}{L_s}\right)\frac{\mathrm{d}\hat{i}_{rb}}{\mathrm{d}t} &= -R_r\hat{i}_{rb} + \frac{MR_s}{L_s}i_{sb} + \omega L_r\hat{i}_{ra} + \omega M i_{sa} + \frac{M}{L_s}u_{sb}\end{aligned} \tag{3.11}$$

whose error dynamics are given by

$$\begin{aligned}\left(L_r - \frac{M^2}{L_s}\right)\frac{\mathrm{d}\tilde{i}_{ra}}{\mathrm{d}t} &= -R_r\tilde{i}_{ra} - \omega L_r\tilde{i}_{rb}\\ \left(L_r - \frac{M^2}{L_s}\right)\frac{\mathrm{d}\tilde{i}_{rb}}{\mathrm{d}t} &= -R_r\tilde{i}_{rb} + \omega L_r\tilde{i}_{ra}\end{aligned} \tag{3.12}$$

in which $\tilde{i}_{ra} = i_{ra} - \hat{i}_{ra}$ and $\tilde{i}_{rb} = i_{rb} - \hat{i}_{rb}$ denote the rotor current estimation errors. Consider the quadratic positive definite function

$$V = \frac{1}{2}\left(L_r - \frac{M^2}{L_s}\right)\left(\tilde{i}_{ra}^2 + \tilde{i}_{rb}^2\right) \tag{3.13}$$

whose time derivative along the trajectories (3.12) is

$$\dot{V} = -R_r\left(\tilde{i}_{ra}^2 + \tilde{i}_{rb}^2\right) = -\frac{2R_rL_s}{L_rL_s - M^2}V\,; \tag{3.14}$$

integrating with respect to time on $[0,t]$, we obtain

$$V(t) = V(0)\mathrm{e}^{-\frac{2R_rL_s}{L_rL_s-M^2}t} \tag{3.15}$$

which implies that the norm of the rotor current error vector $\tilde{i}_r(t) = [\tilde{i}_{ra}(t), \tilde{i}_{rb}(t)]^{\mathrm{T}}$ decays exponentially with rate of convergence $\alpha/[1 - M^2/(L_rL_s)]$, which is greater than α since $L_rL_s > M^2$. Therefore, by using the rotor current observer (3.11) along with the relations

$$\begin{aligned}\hat{\psi}_{ra} &= Mi_{sa} + L_r\hat{i}_{ra} \\ \hat{\psi}_{rb} &= Mi_{sb} + L_r\hat{i}_{rb}\end{aligned} \tag{3.16}$$

converging estimates of the rotor fluxes which are faster than those obtained by (3.2) can be recovered. On the other hand, if we compare the two reduced order flux observers (3.11), (3.16) and (3.2), we note that the knowledge of the additional signals (u_{sa}, u_{sb}) and of the additional parameters (R_s, L_s) is required in (3.11), (3.16).

3.1.3 Rotor Flux Observer with Arbitrary Rate of Convergence

The rotor flux observers presented in the previous sections have a fixed rate of convergence which is related to the motor parameters. We show in this section how a rotor flux observer with arbitrary rate of convergence can be designed. This property is of crucial importance since in Chapter 2 state feedback controls with arbitrary tracking convergence rate are designed, assuming that rotor fluxes are available: if these signals are replaced by estimates given by a flux observer, it is clear that the property of arbitrary tracking convergence can be retained at least locally only if the observer estimation errors decay to zero with arbitrary rate. We begin our design by introducing two terms, w_a and w_b, in the observer equations (3.2),

$$\begin{aligned}\dot{\hat{\psi}}_{ra} &= -\alpha\hat{\psi}_{ra} - \omega\hat{\psi}_{rb} + \alpha M i_{sa} - w_a \\ \dot{\hat{\psi}}_{rb} &= -\alpha\hat{\psi}_{rb} + \omega\hat{\psi}_{ra} + \alpha M i_{sb} - w_b\end{aligned} \tag{3.17}$$

whose choice is postponed. The estimation error dynamics are given by

$$\begin{aligned}\dot{\tilde{\psi}}_{ra} &= -\alpha\tilde{\psi}_{ra} - \omega\tilde{\psi}_{rb} + w_a \\ \dot{\tilde{\psi}}_{rb} &= -\alpha\tilde{\psi}_{rb} + \omega\tilde{\psi}_{ra} + w_b\,.\end{aligned} \tag{3.18}$$

At this point, we note that if we could choose

$$\begin{aligned}w_a &= k\beta\frac{\mathrm{d}\tilde{\psi}_{ra}}{\mathrm{d}t} \\ w_b &= k\beta\frac{\mathrm{d}\tilde{\psi}_{rb}}{\mathrm{d}t}\end{aligned} \tag{3.19}$$

the estimation error dynamics would become

$$\begin{aligned}\dot{\tilde{\psi}}_{ra} &= -\frac{\alpha}{1-k\beta}\tilde{\psi}_{ra} - \frac{\omega}{1-k\beta}\tilde{\psi}_{rb} \\ \dot{\tilde{\psi}}_{rb} &= -\frac{\alpha}{1-k\beta}\tilde{\psi}_{rb} + \frac{\omega}{1-k\beta}\tilde{\psi}_{ra}\,.\end{aligned} \tag{3.20}$$

The time derivative of the Lyapunov function (3.5), in the light of (3.20), is given by

$$\dot{V} = -\frac{\alpha}{1-k\beta}\left(\tilde{\psi}_{ra}^2 + \tilde{\psi}_{rb}^2\right) = -\frac{2\alpha}{1-k\beta}V \tag{3.21}$$

whose time integration leads to

$$V(t) = V(0)\mathrm{e}^{-\frac{2\alpha}{1-k\beta}t} \tag{3.22}$$

which implies the exponential convergence to zero of the norm of the rotor flux error with rate of convergence $\alpha/(1-k\beta)$. By choosing k such that $1-k\beta$ is positive the smaller $1-k\beta$ is, the larger the rate of convergence results. Now, let us reconsider the choice of w_a and w_b by noting that the expressions in (3.19) are not available from measurements and cannot be directly used in the observer equations (3.17). Nevertheless, using

$$\begin{aligned} \beta\frac{\mathrm{d}\psi_{ra}}{\mathrm{d}t} &= \frac{1}{\sigma}u_{sa} - \frac{R_s}{\sigma}i_{sa} - \frac{\mathrm{d}i_{sa}}{\mathrm{d}t} \\ \beta\frac{\mathrm{d}\psi_{rb}}{\mathrm{d}t} &= \frac{1}{\sigma}u_{sb} - \frac{R_s}{\sigma}i_{sb} - \frac{\mathrm{d}i_{sb}}{\mathrm{d}t} \end{aligned} \tag{3.23}$$

from (3.17) and (3.19), we obtain

$$\begin{aligned} \frac{\mathrm{d}\hat{\psi}_{ra}}{\mathrm{d}t} &= -\alpha\hat{\psi}_{ra} - \omega\hat{\psi}_{rb} + \alpha M i_{sa} - k\left(\frac{1}{\sigma}u_{sa} - \frac{R_s}{\sigma}i_{sa} - \frac{\mathrm{d}i_{sa}}{\mathrm{d}t} - \beta\frac{\mathrm{d}\hat{\psi}_{ra}}{\mathrm{d}t}\right) \\ \frac{\mathrm{d}\hat{\psi}_{rb}}{\mathrm{d}t} &= -\alpha\hat{\psi}_{rb} + \omega\hat{\psi}_{ra} + \alpha M i_{sb} - k\left(\frac{1}{\sigma}u_{sb} - \frac{R_s}{\sigma}i_{sb} - \frac{\mathrm{d}i_{sb}}{\mathrm{d}t} - \beta\frac{\mathrm{d}\hat{\psi}_{rb}}{\mathrm{d}t}\right) \end{aligned} \tag{3.24}$$

from which

$$\begin{aligned} \frac{\mathrm{d}}{\mathrm{d}t}\left[(1-k\beta)\hat{\psi}_{ra} - ki_{sa}\right] &= -\alpha\hat{\psi}_{ra} - \omega\hat{\psi}_{rb} + \left(\alpha M + \frac{kR_s}{\sigma}\right)i_{sa} - \frac{k}{\sigma}u_{sa} \\ \frac{\mathrm{d}}{\mathrm{d}t}\left[(1-k\beta)\hat{\psi}_{rb} - ki_{sb}\right] &= -\alpha\hat{\psi}_{rb} + \omega\hat{\psi}_{ra} + \left(\alpha M + \frac{kR_s}{\sigma}\right)i_{sb} - \frac{k}{\sigma}u_{sb}\,. \end{aligned} \tag{3.25}$$

From (3.25), by introducing the variables

$$\begin{aligned} z_a &= (1-k\beta)\hat{\psi}_{ra} - ki_{sa} \\ z_b &= (1-k\beta)\hat{\psi}_{rb} - ki_{sb} \end{aligned} \tag{3.26}$$

we have

$$\begin{aligned} \dot{z}_a = &-\frac{\alpha}{1-k\beta}z_a - \frac{\omega}{1-k\beta}z_b + \left(\alpha M + \frac{kR_s}{\sigma} - \frac{\alpha k}{1-k\beta}\right)i_{sa} \\ &-\frac{\omega}{1-k\beta}ki_{sb} - \frac{k}{\sigma}u_{sa} \end{aligned}$$

$$\begin{aligned}
\dot{z}_b &= -\frac{\alpha}{1-k\beta}z_b + \frac{\omega}{1-k\beta}z_a + \left(\alpha M + \frac{kR_s}{\sigma} - \frac{\alpha k}{1-k\beta}\right)i_{sb} \\
&\quad + \frac{\omega}{1-k\beta}ki_{sa} - \frac{k}{\sigma}u_{sb} \\
\hat{\psi}_{ra} &= \frac{1}{1-k\beta}z_a + \frac{k}{1-k\beta}i_{sa} \\
\hat{\psi}_{rb} &= \frac{1}{1-k\beta}z_b + \frac{k}{1-k\beta}i_{sb} \,.
\end{aligned} \tag{3.27}$$

The observer (3.27) only depends on the measured signals ω, i_{sa}, i_{sb}, u_{sa} and u_{sb}: its estimation error dynamics are given by (3.20) and have an arbitrary exponential rate of convergence. The initial conditions $(\hat{\psi}_{ra}(0), \hat{\psi}_{rb}(0))$ are freely chosen and the initial conditions of (z_a, z_b) are computed as

$$\begin{aligned}
z_a(0) &= (1-k\beta)\hat{\psi}_{ra}(0) - ki_{sa}(0) \\
z_b(0) &= (1-k\beta)\hat{\psi}_{rb}(0) - ki_{sb}(0) \,.
\end{aligned} \tag{3.28}$$

Note that by increasing the convergence rate of (3.27), *i.e.* by making $1-k\beta$ close to zero, the sensitivity of the observer to stator currents and rotor speed measurement errors is also increased, since ω, i_{sa}, and i_{sb} are divided by $1-k\beta$ in (3.27).

The rotor flux observer (3.27) can also be directly obtained by restricting the flux estimates $(\hat{\psi}_{ra}, \hat{\psi}_{rb})$ to be of the form

$$\begin{aligned}
\hat{\psi}_{ra} &= ci_{sa} + \xi_a \\
\hat{\psi}_{rb} &= ci_{sb} + \xi_b
\end{aligned} \tag{3.29}$$

in which the arbitrary constant c and the signals (ξ_a, ξ_b) are to be designed in order to obtain exponential convergence to zero with arbitrary rate for the estimation errors $\tilde{\psi}_{ra} = \psi_{ra} - ci_{sa} - \xi_a$ and $\tilde{\psi}_{rb} = \psi_{rb} - ci_{sb} - \xi_b$. To this end, let us compute

$$\begin{aligned}
\dot{\tilde{\psi}}_{ra} &= -\alpha\psi_{ra} - \omega\psi_{rb} + \alpha M i_{sa} \\
&\quad - c\left(-\gamma i_{sa} + \frac{u_{sa}}{\sigma} + \beta\alpha\psi_{ra} + \beta\omega\psi_{rb}\right) - \dot{\xi}_a \\
\dot{\tilde{\psi}}_{rb} &= -\alpha\psi_{rb} + \omega\psi_{ra} + \alpha M i_{sb} \\
&\quad - c\left(-\gamma i_{sb} + \frac{u_{sb}}{\sigma} + \beta\alpha\psi_{rb} - \beta\omega\psi_{ra}\right) - \dot{\xi}_b \,.
\end{aligned} \tag{3.30}$$

If we design

$$\begin{aligned}
\dot{\xi}_a &= -\alpha(1+c\beta)\hat{\psi}_{ra} - \omega(1+c\beta)\hat{\psi}_{rb} + (\alpha M + c\gamma)i_{sa} - \frac{c}{\sigma}u_{sa} \\
\dot{\xi}_b &= -\alpha(1+c\beta)\hat{\psi}_{rb} + \omega(1+c\beta)\hat{\psi}_{ra} + (\alpha M + c\gamma)i_{sb} - \frac{c}{\sigma}u_{sb}
\end{aligned} \tag{3.31}$$

the equations (3.30) become

$$\begin{aligned}\dot{\tilde{\psi}}_{ra} &= -\alpha(1+c\beta)\tilde{\psi}_{ra} - (1+c\beta)\omega\tilde{\psi}_{rb} \\ \dot{\tilde{\psi}}_{rb} &= -\alpha(1+c\beta)\tilde{\psi}_{rb} + (1+c\beta)\omega\tilde{\psi}_{ra} \,.\end{aligned} \tag{3.32}$$

The time derivative of the positive definite function (3.5), along the trajectories of (3.32), is given by

$$\dot{V} = -\alpha(1+c\beta)\left(\tilde{\psi}_{ra}^2 + \tilde{\psi}_{rb}^2\right) = -2\alpha(1+c\beta)V \,.$$

Integrating with respect to time on $[0,t]$, we obtain

$$V(t) = V(0)\mathrm{e}^{-2\alpha(1+c\beta)t}$$

so that we can conclude that the flux estimates $(\hat{\psi}_{ra}, \hat{\psi}_{rb})$ converge to zero exponentially with arbitrary rate of convergence $\alpha(1+c\beta)$, since c can be freely chosen. The observer (3.29), (3.31) coincides with (3.27) by setting

$$\begin{aligned}\xi_a &= \frac{1}{1-k\beta}z_a \\ \xi_b &= \frac{1}{1-k\beta}z_b \\ c &= \frac{k}{1-k\beta} \,.\end{aligned}$$

Note also that if we set

$$\begin{aligned}c &= M \\ \xi_a &= L_r\hat{i}_{ra} \\ \xi_b &= L_r\hat{i}_{rb}\end{aligned}$$

in (3.29) and (3.31), we reobtain the observer (3.11) and (3.16).

In conclusion: the *rotor flux observer with arbitrary rate of convergence*

$$\begin{aligned}\dot{\xi}_a &= -\alpha(1+c\beta)\xi_a - \omega(1+c\beta)\xi_b + [(\alpha M + c\gamma) - \alpha c(1+c\beta)]i_{sa} \\ &\quad -\omega c(1+c\beta)i_{sb} - \frac{c}{\sigma}u_{sa}\,, \qquad \xi_a(0) = \hat{\psi}_{ra}(0) - ci_{sa}(0) \\ \dot{\xi}_b &= -\alpha(1+c\beta)\xi_b + \omega(1+c\beta)\xi_a + [(\alpha M + c\gamma) - \alpha c(1+c\beta)]i_{sb} \\ &\quad +\omega c(1+c\beta)i_{sa} - \frac{c}{\sigma}u_{sb}\,, \qquad \xi_b(0) = \hat{\psi}_{rb}(0) - ci_{sb}(0) \\ \hat{\psi}_{ra} &= ci_{sa} + \xi_a \\ \hat{\psi}_{rb} &= ci_{sb} + \xi_b\end{aligned} \tag{3.33}$$

guarantees that, for any initial condition $(\hat{\psi}_{ra}(0), \hat{\psi}_{rb}(0))$, the estimation errors $\tilde{\psi}_{ra} = \psi_{ra} - \hat{\psi}_{ra}$ and $\tilde{\psi}_{rb} = \psi_{rb} - \hat{\psi}_{rb}$ are such that

$$\left\| \begin{bmatrix} \tilde{\psi}_{ra}(t) \\ \tilde{\psi}_{rb}(t) \end{bmatrix} \right\| \leq \left\| \begin{bmatrix} \tilde{\psi}_{ra}(0) \\ \tilde{\psi}_{rb}(0) \end{bmatrix} \right\| \mathrm{e}^{-\alpha(1+c\beta)t},$$

i.e. they converge to zero exponentially with arbitrary rate of convergence $\alpha(1+c\beta)$, since c is an arbitrary positive real.

Illustrative Simulations

We tested the rotor flux observer with arbitrary rate of convergence (3.33) by simulations for the three-phase single pole pair 0.6-kW induction motor whose parameters have been reported in Chapter 1. The motor (with initial conditions $\psi_{ra}(0) = \psi_{rb}(0) = 0.1\,\mathrm{Wb}$) is controlled by the input–output feedback linearizing control (with control parameters, rotor speed and flux modulus references, and applied load torque as reported in Section 2.4). The design parameter c is set equal to 0.3 (the value is in SI units) while the observer initial conditions are set to zero. The rotor flux vector (a,b) components along with the corresponding estimation errors are reported in Figures 3.5 and 3.6: exponential estimation is achieved with a rate of convergence which is greater than that obtained by the open-loop rotor flux observer (recall Figures 3.1 and 3.2). The influence of rotor resistance uncertainty is illustrated by Figures 3.7 and 3.8, when the rotor resistance value used by the observer is 50% greater than the motor rotor resistance value: steady-state estimation errors appear, which are however smaller than those obtained by the open-loop rotor flux observer (compare with Figures 3.3 and 3.4).

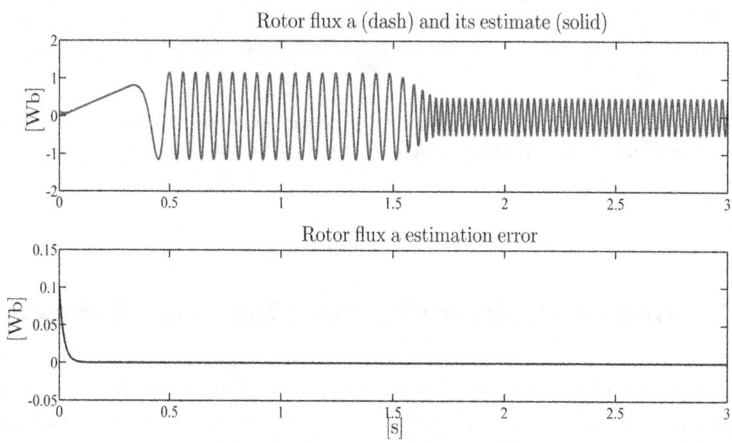

Fig. 3.5 Rotor flux observer with arbitrary rate of convergence: rotor flux ψ_{ra} and its estimate; rotor flux ψ_{ra} estimation error

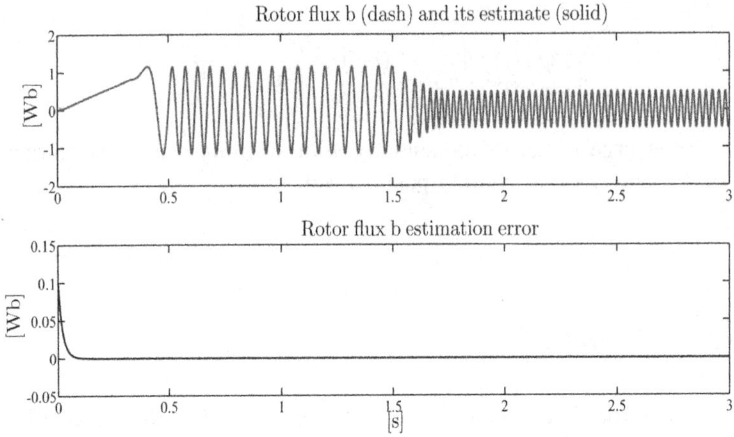

Fig. 3.6 Rotor flux observer with arbitrary rate of convergence: rotor flux ψ_{rb} and its estimate; rotor flux ψ_{rb} estimation error

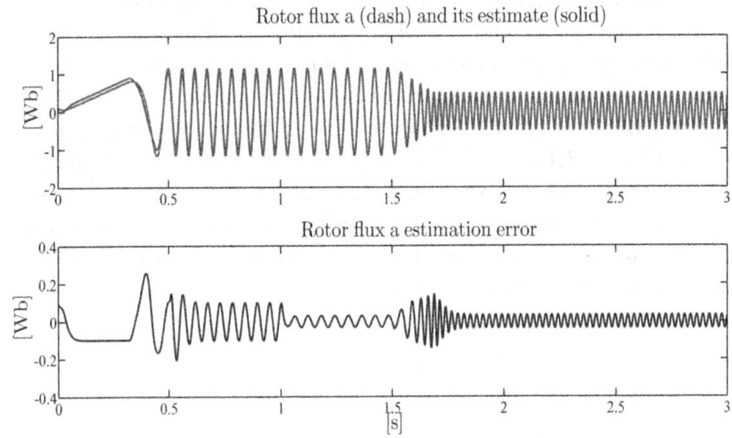

Fig. 3.7 Rotor flux observer with arbitrary rate of convergence in the case of rotor resistance uncertainty: rotor flux ψ_{ra} and its estimate; rotor flux ψ_{ra} estimation error

3.2 Adaptive Flux Observer with Rotor Resistance Estimator

In this section we show how an adaptive rotor flux observer with rotor resistance estimator can be designed on the basis of the measurements of the stator currents i_{sa} and i_{sb}, the rotor speed ω, and the stator voltages u_{sa} and u_{sb}. Recall that, as established in Chapter 1, in some operating conditions the rotor resistance cannot

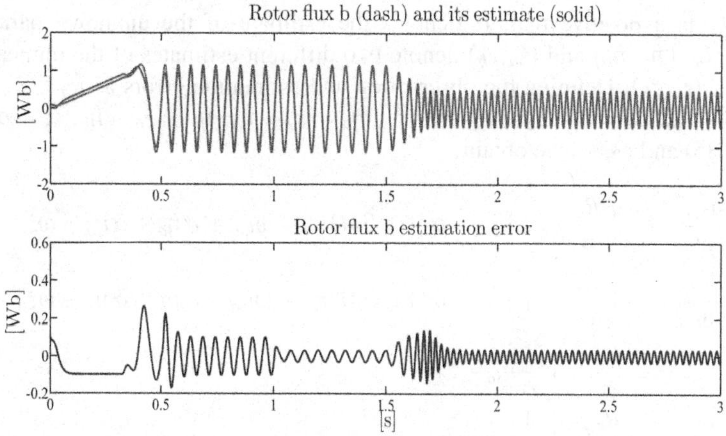

Fig. 3.8 Rotor flux observer with arbitrary rate of convergence in the case of rotor resistance uncertainty: rotor flux ψ_{rb} and its estimate; rotor flux ψ_{rb} estimation error

be identified by stator currents and rotor speed measurements. In order to design the rotor flux observer and rotor resistance estimator, we introduce the variables

$$\begin{aligned} z_a &= i_{sa} + \beta \psi_{ra} \\ z_b &= i_{sb} + \beta \psi_{rb} \end{aligned} \tag{3.34}$$

so that the motor electromagnetic equations in (1.26) become

$$\begin{aligned} \dot{z}_a &= -\frac{R_s}{\sigma} i_{sa} + \frac{1}{\sigma} u_{sa} \\ \dot{z}_b &= -\frac{R_s}{\sigma} i_{sb} + \frac{1}{\sigma} u_{sb} \\ \frac{\mathrm{d}i_{sa}}{\mathrm{d}t} &= -\frac{R_s}{\sigma} i_{sa} - \alpha(1+\beta M) i_{sa} - \omega i_{sb} + \alpha z_a + \omega z_b + \frac{1}{\sigma} u_{sa} \\ \frac{\mathrm{d}i_{sb}}{\mathrm{d}t} &= -\frac{R_s}{\sigma} i_{sb} - \alpha(1+\beta M) i_{sb} + \omega i_{sa} + \alpha z_b - \omega z_a + \frac{1}{\sigma} u_{sb} \,. \end{aligned} \tag{3.35}$$

The advantage of using (z_a, z_b) variables is that their dynamics depend neither on the unmeasured rotor fluxes nor on the uncertain rotor resistance. Now, we begin to design the adaptive observer by defining the dynamics for the estimates of the stator currents

$$\begin{aligned} \frac{\mathrm{d}\hat{i}_{sa}}{\mathrm{d}t} &= -\frac{R_s}{\sigma} \hat{i}_{sa} - \hat{\alpha}(1+\beta M) i_{sa} - \omega \hat{i}_{sb} + \hat{\alpha}\hat{\eta}_a + \omega \hat{z}_b + \frac{1}{\sigma} u_{sa} + k_1 (i_{sa} - \hat{i}_{sa}) \\ \frac{\mathrm{d}\hat{i}_{sb}}{\mathrm{d}t} &= -\frac{R_s}{\sigma} \hat{i}_{sb} - \hat{\alpha}(1+\beta M) i_{sb} + \omega \hat{i}_{sa} + \hat{\alpha}\hat{\eta}_b - \omega \hat{z}_a + \frac{1}{\sigma} u_{sb} + k_1 (i_{sb} - \hat{i}_{sb}) \end{aligned} \tag{3.36}$$

where k_1 is a positive real, $\hat{\alpha}$ denotes the estimate of the unknown parameter $\alpha = R_r/L_r$, $(\hat{\eta}_a, \hat{\eta}_b)$ and $(\hat{z}_a, \hat{z}_b)$ denote two different estimates of the unmeasured variables (z_a, z_b). Defining the observation and estimation errors as $\tilde{i}_{sa} = i_{sa} - \hat{i}_{sa}$, $\tilde{i}_{sb} = i_{sb} - \hat{i}_{sb}$, $\tilde{z}_a = z_a - \hat{z}_a$, $\tilde{z}_b = z_b - \hat{z}_b$, $\tilde{\eta}_a = z_a - \hat{\eta}_a$, $\tilde{\eta}_b = z_b - \hat{\eta}_b$, $\tilde{\alpha} = \alpha - \hat{\alpha}$, from (3.35) and (3.36) we obtain

$$\begin{aligned}
\frac{d\tilde{i}_{sa}}{dt} &= -\left(\frac{R_s}{\sigma} + k_1\right)\tilde{i}_{sa} - \tilde{\alpha}(1+\beta M)i_{sa} - \omega\tilde{i}_{sb} + \alpha\tilde{\eta}_a + \tilde{\alpha}\hat{\eta}_a + \omega\tilde{z}_b \\
\frac{d\tilde{i}_{sb}}{dt} &= -\left(\frac{R_s}{\sigma} + k_1\right)\tilde{i}_{sb} - \tilde{\alpha}(1+\beta M)i_{sb} + \omega\tilde{i}_{sa} + \alpha\tilde{\eta}_b + \tilde{\alpha}\hat{\eta}_b - \omega\tilde{z}_a \\
\dot{\tilde{z}}_a &= -\frac{R_s}{\sigma}i_{sa} + \frac{1}{\sigma}u_{sa} - \dot{\hat{z}}_a \\
\dot{\tilde{z}}_b &= -\frac{R_s}{\sigma}i_{sb} + \frac{1}{\sigma}u_{sb} - \dot{\hat{z}}_b \\
\dot{\tilde{\eta}}_a &= -\frac{R_s}{\sigma}i_{sa} + \frac{1}{\sigma}u_{sa} - \dot{\hat{\eta}}_a \\
\dot{\tilde{\eta}}_b &= -\frac{R_s}{\sigma}i_{sb} + \frac{1}{\sigma}u_{sb} - \dot{\hat{\eta}}_b\,.
\end{aligned} \tag{3.37}$$

Recalling the definition of (z_a, z_b) in (3.34), the rotor flux estimates can be obtained as

$$\begin{aligned}
\hat{\psi}_{ra} &= \frac{1}{\beta}(\hat{z}_a - \hat{i}_{sa}) \\
\hat{\psi}_{rb} &= \frac{1}{\beta}(\hat{z}_b - \hat{i}_{sb})
\end{aligned} \tag{3.38}$$

which imply

$$\begin{aligned}
\tilde{\psi}_{ra} &= \frac{1}{\beta}(\tilde{z}_a - \tilde{i}_{sa}) \\
\tilde{\psi}_{rb} &= \frac{1}{\beta}(\tilde{z}_b - \tilde{i}_{sb})\,.
\end{aligned} \tag{3.39}$$

In order to choose the dynamics of the estimates $(\hat{z}_a, \hat{z}_b)$ and the estimation law for the estimate $\hat{\alpha}$, we consider the positive definite function

$$V = \frac{1}{2}\left(\tilde{i}_{sa}^2 + \tilde{i}_{sb}^2\right) + \frac{1}{2k_2}\left(\tilde{z}_a^2 + \tilde{z}_b^2\right) + \frac{\alpha}{2k_3}\left(\tilde{\eta}_a^2 + \tilde{\eta}_b^2\right) + \frac{1}{2g}\tilde{\alpha}^2 \tag{3.40}$$

in which k_2, k_3 and g are positive reals. The time derivative of the function V along the trajectories of the error system is given by

$$\begin{aligned}
\dot{V} = &-\left(k_1 + \frac{R_s}{\sigma}\right)(\tilde{i}_{sa}^2 + \tilde{i}_{sb}^2) + \tilde{i}_{sa}\tilde{\alpha}[\hat{\eta}_a - (1+\beta M)i_{sa}] \\
&+\alpha\tilde{i}_{sa}\tilde{\eta}_a + \tilde{i}_{sa}\omega\tilde{z}_b + \alpha\tilde{i}_{sb}\tilde{\eta}_b - \tilde{i}_{sb}\omega\tilde{z}_a + \tilde{i}_{sb}\tilde{\alpha}[\hat{\eta}_b - (1+\beta M)i_{sb}]
\end{aligned}$$

$$+\frac{1}{k_2}(\tilde{z}_a\dot{\tilde{z}}_a+\tilde{z}_b\dot{\tilde{z}}_b)+\frac{\alpha}{k_3}(\tilde{\eta}_a\dot{\tilde{\eta}}_a+\tilde{\eta}_b\dot{\tilde{\eta}}_b)+\frac{1}{g}\tilde{\alpha}\dot{\tilde{\alpha}}\,. \tag{3.41}$$

Choose

$$\begin{aligned}
\dot{\hat{z}}_a &= -\frac{R_s}{\sigma}i_{sa}+\frac{1}{\sigma}u_{sa}-k_2\omega\tilde{i}_{sb}\\
\dot{\hat{z}}_b &= -\frac{R_s}{\sigma}i_{sb}+\frac{1}{\sigma}u_{sb}+k_2\omega\tilde{i}_{sa}\\
\dot{\hat{\eta}}_a &= -\frac{R_s}{\sigma}i_{sa}+\frac{1}{\sigma}u_{sa}+k_3\tilde{i}_{sa}\\
\dot{\hat{\eta}}_b &= -\frac{R_s}{\sigma}i_{sb}+\frac{1}{\sigma}u_{sb}+k_3\tilde{i}_{sb}\\
\dot{\tilde{\alpha}} &= -\dot{\hat{\alpha}} = \text{Proj}\left(g\{[(1+\beta M)i_{sa}-\hat{\eta}_a]\tilde{i}_{sa}+[(1+\beta M)i_{sb}-\hat{\eta}_b]\tilde{i}_{sb}\},\hat{\alpha}\right)
\end{aligned} \tag{3.42}$$

where $\text{Proj}(\xi,\hat{\alpha})$ is the smooth projection algorithm given by (A.25) in Appendix A and defined in our case by

$$\begin{aligned}
\text{Proj}(\xi,\hat{\alpha}) &= \xi\ ,\ \text{if } p(\hat{\alpha})\leq 0\\
\text{Proj}(\xi,\hat{\alpha}) &= \xi\ ,\ \text{if } p(\hat{\alpha})\geq 0 \text{ and } \xi\geq 0\\
\text{Proj}(\xi,\hat{\alpha}) &= [1-p(\hat{\alpha})]\xi\ ,\ \text{otherwise}
\end{aligned}$$

in which $p(\hat{\alpha})=\frac{\alpha_m^2-\hat{\alpha}^2}{2\delta\alpha_m-\delta^2}$ with $\alpha_m=R_{rm}/L_r$ the minimum (known) value of α and $\delta>0$ such that $\alpha_m-\delta>0$. The initial condition $\hat{\alpha}_0$ in (3.42) is chosen so that $\hat{\alpha}_0\geq\alpha_m$. The properties of the operator $\text{Proj}(\xi,\hat{\alpha})$ imply that, substituting (3.42) in (3.41), we have the inequality

$$\dot{V}\leq-\left(k_1+\frac{R_s}{\sigma}\right)(\tilde{i}_{sa}^2+\tilde{i}_{sb}^2)\,. \tag{3.43}$$

From (3.40) and (3.43) it follows that $\tilde{i}_{sa}$, $\tilde{i}_{sb}$, $\tilde{z}_a$, $\tilde{z}_b$, $\tilde{\eta}_a$, $\tilde{\eta}_b$, and $\tilde{\alpha}$ are bounded; since $(\omega,\ \psi_{ra},\ \psi_{rb},\ i_{sa},\ i_{sb})$ are assumed to be bounded, from the stator current observation error dynamics in (3.37) it follows that $\dot{\tilde{i}}_{sa}$ and $\dot{\tilde{i}}_{sb}$ are also bounded and, therefore, $\tilde{i}_{sa}$ and $\tilde{i}_{sb}$ are uniformly continuous. Since

$$-\int_0^t\dot{V}(\tau)\mathrm{d}\tau=V(0)-V(t)\geq\left(k_1+\frac{R_s}{\sigma}\right)\int_0^t[\tilde{i}_{sa}^2(\tau)+\tilde{i}_{sb}^2(\tau)]\mathrm{d}\tau$$

it follows that

$$\lim_{t\to\infty}\int_0^t[\tilde{i}_{sa}^2(\tau)+\tilde{i}_{sb}^2(\tau)]\mathrm{d}\tau\leq\left(k_1+\frac{R_s}{\sigma}\right)V(0)$$

so that Barbalat's Lemma A.2 in Appendix A applies and we can conclude that

$$\lim_{t\to\infty}[\tilde{i}_{sa}^2(t)+\tilde{i}_{sb}^2(t)]=0$$

which implies asymptotic convergence to zero of the stator current estimation errors $\tilde{i}_{sa}(t)$ and $\tilde{i}_{sb}(t)$. System (3.37), (3.42) may be written in the form

$$\begin{aligned}\dot{x}(t)&=A(t)x(t)+B(t)z(t)\\ \dot{z}(t)&=D(t)x(t)\end{aligned}\tag{3.44}$$

with $x=[\tilde{i}_{sa},\tilde{i}_{sb}]^{\mathrm{T}}$, $z=[\tilde{z}_a,\tilde{z}_b,\tilde{\eta}_a,\tilde{\eta}_b,\tilde{\alpha}]^{\mathrm{T}}$,

$$\begin{aligned}A&=\begin{bmatrix}-\left(k_1+\frac{R_s}{\sigma}\right) & -\omega\\ \omega & -\left(k_1+\frac{R_s}{\sigma}\right)\end{bmatrix}\\ B&=\begin{bmatrix}0 & \omega & \alpha & 0 & \hat{\eta}_a-(1+\beta M)i_{sa}\\ -\omega & 0 & 0 & \alpha & \hat{\eta}_b-(1+\beta M)i_{sb}\end{bmatrix}\end{aligned}\tag{3.45}$$

and D a suitable matrix. According to the Persistency of Excitation Lemma A.3 in Appendix A, if ω, i_{sa}, i_{sb}, ψ_{ra}, ψ_{rb}, u_{sa}, and u_{sb} are bounded and there exist two positive reals T, k_T such that the persistency of excitation condition

$$\int_t^{t+T}B^{\mathrm{T}}(\tau)B(\tau)\mathrm{d}\tau\geq k_T I,\quad\forall t\geq 0\tag{3.46}$$

is satisfied, then the errors $\tilde{i}_{sa}(t)$, $\tilde{i}_{sb}(t)$, $\tilde{z}_a(t)$, $\tilde{z}_b(t)$, $\tilde{\eta}_a(t)$, $\tilde{\eta}_b(t)$, and $\tilde{\alpha}(t)$ exponentially converge to zero from any motor initial condition: the rotor flux estimation errors (3.39) exponentially converge to zero while the rotor resistance estimate $L_r\hat{\alpha}(t)$ in (3.42) exponentially converges to its true value.

A different, less restrictive, persistency of excitation condition can be obtained in the case of constant rotor speed, including zero speed. In this case it is useful to introduce the new variables

$$\begin{aligned}\tilde{p}_a&=\alpha\tilde{\eta}_a+\omega\tilde{z}_b\\ \tilde{p}_b&=\alpha\tilde{\eta}_b-\omega\tilde{z}_a\end{aligned}\tag{3.47}$$

so that (3.37), (3.42) become

$$\begin{aligned}\frac{\mathrm{d}\tilde{i}_{sa}}{\mathrm{d}t}&=-\left(\frac{R_s}{\sigma}+k_1\right)\tilde{i}_{sa}-\omega\tilde{i}_{sb}+\tilde{\alpha}[\hat{\eta}_a-(1+\beta M)i_{sa}]+\tilde{p}_a\\ \frac{\mathrm{d}\tilde{i}_{sb}}{\mathrm{d}t}&=-\left(\frac{R_s}{\sigma}+k_1\right)\tilde{i}_{sb}+\omega\tilde{i}_{sa}+\tilde{\alpha}[\hat{\eta}_b-(1+\beta M)i_{sb}]+\tilde{p}_b\\ \dot{\tilde{p}}_a&=-(\alpha k_3+k_2\omega^2)\tilde{i}_{sa}\\ \dot{\tilde{p}}_b&=-(\alpha k_3+k_2\omega^2)\tilde{i}_{sb}\\ \dot{\hat{\alpha}}&=\mathrm{Proj}\left(g\{[(1+\beta M)i_{sa}-\hat{\eta}_a]\tilde{i}_{sa}+[(1+\beta M)i_{sb}-\hat{\eta}_b]\tilde{i}_{sb}\},\hat{\alpha}\right).\end{aligned}\tag{3.48}$$

By considering the positive definite function

$$V = \frac{1}{2}\left(\tilde{i}_{sa}^2 + \tilde{i}_{sb}^2\right) + \frac{1}{2g}\tilde{\alpha}^2 + \frac{1}{2(\alpha k_3 + k_2\omega^2)}\left(\tilde{p}_a^2 + \tilde{p}_b^2\right) \tag{3.49}$$

we obtain along the trajectories of the error system

$$\dot{V} \leq -\left(k_1 + \frac{R_s}{\sigma}\right)\left(\tilde{i}_{sa}^2 + \tilde{i}_{sb}^2\right). \tag{3.50}$$

System (3.48) may be written in the compact form

$$\begin{aligned} \dot{x}(t) &= \bar{A}(t)x(t) + \bar{B}(t)z(t) \\ \dot{z}(t) &= \bar{D}(t)x(t) \end{aligned} \tag{3.51}$$

with $x = [\tilde{i}_{sa}, \tilde{i}_{sb}]^{\mathrm{T}}$, $z = [\tilde{p}_a, \tilde{p}_b, \tilde{\alpha}]^{\mathrm{T}}$,

$$\begin{aligned} \bar{A} &= \begin{bmatrix} -\left(k_1 + \frac{R_s}{\sigma}\right) & -\omega \\ \omega & -\left(k_1 + \frac{R_s}{\sigma}\right) \end{bmatrix} \\ \bar{B} &= \begin{bmatrix} 1 & 0 & \hat{\eta}_a - (1+\beta M)i_{sa} \\ 0 & 1 & \hat{\eta}_b - (1+\beta M)i_{sb} \end{bmatrix} \end{aligned} \tag{3.52}$$

and $\bar{D}$ a suitable matrix. According to the Persistency of Excitation Lemma A.3 in Appendix A, if ω, i_{sa}, i_{sb}, ψ_{ra}, ψ_{rb}, u_{sa}, and u_{sb} are bounded and there exist two positive reals T, k_T such that the persistency of excitation condition

$$\int_t^{t+T} \bar{B}^{\mathrm{T}}(\tau)\bar{B}(\tau)\mathrm{d}\tau \geq k_T I, \quad \forall t \geq 0 \tag{3.53}$$

is satisfied, then the errors $\tilde{i}_{sa}(t)$, $\tilde{i}_{sb}(t)$, $\tilde{p}_a(t)$, $\tilde{p}_b(t)$, and $\tilde{\alpha}(t)$ tend exponentially to zero from any initial condition. On the other hand, from (3.47), we have

$$\begin{aligned} (\hat{\alpha} + \tilde{\alpha})(z_a - \hat{\eta}_a) + \omega(z_b - \hat{z}_b) &= \tilde{p}_a \\ (\hat{\alpha} + \tilde{\alpha})(z_b - \hat{\eta}_b) - \omega(z_a - \hat{z}_a) &= \tilde{p}_b \end{aligned} \tag{3.54}$$

from which

$$\begin{bmatrix} \hat{\alpha} & \omega \\ -\omega & \hat{\alpha} \end{bmatrix} \begin{bmatrix} z_a \\ z_b \end{bmatrix} = \begin{bmatrix} \hat{\alpha}\hat{\eta}_a + \omega\hat{z}_b - \tilde{\alpha}(z_a - \hat{\eta}_a) + \tilde{p}_a \\ \hat{\alpha}\hat{\eta}_b - \omega\hat{z}_a - \tilde{\alpha}(z_b - \hat{\eta}_b) + \tilde{p}_b \end{bmatrix}$$

which implies that

$$\begin{bmatrix} i_{sa} + \beta\psi_{ra} \\ i_{sb} + \beta\psi_{rb} \end{bmatrix} - \begin{bmatrix} \hat{\alpha} & \omega \\ -\omega & \hat{\alpha} \end{bmatrix}^{-1} \begin{bmatrix} \hat{\alpha}\hat{\eta}_a + \omega\hat{z}_b \\ \hat{\alpha}\hat{\eta}_b - \omega\hat{z}_a \end{bmatrix} = \begin{bmatrix} \hat{\alpha} & \omega \\ -\omega & \hat{\alpha} \end{bmatrix}^{-1} \begin{bmatrix} -\tilde{\alpha}(z_a - \hat{\eta}_a) + \tilde{p}_a \\ -\tilde{\alpha}(z_b - \hat{\eta}_b) + \tilde{p}_b \end{bmatrix}. \tag{3.55}$$

Since we have established that $\tilde{p}_a(t)$, $\tilde{p}_b(t)$, and $\tilde{\alpha}(t)$ are exponentially decaying signals in (3.55), in the case of constant rotor speed we replace the rotor flux estimates defined in (3.38) by

$$\begin{bmatrix} \hat{\psi}_{ra} \\ \hat{\psi}_{rb} \end{bmatrix} = -\frac{1}{\beta}\begin{bmatrix} i_{sa} \\ i_{sb} \end{bmatrix} + \frac{1}{\beta}\begin{bmatrix} \hat{\alpha} & \omega \\ -\omega & \hat{\alpha} \end{bmatrix}^{-1}\begin{bmatrix} \hat{\alpha}\hat{\eta}_a + \omega\hat{z}_b \\ \hat{\alpha}\hat{\eta}_b - \omega\hat{z}_a \end{bmatrix} \tag{3.56}$$

so that from (3.55) and (3.56) we have

$$\begin{bmatrix} \tilde{\psi}_{ra} \\ \tilde{\psi}_{rb} \end{bmatrix} = \frac{1}{\beta}\begin{bmatrix} \hat{\alpha} & \omega \\ -\omega & \hat{\alpha} \end{bmatrix}^{-1}\begin{bmatrix} -\tilde{\alpha}(z_a - \hat{\eta}_a) + \tilde{p}_a \\ -\tilde{\alpha}(z_b - \hat{\eta}_b) + \tilde{p}_b \end{bmatrix}. \tag{3.57}$$

Therefore, the estimation errors $\tilde{\psi}_{ra}(t)$ and $\tilde{\psi}_{rb}(t)$ exponentially converge to zero from any motor initial condition.

Remarks

1. Note that from (3.54) and (3.57) we obtain

$$\begin{bmatrix} \tilde{\psi}_{ra} \\ \tilde{\psi}_{rb} \end{bmatrix} = \frac{1}{\beta}\begin{bmatrix} \hat{\alpha} & \omega \\ -\omega & \hat{\alpha} \end{bmatrix}^{-1}\begin{bmatrix} \hat{\alpha}\tilde{\eta}_a + \omega\tilde{z}_b \\ \hat{\alpha}\tilde{\eta}_b - \omega\tilde{z}_a \end{bmatrix}.$$

   Hence, if $(\tilde{\eta}_a, \tilde{\eta}_b, \tilde{z}_a, \tilde{z}_b)$ converge exponentially to zero, $(\tilde{\psi}_{ra}, \tilde{\psi}_{rb})$ also converge exponentially to zero: in this case (3.56) provides converging flux estimates which may replace (3.38) when the persistency of excitation condition (3.46) is satisfied.
2. Note that when the rotor speed and the rotor flux modulus are constant and the load torque is zero, so that $\psi_{ra} = Mi_{sa}$ and $\psi_{rb} = Mi_{sb}$, the persistency of excitation condition (3.53) cannot be satisfied: recall that in those operating conditions the rotor resistance cannot be identified by stator currents and rotor speed measurements since in that case the motor equations

$$\begin{aligned}
\frac{d\omega}{dt} &= 0 \\
\frac{d\psi_{ra}}{dt} &= -\omega\psi_{rb} \\
\frac{d\psi_{rb}}{dt} &= \omega\psi_{ra} \\
\frac{di_{sa}}{dt} &= -\omega i_{sb} \\
\frac{di_{sb}}{dt} &= \omega i_{sa}
\end{aligned}$$

do not depend on the rotor resistance R_r. However, the rotor fluxes can be asymptotically estimated even though the rotor resistance estimate does not converge to its true value.

In conclusion: the *adaptive flux observer with rotor resistance estimator*

$$\begin{aligned}
\frac{d\hat{i}_{sa}}{dt} &= -\frac{R_s}{\sigma}\hat{i}_{sa} - \hat{\alpha}(1+\beta M)i_{sa} - \omega\hat{i}_{sb} + \hat{\alpha}\hat{\eta}_a + \omega\hat{z}_b + \frac{1}{\sigma}u_{sa} + k_1(i_{sa} - \hat{i}_{sa})\\
\frac{d\hat{i}_{sb}}{dt} &= -\frac{R_s}{\sigma}\hat{i}_{sb} - \hat{\alpha}(1+\beta M)i_{sb} + \omega\hat{i}_{sa} + \hat{\alpha}\hat{\eta}_b - \omega\hat{z}_a + \frac{1}{\sigma}u_{sb} + k_1(i_{sb} - \hat{i}_{sb})\\
\frac{d\hat{z}_a}{dt} &= -\frac{R_s}{\sigma}i_{sa} + \frac{1}{\sigma}u_{sa} - k_2\omega(i_{sb} - \hat{i}_{sb})\\
\frac{d\hat{z}_b}{dt} &= -\frac{R_s}{\sigma}i_{sb} + \frac{1}{\sigma}u_{sb} + k_2\omega(i_{sa} - \hat{i}_{sa})\\
\frac{d\hat{\eta}_a}{dt} &= -\frac{R_s}{\sigma}i_{sa} + \frac{1}{\sigma}u_{sa} + k_3(i_{sa} - \hat{i}_{sa})\\
\frac{d\hat{\eta}_b}{dt} &= -\frac{R_s}{\sigma}i_{sb} + \frac{1}{\sigma}u_{sb} + k_3(i_{sb} - \hat{i}_{sb})\\
\frac{d\hat{\alpha}}{dt} &= -\mathrm{Proj}\Big(g\{[(1+\beta M)i_{sa} - \hat{\eta}_a](i_{sa} - \hat{i}_{sa})\\
&\qquad + [(1+\beta M)i_{sb} - \hat{\eta}_b](i_{sb} - \hat{i}_{sb})\}, \hat{\alpha}\Big)\\
\begin{bmatrix}\hat{\psi}_{ra}\\ \hat{\psi}_{rb}\end{bmatrix} &= -\frac{1}{\beta}\begin{bmatrix}i_{sa}\\ i_{sb}\end{bmatrix} + \frac{1}{\beta}\begin{bmatrix}\hat{\alpha} & \omega\\ -\omega & \hat{\alpha}\end{bmatrix}^{-1}\begin{bmatrix}\hat{\alpha}\hat{\eta}_a + \omega\hat{z}_b\\ \hat{\alpha}\hat{\eta}_b - \omega\hat{z}_a\end{bmatrix}
\end{aligned} \tag{3.58}$$

guarantees that all the estimation variables ($\hat{i}_{sa}(t)$, $\hat{i}_{sb}(t)$, $\hat{z}_a(t)$, $\hat{z}_b(t)$, $\hat{\eta}_a(t)$, $\hat{\eta}_b(t)$, $\hat{\alpha}(t)$) are bounded for all $t \geq 0$ and that

$$\lim_{t\to\infty}[i_{sa}(t) - \hat{i}_{sa}(t)] = 0$$
$$\lim_{t\to\infty}[i_{sb}(t) - \hat{i}_{sb}(t)] = 0$$

for any initial condition, provided that $(\omega(t), \psi_{ra}(t), \psi_{rb}(t), i_{sa}(t), i_{sb}(t))$ are bounded on $[0,\infty)$; moreover, the estimation errors

$$\begin{aligned}
\tilde{i}_{sa} &= i_{sa} - \hat{i}_{sa}\\
\tilde{i}_{sb} &= i_{sb} - \hat{i}_{sb}\\
\tilde{\psi}_{ra} &= \psi_{ra} - \hat{\psi}_{ra}\\
\tilde{\psi}_{rb} &= \psi_{rb} - \hat{\psi}_{rb}\\
\tilde{\alpha} &= \alpha - \hat{\alpha}
\end{aligned}$$

converge exponentially to zero for any motor initial condition, provided that $\omega(t)$, $i_{sa}(t)$, $i_{sb}(t)$, $\psi_{ra}(t)$, $\psi_{rb}(t)$, $u_{sa}(t)$, and $u_{sb}(t)$ are bounded on $[0,\infty)$ and there exist two positive reals T and k_T such that

$$\int_t^{t+T} B^{\mathrm{T}}(\tau)B(\tau)\mathrm{d}\tau \geq k_T I\,, \quad \forall t \geq 0 \tag{3.59}$$

with

$$B(t) = \begin{bmatrix} 0 & \omega & \alpha & 0 & \hat{\eta}_a - (1+\beta M)i_{sa} \\ -\omega & 0 & 0 & \alpha & \hat{\eta}_b - (1+\beta M)i_{sb} \end{bmatrix}.$$

If (3.59) is not satisfied but the rotor speed is constant then the estimation errors

$$\begin{aligned} \tilde{i}_{sa} &= i_{sa} - \hat{i}_{sa} \\ \tilde{i}_{sb} &= i_{sb} - \hat{i}_{sb} \\ \tilde{\psi}_{ra} &= \psi_{ra} - \hat{\psi}_{ra} \\ \tilde{\psi}_{rb} &= \psi_{rb} - \hat{\psi}_{rb} \\ \tilde{\alpha} &= \alpha - \hat{\alpha} \end{aligned}$$

converge exponentially to zero for any motor initial condition, provided that $\omega(t)$, $i_{sa}(t)$, $i_{sb}(t)$, $\psi_{ra}(t)$, $\psi_{rb}(t)$, $u_{sa}(t)$, and $u_{sb}(t)$ are bounded on $[0,\infty)$ and there exist two positive reals T and k_T such that

$$\int_t^{t+T} \bar{B}^{\mathrm{T}}(\tau)\bar{B}(\tau)\mathrm{d}\tau \geq k_T I\,, \quad \forall t \geq 0 \tag{3.60}$$

with

$$\bar{B} = \begin{bmatrix} 1 & 0 & \hat{\eta}_a - (1+\beta M)i_{sa} \\ 0 & 1 & \hat{\eta}_b - (1+\beta M)i_{sb} \end{bmatrix}.$$

The adaptive observer (3.58) relies on: (1) the measurements of the stator currents (i_{sa}, i_{sb}), the stator voltages (u_{sa}, u_{sb}) and the rotor speed ω; (2) the positive design parameters k_1, k_2, k_3 and g.

Illustrative Simulations

We tested the adaptive rotor flux observer with rotor resistance estimator (3.58) by simulations for the three-phase single pole pair 0.6-kW induction motor whose parameters have been reported in Chapter 1. The motor (with initial conditions

$\psi_{ra}(0) = \psi_{rb}(0) = 0.1\,\mathrm{Wb}$) is controlled by the input–output feedback linearizing control (with control parameters, rotor speed and flux modulus references, and applied load torque as reported in Section 2.4). The design parameters are chosen as (all values are in SI units): $k_1 = 120$, $k_2 = 3$, $k_3 = 270$, $g = 450$. All observer initial conditions are set to zero except for $\hat{\alpha}(0) = 13.2\,\mathrm{s}^{-1}$, which is 50% greater than the corresponding value $\alpha = 8.8\,\mathrm{s}^{-1}$ in the motor model. The rotor flux vector (a,b) components along with the corresponding estimation errors are reported in Figures 3.9 and 3.10, while the estimate of the critical parameter α is reported in Figure 3.11: under persistency of excitation condition, exponentially converging estimates of both the rotor fluxes and the parameter α are achieved. Note that the flux estimation errors $(\tilde{\psi}_{ra}, \tilde{\psi}_{rb})$ converge to zero only when the rotor resistance error $\tilde{\alpha}$ converges to zero.

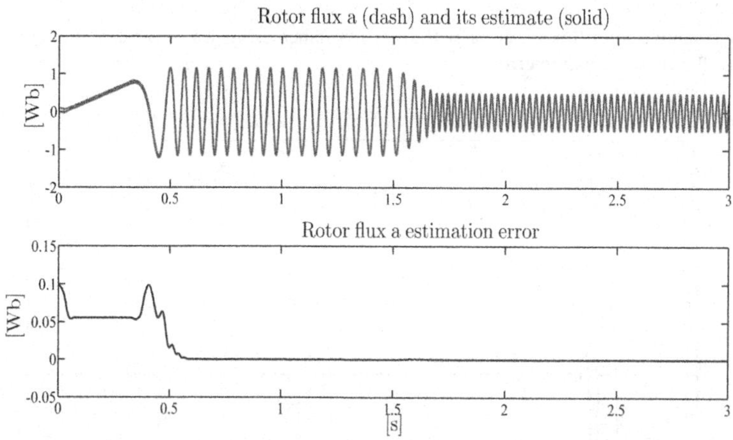

Fig. 3.9 Adaptive rotor flux observer with rotor resistance estimator: rotor flux ψ_{ra} and its estimate; rotor flux ψ_{ra} estimation error

The rotor flux observer designed in this section can be made adaptive with respect to both rotor and stator resistances so that an adaptive rotor flux observer with rotor and stator resistance estimators is achieved. Consider the model (1.26) along with the variables

$$
\begin{aligned}
z_a &= i_{sa} + \beta\psi_{ra} + \frac{R_s}{\sigma}\xi_a \\
z_b &= i_{sb} + \beta\psi_{rb} + \frac{R_s}{\sigma}\xi_b \\
\dot{\xi}_a &= i_{sa} \\
\dot{\xi}_b &= i_{sb}
\end{aligned}
\tag{3.61}
$$

so that we have for the motor electromagnetic equations

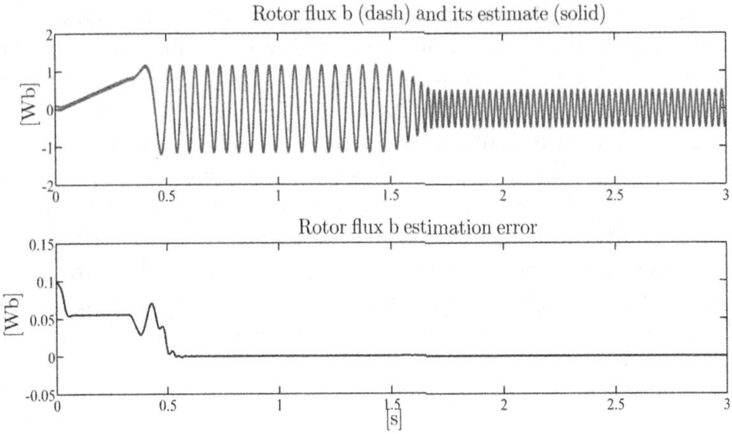

Fig. 3.10 Adaptive rotor flux observer with rotor resistance estimator: rotor flux ψ_{rb} and its estimate; rotor flux ψ_{rb} estimation error

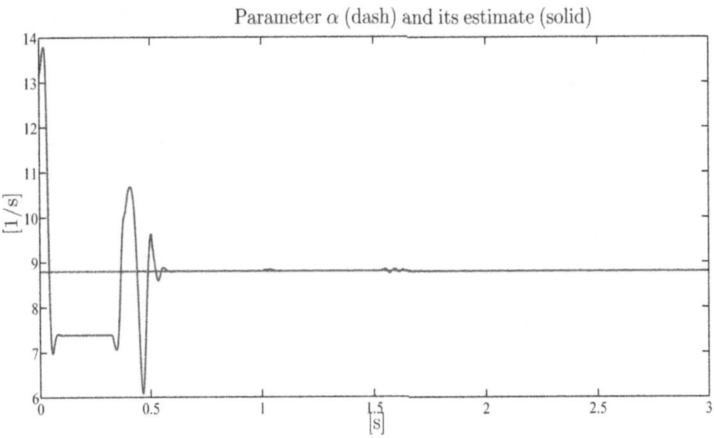

Fig. 3.11 Adaptive rotor flux observer with rotor resistance estimator: parameter α and its estimate

$$
\begin{aligned}
\frac{\mathrm{d}z_a}{\mathrm{d}t} &= \frac{1}{\sigma}u_{sa}\\
\frac{\mathrm{d}z_b}{\mathrm{d}t} &= \frac{1}{\sigma}u_{sb}\\
\frac{\mathrm{d}i_{sa}}{\mathrm{d}t} &= -\frac{R_s}{\sigma}i_{sa} - \alpha(1+\beta M)i_{sa} - \omega i_{sb} + \alpha z_a + \omega z_b\\
&\quad -\alpha\frac{R_s}{\sigma}\xi_a - \omega\frac{R_s}{\sigma}\xi_b + \frac{1}{\sigma}u_{sa}\\
\frac{\mathrm{d}i_{sb}}{\mathrm{d}t} &= -\frac{R_s}{\sigma}i_{sb} - \alpha(1+\beta M)i_{sb} + \omega i_{sa} + \alpha z_b - \omega z_a
\end{aligned}
$$

$$-\alpha\frac{R_s}{\sigma}\xi_b + \omega\frac{R_s}{\sigma}\xi_a + \frac{1}{\sigma}u_{sb}\,. \tag{3.62}$$

Note that even though the new variables z_a and z_b are not available from measurements, their time derivatives do not depend on the unmeasured rotor fluxes and on the uncertain rotor and stator resistances. Now, we design the dynamics of the estimates of the stator currents

$$\begin{aligned}
\frac{\mathrm{d}\hat{i}_{sa}}{\mathrm{d}t} &= -\frac{\hat{R}_s}{\sigma}i_{sa} - \hat{\alpha}(1+\beta M)i_{sa} - \omega\hat{i}_{sb} + \hat{\alpha}\hat{\eta}_a + \omega\hat{z}_b \\
&\quad -\frac{\hat{\theta}}{\sigma}\xi_a - \frac{\hat{R}_s}{\sigma}\omega\xi_b + \frac{1}{\sigma}u_{sa} + k_1(i_{sa} - \hat{i}_{sa}) \\
\frac{\mathrm{d}\hat{i}_{sb}}{\mathrm{d}t} &= -\frac{\hat{R}_s}{\sigma}i_{sb} - \hat{\alpha}(1+\beta M)i_{sb} + \omega\hat{i}_{sa} + \hat{\alpha}\hat{\eta}_b - \omega\hat{z}_a \\
&\quad -\frac{\hat{\theta}}{\sigma}\xi_b + \frac{\hat{R}_s}{\sigma}\omega\xi_a + \frac{1}{\sigma}u_{sb} + k_1(i_{sb} - \hat{i}_{sb})
\end{aligned} \tag{3.63}$$

in which: k_1 is a positive design parameter; $\hat{\alpha}$, $\hat{R}_s$, and $\hat{\theta}$ denote the estimates of the unknown parameters $\alpha = R_r/L_r$, R_s and αR_s, respectively; $(\hat{\eta}_a, \hat{\eta}_b)$ and $(\hat{z}_a, \hat{z}_b)$ denote two pairs of different estimates of the unmeasured variables (z_a, z_b). Defining the observation and estimation errors as $\tilde{i}_{sa} = i_{sa} - \hat{i}_{sa}$, $\tilde{i}_{sb} = i_{sb} - \hat{i}_{sb}$, $\tilde{z}_a = z_a - \hat{z}_a$, $\tilde{z}_b = z_b - \hat{z}_b$, $\tilde{\eta}_a = z_a - \hat{\eta}_a$, $\tilde{\eta}_b = z_b - \hat{\eta}_b$, $\tilde{\alpha} = \alpha - \hat{\alpha}$, $\tilde{R}_s = R_s - \hat{R}_s$, $\tilde{\theta} = \alpha R_s - \hat{\theta}$, from (3.62) and (3.63) we have

$$\begin{aligned}
\frac{\mathrm{d}\tilde{i}_{sa}}{\mathrm{d}t} &= -\frac{\tilde{R}_s}{\sigma}i_{sa} - \tilde{\alpha}(1+\beta M)i_{sa} - \omega\tilde{i}_{sb} + \alpha\tilde{\eta}_a + \tilde{\alpha}\hat{\eta}_a \\
&\quad +\omega\tilde{z}_b - \frac{\tilde{\theta}}{\sigma}\xi_a - \frac{\tilde{R}_s}{\sigma}\omega\xi_b - k_1\tilde{i}_{sa} \\
\frac{\mathrm{d}\tilde{i}_{sb}}{\mathrm{d}t} &= -\frac{\tilde{R}_s}{\sigma}i_{sb} - \tilde{\alpha}(1+\beta M)i_{sb} + \omega\tilde{i}_{sa} + \alpha\tilde{\eta}_b + \tilde{\alpha}\hat{\eta}_b \\
&\quad -\omega\tilde{z}_a - \frac{\tilde{\theta}}{\sigma}\xi_b + \frac{\tilde{R}_s}{\sigma}\omega\xi_a - k_1\tilde{i}_{sb}\,.
\end{aligned} \tag{3.64}$$

The rotor flux estimates are obtained as

$$\begin{aligned}
\hat{\psi}_{ra} &= \frac{1}{\beta}\left(\hat{\eta}_a - \hat{i}_{sa} - \frac{\hat{R}_s}{\sigma}\xi_a\right) \\
\hat{\psi}_{rb} &= \frac{1}{\beta}\left(\hat{\eta}_b - \hat{i}_{sb} - \frac{\hat{R}_s}{\sigma}\xi_b\right)
\end{aligned} \tag{3.65}$$

which, recalling (3.61), imply

$$\tilde{\psi}_{ra} = \frac{1}{\beta}\left(\tilde{\eta}_a - \tilde{i}_{sa} - \frac{\tilde{R}_s}{\sigma}\xi_a\right)$$

$$\tilde{\psi}_{rb} = \frac{1}{\beta}\left(\tilde{\eta}_b - \tilde{i}_{sb} - \frac{\tilde{R}_s}{\sigma}\xi_b\right). \tag{3.66}$$

In order to choose the dynamics of the estimates $(\hat{z}_a, \hat{z}_b)$, $(\hat{\eta}_a, \hat{\eta}_b)$, and of the parameters estimates $(\hat{\alpha}, \hat{R}_s, \hat{\theta})$, we consider the positive definite function (k_2, k_3, g_1, g_2, g_3 are positive design parameters)

$$V = \frac{1}{2}\left(\tilde{i}_{sa}^2 + \tilde{i}_{sb}^2\right) + \frac{1}{2k_2}\left(\tilde{z}_a^2 + \tilde{z}_b^2\right) + \frac{\alpha}{2k_3}\left(\tilde{\eta}_a^2 + \tilde{\eta}_b^2\right) + \frac{1}{2g_1}\tilde{\alpha}^2 + \frac{1}{2g_2}\tilde{R}_s^2 + \frac{1}{2g_3}\tilde{\theta}^2 \tag{3.67}$$

whose time derivative is given by

$$\begin{aligned}
\dot{V} = & -k_1\left(\tilde{i}_{sa}^2 + \tilde{i}_{sb}^2\right) + \tilde{z}_a\left(-\omega\tilde{i}_{sb} + \frac{1}{k_2}\dot{\tilde{z}}_a\right) + \tilde{z}_b\left(\omega\tilde{i}_{sa} + \frac{1}{k_2}\dot{\tilde{z}}_b\right) \\
& + \alpha\tilde{\eta}_a\left(\tilde{i}_{sa} + \frac{1}{k_3}\dot{\tilde{\eta}}_a\right) + \alpha\tilde{\eta}_b\left(\tilde{i}_{sb} + \frac{1}{k_3}\dot{\tilde{\eta}}_b\right) \\
& + \tilde{\alpha}\left\{\tilde{i}_{sa}\left[\hat{\eta}_a - i_{sa}(1+\beta M)\right] + \tilde{i}_{sb}\left[\hat{\eta}_b - i_{sb}(1+\beta M)\right] + \frac{1}{g_1}\dot{\tilde{\alpha}}\right\} \\
& + \tilde{R}_s\left[\tilde{i}_{sa}\left(-\frac{i_{sa}}{\sigma} - \frac{\omega\xi_b}{\sigma}\right) + \tilde{i}_{sb}\left(-\frac{i_{sb}}{\sigma} + \frac{\omega\xi_a}{\sigma}\right) + \frac{1}{g_2}\dot{\tilde{R}}_s\right] \\
& + \tilde{\theta}\left(-\tilde{i}_{sa}\frac{\xi_a}{\sigma} - \tilde{i}_{sb}\frac{\xi_b}{\sigma} + \frac{1}{g_3}\dot{\tilde{\theta}}\right)
\end{aligned} \tag{3.68}$$

so that, by choosing

$$\begin{aligned}
\dot{\hat{z}}_a &= \frac{1}{\sigma}u_{sa} - k_2\omega\tilde{i}_{sb} \\
\dot{\hat{z}}_b &= \frac{1}{\sigma}u_{sb} + k_2\omega\tilde{i}_{sa} \\
\dot{\hat{\eta}}_a &= \frac{1}{\sigma}u_{sa} + k_3\tilde{i}_{sa} \\
\dot{\hat{\eta}}_b &= \frac{1}{\sigma}u_{sb} + k_3\tilde{i}_{sb} \\
\dot{\hat{\alpha}} &= -g_1\left\{\left[(1+\beta M)i_{sa} - \hat{\eta}_a\right]\tilde{i}_{sa} + \left[(1+\beta M)i_{sb} - \hat{\eta}_b\right]\tilde{i}_{sb}\right\} \\
\dot{\hat{R}}_s &= -\frac{g_2}{\sigma}\left[(i_{sa} + \omega\xi_b)\tilde{i}_{sa} + (i_{sb} - \omega\xi_a)\tilde{i}_{sb}\right] \\
\dot{\hat{\theta}} &= -\frac{g_3}{\sigma}\left(\xi_a\tilde{i}_{sa} + \xi_b\tilde{i}_{sb}\right),
\end{aligned} \tag{3.69}$$

we obtain

$$\dot{V} = -k_1\left(\tilde{i}_{sa}^2 + \tilde{i}_{sb}^2\right) \tag{3.70}$$

which implies the boundedness of $\tilde{i}_{sa}$, $\tilde{i}_{sb}$, $\tilde{z}_a$, $\tilde{z}_b$, $\tilde{\eta}_a$, $\tilde{\eta}_b$, $\tilde{\alpha}$, $\tilde{R}_s$, and $\tilde{\theta}$. Denoting by $x = [\tilde{i}_{sa}, \tilde{i}_{sb}]^{\mathrm{T}}$ and $z = [\tilde{z}_a, \tilde{z}_b, \tilde{\eta}_a, \tilde{\eta}_b, \tilde{\alpha}, \tilde{R}_s, \tilde{\theta}]^{\mathrm{T}}$, the error dynamics may be written in the form

$$\begin{aligned} \dot{x}(t) &= A(t)x(t) + B(t)z(t) \\ \dot{z}(t) &= D(t)x(t) \end{aligned} \tag{3.71}$$

with

$$\begin{aligned} A &= \begin{bmatrix} -k_1 & -\omega \\ \omega & -k_1 \end{bmatrix} \\ B &= \begin{bmatrix} 0 & \omega & \alpha & 0 & \hat{\eta}_a - (1+\beta M) i_{sa} & -\frac{1}{\sigma}(i_{sa} + \omega \xi_b) & -\frac{\xi_a}{\sigma} \\ -\omega & 0 & 0 & \alpha & \hat{\eta}_b - (1+\beta M) i_{sb} & -\frac{1}{\sigma}(i_{sb} - \omega \xi_a) & -\frac{\xi_b}{\sigma} \end{bmatrix} \\ D^{\mathrm{T}} &= \begin{bmatrix} 0 & -k_2\omega & -k_3 & 0 & g_1[(1+\beta M) i_{sa} - \hat{\eta}_a] & \frac{g_2}{\sigma}(i_{sa} + \omega \xi_b) & \frac{g_3}{\sigma}\xi_a \\ k_2\omega & 0 & 0 & -k_3 & g_1[(1+\beta M) i_{sb} - \hat{\eta}_b] & \frac{g_2}{\sigma}(i_{sb} - \omega \xi_a) & \frac{g_3}{\sigma}\xi_b \end{bmatrix} . \end{aligned} \tag{3.72}$$

According to the Persistency of Excitation Lemma A.3 in Appendix A, if ω, i_{sa}, i_{sb}, ψ_{ra}, ψ_{rb}, ξ_a, ξ_b, u_{sa}, and u_{sb} are bounded and there exist two positive reals T, k_T such that the persistency of excitation condition

$$\int_t^{t+T} B^{\mathrm{T}}(\tau) B(\tau) \mathrm{d}\tau \geq k_T I\,, \quad \forall t \geq 0 \tag{3.73}$$

is satisfied, then $x(t)$ and $z(t)$ exponentially converge to zero from any initial condition, *i.e.* the rotor flux estimation errors $\tilde{\psi}_{ra}(t)$ and $\tilde{\psi}_{rb}(t)$, the rotor resistance and stator resistance estimation errors $\tilde{\alpha}(t)$ and $\tilde{R}_s(t)$ exponentially converge to zero from any initial condition.

When the rotor speed is constant, a less restrictive persistency of excitation condition can be obtained on the basis of the variables

$$\begin{aligned} \tilde{p}_a &= \alpha \tilde{\eta}_a + \omega \tilde{z}_b \\ \tilde{p}_b &= \alpha \tilde{\eta}_b - \omega \tilde{z}_a \end{aligned} \tag{3.74}$$

which lead to a modification of (3.65) similar to that obtained in (3.56).

3.3 Two Load Torque Estimators

In this section we design two different load torque identifiers which can be used along with any rotor flux observer to provide an online estimation of the load torque: the first one is of second order and also provides an estimate of the rotor speed while the second one is of reduced (first) order and does not rely on a rotor speed estimate. We begin to show how a second order load torque estimator can be designed on the basis of the motor mechanical equation

$$\dot{\omega} = \mu(\psi_{ra}i_{sb} - \psi_{rb}i_{sa}) - \frac{T_L}{J} \tag{3.75}$$

assuming that: (1) the load torque T_L is constant; (2) the parameters μ and J are known; (3) the rotor speed ω and the stator currents (i_{sa}, i_{sb}) are available from measurements; (4) exponentially converging rotor flux estimates are provided by suitable observers. We design the following dynamics for a load torque estimator and rotor speed observer

$$\begin{aligned} \dot{\hat{\omega}} &= \mu(\hat{\psi}_{ra}i_{sb} - \hat{\psi}_{rb}i_{sa}) - \frac{\hat{T}_L}{J} + k(\omega - \hat{\omega}) \\ \dot{\hat{T}}_L &= -\lambda(\omega - \hat{\omega}) \end{aligned} \tag{3.76}$$

in which k and λ are positive reals. Defining the estimation errors

$$\begin{aligned} \tilde{\omega} &= \omega - \hat{\omega} \\ \tilde{T}_L &= T_L - \hat{T}_L \\ \tilde{\psi}_{ra} &= \psi_{ra} - \hat{\psi}_{ra} \\ \tilde{\psi}_{rb} &= \psi_{rb} - \hat{\psi}_{rb} \end{aligned}$$

we obtain for the estimation error dynamics

$$\begin{aligned} \dot{\tilde{T}}_L &= \lambda\tilde{\omega} \\ \dot{\tilde{\omega}} &= \mu(\tilde{\psi}_{ra}i_{sb} - \tilde{\psi}_{rb}i_{sa}) - \frac{\tilde{T}_L}{J} - k\tilde{\omega} \,. \end{aligned} \tag{3.77}$$

Consider the positive definite quadratic function $(0 < \varepsilon < 1/\sqrt{\lambda J})$

$$V = \frac{1}{2\lambda J}\tilde{T}_L^2 + \frac{1}{2}\tilde{\omega}^2 + \varepsilon\tilde{\omega}\tilde{T}_L \,. \tag{3.78}$$

Its time derivative along the trajectories of (3.77) is given by

$$\begin{aligned} \dot{V} = &-(k - \varepsilon\lambda)\tilde{\omega}^2 - \frac{\varepsilon}{J}\tilde{T}_L^2 - \varepsilon k\tilde{T}_L\tilde{\omega} \\ &+\mu\tilde{\omega}(\tilde{\psi}_{ra}i_{sb} - \tilde{\psi}_{rb}i_{sa}) + \varepsilon\mu\tilde{T}_L(\tilde{\psi}_{ra}i_{sb} - \tilde{\psi}_{rb}i_{sa}) \,. \end{aligned} \tag{3.79}$$

By virtue of the inequality

$$ab \le \frac{a^2}{2k} + \frac{kb^2}{2} \tag{3.80}$$

we can write

$$\mu\tilde{\omega}(\tilde{\psi}_{ra}i_{sb} - \tilde{\psi}_{rb}i_{sa}) \le \frac{\mu}{2}\tilde{\omega}^2\varepsilon\frac{|i_{sa}| + |i_{sb}|}{1 + |i_{sa}| + |i_{sb}|}$$

$$
\begin{aligned}
&+\frac{\mu}{2\varepsilon}(1+|i_{sa}|+|i_{sb}|)(|i_{sa}|+|i_{sb}|)(\tilde{\psi}_{ra}^2+\tilde{\psi}_{rb}^2)\\
&\varepsilon\mu\tilde{T}_L(\tilde{\psi}_{ra}i_{sb}-\tilde{\psi}_{rb}i_{sa}) \le \frac{\varepsilon\tilde{T}_L^2}{2J}\frac{|i_{sa}|+|i_{sb}|}{1+|i_{sa}|+|i_{sb}|}\\
&+\frac{\varepsilon\mu^2 J}{2}(|i_{sa}|+|i_{sb}|)(1+|i_{sa}|+|i_{sb}|)(\tilde{\psi}_{ra}^2+\tilde{\psi}_{rb}^2)
\end{aligned} \tag{3.81}
$$

so that from (3.79), we have

$$
\begin{aligned}
\dot{V} \le &-\left(k-\varepsilon\lambda-\frac{\mu\varepsilon}{2}\right)\tilde{\omega}^2-\frac{\varepsilon}{2J}\tilde{T}_L^2-\varepsilon k\tilde{T}_L\tilde{\omega}\\
&+\frac{\mu}{2}\left(\frac{1}{\varepsilon}+\varepsilon\mu J\right)(|i_{sa}|+|i_{sb}|)(1+|i_{sa}|+|i_{sb}|)(\tilde{\psi}_{ra}^2+\tilde{\psi}_{rb}^2)\,.
\end{aligned} \tag{3.82}
$$

By choosing ε such that

$$
\left(k-\varepsilon\lambda-\frac{\mu\varepsilon}{2}\right)\frac{\varepsilon}{2J}-\frac{\varepsilon^2k^2}{4}>0
$$

which implies

$$
\varepsilon < \frac{k}{J\left(\frac{\lambda}{J}+\frac{\mu}{2J}+\frac{k^2}{2}\right)} \tag{3.83}
$$

and assuming that $i_{sa}(t)$ and $i_{sb}(t)$ are bounded on $[0,\infty)$, system (3.77) satisfies assumption (i) in Lemma A.1 in Appendix A, viewing (3.77) as the x_1-subsystem and the observation error dynamics as the x_2-subsystem. Therefore, any globally exponentially convergent rotor flux observer allows us to apply Lemma A.1, proving that the load torque estimation error $\tilde{T}_L(t)$ converges exponentially to zero from any initial condition.

In conclusion: the second order *load torque estimator*

$$
\begin{aligned}
\dot{\hat{\omega}} &= \mu(\hat{\psi}_{ra}i_{sb}-\hat{\psi}_{rb}i_{sa})-\frac{\hat{T}_L}{J}+k(\omega-\hat{\omega})\\
\dot{\hat{T}}_L &= -\lambda(\omega-\hat{\omega})\,,
\end{aligned} \tag{3.84}
$$

with k and λ positive design parameters, guarantees that the estimation errors

$$
\begin{aligned}
\tilde{\omega} &= \omega-\hat{\omega}\\
\tilde{T}_L &= T_L-\hat{T}_L
\end{aligned}
$$

converge exponentially to zero for any initial condition, provided that:

1. The stator currents $i_{sa}(t)$ and $i_{sb}(t)$ are bounded on $[0,\infty)$.

2. $\hat{\psi}_{ra}$ and $\hat{\psi}_{rb}$ are such that the corresponding estimation errors $[\psi_{ra}(t) - \hat{\psi}_{ra}(t)]$ and $[\psi_{rb}(t) - \hat{\psi}_{rb}(t)]$, respectively, exponentially converge to zero for any initial condition.

Illustrative Simulations

We tested the load torque estimator (3.84) by simulations for the three-phase single pole pair 0.6-kW induction motor whose parameters have been reported in Chapter 1. The motor (with initial conditions $\psi_{ra}(0) = \psi_{rb}(0) = 0.1$ Wb) is controlled by the input–output feedback linearizing control (with control parameters, rotor speed and flux modulus references, and applied load torque as reported in Section 2.4). The design parameters are chosen as (all values are in SI units) $k = 200$, $\lambda = 100^2 J$. All initial conditions for the load torque estimator are set to zero while the rotor flux estimates are generated by the adaptive rotor flux observer (3.58) with (all values are in SI units) $k_1 = 120$, $k_2 = 3$, $k_3 = 270$, $g = 450$. All initial conditions for the adaptive rotor flux observer are set to zero except for $\hat{\alpha}(0) = 13.2\,\mathrm{s}^{-1}$, which is 50% greater than the true value $\alpha = 8.8\,\mathrm{s}^{-1}$. The estimate of the load torque T_L is reported in Figure 3.12: exponentially converging load torque estimation is achieved.

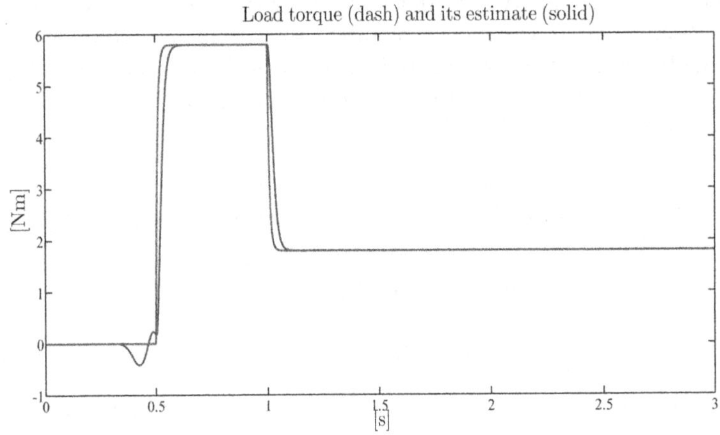

Fig. 3.12 Load torque estimator: load torque T_L and its estimate

We now design a reduced first order load torque estimator by avoiding the use of an estimate for the measured variable ω. Consider the motor mechanical equations

$$\dot{\omega} = \mu(\psi_{ra} i_{sb} - \psi_{rb} i_{sa}) - \frac{T_L}{J}$$

$$\dot{T}_L = 0 \tag{3.85}$$

and make the change of variable ($\lambda > 0$)

$$z = T_L + \lambda\omega \tag{3.86}$$

so that from (3.85) and (3.86), we have

$$\dot{z} = \lambda\dot{\omega} = \lambda\mu(\psi_{ra}i_{sb} - \psi_{rb}i_{sa}) - \frac{\lambda}{J}z + \frac{\lambda^2}{J}\omega\,. \tag{3.87}$$

Let $\hat{T}_L$ be an estimate of T_L generated by

$$\begin{aligned}
\dot{\hat{z}} &= -\frac{\lambda}{J}\hat{z} + \frac{\lambda^2}{J}\omega + \lambda\mu(\hat{\psi}_{ra}i_{sb} - \hat{\psi}_{rb}i_{sa}) \\
\hat{T}_L &= \hat{z} - \lambda\omega\,, \quad \hat{z}(0) = \hat{T}_L(0) + \lambda\omega(0)
\end{aligned} \tag{3.88}$$

and suitable rotor flux estimates. Defining $\tilde{z} = z - \hat{z}$, $\tilde{T}_L = T_L - \hat{T}_L$, $\tilde{\psi}_{ra} = \psi_{ra} - \hat{\psi}_{ra}$, and $\tilde{\psi}_{rb} = \psi_{rb} - \hat{\psi}_{rb}$, we have for the estimation error dynamics

$$\begin{aligned}
\dot{\tilde{z}} &= -\frac{\lambda}{J}\tilde{z} + \lambda\mu(\tilde{\psi}_{ra}i_{sb} - \tilde{\psi}_{rb}i_{sa}) \\
\tilde{T}_L &= \tilde{z}\,.
\end{aligned} \tag{3.89}$$

Consider the Lyapunov function

$$V = \frac{1}{2}\tilde{z}^2 \tag{3.90}$$

whose time derivative along the solutions of (3.89) is such that

$$\begin{aligned}
\dot{V} &= -\frac{\lambda}{J}\tilde{z}^2 + \lambda\mu(\tilde{\psi}_{ra}i_{sb} - \tilde{\psi}_{rb}i_{sa})\tilde{z} \\
&\le -\frac{\lambda}{2J}\tilde{z}^2 + \frac{\lambda\mu^2 J}{2}(|i_{sa}| + |i_{sb}|)(1 + |i_{sa}| + |i_{sb}|)(\tilde{\psi}_{ra}^2 + \tilde{\psi}_{rb}^2)
\end{aligned} \tag{3.91}$$

so that, if $i_{sa}(t)$ and $i_{sb}(t)$ are bounded on $[0,\infty)$, system (3.89) satisfies the conditions of the x_1-subsystem in Lemma A.1 in Appendix A. Hence, any globally exponentially convergent rotor flux observer guarantees that Lemma A.1 applies, demonstrating that $\tilde{\psi}_{ra}(t)$, $\tilde{\psi}_{rb}(t)$, and $\tilde{T}_L(t)$ exponentially converge to zero for any initital condition.

3.4 Experimental Results

The experimental dynamic behavior of the adaptive rotor flux observer (3.98) given in Problem 3.5 (whose parameters are set equal to $k = 200$, $\gamma_1 = 0.3$, $\gamma_2 = 1$, $\gamma_3 = 1000$, and whose initial conditions except $\hat{\alpha}$ are set equal to zero) is illustrated in Figure 3.13 in the case of $\hat{\alpha}(0) = 0.5R_{rN}/L_r$ (which corresponds to an initial error of -50%) and in Figures 3.14 and 3.15 in the case of $\hat{\alpha}(0) = 2R_{rN}/L_r$ (which corresponds to 100% initial error). In both cases the experiments are carried out during the motor start-up with a constant load torque (equal to 66% of the rated value): this corresponds to the most critical situation from the modeling view point. The effects of the high-order harmonics introduced by the inverter on the adaptive observer are illustrated in Figures 3.14 and 3.15: Figure 3.14 corresponds to measured voltage (output of the inverter) while Figure 3.15 refers to the reference voltage (input to the inverter). No significant effects on resistance and flux estimation are noticed. From Figures 3.13–3.15 we also note that the convergence of the rotor resistance and flux estimates is exponential. The results reported in Figure 3.16 are obtained when the

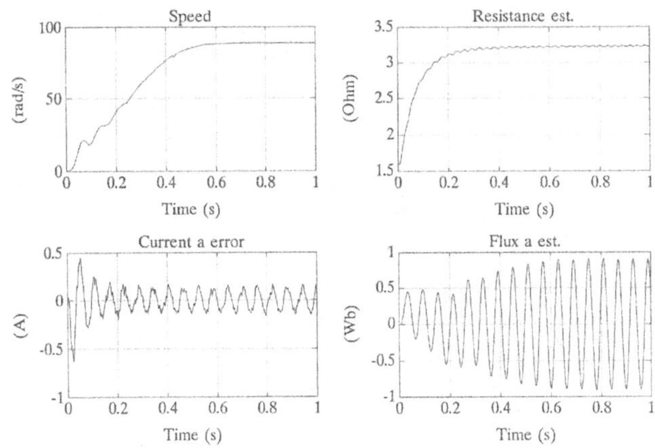

Fig. 3.13 Rotor resistance estimation with measured voltages (-50% initial error)

motor is controlled by an indirect field-oriented control algorithm with a reference speed of 1 rad/s and when the rated load torque is applied at $t = 1.2$ s: the parameters of the estimation algorithm reported in Problem 3.5 are set equal to $k = 60$, $\gamma_1 = 0.006$, $\gamma_2 = 0.1$, and $\gamma_3 = 200$. Note that a slower convergence is obtained when this new set of parameters is used. The experiment reported in Figures 3.17 and 3.18 illustrates the influence of the stator resistance variations on the estimation algorithm when the same set of parameters as in Figure 3.16 are used. A stator resistance variation causes the appearance of additional terms in the estimation algorithm which may be viewed as external periodic disturbances at power supply frequency. The error due to the stator resistance variations becomes greater as the ratio R_s/R_r increases. Note that R_s/R_r reduces when the temperature increases. When the stator

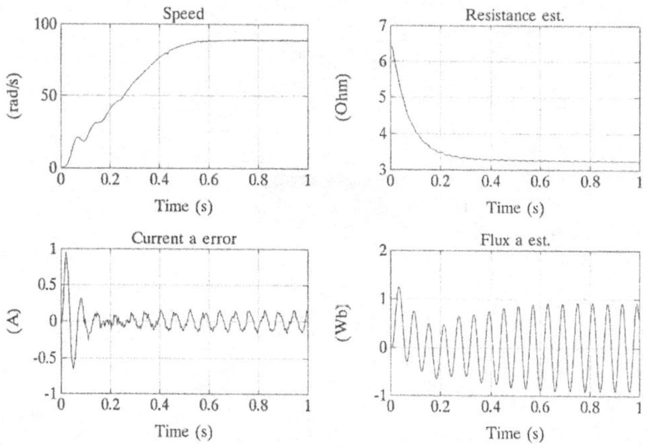

Fig. 3.14 Rotor resistance estimation with measured voltages (+100% initial error)

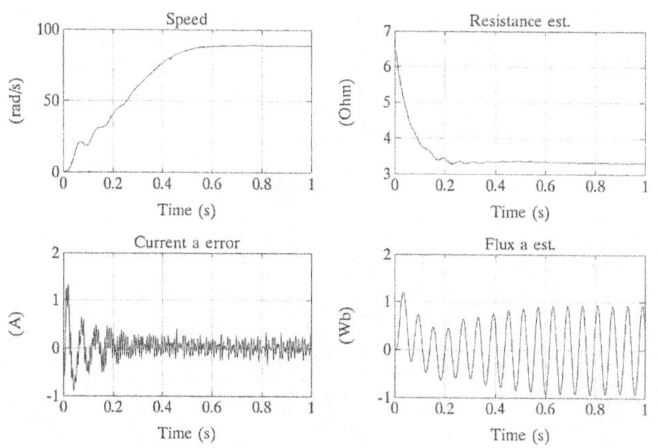

Fig. 3.15 Rotor resistance estimation with reference voltages (+100% initial error)

resistance in the algorithm is $1\,\Omega$ smaller than the true value, a rotor resistance error of about $0.3\,\Omega$ is noticed both at 1 rad/s (Figure 3.17) and at 100rad/s (Figure 3.18).

3.5 Conclusions

Since in Chapter 2 high-performance feedback controls were obtained under the unrealistic assumption that the rotor flux measurements were available, in this chapter the design of online rotor flux estimators has been fully addressed. Experimental results in Chapter 2 clearly indicate that the rotor resistance is a very critical parameter

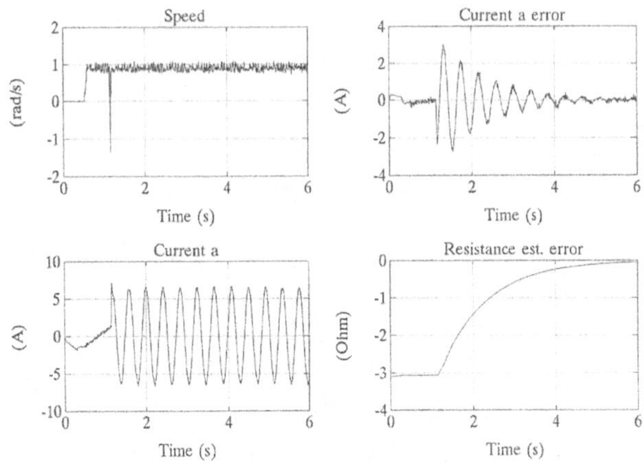

Fig. 3.16 Rotor resistance estimation at low speed ($\omega = 1$ rad/s)

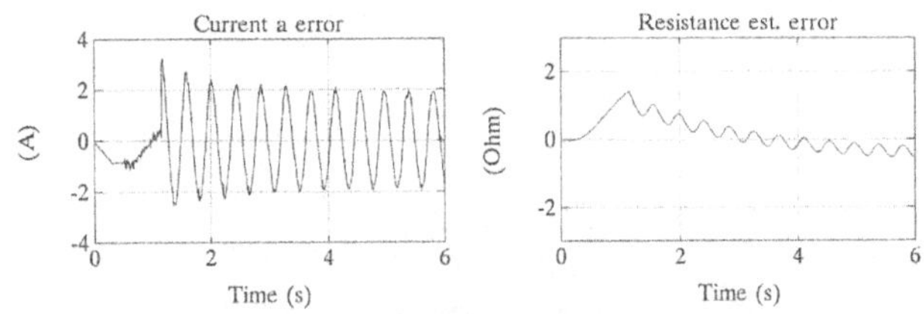

Fig. 3.17 Effect of stator resistance variation at low speed ($\omega = 1$ rad/s)

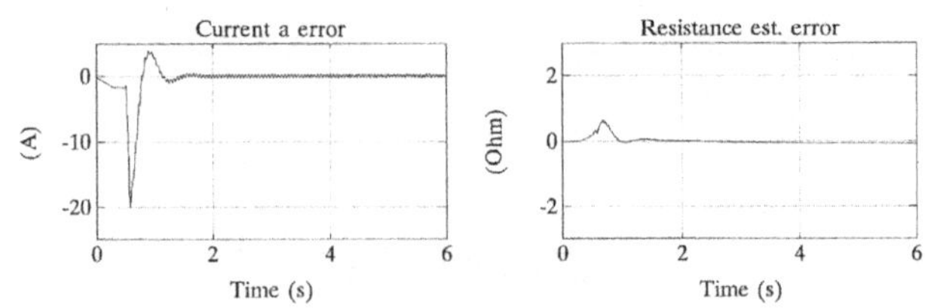

Fig. 3.18 Effect of stator resistance variation at high speed ($\omega = 100$ rad/s)

in the control of induction motors: hence, in this chapter its online estimation has been studied which is linked to rotor flux estimation. In Section 3.1 it is shown how to design globally exponentially converging rotor flux observers: observers with arbitrary exponential rate of convergence can also be obtained. Those observers are very sensitive with respect to rotor resistance, which may indeed vary more than 100% during operations. To overcome this difficult problem adaptive rotor flux observers are designed in Section 3.2 which provide globally converging estimates both of the rotor fluxes and of the rotor resistance during typical operating conditions, that is when the electromagnetic torque is different from zero. The estimation errors cannot converge to zero when the load torque is zero and both the rotor speed and the rotor flux modulus are constant. In Section 3.2 rotor flux observers which are adaptive with respect to both rotor and stator resistances are also presented. In Section 3.3 exponentially convergent load torque estimators are given which are to be used in conjunction with globally exponentially convergent rotor flux estimates. Finally, in Section 3.4 some experimental results are reported for the adaptive rotor flux observer illustrated in Section 3.2 (see also Problem 3.5): the rotor resistance is correctly estimated online both at high speed and at low speed with an initial error of 100%; some sensitivity with respect to stator resistance uncertainty is noted at low speed.

Problems

3.1. Consider the model (1.26) and make the change of variables

$$\begin{aligned} z_a &= i_{sa} + \beta \psi_{ra} \\ z_b &= i_{sb} + \beta \psi_{rb} \end{aligned} \tag{3.92}$$

so that we have the motor electromagnetic equations in the new variables:

$$\begin{aligned} \dot{z}_a &= -\frac{R_s}{\sigma} i_{sa} + \frac{1}{\sigma} u_{sa} \\ \dot{z}_b &= -\frac{R_s}{\sigma} i_{sb} + \frac{1}{\sigma} u_{sb} \\ \frac{\mathrm{d} i_{sa}}{\mathrm{d}t} &= -(\gamma+\alpha) i_{sa} - \omega i_{sb} + \alpha z_a + \omega z_b + \frac{1}{\sigma} u_{sa} \\ \frac{\mathrm{d} i_{sb}}{\mathrm{d}t} &= -(\gamma+\alpha) i_{sb} + \omega i_{sa} + \alpha z_b - \omega z_a + \frac{1}{\sigma} u_{sb} \, . \end{aligned} \tag{3.93}$$

Consider the full order rotor flux observer

$$\begin{aligned} \dot{\hat{z}}_a &= -\frac{R_s}{\sigma} \hat{i}_{sa} + \frac{1}{\sigma} u_{sa} - \frac{R_s}{\alpha\sigma} \omega (i_{sb} - \hat{i}_{sb}) \\ \dot{\hat{z}}_b &= -\frac{R_s}{\sigma} \hat{i}_{sb} + \frac{1}{\sigma} u_{sb} + \frac{R_s}{\alpha\sigma} \omega (i_{sa} - \hat{i}_{sa}) \\ \frac{\mathrm{d} \hat{i}_{sa}}{\mathrm{d}t} &= -(\gamma+\alpha) \hat{i}_{sa} - \omega \hat{i}_{sb} + \alpha \hat{z}_a + \omega \hat{z}_b + \frac{1}{\sigma} u_{sa} + k(i_{sa} - \hat{i}_{sa}) \\ \frac{\mathrm{d} \hat{i}_{sb}}{\mathrm{d}t} &= -(\gamma+\alpha) \hat{i}_{sb} + \omega \hat{i}_{sa} + \alpha \hat{z}_b - \omega \hat{z}_a + \frac{1}{\sigma} u_{sb} + k(i_{sb} - \hat{i}_{sb}) \\ \hat{\psi}_{ra} &= \frac{1}{\beta} (\hat{z}_a - \hat{i}_{sa}) \\ \hat{\psi}_{rb} &= \frac{1}{\beta} (\hat{z}_b - \hat{i}_{sb}) \, . \end{aligned} \tag{3.94}$$

Using the positive definite function

$$V = \frac{1}{2}\left(\tilde{i}_{sa}^2 + \tilde{i}_{sb}^2\right) + \frac{1}{2}\frac{\alpha\sigma}{R_s}\left(\tilde{z}_a^2 + \tilde{z}_b^2\right) \tag{3.95}$$

show that the estimation errors are bounded. Assuming that $\omega(t)$ and $\dot{\omega}(t)$ are bounded on $[0,\infty)$, establish that the estimation errors tend exponentially to zero for every initial condition using the Persistency of Excitation Lemma A.3 in Appendix A.

3.2. Using the notations

$$L = \begin{bmatrix} L_s & 0 & M & 0 \\ 0 & L_s & 0 & M \\ M & 0 & L_r & 0 \\ 0 & M & 0 & L_r \end{bmatrix}, \quad E_1 = \begin{bmatrix} 1 & 0 & 0 & 0 \\ 0 & 1 & 0 & 0 \\ 0 & 0 & 0 & 0 \\ 0 & 0 & 0 & 0 \end{bmatrix}$$

$$E_2 = \begin{bmatrix} 0 & 0 & 0 & 0 \\ 0 & 0 & 0 & 0 \\ 0 & 0 & 1 & 0 \\ 0 & 0 & 0 & 1 \end{bmatrix}, \quad E_3 = \begin{bmatrix} 0 & 0 & 0 & 0 \\ 0 & 0 & 0 & 0 \\ 0 & 0 & 0 & 1 \\ 0 & 0 & -1 & 0 \end{bmatrix}$$

$$E_4 = \begin{bmatrix} 0 & 0 & 0 & 0 \\ 0 & 0 & 0 & 0 \\ 0 & 1 & 0 & 0 \\ -1 & 0 & 0 & 0 \end{bmatrix}$$

$$i = \begin{bmatrix} i_{sa} \\ i_{sb} \\ i_{ra} \\ i_{rb} \end{bmatrix}, \quad u = \begin{bmatrix} u_{sa} \\ u_{sb} \\ 0 \\ 0 \end{bmatrix}, \quad \hat{i} = \begin{bmatrix} \hat{i}_{sa} \\ \hat{i}_{sb} \\ \hat{i}_{ra} \\ \hat{i}_{rb} \end{bmatrix} \tag{3.96}$$

show that the rotor current observer

$$L\frac{\mathrm{d}\hat{i}}{\mathrm{d}t} + R_s E_1 i + R_r E_2 \hat{i} + L_r \omega E_3 \hat{i} + M\omega E_4 i + k_1 E_1 (i - \hat{i}) = u \tag{3.97}$$

in which k_1 is a positive design parameter, is such that the current estimation error $\tilde{i}(t) = i(t) - \hat{i}(t)$ converges exponentially to zero for any initial condition $\hat{i}(0)$. Analyze the influence of the uncertainties on the rotor and stator resistances R_r and R_s. *Suggestion: use the positive definite function* $V = \frac{1}{2}\tilde{i}^{\mathrm{T}} L\tilde{i}$.

3.3. Show that the observer (3.97) can be made adaptive with respect to R_s by replacing R_s with $\hat{R}_s$ in (3.97), with the adaptation dynamics (k is a positive design parameter)

$$\dot{\hat{R}}_s = -k i^{\mathrm{T}} E_1 \tilde{i}\,.$$

Analyze the influence of the uncertainties on the rotor resistance R_r. *Suggestion: use the positive definite function* $V = \frac{1}{2}\tilde{i}^{\mathrm{T}} L\tilde{i} + \frac{1}{2k}(R_s - \hat{R}_s)^2$ *and the Persistency of Excitation Lemma A.3 in Appendix A to show the convergence, provided that* $i(t)$,

$\mathrm{d}i(t)/\mathrm{d}t$, *and* $\omega(t)$ *are bounded on* $[0,\infty)$ *and two positive reals* T *and* k_T *exist such that, for any* $t \geq 0$, $\int_t^{t+T} i^{\mathrm{T}}(\tau)i(\tau)\mathrm{d}\tau \geq k_T I$.

3.4. Show that the third order stator resistance estimator (k is a positive design parameter)

$$\begin{aligned}
\dot{\hat{\psi}}_{ra} &= -\alpha\hat{\psi}_{ra} - \omega\hat{\psi}_{rb} + \alpha M i_{sa} \\
\dot{\hat{\psi}}_{rb} &= -\alpha\hat{\psi}_{rb} + \omega\hat{\psi}_{ra} + \alpha M i_{sb} \\
\dot{\xi} &= -k\frac{\hat{R}_s}{\sigma}(i_{sa}^2 + i_{sb}^2) - k\alpha\beta M(i_{sa}^2 + i_{sb}^2) + k\beta\omega(\hat{\psi}_{rb}i_{sa} - \hat{\psi}_{ra}i_{sb}) \\
&\quad + \frac{k}{\sigma}(u_{sa}i_{sa} + u_{sb}i_{sb}) + k\alpha\beta(\hat{\psi}_{ra}i_{sa} + \hat{\psi}_{rb}i_{sb}) \\
\hat{R}_s &= \xi - \frac{k}{2}(i_{sa}^2 + i_{sb}^2)
\end{aligned}$$

is such that for any initial condition $(\hat{\psi}_{ra}(0), \hat{\psi}_{rb}(0), \xi(0))$ the estimation error $[R_s - \hat{R}_s(t)]$ exponentially converges to zero for any $t \geq 0$, provided that: $i_{sa}^2(t) + i_{sb}^2(t) \geq c > 0$ for any $t \geq 0$; $\omega(t)$, $i_{sa}(t)$ and $i_{sb}(t)$ are bounded on $[0,\infty)$. *Suggestion: compute the dynamics of the estimation error* $R_s - \hat{R}_s$ *and use Lemma A.1 in Appendix A.*

3.5. Consider the fixed frame model (1.26) with

$$\begin{aligned}
\gamma &= \frac{R_s}{\sigma} + \beta(R_{rN} + \theta)M \\
\alpha &= \frac{R_{rN} + \theta}{L_r}
\end{aligned}$$

in which R_{rN} is the nominal value for the rotor resistance and θ is the rotor resistance uncertainty (*i.e.* $R_r = R_{rN} + \theta$). Consider the adaptive observer

$$\begin{aligned}
\frac{\mathrm{d}\hat{\psi}_{ra}}{\mathrm{d}t} &= -\frac{R_{rN} + \hat{\theta}}{L_r}\hat{\psi}_{ra} - \omega\hat{\psi}_{rb} + (R_{rN} + \hat{\theta})\frac{M}{L_r}i_{sa} \\
\frac{\mathrm{d}\hat{\psi}_{rb}}{\mathrm{d}t} &= -\frac{R_{rN} + \hat{\theta}}{L_r}\hat{\psi}_{rb} + \omega\hat{\psi}_{ra} + (R_{rN} + \hat{\theta})\frac{M}{L_r}i_{sb} \\
\frac{\mathrm{d}\hat{i}_{sa}}{\mathrm{d}t} &= -\frac{R_s}{\sigma}i_{sa} + \frac{1}{\sigma}u_{sa} - \beta\dot{\hat{\psi}}_{ra} - \dot{\xi}_a \\
\frac{\mathrm{d}\hat{i}_{sb}}{\mathrm{d}t} &= -\frac{R_s}{\sigma}i_{sb} + \frac{1}{\sigma}u_{sb} - \beta\dot{\hat{\psi}}_{rb} - \dot{\xi}_b \\
\frac{\mathrm{d}\xi_a}{\mathrm{d}t} &= -k\tilde{i}_{sa} - \frac{R_{rN} + \hat{\theta}}{L_r}(\xi_a - \tilde{i}_{sa}) - \omega\xi_b - \hat{C} - \omega\hat{B} \\
\frac{\mathrm{d}\xi_b}{\mathrm{d}t} &= -k\tilde{i}_{sb} + \omega\xi_a - \frac{R_{rN} + \hat{\theta}}{L_r}(\xi_b - \tilde{i}_{sb}) + \omega\hat{A} - \hat{D} \\
\frac{\mathrm{d}\hat{\theta}}{\mathrm{d}t} &= \gamma_1\left[\tilde{i}_{sa}\left(\frac{\beta}{L_r}\hat{\psi}_{ra} - \frac{\beta}{L_r}Mi_{sa} + \frac{(\xi_a - \tilde{i}_{sa})}{L_r}\right)\right.
\end{aligned}$$

$$
\begin{aligned}
& \left. + \tilde{i}_{sb}\left(\frac{\beta}{L_r}\hat{\psi}_{rb} - \frac{\beta}{L_r}Mi_{sb} + \frac{(\xi_b - \hat{i}_{sb})}{L_r}\right)\right] \\
\frac{\mathrm{d}\hat{A}}{\mathrm{d}t} &= -\gamma_2\omega\tilde{i}_{sb} \\
\frac{\mathrm{d}\hat{B}}{\mathrm{d}t} &= \gamma_2\omega\tilde{i}_{sa} \\
\frac{\mathrm{d}\hat{C}}{\mathrm{d}t} &= \gamma_3\tilde{i}_{sa} \\
\frac{\mathrm{d}\hat{D}}{\mathrm{d}t} &= \gamma_3\tilde{i}_{sb}
\end{aligned} \tag{3.98}
$$

in which $\tilde{i}_{sa} = i_{sa} - \hat{i}_{sa}$, $\tilde{i}_{sb} = i_{sb} - \hat{i}_{sb}$, $(\hat{\psi}_{ra}, \hat{\psi}_{rb})$ are estimates of the rotor fluxes (ψ_{ra}, ψ_{rb}), $(\hat{i}_{sa}, \hat{i}_{sb})$ are estimates of the stator currents (i_{sa}, i_{sb}), and $\hat{\theta}$ is the estimate of rotor resistance uncertainty θ. Choose the positive design parameters $(k, \gamma_1, \gamma_2, \gamma_3)$, simulate the adaptive observer (3.98), and compare with the experiments given in Section 3.4.

3.6. Consider the load torque estimator (3.84) and assume that the rotor fluxes (ψ_{ra}, ψ_{rb}) are available from measurements: modify the estimator (3.84) so that global exponential convergence to zero with arbitrary rate of convergence of the estimation errors $\tilde{\omega}$ and $\tilde{T}_L$ is obtained.

3.7. Consider the load torque estimator (3.84): if the rotor flux estimation errors $\tilde{\psi}_{ra}$ and $\tilde{\psi}_{rb}$ converge exponentially to zero with arbitrary rate of convergence, can we also conclude that the estimation errors $\tilde{\omega}$ and $\tilde{T}_L$ converge exponentially to zero with arbitrary rate of convergence? *Suggestion: use Lemma A.4 in Appendix A.*

3.8. Assume that the rotor speed and the rotor flux modulus are constant and that the load torque is zero, so that $\psi_{ra} = Mi_{sa}$ and $\psi_{rb} = Mi_{sb}$; show that the stator resistance estimator

$$
\begin{aligned}
\dot{\xi} &= -\frac{k\hat{R}_s}{\sigma}(i_{sa}^2 + i_{sb}^2) + \frac{k}{\sigma}(u_{sa}i_{sa} + u_{sb}i_{sb}) \\
\hat{R}_s &= \xi - \frac{k}{2}(i_{sa}^2 + i_{sb}^2)
\end{aligned}
$$

guarantees that the stator resistance estimation error $[R_s - \hat{R}_s(t)]$ converges exponentially to zero, provided that $i_{sa}^2(t) + i_{sb}^2(t) \geq c > 0$ for any $t \geq 0$.

3.9. Assume that the rotor speed and the rotor flux modulus are constant and the load torque is zero so that

$$
\begin{aligned}
\psi_{ra} &= Mi_{sa} \\
\psi_{rb} &= Mi_{sb}\,.
\end{aligned}
$$

Show that asymptotic flux estimation is guaranteed by the adaptive observer (3.58) even when the rotor resistance R_r is not correctly estimated. *Suggestion: by using*

Barbalat's Lemma A.2 in Appendix A show that

$$\lim_{t\to\infty}\frac{\mathrm{d}\tilde{i}_{sa}(t)}{\mathrm{d}t}=0$$
$$\lim_{t\to\infty}\frac{\mathrm{d}\tilde{i}_{sb}(t)}{\mathrm{d}t}=0\,.$$

3.10. Find a reparameterization $(\theta_1,\,\theta_2,\,\theta_3,\,\theta_4,\,\theta_5,\,\theta_6)$ for the motor electrical parameters, a matrix $A(\cdot)$ and a vector B such that the induction motor dynamics admits the affine representation

$$\dot{X}=A(u_{sa},u_{sb},i_{sa},i_{sb},\omega)X+B\begin{bmatrix}u_{sa}\\u_{sb}\end{bmatrix}$$
$$Y=\begin{bmatrix}i_{sa}\\i_{sb}\\\omega\end{bmatrix}\triangleq CX$$

with $X=\left[i_{sa},\,i_{sb},\,\psi_{sa},\,\psi_{sb},\,\theta_5\psi_{sa},\,\theta_5\psi_{sb},\,\theta_4\psi_{sa},\,\theta_4\psi_{sb},\,\omega,\,\theta_1,\,\theta_2,\,\theta_3,\,\theta_4,\,\theta_5,\,\theta_6,\,T_L\right]^{\mathrm{T}}$. Consider the observer for stator fluxes, load torque, and induction motor electrical parameters via stator currents, stator voltages, and rotor speed measurements (the rotor inertia is assumed to be known)

$$\begin{aligned}\dot{\hat{X}}&=A(u_{sa},u_{sb},i_{sa},i_{sb},\omega)\hat{X}+B\begin{bmatrix}u_{sa}\\u_{sb}\end{bmatrix}-S^{-1}C^{\mathrm{T}}(C\hat{X}-Y)\\ \dot{S}&=-\lambda S-A^{\mathrm{T}}(u_{sa},u_{sb},i_{sa},i_{sb},\omega)S-SA(u_{sa},u_{sb},i_{sa},i_{sb},\omega)+C^{\mathrm{T}}C\end{aligned}\quad(3.99)$$

with $S(0)>0$. Choose the design parameter λ and simulate the observer (3.99) for sufficiently exciting input and output signals.

3.11. Consider the fixed frame model (1.26) and assume that the rotor speed is constant so that $\dot{\omega}=0$ replaces the first equation in (1.26). Design an observer for rotor flux and rotor speed from stator current and voltage measurements. Analyze the persistency of excitation conditions and compare with the results in Section 1.5.

3.12. Consider the fixed frame model (1.26) with constant rotor speed and unknown rotor resistance R_r. Design an adaptive observer which estimates the rotor speed and flux and identifies the rotor resistance from stator current and voltage measurements. Analyze the persistency of excitation conditions and compare with the results in Section 1.6.

Chapter 4
Output Feedback Control

Abstract In this chapter the state feedback controls presented in Chapter 2 and, in particular, the control given in Section 2.7 are combined with the rotor flux observers presented in Chapter 3: the goal is to obtain global output feedback controls which do not require flux measurements and guarantee rotor speed tracking for any initial condition of the motor. In Section 4.1 the global control with arbitrary rate of convergence which was presented in Section 2.7 is modified to eliminate the need of rotor flux measurements, at the expense of the property of arbitrary exponential rate of convergence. In order to recover this important property, in Section 4.2 the global state feedback control with arbitrary rate of convergence, discussed in Section 2.7, is modified so that the rotor fluxes can be replaced by the estimates provided by the rotor flux observer with arbitrary rate of convergence given in Section 3.1: the resulting observer-based global controller guarantees exponential convergence with arbitrary rate of both the tracking and the estimation errors. In Section 4.3 the control algorithm presented in Section 4.2 is made adaptive with respect to an uncertain load torque by incorporating the load torque estimator presented in Section 3.3. Finally, in Section 4.4 a global output feedback control algorithm is presented which is adaptive with respect to both the unknown load torque and the uncertain rotor resistance and achieves asymptotic rotor speed tracking. Under persistency of excitation conditions, exponentially converging estimates of the unmeasured rotor fluxes and of the uncertain parameters are obtained, while exponential tracking of rotor speed and flux modulus is achieved from any motor initial condition.

4.1 Generalized Indirect Field-oriented Control

In this section we redesign the global state feedback control with arbitrary rate of convergence presented in Section 2.7 without using the rotor flux measurements (ψ_{ra}, ψ_{rb}), which are typically not available from measurements. We shall obtain a generalization of the indirect field-oriented control presented in Section 2.3 whose rate of convergence, however, cannot be larger than $\alpha = R_r/L_r$.

Reconsider the control algorithm (2.48) and modify the reference for the stator current vector d-component and the speed of the rotating (d,q) frame as

$$
\begin{aligned}
i_{sd}^* &= \frac{\psi^*}{M} + \frac{\dot{\psi}^*}{\alpha M} + \frac{\eta_d}{\alpha M} \\
\omega_0 &= \omega + \frac{\alpha M i_{sq}}{\psi^*} - \frac{\eta_q}{\psi^*}
\end{aligned}
\tag{4.1}
$$

i.e. by adding the terms η_d and η_q which will be suitably designed in the following. The reference for the stator current vector q-component is the same as in (2.48), *i.e.*

$$
i_{sq}^* = \frac{1}{\mu \psi^*}\left[-k_\omega(\omega - \omega^*) + \dot{\omega}^* + \frac{T_L}{J}\right] . \tag{4.2}
$$

Introduce the tracking errors

$$
\begin{aligned}
\tilde{\omega} &= \omega - \omega^* \\
\tilde{\psi}_{rd} &= \psi_{rd} - \psi^* \\
\tilde{\psi}_{rq} &= \psi_{rq} \\
\tilde{i}_{sd} &= i_{sd} - i_{sd}^* \\
\tilde{i}_{sq} &= i_{sq} - i_{sq}^*
\end{aligned}
$$

for the rotor speed, rotor flux vector (d,q)-components and stator current vector (d,q)-components, respectively. The dynamics for the tracking errors $\tilde{\omega}$, $\tilde{\psi}_{rd}$ and $\tilde{\psi}_{rq}$ are given by

$$
\begin{aligned}
\dot{\tilde{\omega}} &= -k_\omega \tilde{\omega} + \mu(\tilde{\psi}_{rd} i_{sq} - \tilde{\psi}_{rq} i_{sd}) + \mu \psi^* \tilde{i}_{sq} \\
\dot{\tilde{\psi}}_{rd} &= -\alpha \tilde{\psi}_{rd} + (\omega_0 - \omega)\tilde{\psi}_{rq} + \alpha M \tilde{i}_{sd} + \eta_d \\
\dot{\tilde{\psi}}_{rq} &= -\alpha \tilde{\psi}_{rq} - (\omega_0 - \omega)\tilde{\psi}_{rd} + \eta_q .
\end{aligned}
\tag{4.3}
$$

In order to design the additional terms η_d and η_q, introduce the positive control parameter λ and consider the positive definite quadratic function

$$
W = \frac{1}{2}\left(\lambda \tilde{\omega}^2 + \tilde{\psi}_{rd}^2 + \tilde{\psi}_{rq}^2\right) \tag{4.4}
$$

whose time derivative along the trajectories of the closed-loop system is

$$
\begin{aligned}
\dot{W} = &-\lambda k_\omega \tilde{\omega}^2 + \lambda \mu(\tilde{\psi}_{rd} i_{sq} - \tilde{\psi}_{rq} i_{sd})\tilde{\omega} + \lambda \mu \psi^* \tilde{i}_{sq} \tilde{\omega} \\
&-\alpha(\tilde{\psi}_{rd}^2 + \tilde{\psi}_{rq}^2) + \alpha M \tilde{i}_{sd} \tilde{\psi}_{rd} + \eta_d \tilde{\psi}_{rd} + \eta_q \tilde{\psi}_{rq} .
\end{aligned}
\tag{4.5}
$$

Since $i_{sq} = \tilde{i}_{sq} + i_{sq}^*$, we define

$$
\begin{aligned}
\eta_d &= -\lambda \mu i_{sq}^* \tilde{\omega} \\
\eta_q &= \lambda \mu i_{sd} \tilde{\omega}
\end{aligned}
\tag{4.6}
$$

so that (4.5) becomes

$$\begin{aligned}\dot{W} &= -\lambda k_\omega \tilde{\omega}^2 - \alpha(\tilde{\psi}_{rd}^2 + \tilde{\psi}_{rq}^2) + \lambda\mu\tilde{\psi}_{rd}\tilde{i}_{sq}\tilde{\omega} + \lambda\mu\psi^*\tilde{i}_{sq}\tilde{\omega} \\ &\quad +\alpha M\tilde{i}_{sd}\tilde{\psi}_{rd}\,. \end{aligned} \tag{4.7}$$

Since the rotor fluxes are not available, the feedback terms $-k_\psi(\psi_{rd} - \psi^*)$ and $-k_\psi\psi_{rq}$ which were present in (2.48) disappear. As a consequence, we are losing the property of arbitrary rate of convergence. The influence of the last three terms in (4.7) will be compensated by a suitable choice of the stator voltages (u_{sd}, u_{sq}). To this end, let us compute

$$\frac{\mathrm{d}i_{sq}^*}{\mathrm{d}t} = \Gamma_{q1}\tilde{\psi}_{rd} + \Gamma_{q2}\tilde{\psi}_{rq} + \Gamma_q \tag{4.8}$$

in which the known functions

$$\begin{aligned}\Gamma_{q1} &= -\frac{k_\omega i_{sq}}{\psi^*} \\ \Gamma_{q2} &= \frac{k_\omega i_{sd}}{\psi^*} \\ \Gamma_q &= \frac{1}{\mu\psi^*}\left[k_\omega^2\tilde{\omega} - k_\omega\mu\psi^*\tilde{i}_{sq} + \ddot{\omega}^*\right] - \frac{\dot{\psi}^*}{\mu\psi^{*2}}\left[-k_\omega\tilde{\omega} + \frac{T_L}{J} + \dot{\omega}^*\right]\end{aligned} \tag{4.9}$$

appear. Similarly, let us compute

$$\frac{\mathrm{d}i_{sd}^*}{\mathrm{d}t} = \Gamma_{d1}\tilde{\psi}_{rd} + \Gamma_{d2}\tilde{\psi}_{rq} + \Gamma_d \tag{4.10}$$

in which the known functions

$$\begin{aligned}\Gamma_{d1} &= -\frac{\lambda\mu}{\alpha M}\left[\mu i_{sq} i_{sq}^* + \tilde{\omega}\Gamma_{q1}\right] \\ \Gamma_{d2} &= \frac{\lambda\mu}{\alpha M}\left[\mu i_{sd} i_{sq}^* - \tilde{\omega}\Gamma_{q2}\right] \\ \Gamma_d &= \frac{\dot{\psi}^*}{M} + \frac{\ddot{\psi}^*}{\alpha M} + \frac{\lambda\mu k_\omega\tilde{\omega} i_{sq}^*}{\alpha M} - \frac{\lambda\mu^2\psi^*\tilde{i}_{sq} i_{sq}^*}{\alpha M} - \frac{\lambda\mu\Gamma_q\tilde{\omega}}{\alpha M}\end{aligned} \tag{4.11}$$

appear. According to (4.8) and (4.10) the dynamics for the stator current tracking errors $\tilde{i}_{sd}$ and $\tilde{i}_{sq}$ are

$$\begin{aligned}\frac{\mathrm{d}\tilde{i}_{sd}}{\mathrm{d}t} &= -\gamma i_{sd} + \omega_0 i_{sq} + \alpha\beta\psi_{rd} + \beta\omega\psi_{rq} + \frac{1}{\sigma}u_{sd} - \Gamma_{d1}\tilde{\psi}_{rd} \\ &\quad -\Gamma_{d2}\tilde{\psi}_{rq} - \Gamma_d \\ \frac{\mathrm{d}\tilde{i}_{sq}}{\mathrm{d}t} &= -\gamma i_{sq} - \omega_0 i_{sd} + \alpha\beta\psi_{rq} - \beta\omega\psi_{rd} + \frac{1}{\sigma}u_{sq} - \Gamma_{q1}\tilde{\psi}_{rd} \\ &\quad -\Gamma_{q2}\tilde{\psi}_{rq} - \Gamma_q\,. \end{aligned} \tag{4.12}$$

Design the control inputs (u_{sd}, u_{sq}) as

$$\begin{aligned} u_{sd} &= \sigma\gamma i^*_{sd} - \sigma\omega_0 i_{sq} - \sigma\alpha\beta\psi^* + \sigma\Gamma_d - \sigma k_i \tilde{i}_{sd} + \sigma v_d \\ u_{sq} &= \sigma\gamma i^*_{sq} + \sigma\omega_0 i_{sd} + \sigma\beta\omega\psi^* + \sigma\Gamma_q - \sigma k_i \tilde{i}_{sq} + \sigma v_q \end{aligned} \tag{4.13}$$

in which the additional terms v_d and v_q are yet to be designed and k_i is a positive control parameter, so that (4.12) become

$$\begin{aligned} \frac{d\tilde{i}_{sd}}{dt} &= -(\gamma + k_i)\tilde{i}_{sd} + \alpha\beta\tilde{\psi}_{rd} + \beta\omega\tilde{\psi}_{rq} - \Gamma_{d1}\tilde{\psi}_{rd} - \Gamma_{d2}\tilde{\psi}_{rq} + v_d \\ \frac{d\tilde{i}_{sq}}{dt} &= -(\gamma + k_i)\tilde{i}_{sq} + \alpha\beta\tilde{\psi}_{rq} - \beta\omega\tilde{\psi}_{rd} - \Gamma_{q1}\tilde{\psi}_{rd} - \Gamma_{q2}\tilde{\psi}_{rq} + v_q\,. \end{aligned} \tag{4.14}$$

In order to design the yet undefined terms v_d and v_q in (4.13), consider the positive definite function

$$V = W + \frac{1}{2}\left(\tilde{i}^2_{sd} + \tilde{i}^2_{sq}\right) \tag{4.15}$$

whose time derivative along the trajectories of the closed-loop system, according to (4.7), satisfies

$$\begin{aligned} \dot{V} &= -\lambda k_\omega \tilde{\omega}^2 + \lambda\mu\tilde{\psi}_{rd}\tilde{i}_{sq}\tilde{\omega} + \lambda\mu\psi^*\tilde{i}_{sq}\tilde{\omega} - \alpha(\tilde{\psi}^2_{rd} + \tilde{\psi}^2_{rq}) + \alpha M\tilde{i}_{sd}\tilde{\psi}_{rd} \\ &\quad -(\gamma + k_i)(\tilde{i}^2_{sd} + \tilde{i}^2_{sq}) + \alpha\beta\tilde{\psi}_{rd}\tilde{i}_{sd} + \beta\omega\tilde{\psi}_{rq}\tilde{i}_{sd} - \Gamma_{d1}\tilde{\psi}_{rd}\tilde{i}_{sd} - \Gamma_{d2}\tilde{\psi}_{rq}\tilde{i}_{sd} \\ &\quad + v_d\tilde{i}_{sd} + \alpha\beta\tilde{\psi}_{rq}\tilde{i}_{sq} - \beta\omega\tilde{\psi}_{rd}\tilde{i}_{sq} - \Gamma_{q1}\tilde{\psi}_{rd}\tilde{i}_{sq} - \Gamma_{q2}\tilde{\psi}_{rq}\tilde{i}_{sq} + v_q\tilde{i}_{sq} \\ &\triangleq -\lambda k_\omega \tilde{\omega}^2 + \lambda\mu\psi^*\tilde{i}_{sq}\tilde{\omega} - \alpha(\tilde{\psi}^2_{rd} + \tilde{\psi}^2_{rq}) - (\gamma + k_i)(\tilde{i}^2_{sd} + \tilde{i}^2_{sq}) + v_d\tilde{i}_{sd} \\ &\quad + v_q\tilde{i}_{sq} + \Delta_1\tilde{i}_{sd}\tilde{\psi}_{rd} + \Delta_2\tilde{i}_{sd}\tilde{\psi}_{rq} + \Delta_3\tilde{i}_{sq}\tilde{\psi}_{rd} + \Delta_4\tilde{i}_{sq}\tilde{\psi}_{rq} \end{aligned} \tag{4.16}$$

in which the known functions

$$\begin{aligned} \Delta_1 &= \alpha M + \alpha\beta - \Gamma_{d1} \\ \Delta_2 &= \beta\omega - \Gamma_{d2} \\ \Delta_3 &= \lambda\mu\tilde{\omega} - \beta\omega - \Gamma_{q1} \\ \Delta_4 &= \alpha\beta - \Gamma_{q2} \end{aligned} \tag{4.17}$$

appear. Since

$$\begin{aligned} \left(\frac{\sqrt{\alpha}}{2}\tilde{\psi}_{rd} - \frac{\Delta_1}{\sqrt{\alpha}}\tilde{i}_{sd}\right)^2 &\geq 0 \\ \left(\frac{\sqrt{\alpha}}{2}\tilde{\psi}_{rq} - \frac{\Delta_2}{\sqrt{\alpha}}\tilde{i}_{sd}\right)^2 &\geq 0 \\ \left(\frac{\sqrt{\alpha}}{2}\tilde{\psi}_{rd} - \frac{\Delta_3}{\sqrt{\alpha}}\tilde{i}_{sq}\right)^2 &\geq 0 \end{aligned}$$

$$\left(\frac{\sqrt{\alpha}}{2}\tilde{\psi}_{rq} - \frac{\Delta_4}{\sqrt{\alpha}}\tilde{i}_{sq}\right)^2 \geq 0 \tag{4.18}$$

the following inequalities hold:

$$\begin{aligned}
\Delta_1 \tilde{i}_{sd}\tilde{\psi}_{rd} &\leq \frac{\alpha}{4}\tilde{\psi}_{rd}^2 + \frac{\Delta_1^2}{\alpha}\tilde{i}_{sd}^2 \\
\Delta_2 \tilde{i}_{sd}\tilde{\psi}_{rq} &\leq \frac{\alpha}{4}\tilde{\psi}_{rq}^2 + \frac{\Delta_2^2}{\alpha}\tilde{i}_{sd}^2 \\
\Delta_3 \tilde{i}_{sq}\tilde{\psi}_{rd} &\leq \frac{\alpha}{4}\tilde{\psi}_{rd}^2 + \frac{\Delta_3^2}{\alpha}\tilde{i}_{sq}^2 \\
\Delta_4 \tilde{i}_{sq}\tilde{\psi}_{rq} &\leq \frac{\alpha}{4}\tilde{\psi}_{rq}^2 + \frac{\Delta_4^2}{\alpha}\tilde{i}_{sq}^2
\end{aligned} \tag{4.19}$$

and therefore from (4.16) we have

$$\begin{aligned}
\dot{V} \leq &-\lambda k_\omega \tilde{\omega}^2 + \lambda\mu\psi^*\tilde{i}_{sq}\tilde{\omega} - \frac{\alpha}{2}(\tilde{\psi}_{rd}^2 + \tilde{\psi}_{rq}^2) - (\gamma + k_i)(\tilde{i}_{sd}^2 + \tilde{i}_{sq}^2) + v_d\tilde{i}_{sd} + v_q\tilde{i}_{sq} \\
&+ \left(\frac{\Delta_1^2}{\alpha} + \frac{\Delta_2^2}{\alpha}\right)\tilde{i}_{sd}^2 + \left(\frac{\Delta_3^2}{\alpha} + \frac{\Delta_4^2}{\alpha}\right)\tilde{i}_{sq}^2 .
\end{aligned} \tag{4.20}$$

If we design the yet undefined terms v_d and v_q as

$$\begin{aligned}
v_d &= -\left(\frac{\Delta_1^2}{\alpha} + \frac{\Delta_2^2}{\alpha}\right)\tilde{i}_{sd} \\
v_q &= -\left(\frac{\Delta_3^2}{\alpha} + \frac{\Delta_4^2}{\alpha}\right)\tilde{i}_{sq} - \lambda\mu\psi^*\tilde{\omega}
\end{aligned} \tag{4.21}$$

from (4.20) we obtain

$$\dot{V} \leq -\lambda k_\omega \tilde{\omega}^2 - \frac{\alpha}{2}(\tilde{\psi}_{rd}^2 + \tilde{\psi}_{rq}^2) - (\gamma + k_i)(\tilde{i}_{sd}^2 + \tilde{i}_{sq}^2) . \tag{4.22}$$

Therefore we have

$$\dot{V} \leq -\min\{2k_\omega, \alpha, 2(\gamma + k_i)\}V \triangleq -cV . \tag{4.23}$$

By integrating (4.23), we have

$$V(t) \leq \mathrm{e}^{-ct}V(0) . \tag{4.24}$$

Thus the origin of the closed-loop error system is exponentially stable with rate of decay

$$c = \min\{2k_\omega, \alpha, 2(\gamma + k_i)\} \tag{4.25}$$

which, however, cannot be made larger than α and it is, therefore, not arbitrary.

In conclusion, the *generalized indirect field-oriented control*

$$
\begin{aligned}
\begin{bmatrix} u_{sa} \\ u_{sb} \end{bmatrix} &= \begin{bmatrix} \cos\varepsilon_0 & -\sin\varepsilon_0 \\ \sin\varepsilon_0 & \cos\varepsilon_0 \end{bmatrix} \begin{bmatrix} u_{sd} \\ u_{sq} \end{bmatrix} \\
u_{sd} &= \sigma\gamma i_{sd}^* - \sigma\omega_0 i_{sq} - \sigma\alpha\beta\psi^* + \sigma\Gamma_d - \sigma k_i(i_{sd} - i_{sd}^*) + \sigma v_d \\
u_{sq} &= \sigma\gamma i_{sq}^* + \sigma\omega_0 i_{sd} + \sigma\beta\omega\psi^* + \sigma\Gamma_q - \sigma k_i(i_{sq} - i_{sq}^*) + \sigma v_q \\
i_{sd}^* &= \frac{\psi^*}{M} + \frac{\dot{\psi}^*}{\alpha M} + \frac{\eta_d}{\alpha M} \\
i_{sq}^* &= \frac{1}{\mu\psi^*}\left[-k_\omega(\omega - \omega^*) + \dot{\omega}^* + \frac{T_L}{J}\right] \\
\dot{\varepsilon}_0 &= \omega_0 = \omega + \frac{\alpha M i_{sq}}{\psi^*} - \frac{\eta_q}{\psi^*} \\
\eta_d &= -\lambda\mu i_{sq}^*(\omega - \omega^*) \\
\eta_q &= \lambda\mu i_{sd}(\omega - \omega^*) \\
v_d &= -\left(\frac{\Delta_1^2}{\alpha} + \frac{\Delta_2^2}{\alpha}\right)(i_{sd} - i_{sd}^*) \\
v_q &= -\left(\frac{\Delta_3^2}{\alpha} + \frac{\Delta_4^2}{\alpha}\right)(i_{sq} - i_{sq}^*) - \lambda\mu\psi^*(\omega - \omega^*) \\
\begin{bmatrix} i_{sd} \\ i_{sq} \end{bmatrix} &= \begin{bmatrix} \cos\varepsilon_0 & \sin\varepsilon_0 \\ -\sin\varepsilon_0 & \cos\varepsilon_0 \end{bmatrix} \begin{bmatrix} i_{sa} \\ i_{sb} \end{bmatrix} \\
\Gamma_{d1} &= -\frac{\lambda\mu}{\alpha M}\left[\mu i_{sq} i_{sq}^* + (\omega - \omega^*)\Gamma_{q1}\right] \\
\Gamma_{d2} &= \frac{\lambda\mu}{\alpha M}\left[\mu i_{sd} i_{sq}^* - (\omega - \omega^*)\Gamma_{q2}\right] \\
\Gamma_d &= \frac{\dot{\psi}^*}{M} + \frac{\ddot{\psi}^*}{\alpha M} + \frac{\lambda\mu k_\omega(\omega - \omega^*) i_{sq}^*}{\alpha M} - \frac{\lambda\mu^2\psi^*(i_{sq} - i_{sq}^*) i_{sq}^*}{\alpha M} \\
&\quad - \frac{\lambda\mu\Gamma_q(\omega - \omega^*)}{\alpha M} \\
\Gamma_{q1} &= -\frac{k_\omega i_{sq}}{\psi^*} \\
\Gamma_{q2} &= \frac{k_\omega i_{sd}}{\psi^*} \\
\Gamma_q &= \frac{1}{\mu\psi^*}\left[k_\omega^2(\omega - \omega^*) - k_\omega\mu\psi^*(i_{sq} - i_{sq}^*) + \ddot{\omega}^*\right] \\
&\quad - \frac{\dot{\psi}^*}{\mu\psi^{*2}}\left[-k_\omega(\omega - \omega^*) + \frac{T_L}{J} + \dot{\omega}^*\right] \\
\Delta_1 &= \alpha M + \alpha\beta - \Gamma_{d1} \\
\Delta_2 &= \beta\omega - \Gamma_{d2}
\end{aligned}
$$

$$\Delta_3 = \lambda\mu(\omega-\omega^*)-\beta\omega-\Gamma_{q1}$$
$$\Delta_4 = \alpha\beta-\Gamma_{q2} \tag{4.26}$$

is a first order dynamic control algorithm which depends on the measurements of the rotor speed ω and stator currents (i_{sa}, i_{sb}), on the reference signals (ω^*, ψ^*), on the positive control parameters k_ω, λ, k_i, on the load torque T_L, and on the machine parameters M, R_r, L_r, J, R_s, L_s: it guarantees that for any motor initial condition the tracking errors $(\omega-\omega^*)$ and $(\sqrt{\psi_{ra}^2+\psi_{rb}^2}-\psi^*)$ tend exponentially to zero with a rate of convergence no larger than α.

Illustrative Simulations

We tested the generalized indirect field-oriented control by simulations for the three-phase single pole pair 0.6-kW induction motor whose parameters have been reported in Chapter 1. All the motor initial conditions have been set to zero except for $\psi_{ra}(0) = \psi_{rb}(0) = 0.1\,\text{Wb}$. A simplified version of the controller (in which the robustifying terms have been set to zero, *i.e.* $v_d = 0$, $v_q = -\lambda\mu\psi^*(\omega-\omega^*)$) has been simulated with zero controller initial conditions and control parameters (all the values are in SI units) $\lambda = 0.005$, $k_\omega = 450$, $k_i = 800$. The references for the rotor speed and flux modulus along with the applied load torque are reported in Figures 4.1–4.3. The rotor flux modulus reference signal starts from 0.01 Wb at $t = 0\,\text{s}$ and grows up to the constant value 1.16 Wb. The speed reference is zero until $t = 0.32\,\text{s}$ and grows up to the constant value 100 rad/s; at $t = 1.5\,\text{s}$ the speed is required to go up to the value 200 rad/s, while the reference for the flux modulus is reduced to 0.5 Wb. A 5.8-Nm load torque is applied to the motor and is reduced to 1.8 Nm. Figures 4.2 and 4.3 show the time histories of rotor speed and flux modulus along with the corresponding tracking errors: the rotor speed and the flux modulus track their references tightly. Finally, the stator current and voltage profiles (which are within physical saturation limits) are reported in Figures 4.4 and 4.5.

4.2 Observer-based Control

We have shown in Section 3.1 that exponential estimation of rotor fluxes can be achieved by the rotor flux observer (3.33), for any flux initial condition and with an arbitrarily large rate of convergence, provided that the rotor speed and the stator current measurements are available for feedback and the parameters of the electromagnetic subsystem (in particular the rotor resistance) are known. On the other hand, the indirect field-oriented control (2.48) was modified in Section 2.7 in order

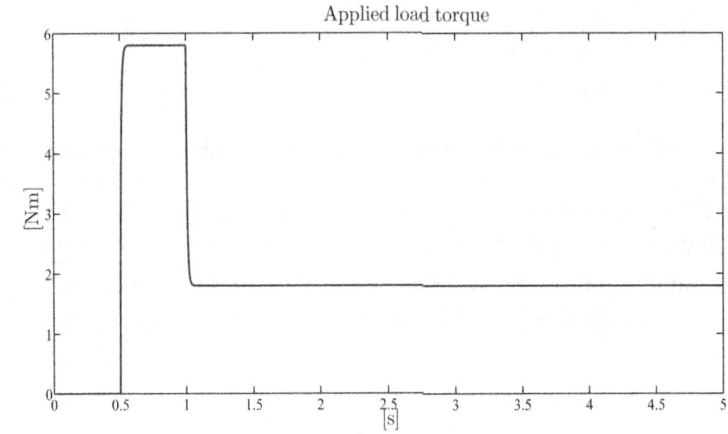

Fig. 4.1 Generalized indirect field-oriented control: applied load torque T_L

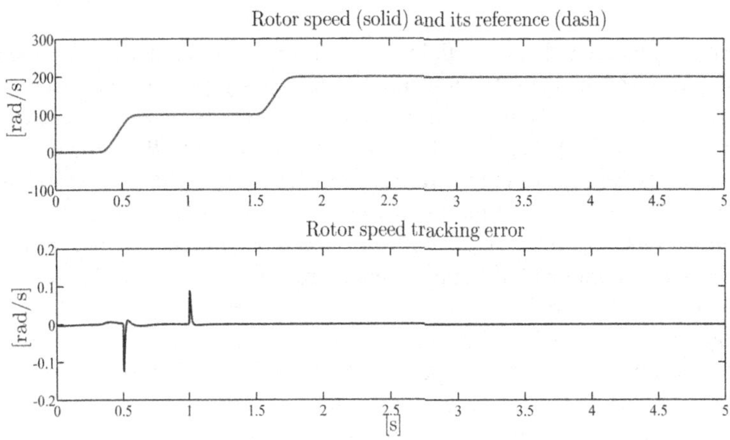

Fig. 4.2 Generalized indirect field-oriented control: rotor speed ω and its reference ω^*; rotor speed tracking error

to guarantee, for any initial condition of the motor (1.26) and with an arbitrarily large rate of convergence, exponential rotor speed and flux modulus tracking provided that the measurements of all the state variables are available for feedback and the model parameters (in particular, the rotor resistance and the load torque) are known. The question which then naturally arises and will be addressed in this section is the following: how to design an output feedback control from rotor speed and stator current measurements only (*i.e.* without feeding back the rotor flux measurements), which is able to guarantee exponential rotor speed and flux modulus tracking for any motor initial condition and with an arbitrarily large rate of convergence? We provide in this section an affirmative answer to the above question by reconsidering the global rotor flux observer (3.33) and the global state feedback

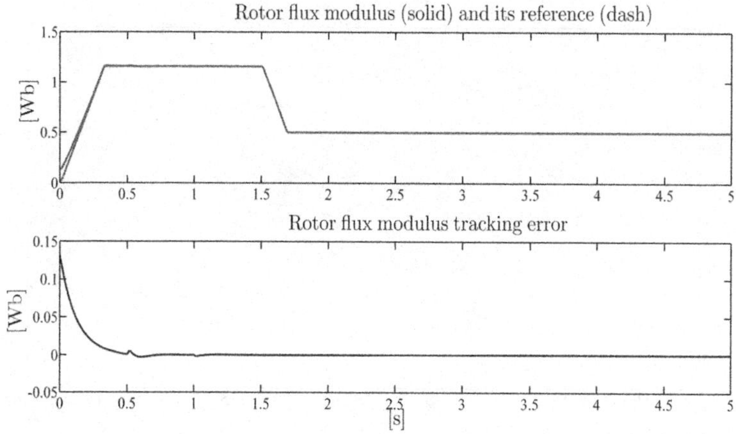

Fig. 4.3 Generalized indirect field-oriented control: rotor flux modulus $\sqrt{\psi_{ra}^2 + \psi_{rb}^2}$ and its reference ψ^*; rotor flux modulus tracking error

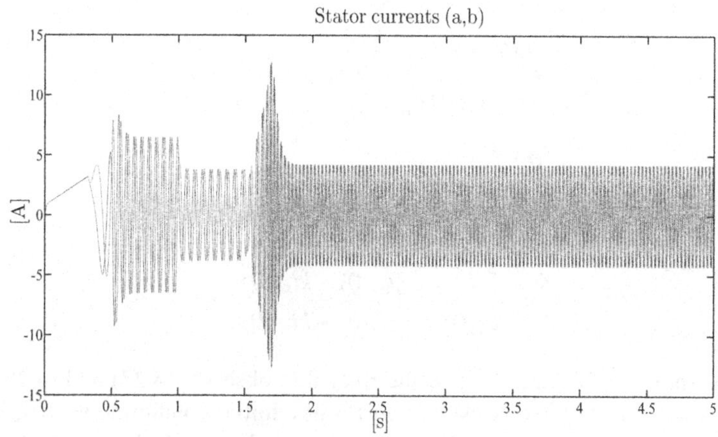

Fig. 4.4 Generalized indirect field-oriented control: stator current vector (a,b)-components (i_{sa}, i_{sb})

control with arbitrary rate of convergence (2.113): the two algorithms are modified and integrated so that an overall stability analysis can be performed. First we recall the observer (3.33), namely

$$\begin{aligned}\hat{\psi}_{ra} &= ki_{sa} + \xi_a \\ \hat{\psi}_{rb} &= ki_{sb} + \xi_b\end{aligned} \tag{4.27}$$

in which k is a positive control parameter and ξ_a, ξ_b satisfy the differential equations

$$\dot{\xi}_a = -\alpha(1+k\beta)(ki_{sa}+\xi_a) - \omega(1+k\beta)(ki_{sb}+\xi_b)$$

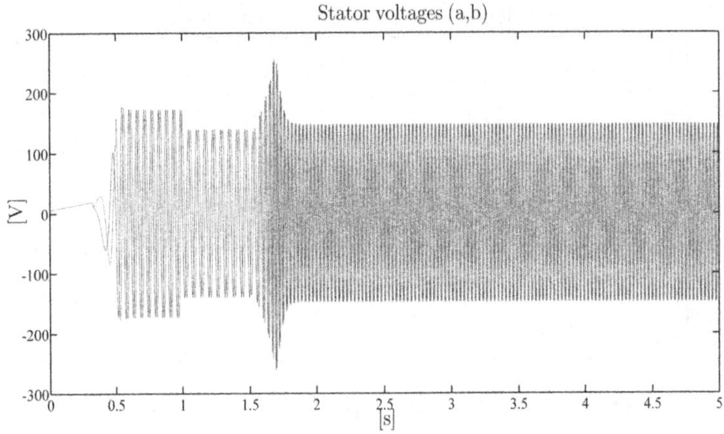

Fig. 4.5 Generalized indirect field-oriented control: stator voltage vector (a,b)-components (u_{sa}, u_{sb})

$$
\begin{aligned}
&+(\alpha M + k\gamma)i_{sa} - \frac{k u_{sa}}{\sigma} \\
\dot{\xi}_b &= -\alpha(1+k\beta)(k i_{sb} + \xi_b) + \omega(1+k\beta)(k i_{sa} + \xi_a) \\
&+(\alpha M + k\gamma)i_{sb} - \frac{k u_{sb}}{\sigma}
\end{aligned}
\tag{4.28}
$$

with initial conditions

$$
\begin{aligned}
\xi_a(0) &= \hat{\psi}_{ra}(0) - k i_{sa}(0) \\
\xi_b(0) &= \hat{\psi}_{rb}(0) - k i_{sb}(0)\,.
\end{aligned}
$$

We have shown in Section 3.1 that the rotor flux observer (4.27) and (4.28) guarantees exponential rotor flux estimation for any initial condition $(\psi_{ra}(0), \psi_{rb}(0))$ and with rate of convergence $\alpha(1+k\beta)$, which can be made arbitrarily large by increasing the control parameter k. Now reconsider the control algorithm (2.113) and, in particular, the references for the stator current vector (d,q)-components and the speed of the rotating (d,q) frame in (2.113):

$$
\begin{aligned}
i_{sd}^* &= \frac{\psi^*}{M} + \frac{\dot{\psi}^*}{\alpha M} + \frac{\eta_d}{\alpha M} \\
i_{sq}^* &= \frac{1}{\mu\psi^*}\left[-k_\omega(\omega - \omega^*) + \dot{\omega}^* + \frac{T_L}{J}\right] \\
\omega_0 &= \omega + \frac{\alpha M i_{sq}}{\psi^*} - \frac{\eta_q}{\psi^*}\,.
\end{aligned}
\tag{4.29}
$$

Introduce the tracking errors

$$\tilde{\omega} = \omega - \omega^*$$

$$\begin{aligned}\tilde{\psi}_{rd} &= \psi_{rd} - \psi^* \\ \tilde{\psi}_{rq} &= \psi_{rq} \\ \tilde{i}_{sd} &= i_{sd} - i^*_{sd} \\ \tilde{i}_{sq} &= i_{sq} - i^*_{sq}\end{aligned}$$

for the rotor speed, rotor flux vector (d,q)-components and stator current vector (d,q)-components, respectively. The dynamics for $\tilde{\omega}$, $\tilde{\psi}_{rd}$ and $\tilde{\psi}_{rq}$ are given by

$$\begin{aligned}\dot{\tilde{\omega}} &= -k_\omega \tilde{\omega} + \mu(\tilde{\psi}_{rd} i_{sq} - \tilde{\psi}_{rq} i_{sd}) + \mu \psi^* \tilde{i}_{sq} \\ \dot{\tilde{\psi}}_{rd} &= -\alpha \tilde{\psi}_{rd} + (\omega_0 - \omega)\tilde{\psi}_{rq} + \alpha M \tilde{i}_{sd} + \eta_d \\ \dot{\tilde{\psi}}_{rq} &= -\alpha \tilde{\psi}_{rq} - (\omega_0 - \omega)\tilde{\psi}_{rd} + \eta_q \,.\end{aligned} \tag{4.30}$$

In order to design the undetermined terms η_d and η_q, introduce the positive control parameter λ and consider the Lyapunov function

$$W = \frac{1}{2}\left(\lambda \tilde{\omega}^2 + \tilde{\psi}_{rd}^2 + \tilde{\psi}_{rq}^2\right) \tag{4.31}$$

whose time derivative along the trajectories of the closed-loop system is

$$\begin{aligned}\dot{W} = &-\lambda k_\omega \tilde{\omega}^2 + \lambda\mu(\tilde{\psi}_{rd} i_{sq} - \tilde{\psi}_{rq} i_{sd})\tilde{\omega} + \lambda\mu\psi^* \tilde{i}_{sq}\tilde{\omega} \\ &-\alpha(\tilde{\psi}_{rd}^2 + \tilde{\psi}_{rq}^2) + \alpha M \tilde{i}_{sd}\tilde{\psi}_{rd} + \eta_d \tilde{\psi}_{rd} + \eta_q \tilde{\psi}_{rq} \,.\end{aligned} \tag{4.32}$$

Since $i_{sq} = \tilde{i}_{sq} + i^*_{sq}$, we define

$$\begin{aligned}\eta_d &= -k_\psi(\hat{\psi}_{rd} - \psi^*) - \lambda\mu i^*_{sq}\tilde{\omega} \\ \eta_q &= -k_\psi \hat{\psi}_{rq} + \lambda\mu i_{sd}\tilde{\omega}\end{aligned} \tag{4.33}$$

in which k_ψ is a positive control parameter and $(\hat{\psi}_{rd}, \hat{\psi}_{rq})$ are the estimates of the rotor flux vector (d,q)-components (ψ_{rd}, ψ_{rq}) computed on the basis of (4.27), namely

$$\begin{aligned}\hat{\psi}_{rd} &= \hat{\psi}_{ra}\cos\varepsilon_0 + \hat{\psi}_{rb}\sin\varepsilon_0 \\ \hat{\psi}_{rq} &= -\hat{\psi}_{ra}\sin\varepsilon_0 + \hat{\psi}_{rb}\cos\varepsilon_0 \,.\end{aligned} \tag{4.34}$$

If we compare (4.33) with (2.100) designed in Section 2.7, we note that the rotor fluxes are now replaced by their estimates $(\hat{\psi}_{ra}, \hat{\psi}_{rb})$. By substituting (4.33) in (4.32) and noting that

$$\begin{aligned}\hat{\psi}_{rd} - \psi^* &= \tilde{\psi}_{rd} - e_{\psi d} \\ \hat{\psi}_{rq} &= \tilde{\psi}_{rq} - e_{\psi q}\end{aligned}$$

with the rotor flux estimation errors denoted by

$$\begin{aligned}e_{\psi d} &= \psi_{rd} - \hat{\psi}_{rd} \\ e_{\psi q} &= \psi_{rq} - \hat{\psi}_{rq}\end{aligned} \tag{4.35}$$

we obtain

$$\dot{W} = -\lambda k_\omega \tilde{\omega}^2 - (\alpha + k_\psi)(\tilde{\psi}_{rd}^2 + \tilde{\psi}_{rq}^2) + k_\psi \tilde{\psi}_{rd} e_{\psi d} + k_\psi \tilde{\psi}_{rq} e_{\psi q}$$
$$+ \lambda \mu \tilde{\psi}_{rd} \tilde{i}_{sq} \tilde{\omega} + \lambda \mu \psi^* \tilde{i}_{sq} \tilde{\omega} + \alpha M \tilde{i}_{sd} \tilde{\psi}_{rd} \,. \tag{4.36}$$

By adding and subtracting $\frac{k_\psi}{2}\tilde{\psi}_{rd}^2$ and $\frac{k_\psi}{2}\tilde{\psi}_{rq}^2$ and completing the squares, from (4.36) we have

$$\dot{W} = -\lambda k_\omega \tilde{\omega}^2 - \left(\alpha + \frac{k_\psi}{2}\right)(\tilde{\psi}_{rd}^2 + \tilde{\psi}_{rq}^2) + \frac{k_\psi}{2}\left(e_{\psi d}^2 + e_{\psi q}^2\right)$$
$$+ \lambda \mu \tilde{\psi}_{rd} \tilde{i}_{sq} \tilde{\omega} + \lambda \mu \psi^* \tilde{i}_{sq} \tilde{\omega} + \alpha M \tilde{i}_{sd} \tilde{\psi}_{rd} \,. \tag{4.37}$$

The influence of the last three terms in (4.37) will then be compensated by a suitable choice of the stator voltages (u_{sd}, u_{sq}). To this purpose, let us compute

$$\frac{\mathrm{d}i_{sq}^*}{\mathrm{d}t} = \Gamma_{q1} \tilde{\psi}_{rd} + \Gamma_{q2} \tilde{\psi}_{rq} + \Gamma_q \tag{4.38}$$

in which the known terms

$$\begin{aligned} \Gamma_{q1} &= -\frac{k_\omega i_{sq}}{\psi^*} \\ \Gamma_{q2} &= \frac{k_\omega i_{sd}}{\psi^*} \\ \Gamma_q &= \frac{1}{\mu \psi^*}\left[k_\omega^2 \tilde{\omega} - k_\omega \mu \psi^* \tilde{i}_{sq} + \ddot{\omega}^*\right] - \frac{\dot{\psi}^*}{\mu \psi^{*2}}\left[-k_\omega \tilde{\omega} + \frac{T_L}{J} + \dot{\omega}^*\right] \end{aligned} \tag{4.39}$$

appear. Similarly, let us compute

$$\frac{\mathrm{d}i_{sd}^*}{\mathrm{d}t} = \Gamma_{d1} \tilde{\psi}_{rd} + \Gamma_{d2} \tilde{\psi}_{rq} + \Gamma_d - \frac{k_\psi k}{\alpha M}\left[\Lambda_1 \cos \varepsilon_0 + \Lambda_2 \sin \varepsilon_0\right] \tag{4.40}$$

in which the unavailable terms are collected in (recall that the rotor fluxes are not available)

$$\begin{aligned} \Lambda_1 &= \beta \alpha [\tilde{\psi}_{rd} \cos \varepsilon_0 - \tilde{\psi}_{rq} \sin \varepsilon_0] + \beta \omega [\tilde{\psi}_{rd} \sin \varepsilon_0 + \tilde{\psi}_{rq} \cos \varepsilon_0] \\ \Lambda_2 &= \beta \alpha [\tilde{\psi}_{rd} \sin \varepsilon_0 + \tilde{\psi}_{rq} \cos \varepsilon_0] - \beta \omega [\tilde{\psi}_{rd} \cos \varepsilon_0 - \tilde{\psi}_{rq} \sin \varepsilon_0] \end{aligned} \tag{4.41}$$

while the known terms are collected in

$$\begin{aligned} \Gamma_{d1} &= -\frac{\lambda \mu}{\alpha M}\left[\mu i_{sq} i_{sq}^* + \tilde{\omega} \Gamma_{q1}\right] \\ \Gamma_{d2} &= \frac{\lambda \mu}{\alpha M}\left[\mu i_{sd} i_{sq}^* - \tilde{\omega} \Gamma_{q2}\right] \\ \Gamma_d &= \frac{\dot{\psi}^*}{M} + \frac{\ddot{\psi}^*}{\alpha M} - \frac{\lambda \mu^2 \psi^* \tilde{i}_{sq} i_{sq}^*}{\alpha M} - \frac{\lambda \mu \Gamma_q \tilde{\omega}}{\alpha M} + \frac{\lambda \mu k_\omega \tilde{\omega} i_{sq}^*}{\alpha M} + \frac{k_\psi \dot{\psi}^*}{\alpha M} \end{aligned}$$

$$-\frac{k_\psi}{\alpha M}\left[\omega_0\hat{\psi}_{rb}\cos\varepsilon_0 - \omega_0\hat{\psi}_{ra}\sin\varepsilon_0 + \Gamma_a\cos\varepsilon_0 + \Gamma_b\sin\varepsilon_0\right] \tag{4.42}$$

with

$$\begin{aligned}
\Gamma_a &= k\beta\alpha\psi^*\cos\varepsilon_0 + k\beta\omega\psi^*\sin\varepsilon_0 - \alpha(1+k\beta)\hat{\psi}_{ra} \\
&\quad -\omega(1+k\beta)\hat{\psi}_{rb} + \alpha M i_{sa} \\
\Gamma_b &= k\beta\alpha\psi^*\sin\varepsilon_0 - k\beta\omega\psi^*\cos\varepsilon_0 - \alpha(1+k\beta)\hat{\psi}_{rb} \\
&\quad +\omega(1+k\beta)\hat{\psi}_{ra} + \alpha M i_{sb}\,.
\end{aligned}$$

According to (4.38) and (4.40) the dynamics for the stator current tracking errors $\tilde{i}_{sd}$ and $\tilde{i}_{sq}$ are

$$\begin{aligned}
\frac{d\tilde{i}_{sd}}{dt} &= -\gamma i_{sd} + \omega_0 i_{sq} + \alpha\beta\psi_{rd} + \beta\omega\psi_{rq} + \frac{u_{sd}}{\sigma} - \Gamma_{d1}\tilde{\psi}_{rd} \\
&\quad -\Gamma_{d2}\tilde{\psi}_{rq} - \Gamma_d + \frac{k_\psi k}{\alpha M}\left[\Lambda_1\cos\varepsilon_0 + \Lambda_2\sin\varepsilon_0\right] \\
\frac{d\tilde{i}_{sq}}{dt} &= -\gamma i_{sq} - \omega_0 i_{sd} + \alpha\beta\psi_{rq} - \beta\omega\psi_{rd} + \frac{u_{sq}}{\sigma} - \Gamma_{q1}\tilde{\psi}_{rd} \\
&\quad -\Gamma_{q2}\tilde{\psi}_{rq} - \Gamma_q\,.
\end{aligned} \tag{4.43}$$

Design the control inputs (u_{sd}, u_{sq}) as

$$\begin{aligned}
u_{sd} &= \sigma\left[\gamma i_{sd}^* - \omega_0 i_{sq} - \alpha\beta\psi^* + \Gamma_d - k_i\tilde{i}_{sd} + v_d\right] \\
u_{sq} &= \sigma\left[\gamma i_{sq}^* + \omega_0 i_{sd} + \beta\omega\psi^* + \Gamma_q - k_i\tilde{i}_{sq} + v_q\right]
\end{aligned} \tag{4.44}$$

where v_d and v_q are yet to be defined and k_i is a positive control parameter, so that (4.43) become

$$\begin{aligned}
\frac{d\tilde{i}_{sd}}{dt} &= -(\gamma + k_i)\tilde{i}_{sd} + \alpha\beta\tilde{\psi}_{rd} + \beta\omega\tilde{\psi}_{rq} - \Gamma_{d1}\tilde{\psi}_{rd} - \Gamma_{d2}\tilde{\psi}_{rq} + v_d \\
&\quad + \frac{k_\psi k}{\alpha M}\left[\Lambda_1\cos\varepsilon_0 + \Lambda_2\sin\varepsilon_0\right] \\
\frac{d\tilde{i}_{sq}}{dt} &= -(\gamma + k_i)\tilde{i}_{sq} + \alpha\beta\tilde{\psi}_{rq} - \beta\omega\tilde{\psi}_{rd} - \Gamma_{q1}\tilde{\psi}_{rd} - \Gamma_{q2}\tilde{\psi}_{rq} + v_q\,.
\end{aligned} \tag{4.45}$$

In order to choose the yet undefined terms v_d and v_q, consider the positive definite function

$$V = W + \frac{1}{2}\left(\tilde{i}_{sd}^2 + \tilde{i}_{sq}^2\right) \tag{4.46}$$

whose time derivative along the trajectories of the closed-loop system, according to (4.37) and (4.41), satisfies

$$\dot{V} = -\lambda k_\omega\tilde{\omega}^2 - \left(\alpha + \frac{k_\psi}{2}\right)\left(\tilde{\psi}_{rd}^2 + \tilde{\psi}_{rq}^2\right) + \frac{k_\psi}{2}\left(e_{\psi d}^2 + e_{\psi q}^2\right)$$

$$
\begin{aligned}
&+\lambda\mu\tilde{\psi}_{rd}\tilde{i}_{sq}\tilde{\omega}+\lambda\mu\psi^{*}\tilde{i}_{sq}\tilde{\omega}+\alpha M\tilde{i}_{sd}\tilde{\psi}_{rd}\\
&-(\gamma+k_i)(\tilde{i}_{sd}^2+\tilde{i}_{sq}^2)+\alpha\beta\tilde{\psi}_{rd}\tilde{i}_{sd}+\beta\omega\tilde{\psi}_{rq}\tilde{i}_{sd}-\Gamma_{d1}\tilde{\psi}_{rd}\tilde{i}_{sd}-\Gamma_{d2}\tilde{\psi}_{rq}\tilde{i}_{sd}\\
&+v_d\tilde{i}_{sd}+\frac{k_\psi k}{\alpha M}\left[\Lambda_1\cos\varepsilon_0+\Lambda_2\sin\varepsilon_0\right]\tilde{i}_{sd}\\
&+\alpha\beta\tilde{\psi}_{rq}\tilde{i}_{sq}-\beta\omega\tilde{\psi}_{rd}\tilde{i}_{sq}-\Gamma_{q1}\tilde{\psi}_{rd}\tilde{i}_{sq}-\Gamma_{q2}\tilde{\psi}_{rq}\tilde{i}_{sq}+v_q\tilde{i}_{sq}\\
\triangleq\ &-\lambda k_\omega\tilde{\omega}^2-\left(\alpha+\frac{k_\psi}{2}\right)(\tilde{\psi}_{rd}^2+\tilde{\psi}_{rq}^2)-(\gamma+k_i)(\tilde{i}_{sd}^2+\tilde{i}_{sq}^2)+\frac{k_\psi}{2}\left(e_{\psi d}^2+e_{\psi q}^2\right)\\
&+\lambda\mu\psi^{*}\tilde{i}_{sq}\tilde{\omega}+v_d\tilde{i}_{sd}+v_q\tilde{i}_{sq}+\Delta_1\tilde{i}_{sd}\tilde{\psi}_{rd}+\Delta_2\tilde{i}_{sd}\tilde{\psi}_{rq}+\Delta_3\tilde{i}_{sq}\tilde{\psi}_{rd}+\Delta_4\tilde{i}_{sq}\tilde{\psi}_{rq}\\
&+\frac{2k_\psi k\beta}{\alpha M}(\alpha+|\omega|)(|\tilde{\psi}_{rd}|+|\tilde{\psi}_{rq}|)|\tilde{i}_{sd}| \qquad (4.47)
\end{aligned}
$$

in which the known terms

$$
\begin{aligned}
\Delta_1 &= \alpha M+\alpha\beta-\Gamma_{d1}\\
\Delta_2 &= \beta\omega-\Gamma_{d2}\\
\Delta_3 &= \lambda\mu\tilde{\omega}-\beta\omega-\Gamma_{q1}\\
\Delta_4 &= \alpha\beta-\Gamma_{q2} \qquad (4.48)
\end{aligned}
$$

appear. By noting that $(\alpha+|\omega|)^2\le 2\alpha^2+2\omega^2$, by adding and subtracting the terms $\frac{\alpha}{4}\tilde{\psi}_{rd}^2$, $\frac{\alpha}{4}\tilde{\psi}_{rq}^2$, and by completing the squares, from (4.47) we obtain

$$
\begin{aligned}
\dot{V}\le\ &-\lambda k_\omega\tilde{\omega}^2-\left(\frac{\alpha}{4}+\frac{k_\psi}{2}\right)(\tilde{\psi}_{rd}^2+\tilde{\psi}_{rq}^2)-(\gamma+k_i)(\tilde{i}_{sd}^2+\tilde{i}_{sq}^2)+\frac{k_\psi}{2}\left(e_{\psi d}^2+e_{\psi q}^2\right)\\
&+\lambda\mu\psi^{*}\tilde{i}_{sq}\tilde{\omega}+v_d\tilde{i}_{sd}+v_q\tilde{i}_{sq}+\left(\frac{\Delta_1^2}{\alpha}+\frac{\Delta_2^2}{\alpha}\right)\tilde{i}_{sd}^2+\left(\frac{\Delta_3^2}{\alpha}+\frac{\Delta_4^2}{\alpha}\right)\tilde{i}_{sq}^2\\
&+\frac{8k_\psi^2k^2\beta^2}{\alpha^3M^2}(\alpha^2+\omega^2)\tilde{i}_{sd}^2\,. \qquad (4.49)
\end{aligned}
$$

If we choose the undefined terms v_d and v_q as

$$
\begin{aligned}
v_d &= -\left(\frac{\Delta_1^2}{\alpha}+\frac{\Delta_2^2}{\alpha}+\frac{8k_\psi^2k^2\beta^2}{\alpha^3M^2}(\alpha^2+\omega^2)\right)\tilde{i}_{sd}\\
v_q &= -\left(\frac{\Delta_3^2}{\alpha}+\frac{\Delta_4^2}{\alpha}\right)\tilde{i}_{sq}-\lambda\mu\psi^{*}\tilde{\omega} \qquad (4.50)
\end{aligned}
$$

then from (4.49) we obtain

$$
\begin{aligned}
\dot{V}\le\ &-\lambda k_\omega\tilde{\omega}^2-\left(\frac{\alpha}{4}+\frac{k_\psi}{2}\right)(\tilde{\psi}_{rd}^2+\tilde{\psi}_{rq}^2)-(\gamma+k_i)(\tilde{i}_{sd}^2+\tilde{i}_{sq}^2)\\
&+\frac{k_\psi}{2}\left(e_{\psi d}^2+e_{\psi q}^2\right)\,. \qquad (4.51)
\end{aligned}
$$

From (4.34) and the definition of $(e_{\psi d},e_{\psi q})$ in (4.35), we have

$$e_{\psi d}^2 + e_{\psi q}^2 = e_{\psi a}^2 + e_{\psi b}^2 = 2V_\psi$$

so that (4.51) can be rewritten as

$$\dot{V} \le -\lambda k_\omega \tilde{\omega}^2 - \left(\frac{\alpha}{4} + \frac{k_\psi}{2}\right)(\tilde{\psi}_{rd}^2 + \tilde{\psi}_{rq}^2) - (\gamma + k_i)(\tilde{i}_{sd}^2 + \tilde{i}_{sq}^2) + k_\psi V_\psi \quad (4.52)$$

and therefore, according to (4.31) and (4.46), we have

$$\begin{aligned} \dot{V}(t) &\le -\min\left\{2k_\omega, \frac{\alpha}{2} + k_\psi, 2(\gamma + k_i)\right\} V(t) + k_\psi V_\psi \\ V_\psi(t) &= \mathrm{e}^{-2\alpha(1+k\beta)t} V_\psi(0) . \end{aligned} \quad (4.53)$$

According to Lemma A.4 in Appendix A, (4.46) and (4.53) guarantee exponential rotor speed and flux modulus tracking for any motor initial condition and with a rate of convergence which can be made arbitrarily large by suitably choosing the control parameters $(k_\omega, k_\psi, k_i, k)$.

In conclusion, the third order output feedback *observer-based control*

$$\begin{aligned}
\begin{bmatrix} u_{sa} \\ u_{sb} \end{bmatrix} &= \begin{bmatrix} \cos\varepsilon_0 & -\sin\varepsilon_0 \\ \sin\varepsilon_0 & \cos\varepsilon_0 \end{bmatrix} \begin{bmatrix} u_{sd} \\ u_{sq} \end{bmatrix} \\
\dot{\varepsilon}_0 &= \omega_0 = \omega + \frac{\alpha M i_{sq}}{\psi^*} - \frac{\eta_q}{\psi^*} \\
u_{sd} &= \sigma\left[\gamma i_{sd}^* - \omega_0 i_{sq} - \alpha\beta\psi^* + \Gamma_d - k_i(i_{sd} - i_{sd}^*) + v_d\right] \\
\begin{bmatrix} i_{sd} \\ i_{sq} \end{bmatrix} &= \begin{bmatrix} \cos\varepsilon_0 & \sin\varepsilon_0 \\ -\sin\varepsilon_0 & \cos\varepsilon_0 \end{bmatrix} \begin{bmatrix} i_{sa} \\ i_{sb} \end{bmatrix} \\
u_{sq} &= \sigma\left[\gamma i_{sq}^* + \omega_0 i_{sd} + \beta\omega\psi^* + \Gamma_q - k_i(i_{sq} - i_{sq}^*) + v_q\right] \\
i_{sd}^* &= \frac{\psi^*}{M} + \frac{\dot{\psi}^*}{\alpha M} + \frac{\eta_d}{\alpha M} \\
i_{sq}^* &= \frac{1}{\mu\psi^*}\left[-k_\omega(\omega - \omega^*) + \dot{\omega}^* + \frac{T_L}{J}\right] \\
\eta_d &= -k_\psi(\hat{\psi}_{rd} - \psi^*) - \lambda\mu i_{sq}^*(\omega - \omega^*) \\
\eta_q &= -k_\psi\hat{\psi}_{rq} + \lambda\mu i_{sd}(\omega - \omega^*) \\
\hat{\psi}_{rd} &= \hat{\psi}_{ra}\cos\varepsilon_0 + \hat{\psi}_{rb}\sin\varepsilon_0 \\
\hat{\psi}_{rq} &= -\hat{\psi}_{ra}\sin\varepsilon_0 + \hat{\psi}_{rb}\cos\varepsilon_0 \\
\hat{\psi}_{ra} &= k i_{sa} + \xi_a \\
\hat{\psi}_{rb} &= k i_{sb} + \xi_b \\
\dot{\xi}_a &= -\alpha(1 + k\beta)(k i_{sa} + \xi_a) - \omega(1 + k\beta)(k i_{sb} + \xi_b) + (\alpha M + k\gamma)i_{sa} - \frac{k u_{sa}}{\sigma}
\end{aligned}$$

$$
\begin{aligned}
\dot{\xi}_b &= -\alpha(1+k\beta)(ki_{sb}+\xi_b)+\omega(1+k\beta)(ki_{sa}+\xi_a)+(\alpha M+k\gamma)i_{sb}-\frac{ku_{sb}}{\sigma}\\
\xi_a(0) &= \hat{\psi}_{ra}(0)-ki_{sa}(0)\\
\xi_b(0) &= \hat{\psi}_{rb}(0)-ki_{sb}(0)\\
v_d &= -\left(\frac{\Delta_1^2}{\alpha}+\frac{\Delta_2^2}{\alpha}+\frac{8k_\psi^2k^2\beta^2}{\alpha^3M^2}(\alpha^2+\omega^2)\right)(i_{sd}-i_{sd}^*)\\
v_q &= -\left(\frac{\Delta_3^2}{\alpha}+\frac{\Delta_4^2}{\alpha}\right)(i_{sq}-i_{sq}^*)-\lambda\mu\psi^*(\omega-\omega^*)\\
\Gamma_d &= \frac{\dot{\psi}^*}{M}+\frac{\ddot{\psi}^*}{\alpha M}-\frac{\lambda\mu^2\psi^*(i_{sq}-i_{sq}^*)i_{sq}^*}{\alpha M}-\frac{\lambda\mu\Gamma_q(\omega-\omega^*)}{\alpha M}\\
&\quad+\frac{\lambda\mu k_\omega(\omega-\omega^*)i_{sq}^*}{\alpha M}+\frac{k_\psi\dot{\psi}^*}{\alpha M}\\
&\quad-\frac{k_\psi}{\alpha M}\left[\omega_0\hat{\psi}_{rb}\cos\varepsilon_0-\omega_0\hat{\psi}_{ra}\sin\varepsilon_0+\Gamma_a\cos\varepsilon_0+\Gamma_b\sin\varepsilon_0\right]\\
\Gamma_q &= \frac{1}{\mu\psi^*}\left[k_\omega^2(\omega-\omega^*)-k_\omega\mu\psi^*(i_{sq}-i_{sq}^*)+\ddot{\omega}^*\right]\\
&\quad-\frac{\dot{\psi}^*}{\mu\psi^{*2}}\left[-k_\omega(\omega-\omega^*)+\frac{T_L}{J}+\dot{\omega}^*\right]\\
\Gamma_a &= k\beta\alpha\cos(\varepsilon_0)\psi^*+k\beta\omega\sin(\varepsilon_0)\psi^*-\alpha(1+k\beta)\hat{\psi}_{ra}\\
&\quad-\omega(1+k\beta)\hat{\psi}_{rb}+\alpha Mi_{sa}\\
\Gamma_b &= k\beta\alpha\sin(\varepsilon_0)\psi^*-k\beta\omega\cos(\varepsilon_0)\psi^*-\alpha(1+k\beta)\hat{\psi}_{rb}\\
&\quad+\omega(1+k\beta)\hat{\psi}_{ra}+\alpha Mi_{sb}\\
\Delta_1 &= \alpha M+\alpha\beta-\Gamma_{d1}\\
\Delta_2 &= \beta\omega-\Gamma_{d2}\\
\Gamma_{d1} &= -\frac{\lambda\mu}{\alpha M}\left[\mu i_{sq}i_{sq}^*+(\omega-\omega^*)\Gamma_{q1}\right]\\
\Gamma_{d2} &= \frac{\lambda\mu}{\alpha M}\left[\mu i_{sd}i_{sq}^*-(\omega-\omega^*)\Gamma_{q2}\right]\\
\Delta_3 &= \lambda\mu(\omega-\omega^*)-\beta\omega-\Gamma_{q1}\\
\Delta_4 &= \alpha\beta-\Gamma_{q2}\\
\Gamma_{q1} &= -\frac{k_\omega i_{sq}}{\psi^*}\\
\Gamma_{q2} &= \frac{k_\omega i_{sd}}{\psi^*}
\end{aligned}
\tag{4.54}
$$

which depends on the measurements of the rotor speed ω and stator currents (i_{sa}, i_{sb}), on the reference signals (ω^*, ψ^*), on the positive control parameters $(k_\omega, \lambda, k_i, k)$, on the load torque T_L, and on the machine parameters M, R_r, L_r, J, R_s, L_s, since $\mu = \frac{M}{JL_r}$, $\alpha = \frac{R_r}{L_r}$, $\sigma = L_s\left(1 - \frac{M^2}{L_s L_r}\right)$, $\beta = \frac{M}{\sigma L_r}$, $\gamma = \frac{R_s}{\sigma} + \beta\alpha M$, guarantees exponential rotor speed and flux modulus tracking for any motor initial condition $(\omega(0), \psi_{ra}(0), \psi_{rb}(0), i_{sa}(0), i_{sb}(0))$, and with a rate of convergence which can be made arbitrarily large by suitably choosing the control parameters $(k_\omega, k_\psi, k_i, k)$.

Illustrative Simulations

We tested the observer-based control by simulations for the three-phase single pole pair 0.6-kW induction motor whose parameters have been reported in Chapter 1. All motor initial conditions have been set to zero except for $\psi_{ra}(0) = \psi_{rb}(0) = 0.1$ Wb. A simplified version of the control algorithm, in which the robustifying term v_d has been set to zero and $v_q = -\lambda\mu\psi^*\tilde{\omega}$, has been tested with zero initial conditions and control parameters (all values are in SI units) $\lambda = 0.005$, $k_\omega = 450$, $k_i = 800$, $k_\psi = 12$, $k = 0.3$. The references for the rotor speed and flux modulus along with the applied load torque are reported in Figures 4.6–4.8. The rotor flux modulus reference signal starts from 0.001 Wb at $t = 0$ s and grows up to the constant value 1.16 Wb. The speed reference is zero until $t = 0.32$ s and grows up to the constant value 100 rad/s; at $t = 1.5$ s the speed is required to go up to the value 200 rad/s, while the reference for the flux modulus is reduced to 0.5 Wb. A 5.8-Nm load torque is applied to the motor and is reduced to 1.8 Nm. Figures 4.7 and 4.8 show the time histories of rotor speed and flux modulus along with the corresponding tracking errors: the rotor speed and flux modulus track their references tightly. Finally, the stator current and voltage profiles (which are within physical saturation limits) are reported in Figures 4.9 and 4.10.

4.3 Adaptive Observer-based Control with Uncertain Load Torque

We have shown in the previous section that exponential rotor speed and flux modulus tracking can be achieved, for any motor initial condition and with an arbitrarily large rate of convergence, by the output feedback control (4.54) which feeds back only the rotor speed and stator current measurements. However, it requires the knowledge of the load torque, which is typically uncertain. On the other hand, we designed in Section 3.3 the load torque identifier (3.76) which can be used in conjunction with the rotor flux observer (4.27) and (4.28). The aim of this section is to provide an adaptive version of the controller (4.54) which, by incorporating the load torque identifier (3.76), allows for uncertainties on the load torque T_L within known bounds

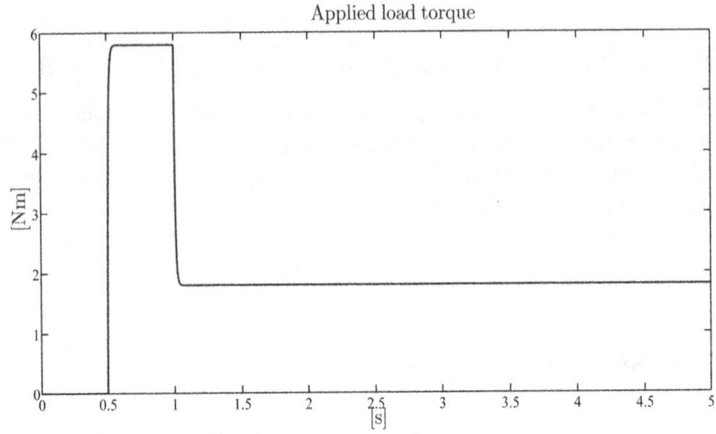

Fig. 4.6 Observer-based control: applied load torque T_L

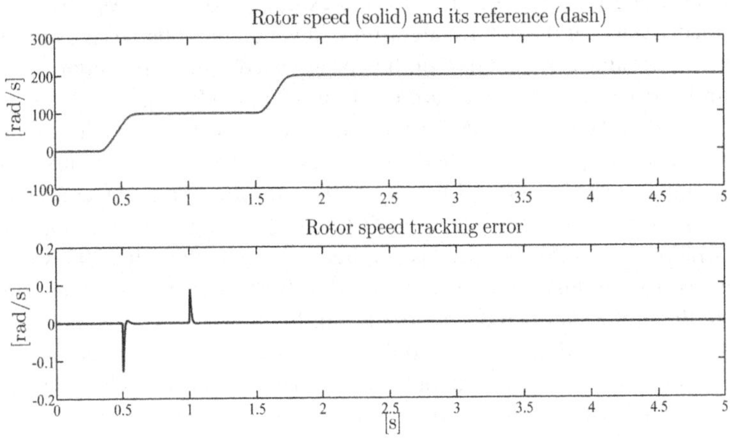

Fig. 4.7 Observer-based control: rotor speed ω and its reference ω^*; rotor speed tracking error

T_{Lm} and T_{LM}. Reconsider the control algorithm (4.54) and replace the load torque T_L in (4.54) by its saturated estimate $\text{sat}(\hat{T}_L)$, *i.e.*

$$i_{sq}^* = \frac{1}{\mu\psi^*}\left[-k_\omega(\omega-\omega^*)+\dot{\omega}^*+\frac{\text{sat}(\hat{T}_L)}{J}\right]. \tag{4.55}$$

where $\text{sat}(\eta)$ is the saturation function, *i.e.* a class C^1 odd function on $\mathbb{R}$ which is linear in the closed set $[T_{Lm}, T_{LM}]$ and admits a finite limit as $|\eta|$ goes to infinity, while $\hat{T}_L$ is the load torque estimate provided by the load torque estimator (3.76), namely (g and k_o are positive control parameters)

$$\dot{\hat{T}}_L = -g(\omega-\hat{\omega})$$

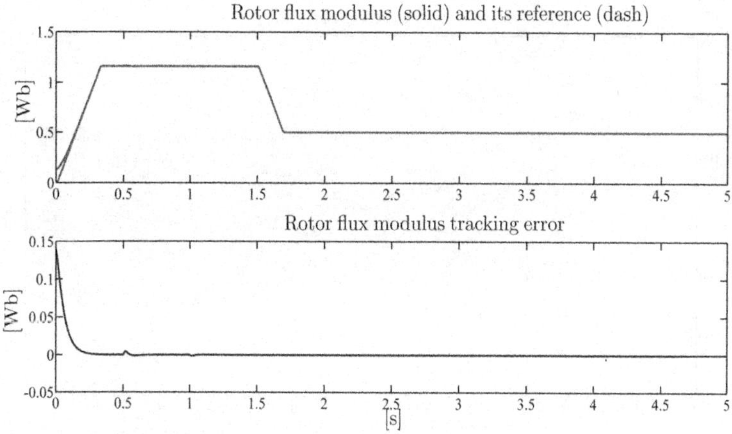

Fig. 4.8 Observer-based control: rotor flux modulus $\sqrt{\psi_{ra}^2+\psi_{rb}^2}$ and its reference ψ^*; rotor flux modulus tracking error

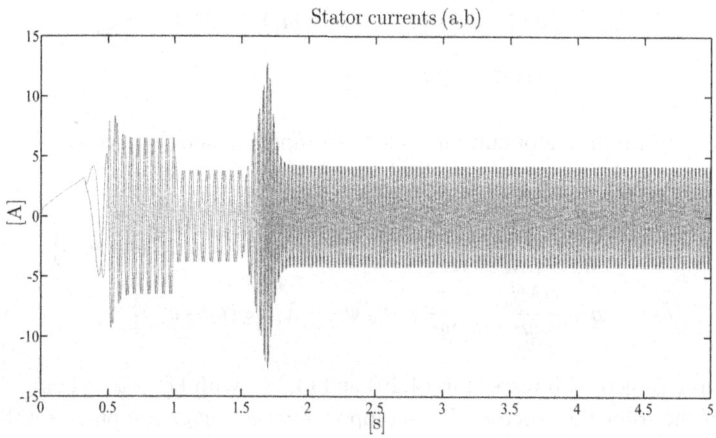

Fig. 4.9 Observer-based control: stator current vector (a,b)-components (i_{sa}, i_{sb})

$$\dot{\hat{\omega}} = \mu(\hat{\psi}_{ra} i_{sb} - \hat{\psi}_{rb} i_{sa}) - \frac{\hat{T}_L}{J} + k_o(\omega - \hat{\omega}) \tag{4.56}$$

in conjunction with the rotor flux observer (4.27) and (4.28), namely (k is a positive control parameter)

$$\begin{aligned}
\hat{\psi}_{ra} &= k i_{sa} + \xi_a \\
\hat{\psi}_{rb} &= k i_{sb} + \xi_b \\
\dot{\xi}_a &= -\alpha(1+k\beta)(k i_{sa} + \xi_a) - \omega(1+k\beta)(k i_{sb} + \xi_b) \\
&\quad + (\alpha M + k\gamma) i_{sa} - \frac{k u_{sa}}{\sigma}
\end{aligned}$$

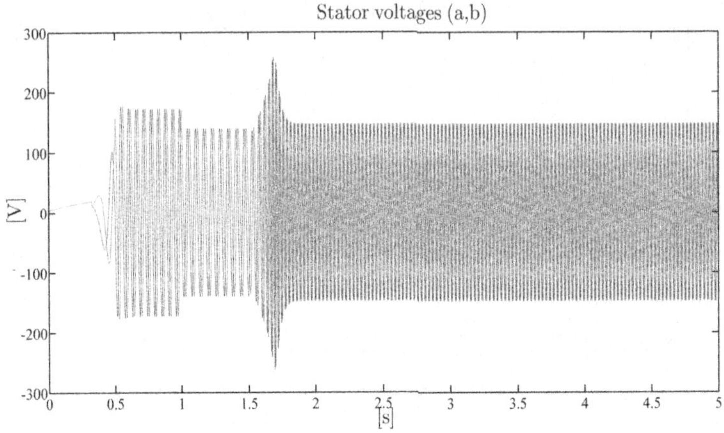

Fig. 4.10 Observer-based control: stator voltage vector (a,b)-components (u_{sa}, u_{sb})

$$\begin{aligned}\dot{\xi}_b &= -\alpha(1+k\beta)(ki_{sb}+\xi_b)+\omega(1+k\beta)(ki_{sa}+\xi_a)\\&\quad+(\alpha M+k\gamma)i_{sb}-\frac{ku_{sb}}{\sigma}\,.\end{aligned}\tag{4.57}$$

The reference for the stator current vector d-component and the speed of the rotating (d,q) frame (λ, k_ψ are positive control parameters)

$$\begin{aligned}i^*_{sd} &= \frac{\psi^*}{M}+\frac{\dot{\psi}^*}{\alpha M}+\frac{1}{\alpha M}\left[-k_\psi(\hat{\psi}_{rd}-\psi^*)-\lambda\mu i^*_{sq}(\omega-\omega^*)\right]\\ \omega_0 &= \omega+\frac{\alpha M i_{sq}}{\psi^*}-\frac{1}{\psi^*}\left[-k_\psi\hat{\psi}_{rq}+\lambda\mu i_{sd}(\omega-\omega^*)\right]\end{aligned}\tag{4.58}$$

remain unchanged with respect to (4.29) and (4.33), with $(\hat{\psi}_{rd}, \hat{\psi}_{rq})$ being the estimates of the rotor flux vector (d,q)-components (ψ_{rd}, ψ_{rq}) computed on the basis of (4.57) as

$$\begin{aligned}\hat{\psi}_{rd} &= \hat{\psi}_{ra}\cos\varepsilon_0+\hat{\psi}_{rb}\sin\varepsilon_0\\ \hat{\psi}_{rq} &= -\hat{\psi}_{ra}\sin\varepsilon_0+\hat{\psi}_{rb}\cos\varepsilon_0\,.\end{aligned}\tag{4.59}$$

Introduce the tracking and the estimation errors

$$\begin{aligned}\tilde{\omega} &= \omega-\omega^*\\ \tilde{\psi}_{rd} &= \psi_{rd}-\psi^*\\ \tilde{\psi}_{rq} &= \psi_{rq}\\ \tilde{i}_{sd} &= i_{sd}-i^*_{sd}\\ \tilde{i}_{sq} &= i_{sq}-i^*_{sq}\\ e_{\psi d} &= \psi_{rd}-\hat{\psi}_{rd}\end{aligned}$$

$$
\begin{aligned}
e_{\psi q} &= \psi_{rq} - \hat{\psi}_{rq} \\
e_T &= T_L - \mathrm{sat}(\hat{T}_L) \,.
\end{aligned}
$$

The dynamics for $\tilde{\omega}$, $\tilde{\psi}_{rd}$ and $\tilde{\psi}_{rq}$ are given by

$$
\begin{aligned}
\dot{\tilde{\omega}} &= -k_\omega \tilde{\omega} + \mu(\tilde{\psi}_{rd} i_{sq} - \tilde{\psi}_{rq} i_{sd}) + \mu \psi^* \tilde{i}_{sq} - \frac{e_T}{J} \\
\dot{\tilde{\psi}}_{rd} &= -(\alpha + k_\psi)\tilde{\psi}_{rd} + (\omega_0 - \omega)\tilde{\psi}_{rq} + \alpha M \tilde{i}_{sd} - \lambda \mu i^*_{sq} \tilde{\omega} + k_\psi e_{\psi d} \\
\dot{\tilde{\psi}}_{rq} &= -(\alpha + k_\psi)\tilde{\psi}_{rq} - (\omega_0 - \omega)\tilde{\psi}_{rd} + \lambda \mu i_{sd} \tilde{\omega} + k_\psi e_{\psi q} \,.
\end{aligned} \tag{4.60}
$$

Let us compute

$$
\frac{\mathrm{d}i^*_{sq}}{\mathrm{d}t} = \Gamma_{q1} \tilde{\psi}_{rd} + \Gamma_{q2} \tilde{\psi}_{rq} + \Gamma_{q3} e_T + \Gamma_q \tag{4.61}
$$

in which the known terms

$$
\begin{aligned}
\Gamma_{q1} &= -\frac{k_\omega i_{sq}}{\psi^*} \\
\Gamma_{q2} &= \frac{k_\omega i_{sd}}{\psi^*} \\
\Gamma_{q3} &= \frac{k_\omega}{J \mu \psi^*} \\
\Gamma_q &= \frac{1}{\mu \psi^*} \left[k_\omega^2 \tilde{\omega} - k_\omega \mu \psi^* \tilde{i}_{sq} + \ddot{\omega}^* + \frac{\mathrm{dsat}(\hat{T}_L)}{\mathrm{d}\hat{T}_L} \frac{\dot{\hat{T}}_L}{J} \right] \\
&\quad - \frac{\dot{\psi}^*}{\mu \psi^{*2}} \left[-k_\omega \tilde{\omega} + \frac{\mathrm{sat}(\hat{T}_L)}{J} + \dot{\omega}^* \right]
\end{aligned} \tag{4.62}
$$

appear. Let us compute

$$
\frac{\mathrm{d}i^*_{sd}}{\mathrm{d}t} = \Gamma_{d1} \tilde{\psi}_{rd} + \Gamma_{d2} \tilde{\psi}_{rq} + \Gamma_{d3} e_T + \Gamma_d - \frac{k_\psi k}{\alpha M} [\Lambda_1 \cos \varepsilon_0 + \Lambda_2 \sin \varepsilon_0] \tag{4.63}
$$

in which the unavailable terms are

$$
\begin{aligned}
\Lambda_1 &= \beta \alpha [\tilde{\psi}_{rd} \cos \varepsilon_0 - \tilde{\psi}_{rq} \sin \varepsilon_0] + \beta \omega [\tilde{\psi}_{rd} \sin \varepsilon_0 + \tilde{\psi}_{rq} \cos \varepsilon_0] \\
\Lambda_2 &= \beta \alpha [\tilde{\psi}_{rd} \sin \varepsilon_0 + \tilde{\psi}_{rq} \cos \varepsilon_0] - \beta \omega [\tilde{\psi}_{rd} \cos \varepsilon_0 - \tilde{\psi}_{rq} \sin \varepsilon_0]
\end{aligned} \tag{4.64}
$$

while the known terms are

$$
\begin{aligned}
\Gamma_{d1} &= -\frac{\lambda \mu}{\alpha M} [\mu i_{sq} i^*_{sq} + \tilde{\omega} \Gamma_{q1}] \\
\Gamma_{d2} &= \frac{\lambda \mu}{\alpha M} [\mu i_{sd} i^*_{sq} - \tilde{\omega} \Gamma_{q2}]
\end{aligned}
$$

$$\begin{aligned}
\Gamma_{d3} &= \frac{\lambda\mu}{\alpha M}\left[\frac{i_{sq}^*}{J} - \Gamma_{q3}\tilde{\omega}\right] \\
\Gamma_d &= \frac{\dot{\psi}^*}{M} + \frac{\ddot{\psi}^*}{\alpha M} - \frac{\lambda\mu^2\psi^*\tilde{i}_{sq}i_{sq}^*}{\alpha M} - \frac{\lambda\mu\Gamma_q\tilde{\omega}}{\alpha M} + \frac{\lambda\mu k_\omega\tilde{\omega}i_{sq}^*}{\alpha M} + \frac{k_\psi\dot{\psi}^*}{\alpha M} \\
&\quad - \frac{k_\psi}{\alpha M}\left[\omega_0\hat{\psi}_{rb}\cos\varepsilon_0 - \omega_0\hat{\psi}_{ra}\sin\varepsilon_0 + \Gamma_a\cos\varepsilon_0 + \Gamma_b\sin\varepsilon_0\right] \qquad (4.65)
\end{aligned}$$

with

$$\begin{aligned}
\Gamma_a &= k\beta\alpha\psi^*\cos\varepsilon_0 + k\beta\omega\psi^*\sin\varepsilon_0 - \alpha(1+k\beta)\hat{\psi}_{ra} \\
&\quad -\omega(1+k\beta)\hat{\psi}_{rb} + \alpha M i_{sa} \\
\Gamma_b &= k\beta\alpha\psi^*\sin\varepsilon_0 - k\beta\omega\psi^*\cos\varepsilon_0 - \alpha(1+k\beta)\hat{\psi}_{rb} \\
&\quad +\omega(1+k\beta)\hat{\psi}_{ra} + \alpha M i_{sb}\,. \qquad (4.66)
\end{aligned}$$

According to (4.61) and (4.63) the dynamics for the stator currents tracking errors $\tilde{i}_{sd}$ and $\tilde{i}_{sq}$ are

$$\begin{aligned}
\frac{\mathrm{d}\tilde{i}_{sd}}{\mathrm{d}t} &= -\gamma i_{sd} + \omega_0 i_{sq} + \alpha\beta\psi_{rd} + \beta\omega\psi_{rq} + \frac{u_{sd}}{\sigma} - \Gamma_{d1}\tilde{\psi}_{rd} \\
&\quad -\Gamma_{d2}\tilde{\psi}_{rq} - \Gamma_{d3}e_T - \Gamma_d + \frac{k_\psi k}{\alpha M}\left[\Lambda_1\cos\varepsilon_0 + \Lambda_2\sin\varepsilon_0\right] \\
\frac{\mathrm{d}\tilde{i}_{sq}}{\mathrm{d}t} &= -\gamma i_{sq} - \omega_0 i_{sd} + \alpha\beta\psi_{rq} - \beta\omega\psi_{rd} + \frac{u_{sq}}{\sigma} - \Gamma_{q1}\tilde{\psi}_{rd} \\
&\quad -\Gamma_{q2}\tilde{\psi}_{rq} - \Gamma_{q3}e_T - \Gamma_q\,. \qquad (4.67)
\end{aligned}$$

Design the control inputs (u_{sd}, u_{sq}) as

$$\begin{aligned}
u_{sd} &= \sigma\left[\gamma i_{sd}^* - \omega_0 i_{sq} - \alpha\beta\psi^* + \Gamma_d - k_i\tilde{i}_{sd} + v_d\right] \\
u_{sq} &= \sigma\left[\gamma i_{sq}^* + \omega_0 i_{sd} + \beta\omega\psi^* + \Gamma_q - k_i\tilde{i}_{sq} + v_q\right] \qquad (4.68)
\end{aligned}$$

where v_d and v_q are yet to be designed and k_i is a positive control parameter, so that (4.67) become

$$\begin{aligned}
\frac{\mathrm{d}\tilde{i}_{sd}}{\mathrm{d}t} &= -(\gamma+k_i)\tilde{i}_{sd} + \alpha\beta\tilde{\psi}_{rd} + \beta\omega\tilde{\psi}_{rq} - \Gamma_{d1}\tilde{\psi}_{rd} - \Gamma_{d2}\tilde{\psi}_{rq} - \Gamma_{d3}e_T + v_d \\
&\quad + \frac{k_\psi k}{\alpha M}\left[\Lambda_1\cos\varepsilon_0 + \Lambda_2\sin\varepsilon_0\right] \\
\frac{\mathrm{d}\tilde{i}_{sq}}{\mathrm{d}t} &= -(\gamma+k_i)\tilde{i}_{sq} + \alpha\beta\tilde{\psi}_{rq} - \beta\omega\tilde{\psi}_{rd} - \Gamma_{q1}\tilde{\psi}_{rd} - \Gamma_{q2}\tilde{\psi}_{rq} - \Gamma_{q3}e_T + v_q\,.
\end{aligned} \qquad (4.69)$$

In order to choose the undefined terms v_d and v_q, consider the positive definite function

$$V = \frac{1}{2}\left(\lambda \tilde{\omega}^2 + \tilde{\psi}_{rd}^2 + \tilde{\psi}_{rq}^2\right) + \frac{1}{2}\left(\tilde{i}_{sd}^2 + \tilde{i}_{sq}^2\right) \tag{4.70}$$

whose time derivative along the trajectories of the closed-loop system, according to (4.60) and (4.69), satisfies

$$\begin{aligned} \dot{V} \leq & -\frac{\lambda k_\omega}{2}\tilde{\omega}^2 - \left(\frac{\alpha}{4} + \frac{k_\psi}{2}\right)\left(\tilde{\psi}_{rd}^2 + \tilde{\psi}_{rq}^2\right) - (\gamma + k_i)(\tilde{i}_{sd}^2 + \tilde{i}_{sq}^2) + \frac{k_\psi}{2}\left(e_{\psi d}^2 + e_{\psi q}^2\right) \\ & + \lambda\mu\psi^* \tilde{i}_{sq}\tilde{\omega} + v_d \tilde{i}_{sd} + v_q \tilde{i}_{sq} + \left(\frac{\Delta_1^2}{\alpha} + \frac{\Delta_2^2}{\alpha} + \frac{\Gamma_{d3}^2}{2}\right)\tilde{i}_{sd}^2 + \left(\frac{\Delta_3^2}{\alpha} + \frac{\Delta_4^2}{\alpha} + \frac{\Gamma_{q3}^2}{2}\right)\tilde{i}_{sq}^2 \\ & + \frac{8k_\psi^2 k^2 \beta^2}{\alpha^3 M^2}(\alpha^2 + \omega^2)\tilde{i}_{sd}^2 + \left(1 + \frac{\lambda}{2k_\omega J^2}\right)e_T^2 \end{aligned} \tag{4.71}$$

in which the known terms

$$\begin{aligned} \Delta_1 &= \alpha M + \alpha\beta - \Gamma_{d1} \\ \Delta_2 &= \beta\omega - \Gamma_{d2} \\ \Delta_3 &= \lambda\mu\tilde{\omega} - \beta\omega - \Gamma_{q1} \\ \Delta_4 &= \alpha\beta - \Gamma_{q2} \end{aligned} \tag{4.72}$$

appear. If we choose the yet undefined terms v_d and v_q as in the previous sections

$$\begin{aligned} v_d &= -\left(\frac{\Delta_1^2}{\alpha} + \frac{\Delta_2^2}{\alpha} + \frac{\Gamma_{d3}^2}{2} + \frac{8k_\psi^2 k^2\beta^2}{\alpha^3 M^2}(\alpha^2 + \omega^2)\right)\tilde{i}_{sd} \\ v_q &= -\left(\frac{\Delta_3^2}{\alpha} + \frac{\Delta_4^2}{\alpha} + \frac{\Gamma_{q3}^2}{2}\right)\tilde{i}_{sq} - \lambda\mu\psi^*\tilde{\omega} \end{aligned} \tag{4.73}$$

then from (4.71) we obtain

$$\begin{aligned} \dot{V} \leq & -\min\left\{k_\omega, \frac{\alpha}{2} + k_\psi, 2(\gamma + k_i)\right\}V(t) + \frac{k_\psi}{2}(e_{\psi d}^2 + e_{\psi q}^2) \\ & + \left(1 + \frac{\lambda}{2k_\omega J^2}\right)e_T^2 \, . \end{aligned} \tag{4.74}$$

According to the analysis performed in Section 3.1, the rotor flux observer (3.33) guarantees exponential rotor flux estimation for any initial condition $(\psi_{ra}(0), \psi_{rb}(0))$ and with an arbitrarily large rate of convergence. Thus inequality (4.74) along with (4.70) imply that $(i_{sd}(t), i_{sq}(t))$ and therefore $(i_{sa}(t), i_{sb}(t))$ are bounded functions on $[0, \infty)$. On the other hand, according to the analysis performed in Section 3.3, the load torque estimator (4.56) guarantees exponential load torque estimation for any initial condition $(\omega(0), \psi_{ra}(0), \psi_{rb}(0))$, provided that the stator currents $i_{sa}(t)$ and $i_{sb}(t)$ are bounded time functions on $[0, \infty)$. Therefore, according to (4.70), (4.74), and Lemma A.4 in Appendix A, exponential rotor speed and flux modulus tracking are guaranteed for any motor initial condition and for any uncertain load torque T_L within known bounds (T_{Lm}, T_{LM}).

In conclusion, the fifth order *adaptive observer-based control* with uncertain load torque

$$
\begin{aligned}
\begin{bmatrix} u_{sa} \\ u_{sb} \end{bmatrix} &= \begin{bmatrix} \cos\varepsilon_0 & -\sin\varepsilon_0 \\ \sin\varepsilon_0 & \cos\varepsilon_0 \end{bmatrix} \begin{bmatrix} u_{sd} \\ u_{sq} \end{bmatrix} \\
\dot{\varepsilon}_0 &= \omega_0 = \omega + \frac{\alpha M i_{sq}}{\psi^*} - \frac{\eta_q}{\psi^*} \\
u_{sd} &= \sigma\left[\gamma i_{sd}^* - \omega_0 i_{sq} - \alpha\beta\psi^* + \Gamma_d - k_i(i_{sd} - i_{sd}^*) + v_d\right] \\
u_{sq} &= \sigma\left[\gamma i_{sq}^* + \omega_0 i_{sd} + \beta\omega\psi^* + \Gamma_q - k_i(i_{sq} - i_{sq}^*) + v_q\right] \\
\begin{bmatrix} i_{sd} \\ i_{sq} \end{bmatrix} &= \begin{bmatrix} \cos\varepsilon_0 & \sin\varepsilon_0 \\ -\sin\varepsilon_0 & \cos\varepsilon_0 \end{bmatrix} \begin{bmatrix} i_{sa} \\ i_{sb} \end{bmatrix} \\
i_{sd}^* &= \frac{\psi^*}{M} + \frac{\dot{\psi}^*}{\alpha M} + \frac{\eta_d}{\alpha M} \\
i_{sq}^* &= \frac{1}{\mu\psi^*}\left[-k_\omega(\omega - \omega^*) + \dot{\omega}^* + \frac{\mathrm{sat}(\hat{T}_L)}{J}\right] \\
\eta_d &= -k_\psi(\hat{\psi}_{rd} - \psi^*) - \lambda\mu i_{sq}^*(\omega - \omega^*) \\
\eta_q &= -k_\psi\hat{\psi}_{rq} + \lambda\mu i_{sd}(\omega - \omega^*) \\
\hat{\psi}_{rd} &= \hat{\psi}_{ra}\cos\varepsilon_0 + \hat{\psi}_{rb}\sin\varepsilon_0 \\
\hat{\psi}_{rq} &= -\hat{\psi}_{ra}\sin\varepsilon_0 + \hat{\psi}_{rb}\cos\varepsilon_0 \\
\hat{\psi}_{ra} &= k i_{sa} + \xi_a \\
\hat{\psi}_{rb} &= k i_{sb} + \xi_b \\
\dot{\xi}_a &= -\alpha(1 + k\beta)(k i_{sa} + \xi_a) - \omega(1 + k\beta)(k i_{sb} + \xi_b) \\
&\quad + (\alpha M + k\gamma) i_{sa} - \frac{k u_{sa}}{\sigma} \\
\dot{\xi}_b &= -\alpha(1 + k\beta)(k i_{sb} + \xi_b) + \omega(1 + k\beta)(k i_{sa} + \xi_a) \\
&\quad + (\alpha M + k\gamma) i_{sb} - \frac{k u_{sb}}{\sigma} \\
\xi_a(0) &= \hat{\psi}_{ra}(0) - k i_{sa}(0) \\
\xi_b(0) &= \hat{\psi}_{rb}(0) - k i_{sb}(0) \\
\dot{\hat{T}}_L &= -g(\omega - \hat{\omega}) \\
\dot{\hat{\omega}} &= \mu(\hat{\psi}_{ra} i_{sb} - \hat{\psi}_{rb} i_{sa}) - \frac{\hat{T}_L}{J} + k_o(\omega - \hat{\omega}) \\
v_d &= -\left(\frac{\Delta_1^2}{\alpha} + \frac{\Delta_2^2}{\alpha} + \frac{\Gamma_{d3}^2}{2} + \frac{8k_\psi^2 k^2\beta^2}{\alpha^3 M^2}(\alpha^2 + \omega^2)\right)(i_{sd} - i_{sd}^*)
\end{aligned}
$$

$$
\begin{aligned}
v_q &= -\left(\frac{\Delta_3^2}{\alpha}+\frac{\Delta_4^2}{\alpha}+\frac{\Gamma_{q3}^2}{2}\right)(i_{sq}-i_{sq}^*)-\lambda\mu\psi^*(\omega-\omega^*)\\
\Gamma_{q3} &= \frac{k_\omega}{J\mu\psi^*}\\
\Gamma_q &= \frac{1}{\mu\psi^*}\left[k_\omega^2(\omega-\omega^*)-k_\omega\mu\psi^*(i_{sq}-i_{sq}^*)+\ddot{\omega}^*+\frac{\mathrm{dsat}(\hat{T}_L)}{\mathrm{d}\hat{T}_L}\frac{\dot{\hat{T}}_L}{J}\right]\\
&\quad -\frac{\dot{\psi}^*}{\mu\psi^{*2}}\left[-k_\omega(\omega-\omega^*)+\frac{\mathrm{sat}(\hat{T}_L)}{J}+\dot{\omega}^*\right]\\
\Gamma_{d1} &= -\frac{\lambda\mu}{\alpha M}\left[\mu i_{sq}i_{sq}^*+(\omega-\omega^*)\Gamma_{q1}\right]\\
\Gamma_{d2} &= \frac{\lambda\mu}{\alpha M}\left[\mu i_{sd}i_{sq}^*-(\omega-\omega^*)\Gamma_{q2}\right]\\
\Gamma_{d3} &= \frac{\lambda\mu}{\alpha M}\left[\frac{i_{sq}^*}{J}-\Gamma_{q3}(\omega-\omega^*)\right]\\
\Gamma_d &= \frac{\dot{\psi}^*}{M}+\frac{\ddot{\psi}^*}{\alpha M}-\frac{\lambda\mu^2\psi^*(i_{sq}-i_{sq}^*)i_{sq}^*}{\alpha M}\\
&\quad -\frac{\lambda\mu\Gamma_q(\omega-\omega^*)}{\alpha M}+\frac{\lambda\mu k_\omega(\omega-\omega^*)i_{sq}^*}{\alpha M}+\frac{k_\psi\dot{\psi}^*}{\alpha M}\\
&\quad -\frac{k_\psi}{\alpha M}\left[-\omega_0\hat{\psi}_{ra}\sin\varepsilon_0+\omega_0\hat{\psi}_{rb}\cos\varepsilon_0+\Gamma_a\cos\varepsilon_0+\Gamma_b\sin\varepsilon_0\right]\\
\Gamma_a &= k\beta\alpha\psi^*\cos\varepsilon_0+k\beta\omega\psi^*\sin\varepsilon_0-\alpha(1+k\beta)\hat{\psi}_{ra}\\
&\quad -\omega(1+k\beta)\hat{\psi}_{rb}+\alpha M i_{sa}\\
\Gamma_b &= k\beta\alpha\psi^*\sin\varepsilon_0-k\beta\omega\psi^*\cos\varepsilon_0-\alpha(1+k\beta)\hat{\psi}_{rb}\\
&\quad +\omega(1+k\beta)\hat{\psi}_{ra}+\alpha M i_{sb}\\
\Delta_1 &= \alpha M+\alpha\beta-\Gamma_{d1}\\
\Delta_2 &= \beta\omega-\Gamma_{d2}\\
\Delta_3 &= \lambda\mu(\omega-\omega^*)-\beta\omega-\Gamma_{q1}\\
\Delta_4 &= \alpha\beta-\Gamma_{q2}\\
\Gamma_{q1} &= -\frac{k_\omega i_{sq}}{\psi^*}\\
\Gamma_{q2} &= \frac{k_\omega i_{sd}}{\psi^*}
\end{aligned}
\tag{4.75}
$$

which depends on the measurements of the rotor speed ω and stator currents (i_{sa}, i_{sb}), on the reference signals (ω^*, ψ^*), on the positive control parameters k_ω, λ, k_i, k, g, k_o, and on the machine parameters M, R_r, L_r, J, R_s, L_s, since $\mu = \frac{M}{JL_r}$, $\alpha = \frac{R_r}{L_r}$, $\sigma = L_s\left(1-\frac{M^2}{L_sL_r}\right)$, $\beta = \frac{M}{\sigma L_r}$, $\gamma = \frac{R_s}{\sigma}+\beta\alpha M$, guarantees exponential rotor speed and flux modulus tracking, for any motor initial condition $(\omega(0), \psi_{ra}(0), \psi_{rb}(0), i_{sa}(0), i_{sb}(0))$ and for any uncertain load torque T_L within known bounds T_{Lm}, T_{LM}.

Remarks

1. The controller (4.75) relies on an observer for the unmeasured rotor fluxes and on an identifier for the uncertain load torque while, as in (2.48), singularities are avoided.
2. In the case of known load torque, by setting $\hat{T}_L = T_L$, the controller (4.75) reduces to the nonadaptive controller (4.54).
3. The knowledge of the critical parameter R_r is required.

Illustrative Simulations

We tested the adaptive observer-based control by simulations for the three-phase single pole pair 0.6-kW induction motor whose parameters have been reported in Chapter 1. All motor initial conditions have been set to zero except for $\psi_{ra}(0) = \psi_{rb}(0) = 0.1$ Wb. A simplified version of the control algorithm, in which the robustifying term v_d has been set to zero and $v_q = -\lambda\mu\psi^*\tilde{\omega}$, has been tested with zero initial conditions and control parameters (all values are in SI units) $\lambda = 0.005$, $k_\omega = 450$, $k_i = 800$, $k_\psi = 12$, $k = 0.3$, $g = 100^2 J$, $k_o = 200$. The references for the rotor speed and flux modulus along with the applied load torque are reported in Figures 4.11–4.13. The rotor flux modulus reference signal starts from 0.001 Wb at $t = 0$ s and grows up to the constant value 1.16 Wb. The speed reference is zero until $t = 0.32$ s and grows up to the constant value 100 rad/s; at $t = 1.5$ s the speed is required to go up to the value 200 rad/s, while the reference for the flux modulus is reduced to 0.5 Wb. A 5.8-Nm load torque is applied to the motor and is reduced to 1.8 Nm. Figures 4.11 and 4.12 show the time histories of rotor speed and flux modulus along with the corresponding tracking errors: the rotor speed and flux modulus track their references tightly. Finally, the stator current and voltage profiles (which are within physical saturation limits) are reported in Figures 4.14 and 4.15.

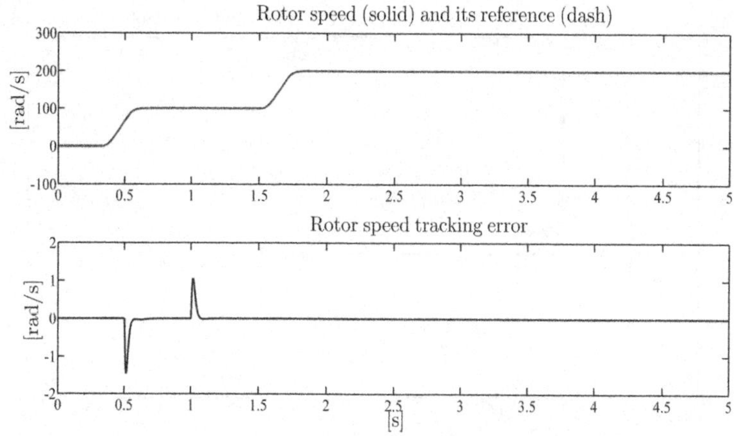

Fig. 4.11 Adaptive observer-based control: rotor speed ω and its reference ω^*; rotor speed tracking error

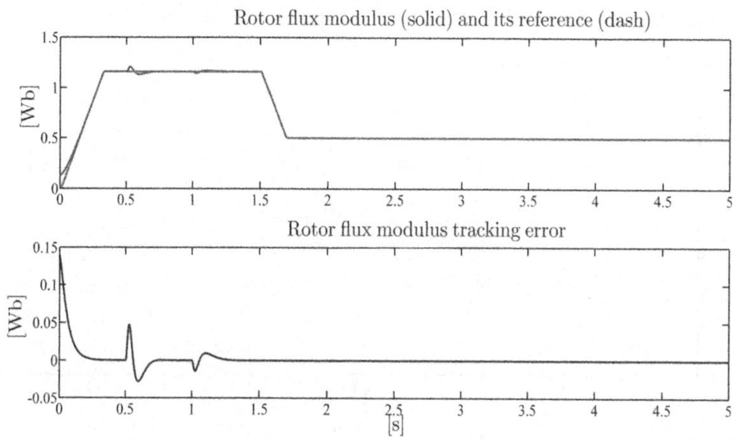

Fig. 4.12 Adaptive observer-based control: rotor flux modulus $\sqrt{\psi_{ra}^2 + \psi_{rb}^2}$ and its reference ψ^*; rotor flux modulus tracking error

4.4 Adaptive Control with Uncertain Load Torque and Rotor Resistance

The aim of this section is to generalize the output feedback control presented in the previous section to the case in which, in addition to load torque, the rotor resistance R_r is also an uncertain parameter. This is not an easy task since the parameter R_r affects both the estimation of the rotor fluxes and directly the control signals. We first reconsider the generalized indirect field-oriented control (4.26) and we show that it is sufficient to replace the true values of α and T_L by the estimates provided by the adaptive observer (3.58) and by the load torque estimator (3.84), in order to

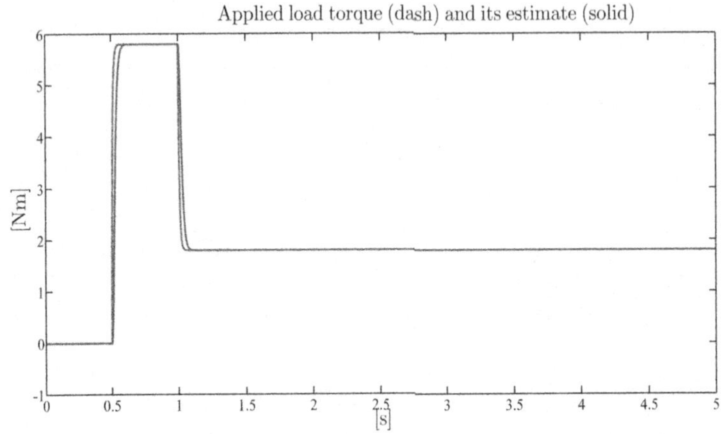

Fig. 4.13 Adaptive observer-based control: applied load torque T_L and its estimate $\hat{T}_L$

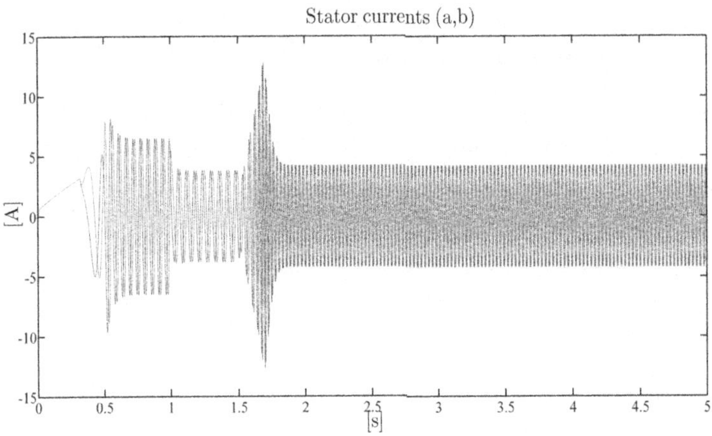

Fig. 4.14 Adaptive observer-based control: stator current vector (a,b)-components (i_{sa}, i_{sb})

achieve exponential rotor speed and flux modulus tracking, despite load torque and rotor resistance uncertainties, provided that the initial estimation and tracking errors are sufficiently small. Then we show that global results, which hold for any initial tracking and estimation errors, can be obtained if a dynamic output feedback control is entirely redesigned by incorporating in the algorithm the rotor flux observer and the estimators of both the rotor resistance and the load torque.

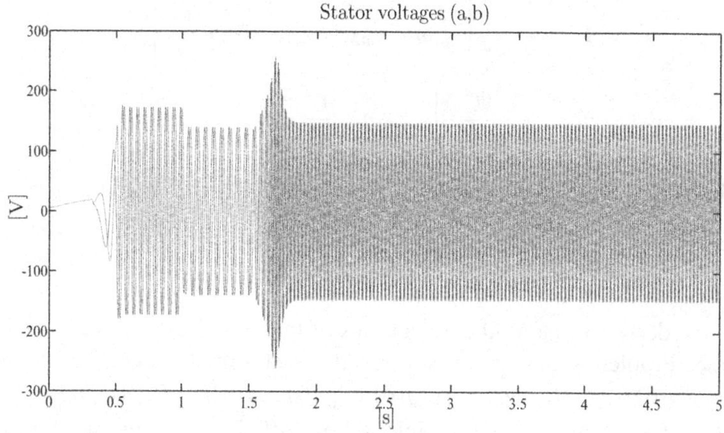

Fig. 4.15 Adaptive observer-based control: stator voltage vector (a,b)-components (u_{sa}, u_{sb})

4.4.1 Observer-based Control

Consider the generalized indirect field-oriented control (4.26) and set $k_\psi = 0$, $v_d = 0$ and $v_q = -\lambda\mu\psi^*\tilde{\omega}$. Moreover, replace T_L and α by the corresponding estimates $\hat{T}_L$ and $\hat{\alpha}$ provided by (along with their time derivatives $\dot{\hat{T}}_L$ and $\dot{\hat{\alpha}}$) the adaptive observer (3.58) and by the load torque identifier (3.84) so that the following adaptive observer-based control algorithm (k_ω, λ and k_i are positive control parameters) is obtained:

$$
\begin{aligned}
i_{sd}^* &= \frac{\psi^*}{M} + \frac{\dot{\psi}^*}{\hat{\alpha}M} - \frac{\lambda\mu i_{sq}^*\tilde{\omega}}{\hat{\alpha}M} \\
i_{sq}^* &= \frac{1}{\mu\psi^*}\left[-k_\omega\tilde{\omega} + \dot{\omega}^* + \frac{\hat{T}_L}{J}\right] \\
\omega_0 &= \omega + \frac{\hat{\alpha}Mi_{sq}}{\psi^*} - \frac{\lambda\mu i_{sd}\tilde{\omega}}{\psi^*} \\
u_{sd} &= \sigma\left[\left(\frac{R_s}{\sigma} + \hat{\alpha}\beta M\right)i_{sd}^* - \omega_0 i_{sq} - \hat{\alpha}\beta\psi^* + \Gamma_d - k_i\tilde{i}_{sd}\right] \\
u_{sq} &= \sigma\left[\left(\frac{R_s}{\sigma} + \hat{\alpha}\beta M\right)i_{sq}^* + \omega_0 i_{sd} + \beta\omega\psi^* + \Gamma_q - k_i\tilde{i}_{sq} - \lambda\mu\psi^*\tilde{\omega}\right]
\end{aligned}
\tag{4.76}
$$

with $\tilde{\omega} = \omega - \omega^*$, $\tilde{i}_{sd} = i_{sd} - i_{sd}^*$, $\tilde{i}_{sq} = i_{sq} - i_{sq}^*$ and

$$
\begin{aligned}
\Gamma_d = \frac{\dot{\psi}^*}{M} + \frac{\ddot{\psi}^*}{\hat{\alpha}M} - \frac{\lambda\mu^2\psi^*\tilde{i}_{sq}i_{sq}^*}{\hat{\alpha}M} - \frac{\lambda\mu\Gamma_q\tilde{\omega}}{\hat{\alpha}M} \\
+ \frac{\lambda\mu k_\omega\tilde{\omega}i_{sq}^*}{\hat{\alpha}M} - \frac{\dot{\psi}^*\dot{\hat{\alpha}}}{\hat{\alpha}^2M} + \frac{\lambda\mu i_{sq}^*\tilde{\omega}\dot{\hat{\alpha}}}{\hat{\alpha}^2M}
\end{aligned}
$$

$$\Gamma_q = \frac{1}{\mu \psi^*} \left[k_\omega^2 \tilde{\omega} - k_\omega \mu \psi^* \tilde{i}_{sq} + \dot{\omega}^* + \frac{\dot{\hat{T}}_L}{J} \right] - \frac{\dot{\psi}^*}{\mu \psi^{*2}} \left[-k_\omega \tilde{\omega} + \frac{\hat{T}_L}{J} + \dot{\omega}^* \right] .$$

On the basis of the Lyapunov function ($\tilde{\psi}_{rd} = \psi_{rd} - \psi^*$, $\tilde{\psi}_{rq} = \psi_{rq}$)

$$V = \frac{1}{2} \left(\lambda \tilde{\omega}^2 + \tilde{\psi}_{rd}^2 + \tilde{\psi}_{rq}^2 + \tilde{i}_{sd}^2 + \tilde{i}_{sq}^2 \right)$$

and its time derivative along the trajectories of the closed-loop system, we can establish (see Problem 4.11) that for sufficiently small initial tracking errors $(\omega(0) - \omega^*(0))$, $(\psi_{rd}(0) - \psi^*(0))$, $\psi_{rq}(0)$, $(i_{sd}(0) - i_{sd}^*(0))$, $(i_{sq}(0) - i_{sq}^*(0))$ and sufficiently small $(T_L - \hat{T}_L(t))$ and $(\alpha - \hat{\alpha}(t))$ (implied by sufficiently small initial estimation errors for the adaptive observer and the load torque identifier), the tracking errors $(\omega(t) - \omega^*(t))$, $(\psi_{rd}(t) - \psi^*(t))$, $\psi_{rq}(t)$, $(i_{sd}(t) - i_{sd}^*(t))$, $(i_{sq}(t) - i_{sq}^*(t))$ are bounded on $[0, \infty)$, provided that the control parameter k_i is sufficiently large. Hence, if persistency of excitation conditions are satisfied for the adaptive observer (3.58), then exponential rotor speed and flux modulus tracking are achieved with a rate of convergence which cannot be made arbitrarily large by a suitable choice of the control parameters (as in the generalized indirect field-oriented control). Note that the result holds under persistency of excitation and for sufficiently small initial tracking and estimation errors.

Illustrative Simulations

We tested the control (4.76), (3.58), (3.84) by simulations for the three-phase single pole pair 0.6-kW induction motor whose parameters have been reported in Chapter 1. All the motor initial conditions have been set to zero except for $\psi_{ra}(0) = \psi_{rb}(0) = 0.1\,\text{Wb}$. The initial conditions are set to zero except for $\hat{\alpha}(0) = 13.2\,\text{s}^{-1}$ while the control parameters are (all the values are in SI units): $\lambda = 0.005$, $k_\omega = 450$, $k_i = 800$, $k_1 = 120$, $k_2 = 3$, $k_3 = 270$, $g_\alpha = 450$, $k = 0.3$, $g = 100^2 J$, $k_o = 200$. The references for the rotor speed and flux modulus along with the applied load torque are reported in Figures 4.16–4.18. The rotor flux modulus reference signal starts from 0.001 Wb at $t = 0\,\text{s}$ and grows up to the constant value 1.16 Wb. The rotor speed reference is zero until $t = 0.32\,\text{s}$ and grows up to the constant value 100 rad/s; at $t = 1.5\,\text{s}$ the speed is required to go up to the value 200 rad/s, while the reference for the flux modulus is reduced to 0.5 Wb. A 5.8-Nm load torque is applied to the motor and is reduced to 1.8 Nm. Figures 4.16 and 4.17 show the time histories of the rotor speed and the flux modulus along with the corresponding tracking errors: the rotor speed and the flux modulus track their references tightly. The load torque and parameter α estimates provided by the load torque identifier (3.84) and by the adaptive observer (3.58), respectively, are reported in Figures 4.18 and 4.19. Finally, the stator current and voltage profiles (which are within physical saturation limits) are reported in Figures 4.20 and 4.21.

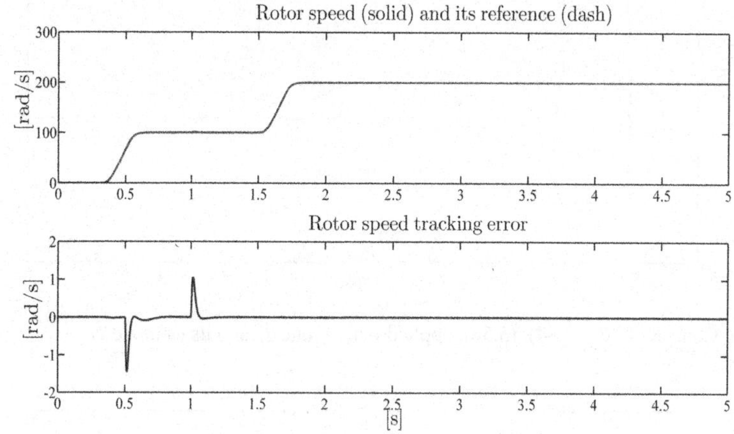

Fig. 4.16 Control (4.76), (3.84), (3.58): rotor speed ω and its reference ω^*; rotor speed tracking error

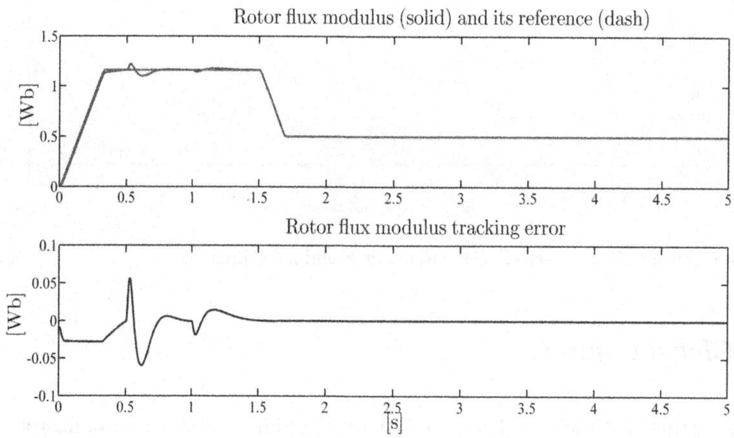

Fig. 4.17 Control (4.76), (3.84), (3.58): rotor flux modulus $\sqrt{\psi_{ra}^2 + \psi_{rb}^2}$ and its reference ψ^*; rotor flux modulus tracking error

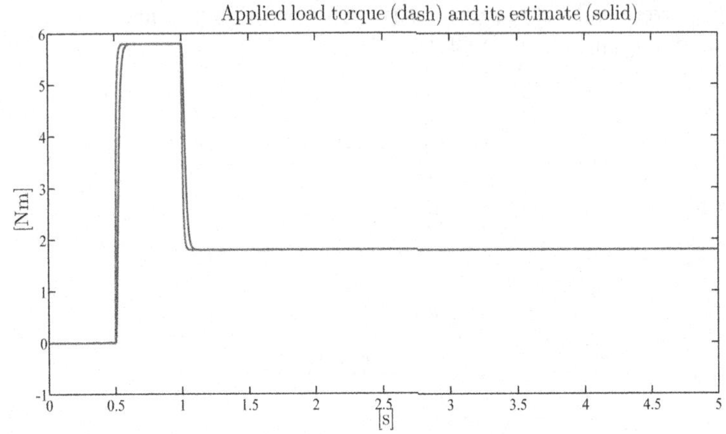

Fig. 4.18 Control (4.76), (3.84), (3.58): applied load torque T_L and its estimate $\hat{T}_L$

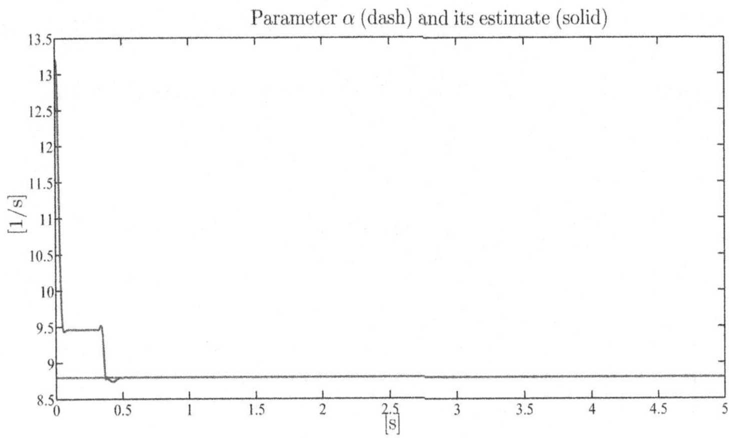

Fig. 4.19 Control (4.76), (3.84), (3.58): parameter α and its estimate $\hat{\alpha}$

4.4.2 Global Control

The aim of this section is to design a dynamic output feedback compensator on the basis of the measured outputs (ω, i_{sa}, i_{sb})

$$\begin{aligned} \dot{\varepsilon}_0 &= \omega_0 \\ \begin{bmatrix} u_{sa} \\ u_{sb} \end{bmatrix} &= \begin{bmatrix} \cos\varepsilon_0 & -\sin\varepsilon_0 \\ \sin\varepsilon_0 & \cos\varepsilon_0 \end{bmatrix} \begin{bmatrix} u_{sd} \\ u_{sq} \end{bmatrix} \end{aligned} \tag{4.77}$$

by choosing $(\omega_0, u_{sd}, u_{sq})$ so that, for any unknown T_L and uncertain α and for any initial condition $(\omega(0), \psi_{ra}(0), \psi_{rb}(0), i_{sa}(0), i_{sb}(0), \varepsilon_0(0))$ we have

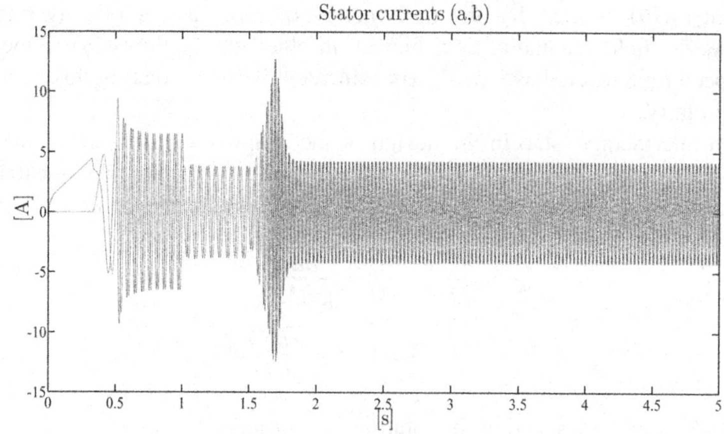

Fig. 4.20 Control (4.76), (3.84), (3.58): stator current vector (a,b)-components (i_{sa}, i_{sb})

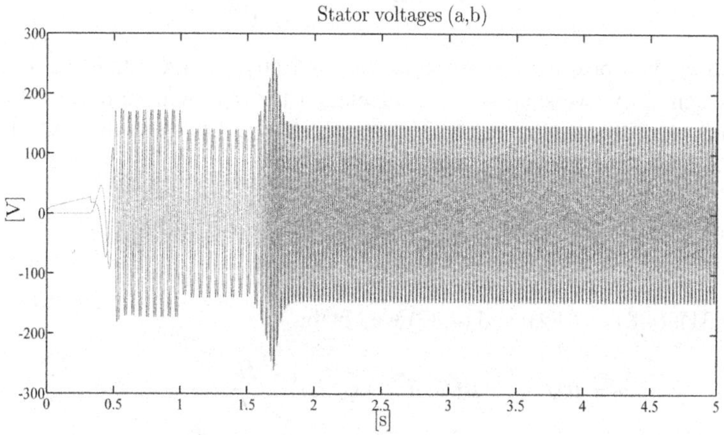

Fig. 4.21 Control (4.76), (3.84), (3.58): stator voltage vector (a,b)-components (u_{sa}, u_{sb})

$$\lim_{t\to\infty}[\omega(t) - \omega^*(t)] = 0 \tag{4.78}$$

and

$$\begin{aligned} \lim_{t\to\infty}[\psi_{rd}(t) - \psi^*(t)] &= 0 \\ \lim_{t\to\infty} \psi_{rq}(t) &= 0 \end{aligned} \tag{4.79}$$

which imply that

$$\lim_{t\to\infty}\left[\sqrt{\psi_{ra}^2(t) + \psi_{rb}^2(t)} - \psi^*(t)\right] = 0\,. \tag{4.80}$$

Note that (4.79) implies that the flux vector (ψ_{ra}, ψ_{rb}) asymptotically rotates at speed ω_0, *i.e.* field orientation is achieved: in other words, the (d,q) frame rotating at speed ω_0 tends to have the d-axis coincident with the rotating flux vector as t goes to infinity.

As an intermediate step in the design of the adaptive control, we define reference signals for the currents (i_{sd}, i_{sq}): since T_L and α are unknown constants, we introduce the estimates $\hat{T}_L$ and $\hat{\alpha}$. Define the reference current signals

$$\begin{aligned} i_{sd}^* &= \frac{\psi^*}{M} + \frac{\dot{\psi}^*}{\hat{\alpha} M} + \frac{\eta_d}{\beta \hat{\alpha} M} \\ i_{sq}^* &= \frac{1}{\mu \psi^*}\left(-k_\omega \tilde{\omega} + \frac{\hat{T}_L}{J} + \dot{\omega}^*\right) \end{aligned} \tag{4.81}$$

for i_{sd} and i_{sq}, and the speed of the rotating (d,q) frame

$$\dot{\varepsilon}_0 = \omega_0 = \omega + \frac{\hat{\alpha} M i_{sq}}{\psi^*} - \frac{\eta_q}{\beta \psi^*} \tag{4.82}$$

in which k_ω is a positive design parameter and (η_d, η_q) are additional terms yet to be designed. As we shall see, the adaptation for $\hat{\alpha}(t)$ will be designed using a projection algorithm guaranteeing $\hat{\alpha}(t) \geq c_\alpha > 0$ for all $t \geq 0$ so that (4.81) is well defined. Introducing the current tracking errors

$$\begin{aligned} \tilde{i}_{sd} &= i_{sd} - i_{sd}^* \\ \tilde{i}_{sq} &= i_{sq} - i_{sq}^* \end{aligned} \tag{4.83}$$

from (1.31), (4.81), (4.82), and (4.83) we obtain

$$\begin{aligned} \dot{\omega} &= \mu \psi^* i_{sq} + \mu(\tilde{\psi}_{rd} i_{sq} - \tilde{\psi}_{rq} i_{sd}) - \frac{T_L}{J} \\ &= -k_\omega \tilde{\omega} + \dot{\omega}^* + \mu(\tilde{\psi}_{rd} i_{sq} - \tilde{\psi}_{rq} i_{sd}) - \frac{\tilde{T}_L}{J} + \mu \psi^* \tilde{i}_{sq} \\ \dot{\psi}_{rd} &= -\alpha \psi_{rd} + (\omega_0 - \omega)\psi_{rq} + \hat{\alpha}\psi^* + \dot{\psi}^* + \frac{\eta_d}{\beta} \\ &\quad + \hat{\alpha} M \tilde{i}_{sd} + \tilde{\alpha} M i_{sd} \\ \dot{\psi}_{rq} &= -\alpha \psi_{rq} - (\omega_0 - \omega)\tilde{\psi}_{rd} - \hat{\alpha} M i_{sq} + \frac{\eta_q}{\beta} + \alpha M i_{sq}\,. \end{aligned} \tag{4.84}$$

The speed and flux tracking errors dynamics are

$$\begin{aligned} \dot{\tilde{\omega}} &= -k_\omega \tilde{\omega} + \mu(\tilde{\psi}_{rd} i_{sq} - \tilde{\psi}_{rq} i_{sd}) - \frac{\tilde{T}_L}{J} + \mu \psi^* \tilde{i}_{sq} \\ \dot{\tilde{\psi}}_{rd} &= -\alpha \tilde{\psi}_{rd} + (\omega_0 - \omega)\tilde{\psi}_{rq} + \frac{\eta_d}{\beta} + \tilde{\alpha}(M i_{sd} - \psi^*) + \hat{\alpha} M \tilde{i}_{sd} \\ \dot{\tilde{\psi}}_{rq} &= -\alpha \tilde{\psi}_{rq} - (\omega_0 - \omega)\tilde{\psi}_{rd} + \frac{\eta_q}{\beta} + \tilde{\alpha} M i_{sq} \end{aligned} \tag{4.85}$$

where $\tilde{\alpha} = \alpha - \hat{\alpha}$ and $\tilde{T}_L = T_L - \hat{T}_L$ denote the parameter estimation errors. It will be the goal of the adaptation law for the parameter estimates $(\hat{T}_L, \hat{\alpha})$ to drive $\tilde{T}_L$ and $\tilde{\alpha}$ to zero while the objective of the control inputs u_{sd} and u_{sq} will be to drive $\tilde{i}_{sd}$ and $\tilde{i}_{sq}$ to zero along with $\tilde{\omega}$, $\tilde{\psi}_{rd}$, and $\tilde{\psi}_{rq}$. As in the adaptive observer presented in Section 3.2, we introduce the stator current estimates $(\hat{i}_{sd}, \hat{i}_{sq})$ which will be auxiliary signals for the estimation of the rotor fluxes and of the rotor resistance: the stator current observer is designed as

$$\begin{aligned}
\frac{\mathrm{d}\hat{i}_{sd}}{\mathrm{d}t} &= -\frac{R_s}{\sigma} i_{sd} + \omega_0 i_{sq} + \hat{\alpha}\beta\psi^* - \hat{\alpha}M\beta i_{sd} + \frac{1}{\sigma}u_{sd} \\
&\quad + k_i e_d + w_{d1} + w_{d2} \\
\frac{\mathrm{d}\hat{i}_{sq}}{\mathrm{d}t} &= -\frac{R_s}{\sigma} i_{sq} - \omega_0 i_{sd} - \beta\omega\psi^* - \hat{\alpha}M\beta i_{sq} + \frac{1}{\sigma}u_{sq} \\
&\quad + k_i e_q + w_{q1} + w_{q2}
\end{aligned} \tag{4.86}$$

in which k_i is a positive design parameter and $(w_{d1}, w_{d2}, w_{q1}, w_{q2})$ are additional terms yet to be chosen. We denote by

$$\begin{aligned}
e_d &= i_{sd} - \hat{i}_{sd} \\
e_q &= i_{sq} - \hat{i}_{sq}
\end{aligned}$$

the current estimation errors. The stator current estimation error dynamics may be written as

$$\begin{aligned}
\dot{e}_d &= -k_i e_d + \alpha\beta\tilde{\psi}_{rd} + \beta\omega\tilde{\psi}_{rq} - \beta\tilde{\alpha}(Mi_{sd} - \psi^*) - w_{d1} - w_{d2} \\
\dot{e}_q &= -k_i e_q + \alpha\beta\tilde{\psi}_{rq} - \beta\omega\tilde{\psi}_{rd} - \beta\tilde{\alpha}Mi_{sq} - w_{q1} - w_{q2}
\end{aligned} \tag{4.87}$$

and used in the design of the adaptive law for $\hat{\alpha}$. We replace the unknown variables $(\tilde{\psi}_{rd}, \tilde{\psi}_{rq})$ by the new error variables

$$\begin{aligned}
z_d &= e_d + \beta\tilde{\psi}_{rd} \\
z_q &= e_q + \beta\tilde{\psi}_{rq}
\end{aligned} \tag{4.88}$$

so that the error equations (4.85) and (4.87) are expressed in new coordinates as

$$\begin{aligned}
\dot{\tilde{\omega}} &= -k_\omega\tilde{\omega} + \frac{\mu}{\beta}(z_d i_{sq} - z_q i_{sd}) + \frac{\mu}{\beta}(-e_d i_{sq} + e_q i_{sd}) - \frac{\tilde{T}_L}{J} + \mu\psi^*\tilde{i}_{sq} \\
\dot{z}_d &= -k_i e_d + \omega_0(z_q - e_q) + \eta_d - w_{d1} - w_{d2} + \beta\hat{\alpha}M\tilde{i}_{sd} \\
\dot{z}_q &= -k_i e_q - \omega_0(z_d - e_d) + \eta_q - w_{q1} - w_{q2} \\
\dot{e}_d &= -(k_i + \alpha)e_d - \omega e_q + \alpha z_d + \omega z_q - \beta\tilde{\alpha}(Mi_{sd} - \psi^*) - w_{d1} - w_{d2} \\
\dot{e}_q &= -(k_i + \alpha)e_q + \omega e_d + \alpha z_q - \omega z_d - \beta\tilde{\alpha}Mi_{sq} - w_{q1} - w_{q2} \\
\dot{\varepsilon}_0 &= \omega_0 = \omega + \frac{\hat{\alpha}Mi_{sq}}{\psi^*} - \frac{\eta_q}{\beta\psi^*}\,.
\end{aligned} \tag{4.89}$$

The advantage of using the unknown variables (z_d, z_q) instead of $(\tilde{\psi}_{rd}, \tilde{\psi}_{rq})$ relies on the fact that their dynamics no longer depend on α (compare (4.89) with (4.85)). We now define some of the yet undetermined terms in (4.81), (4.82), (4.86) as (w_{d2} and w_{q2} in (4.86) are still to be chosen)

$$
\begin{aligned}
\eta_d &= (k_i - k)e_d - 2\gamma_1 \frac{\mu}{\beta}\tilde{\omega} i_{sq} + (\hat{\alpha} - k)\hat{z}_d \\
\eta_q &= (k_i - k)e_q + 2\gamma_1 \frac{\mu}{\beta}\tilde{\omega} i_{sd} + (\hat{\alpha} - k)\hat{z}_q \\
w_{d1} &= -\gamma_1 \frac{\mu}{\beta}\tilde{\omega} i_{sq} + (\hat{\alpha} - k)\hat{z}_d - (\omega_0 + \omega)e_q \\
w_{q1} &= \gamma_1 \frac{\mu}{\beta}\tilde{\omega} i_{sd} + (\hat{\alpha} - k)\hat{z}_q + (\omega_0 + \omega)e_d
\end{aligned}
\qquad (4.90)
$$

with k, γ_1 positive design parameters and $(\hat{z}_d,\ \hat{z}_q)$ estimates of the unknown error variables $(z_d,\ z_q)$ defined by (4.88). Note that the dynamics of $(\hat{z}_d, \hat{z}_q)$ are yet to be defined. If $(\hat{z}_d, \hat{z}_q)$ converge to (z_d, z_q), they allow us to recover the rotor flux vector (ψ_{rd}, ψ_{rq}) since $(\hat{i}_{sd},\ \hat{i}_{sq})$ are known variables and β is a known parameter. Hence, the new variables $(\hat{z}_d, \hat{z}_q)$ may be viewed as rotor flux estimates. Substituting (4.90) in (4.89), we obtain

$$
\begin{aligned}
\dot{\tilde{\omega}} &= -k_\omega \tilde{\omega} + \frac{\mu}{\beta}(z_d i_{sq} - z_q i_{sd}) + \frac{\mu}{\beta}(-e_d i_{sq} + e_q i_{sd}) - \frac{\tilde{T}_L}{J} + \mu \psi^* \tilde{i}_{sq} \\
\dot{z}_d &= -k e_d + \omega_0 z_q - \gamma_1 \frac{\mu}{\beta}\tilde{\omega} i_{sq} + \omega e_q - w_{d2} + \beta \hat{\alpha} M \tilde{i}_{sd} \\
\dot{z}_q &= -k e_q - \omega_0 z_d + \gamma_1 \frac{\mu}{\beta}\tilde{\omega} i_{sd} - \omega e_d - w_{q2} \\
\dot{e}_d &= -(k_i + \alpha)e_d - \omega e_q + \alpha z_d + \omega z_q - \beta \tilde{\alpha}(M i_{sd} - \psi^*) \\
&\quad + \gamma_1 \frac{\mu}{\beta}\tilde{\omega} i_{sq} - (\hat{\alpha} - k)\hat{z}_d + (\omega_0 + \omega)e_q - w_{d2} \\
\dot{e}_q &= -(k_i + \alpha)e_q + \omega e_d + \alpha z_q - \omega z_d - \beta \tilde{\alpha} M i_{sq} \\
&\quad - \gamma_1 \frac{\mu}{\beta}\tilde{\omega} i_{sd} - (\hat{\alpha} - k)\hat{z}_q - (\omega_0 + \omega)e_d - w_{q2}\,.
\end{aligned}
\qquad (4.91)
$$

We compute from (1.31), (4.81), (4.89), and (4.90) the dynamics of the stator current tracking errors $(\tilde{i}_{sd}, \tilde{i}_{sq})$:

$$
\begin{aligned}
\frac{\mathrm{d}\tilde{i}_{sd}}{\mathrm{d}t} &= \frac{1}{\sigma}u_{sd} + \frac{2\gamma_1 \mu \tilde{\omega}}{\sigma \beta^2 \hat{\alpha} M}u_{sq} + \phi_{d0} + z_d \phi_{d1} + z_q \phi_{d2} \\
&\quad + \alpha \tilde{z}_d \phi_{d3} + \alpha \tilde{z}_q \phi_{d4} + \tilde{T}_L \phi_{d5} + \tilde{\alpha} \phi_{d6} + \frac{k_i - k}{\beta \hat{\alpha} M} w_{d2} \\
\frac{\mathrm{d}\tilde{i}_{sq}}{\mathrm{d}t} &= \frac{1}{\sigma}u_{sq} + \phi_{q0} + z_d \phi_{q1} + z_q \phi_{q2} + \alpha \tilde{z}_d \phi_{q3} + \alpha \tilde{z}_q \phi_{q4} \\
&\quad + \tilde{T}_L \phi_{q5} + \tilde{\alpha} \phi_{q6}
\end{aligned}
\qquad (4.92)
$$

with

$$\begin{aligned}\tilde{z}_d &= z_d - \hat{z}_d \\ \tilde{z}_q &= z_q - \hat{z}_q\end{aligned}$$

and

$$\begin{aligned}
\phi_{d0} &= -\frac{R_s}{\sigma} i_{sd} + \omega_0 i_{sq} + \hat{\alpha}\beta\psi^* + \hat{\alpha}\hat{z}_d - \hat{\alpha} e_d - \omega e_q \\
&\quad -\hat{\alpha}\beta M i_{sd} - \frac{\dot{\psi}^*}{M} - \frac{\ddot{\psi}^*}{\hat{\alpha}M} + \frac{\dot{\hat{\alpha}}}{\hat{\alpha}^2 M}\left(\dot{\psi}^* + \frac{\eta_d}{\beta}\right) \\
&\quad -\frac{k_i - k}{\beta\hat{\alpha}M}\Big[-(k_i + \hat{\alpha})e_d - \omega e_q + \hat{\alpha}\hat{z}_d + \gamma_1 \frac{\mu}{\beta}\tilde{\omega} i_{sq} \\
&\quad -(\hat{\alpha} - k)\hat{z}_d + \left(\omega_0 + \omega\right) e_q\Big] \\
&\quad +2\frac{\gamma_1 \mu i_{sq}}{\beta^2 \hat{\alpha} M}\left[-k_\omega \tilde{\omega} + \frac{\mu}{\beta}(-e_d i_{sq} + e_q i_{sd}) + \mu\psi^* \tilde{i}_{sq}\right] \\
&\quad +2\frac{\gamma_1 \mu \tilde{\omega}}{\beta^2 \hat{\alpha} M}\Big[-\frac{R_s}{\sigma} i_{sq} - \omega_0 i_{sd} + \hat{\alpha}\hat{z}_q - \hat{\alpha} e_q + \omega e_d \\
&\quad -\hat{\alpha} M \beta i_{sq} - \beta\omega\psi^*\Big] - \frac{1}{\beta\hat{\alpha}M}[\dot{\hat{\alpha}}\hat{z}_d + (\hat{\alpha} - k)\dot{\hat{z}}_d] \\
\phi_{d1} &= 2\left(\frac{\gamma_1 \mu^2 i_{sq}^2}{\beta^3 \hat{\alpha} M} - \frac{\gamma_1 \mu \tilde{\omega}}{\beta^2 \hat{\alpha} M}\omega\right) \\
\phi_{d2} &= -2\frac{\gamma_1 \mu^2 i_{sq} i_{sd}}{\beta^3 \hat{\alpha} M} + \omega\left(1 - \frac{k_i - k}{\beta\hat{\alpha}M}\right) \\
\phi_{d3} &= 1 - \frac{k_i - k}{\beta\hat{\alpha}M} \\
\phi_{d4} &= 2\frac{\gamma_1 \mu \tilde{\omega}}{\beta^2 \hat{\alpha} M} \\
\phi_{d5} &= -2\frac{\gamma_1 \mu i_{sq}}{J\beta^2 \hat{\alpha} M} \\
\phi_{d6} &= \beta\psi^* + \hat{z}_d - e_d - \beta M i_{sd} + \frac{k_i - k}{\beta\hat{\alpha}M} e_d - \frac{k_i - k}{\beta\hat{\alpha}M}\hat{z}_d \\
&\quad + \frac{k_i - k}{\hat{\alpha}M}(M i_{sd} - \psi^*) + 2\frac{\gamma_1 \mu}{\beta^2 \hat{\alpha} M}\tilde{\omega}(\hat{z}_q - e_q - \beta M i_{sq}) \\
\phi_{q0} &= -\frac{R_s}{\sigma} i_{sq} - \omega_0 i_{sd} + \hat{\alpha}\hat{z}_q - \hat{\alpha} e_q - \beta\omega\psi^* - \hat{\alpha}\beta M i_{sq} \\
&\quad + \frac{\dot{\psi}^*}{\mu\psi^{*2}}\left(-k_\omega \tilde{\omega} + \frac{\hat{T}_L}{J} + \dot{\omega}^*\right) - \frac{1}{\mu\psi^*}\left[\frac{\dot{\hat{T}}_L}{J} + \ddot{\omega}^*\right.
\end{aligned}$$

$$
\begin{aligned}
&\left. +k_\omega^2\tilde{\omega} - \frac{k_\omega\mu}{\beta}(-e_d i_{sq} + e_q i_{sd}) - k_\omega\mu\psi^*\tilde{i}_{sq}\right] + \omega e_d \\
\phi_{q1} &= -\omega + \frac{k_\omega}{\psi^*\beta} i_{sq} \\
\phi_{q2} &= -\frac{k_\omega}{\psi^*\beta} i_{sd} \\
\phi_{q3} &= 0 \\
\phi_{q4} &= 1 \\
\phi_{q5} &= \frac{-k_\omega}{\mu\psi^* J} \\
\phi_{q6} &= \hat{z}_q - e_q - \beta M i_{sq} \,.
\end{aligned}
\tag{4.93}
$$

In order to determine (w_{d2}, w_{q2}), the dynamics of the estimation variables $\hat{z}_d$, $\hat{z}_q$, $\hat{\alpha}$, and $\hat{T}_L$ and of the feedback control inputs (u_{sd}, u_{sq}), we consider the function

$$
V_0 = \frac{1}{2}[\gamma_1\tilde{\omega}^2 + e_d^2 + e_q^2 + \gamma_2(\tilde{z}_d^2 + \tilde{z}_q^2) + \gamma_3\tilde{T}_L^2 + \gamma_4\tilde{\alpha}^2 + z_d^2 + z_q^2] \tag{4.94}
$$

in which γ_i, $2 \le i \le 4$, are positive parameters. We define $\gamma_2 = (\alpha - k)\gamma_0$ with $\gamma_0 > 0$ and, to guarantee that $\gamma_2 > 0$, k is chosen so that $0 < k < \alpha_m$ with α_m the minimum value of α (assumed to be known). Alternatively, we may choose $\gamma_0 < 0$ and $k \ge \alpha_M$ with α_M the maximum value of α (assumed to be known). From (4.91) and (4.94), the time derivative of (4.94) results:

$$
\begin{aligned}
\dot{V}_0 = {} & -\gamma_1 k_\omega\tilde{\omega}^2 - (k_i + \alpha)(e_d^2 + e_q^2) - z_d w_{d2} - z_q w_{q2} \\
& - e_d w_{d2} - e_q w_{q2} + (\alpha - k)\tilde{z}_d(\gamma_0\dot{\tilde{z}}_d + e_d) \\
& + (\alpha - k)\tilde{z}_q(\gamma_0\dot{\tilde{z}}_q + e_q) + \tilde{\alpha}[-\beta e_d(M i_{sd} - \psi^*) \\
& - \beta e_q M i_{sq} + e_d\hat{z}_d + e_q\hat{z}_q + \gamma_4\dot{\tilde{\alpha}}] \\
& + \tilde{T}_L\left(-\gamma_1\frac{\tilde{\omega}}{J} + \gamma_3\dot{\tilde{T}}_L\right) + \gamma_1\mu\psi^*\tilde{\omega}\tilde{i}_{sq} + \hat{\alpha}\beta M z_d\tilde{i}_{sd} \,.
\end{aligned}
\tag{4.95}
$$

We now consider the function

$$
\begin{aligned}
V &= V_0 + \frac{1}{2}(\tilde{i}_{sd}^2 + \tilde{i}_{sq}^2) \\
&= \frac{1}{2}[\gamma_1\tilde{\omega}^2 + e_d^2 + e_q^2 + \gamma_2(\tilde{z}_d^2 + \tilde{z}_q^2) + \gamma_3\tilde{T}_L^2 + \gamma_4\tilde{\alpha}^2 \\
&\quad + z_d^2 + z_q^2 + \tilde{i}_{sd}^2 + \tilde{i}_{sq}^2] \,.
\end{aligned}
\tag{4.96}
$$

From (4.95), (4.91), and (4.92), its time derivative is

$$
\begin{aligned}
\dot{V} = {} & -\gamma_1 k_\omega\tilde{\omega}^2 - (k_i + \alpha)(e_d^2 + e_q^2) - e_d w_{d2} - e_q w_{q2} \\
& + z_d[-w_{d2} + \tilde{i}_{sd}(\phi_{d1} + \hat{\alpha}\beta M) + \tilde{i}_{sq}\phi_{q1} + k(\tilde{i}_{sd}\phi_{d3} + \tilde{i}_{sq}\phi_{q3})] \\
& + z_q[-w_{q2} + \tilde{i}_{sd}\phi_{d2} + \tilde{i}_{sq}\phi_{q2} + k(\tilde{i}_{sd}\phi_{d4} + \tilde{i}_{sq}\phi_{q4})]
\end{aligned}
$$

$$
\begin{aligned}
&+(\alpha-k)\tilde{z}_d(\gamma_0\dot{\tilde{z}}_d+e_d+\tilde{i}_{sd}\phi_{d3}+\tilde{i}_{sq}\phi_{q3})\\
&+(\alpha-k)\tilde{z}_q(\gamma_0\dot{\tilde{z}}_q+e_q+\tilde{i}_{sd}\phi_{d4}+\tilde{i}_{sq}\phi_{q4})\\
&+\tilde{i}_{sd}\left(\frac{1}{\sigma}u_{sd}+\frac{2\gamma_1\mu\tilde{\omega}}{\sigma\beta^2\hat{\alpha}M}u_{sq}+\phi_{d0}-k\hat{z}_d\phi_{d3}-k\hat{z}_q\phi_{d4}+\frac{k_i-k}{\beta\hat{\alpha}M}w_{d2}\right)\\
&+\tilde{i}_{sq}\left(\frac{1}{\sigma}u_{sq}+\phi_{q0}+\gamma_1\mu\psi^*\tilde{\omega}-k\hat{z}_d\phi_{q3}-k\hat{z}_q\phi_{q4}\right)\\
&+\tilde{\alpha}\Big(-\beta e_d(Mi_{sd}-\psi^*)-\beta e_qMi_{sq}+e_d\hat{z}_d+e_q\hat{z}_q\\
&+\gamma_4\dot{\tilde{\alpha}}+\tilde{i}_{sd}\phi_{d6}+\tilde{i}_{sq}\phi_{q6}\Big)\\
&+\tilde{T}_L\left(-\gamma_1\frac{\tilde{\omega}}{J}+\gamma_3\dot{\tilde{T}}_L+\tilde{i}_{sd}\phi_{d5}+\tilde{i}_{sq}\phi_{q5}\right).
\end{aligned}
\tag{4.97}
$$

The terms (w_{d2}, w_{q2}), the feedback controls (u_{sd}, u_{sq}), and the dynamics of $(\hat{z}_d, \hat{z}_q, \hat{\alpha}, \hat{T}_L)$ are now chosen in order to force $\dot{V}$ to be negative semidefinite. Choosing the yet undetermined terms in (4.86) as

$$
\begin{aligned}
w_{d2}&=[\tilde{i}_{sd}(\phi_{d1}+\hat{\alpha}\beta M)+\tilde{i}_{sq}\phi_{q1}+k(\tilde{i}_{sd}\phi_{d3}+\tilde{i}_{sq}\phi_{q3})]\\
w_{q2}&=[\tilde{i}_{sd}\phi_{d2}+\tilde{i}_{sq}\phi_{q2}+k(\tilde{i}_{sd}\phi_{d4}+\tilde{i}_{sq}\phi_{q4})]
\end{aligned}
\tag{4.98}
$$

we obtain

$$
\begin{aligned}
\dot{V}=&-\gamma_1k_\omega\tilde{\omega}^2-(k_i+\alpha)(e_d^2+e_q^2)\\
&+(\alpha-k)\tilde{z}_d(\gamma_0\dot{\tilde{z}}_d+e_d+\tilde{i}_{sd}\phi_{d3}+\tilde{i}_{sq}\phi_{q3})\\
&+(\alpha-k)\tilde{z}_q(\gamma_0\dot{\tilde{z}}_q+e_q+\tilde{i}_{sd}\phi_{d4}+\tilde{i}_{sq}\phi_{q4})\\
&+\tilde{i}_{sd}\Big(\frac{1}{\sigma}u_{sd}+\frac{2\gamma_1\mu\tilde{\omega}}{\sigma\beta^2\hat{\alpha}M}u_{sq}+\phi_{d0}-k\hat{z}_d\phi_{d3}-k\hat{z}_q\phi_{d4}\\
&-e_d(\phi_{d1}+\hat{\alpha}\beta M+k\phi_{d3})-e_q(\phi_{d2}+k\phi_{d4})+\frac{k_i-k}{\beta\hat{\alpha}M}w_{d2}\Big)\\
&+\tilde{i}_{sq}\Big(\frac{1}{\sigma}u_{sq}+\phi_{q0}+\gamma_1\mu\psi^*\tilde{\omega}-k\hat{z}_d\phi_{q3}-k\hat{z}_q\phi_{q4}\\
&-e_d(\phi_{q1}+k\phi_{q3})-e_q(\phi_{q2}+k\phi_{q4})\Big)\\
&+\tilde{\alpha}\Big(-\beta e_d(Mi_{sd}-\psi^*)-\beta e_qMi_{sq}+e_d\hat{z}_d+e_q\hat{z}_q\\
&+\tilde{i}_{sd}\phi_{d6}+\tilde{i}_{sq}\phi_{q6}+\gamma_4\dot{\tilde{\alpha}}\Big)\\
&+\tilde{T}_L\left(-\gamma_1\frac{\tilde{\omega}}{J}+\tilde{i}_{sd}\phi_{d5}+\tilde{i}_{sq}\phi_{q5}+\gamma_3\dot{\tilde{T}}_L\right).
\end{aligned}
\tag{4.99}
$$

We finally design the control inputs and the adaptation dynamics as

$$\begin{aligned}
\dot{\hat{z}}_d &= -ke_d + \omega_0 \hat{z}_q - \gamma_1 \frac{\mu}{\beta} \tilde{\omega} i_{sq} + \omega e_q \\
&\quad - [\tilde{i}_{sd} \phi_{d1} + \tilde{i}_{sq} \phi_{q1} + k(\tilde{i}_{sd} \phi_{d3} + \tilde{i}_{sq} \phi_{q3})] \\
&\quad + \frac{1}{\gamma_0}(e_d + \tilde{i}_{sd} \phi_{d3} + \tilde{i}_{sq} \phi_{q3}) \\
\dot{\hat{z}}_q &= -ke_q - \omega_0 \hat{z}_d + \gamma_1 \frac{\mu}{\beta} \tilde{\omega} i_{sd} - \omega e_d \\
&\quad - [\tilde{i}_{sd} \phi_{d2} + \tilde{i}_{sq} \phi_{q2} + k(\tilde{i}_{sd} \phi_{d4} + \tilde{i}_{sq} \phi_{q4})] \\
&\quad + \frac{1}{\gamma_0}(e_q + \tilde{i}_{sd} \phi_{d4} + \tilde{i}_{sq} \phi_{q4}) \\
u_{sq} &= \sigma[-\phi_{q0} - \gamma_1 \mu \psi^* \tilde{\omega} + k\hat{z}_d \phi_{q3} + k\hat{z}_q \phi_{q4} \\
&\quad + e_d(\phi_{q1} + k\phi_{q3}) + e_q(\phi_{q2} + k\phi_{q4}) - k_e \tilde{i}_{sq}] \\
u_{sd} &= \sigma\left[-\frac{2\gamma_1 \mu \tilde{\omega}}{\sigma \beta^2 \hat{\alpha} M} u_{sq} - \phi_{d0} + k\hat{z}_d \phi_{d3} + k\hat{z}_q \phi_{d4} \right. \\
&\quad + e_d(\phi_{d1} + \hat{\alpha}\beta M + k\phi_{d3}) + e_q(\phi_{d2} + k\phi_{d4}) \\
&\quad \left. - k_e \tilde{i}_{sd} - \frac{k_i - k}{\beta \hat{\alpha} M} w_{d2} \right] \\
\dot{\hat{T}}_L &= \frac{1}{\gamma_3}\left[-\frac{\gamma_1}{J}\tilde{\omega} + \tilde{i}_{sd} \phi_{d5} + \tilde{i}_{sq} \phi_{q5}\right]
\end{aligned} \tag{4.100}$$

and

$$\begin{aligned}
\dot{\hat{\alpha}} &= \frac{1}{\gamma_4}[-\beta e_d(M i_{sd} - \psi^*) - \beta e_q M i_{sq} + e_d \hat{z}_d \\
&\quad + e_q \hat{z}_q + \tilde{i}_{sd} \phi_{d6} + \tilde{i}_{sq} \phi_{q6}]
\end{aligned} \tag{4.101}$$

so that (4.99) becomes

$$\dot{V} = -\gamma_1 k_\omega \tilde{\omega}^2 - (k_i + \alpha)(e_d^2 + e_q^2) - k_e(\tilde{i}_{sd}^2 + \tilde{i}_{sq}^2) \, . \tag{4.102}$$

Equations (4.100), (4.101), (4.93), (4.86), (4.90), (4.98), (4.81), (4.82) and (4.77) define a seventh order dynamic feedback compensator, whose state variables are $(\varepsilon_0, \hat{i}_d, \hat{i}_q, \hat{z}_d, \hat{z}_q, \hat{T}_L, \hat{\alpha})$, which generates the control signals (u_{sa}, u_{sb}) on the basis of the measurements (ω, i_{sa}, i_{sb}), the reference signals (ω^*, ψ^*), and their time derivatives $(\dot{\omega}^*, \ddot{\omega}^*, \dot{\psi}^*, \ddot{\psi}^*)$. In order to guarantee that $\hat{\alpha}(t) \geq c_\alpha > 0$ for every $t \geq 0$ we modify the dynamics (4.101) according to

$$\begin{aligned}
\dot{\hat{\alpha}} &= \mathrm{Proj}\left(\frac{1}{\gamma_4}[-\beta e_d(M i_{sd} - \psi^*) - \beta e_q M i_{sq} + e_d \hat{z}_d \right. \\
&\quad \left. + e_q \hat{z}_q + \tilde{i}_{sd} \phi_{d6} + \tilde{i}_{sq} \phi_{q6}], \hat{\alpha}\right) \, , \; \hat{\alpha}(0) = \hat{\alpha}_0
\end{aligned} \tag{4.103}$$

where $\mathrm{Proj}(\xi, \hat{\alpha})$ is the smooth projection algorithm given by (A.25) in Appendix A and defined in our case by

$$\begin{aligned}\text{Proj}(\xi,\hat{\alpha}) &= \xi \text{ , if } p(\hat{\alpha}) \leq 0 \\ \text{Proj}(\xi,\hat{\alpha}) &= \xi \text{ , if } p(\hat{\alpha}) \geq 0 \text{ and } \xi \geq 0 \\ \text{Proj}(\xi,\hat{\alpha}) &= [1-p(\hat{\alpha})]\xi \text{ , otherwise}\end{aligned}$$

in which $p(\hat{\alpha}) = \frac{\alpha_m^2 - \hat{\alpha}^2}{2\delta\alpha_m - \delta^2}$ with $\alpha_m = R_{rm}/L_r$ the minimum (known) value of α and $\delta > 0$ such that $\alpha_m - \delta > 0$. The initial condition $\hat{\alpha}_0$ in (4.103) is chosen so that $\hat{\alpha}_0 \geq \alpha_m$. The properties of the operator $\text{Proj}(\xi,\hat{\alpha})$ imply that, by substituting (4.103) instead of (4.101) in (4.99), we obtain instead of (4.102) the inequality

$$\dot{V} \leq -\gamma_1 k_\omega \tilde{\omega}^2 - (k_i+\alpha)(e_d^2+e_q^2) - k_e(\tilde{i}_{sd}^2+\tilde{i}_{sq}^2)\,. \tag{4.104}$$

From (4.96) and (4.104), it follows that $(\tilde{\omega}, e_d, e_q, z_d, z_q, \tilde{z}_d, \tilde{z}_q, \tilde{i}_{sd}, \tilde{i}_{sq}, \tilde{\alpha}, \tilde{T}_L)$ are bounded: their bounds depend on the initial errors. Therefore, according to (4.100), (u_{sd}, u_{sq}) and, consequently, ($\dot{\tilde{\omega}}$, $\dot{e}_d$, e_q, $\dot{z}_d$, $\dot{z}_q$, $\dot{\tilde{z}}_d$, $\dot{\tilde{z}}_q$, $\dot{\tilde{i}}_{sd}$, $\dot{\tilde{i}}_{sq}$, $\dot{\tilde{\alpha}}$, $\dot{\tilde{T}}_L$) are bounded. Hence, $(\ddot{\tilde{\omega}}, \ddot{e}_d, \ddot{e}_q, \ddot{\tilde{i}}_{sd}, \ddot{\tilde{i}}_{sq})$ are bounded and therefore ($\dot{\tilde{\omega}}$, $\dot{e}_d$, $\dot{e}_q$, $\dot{\tilde{i}}_{sd}$, $\dot{\tilde{i}}_{sq}$) are uniformly continuous. On the other hand, integrating (4.104) on the time interval $[0,t]$ we have

$$\begin{aligned}&\int_0^t [\gamma_1 k_\omega \tilde{\omega}^2(\tau) + (k_i+\alpha)(e_d^2(\tau)+e_q^2(\tau)) + k_e(\tilde{i}_{sd}^2(\tau)+\tilde{i}_{sq}^2(\tau))]\,\mathrm{d}\tau \\ &\leq \int_0^t -\dot{V}(\tau)\,\mathrm{d}\tau \leq V(0)\end{aligned}$$

which implies by Barbalat's Lemma A.2 in Appendix A that

$$\lim_{t\to\infty} \|[\tilde{\omega}(t), e_d(t), e_q(t), \tilde{i}_{sd}(t), \tilde{i}_{sq}(t)]\| = 0\,.$$

This shows that asymptotic speed tracking is achieved from any initial condition provided that $\hat{\alpha}_0 \geq \alpha_m - \delta$ in (4.103). Moreover, the current estimation errors $(e_d(t), e_q(t))$ and the current tracking errors $(\tilde{i}_{sd}(t), \tilde{i}_{sq}(t))$ asymptotically tend to zero.

We now analyze under which conditions $\tilde{z}_d(t)$, $\tilde{z}_q(t)$, $\tilde{T}_L(t)$, $\tilde{\alpha}(t)$, $z_d(t)$, and $z_q(t)$ also tend asymptotically to zero. The error equations are

$$\begin{aligned}\frac{\mathrm{d}\tilde{\omega}}{\mathrm{d}t} &= -k_\omega\tilde{\omega} + \frac{\mu}{\beta}(z_d i_{sq} - z_q i_{sd}) + \frac{\mu}{\beta}(-e_d i_{sq} + e_q i_{sd}) \\ &\quad -\frac{\tilde{T}_L}{J} + \mu\psi^*\tilde{i}_{sq} \\ \frac{\mathrm{d}e_d}{\mathrm{d}t} &= -(k_i+\alpha)e_d - \omega e_q + \omega z_q + \tilde{\alpha}[\hat{z}_d - \beta(M i_{sd} - \psi^*)] \\ &\quad +(\alpha-k)\tilde{z}_d + k z_d + \gamma_1\frac{\mu}{\beta}\tilde{\omega} i_{sq} + (\omega_0+\omega)e_q \\ &\quad -(\phi_{d1} + \hat{\alpha}\beta M + k\phi_{d3})\tilde{i}_{sd} - (\phi_{q1} + k\phi_{q3})\tilde{i}_{sq} \\ \frac{\mathrm{d}e_q}{\mathrm{d}t} &= -(k_i+\alpha)e_q + \omega e_d - \omega z_d + \tilde{\alpha}[\hat{z}_q - \beta M i_{sq}]\end{aligned}$$

$$
\begin{aligned}
&+(\alpha-k)\tilde{z}_q+kz_q-\gamma_1\frac{\mu}{\beta}\tilde{\omega}i_{sd}-(\omega_0+\omega)e_d \\
&-(\phi_{d2}+k\phi_{d4})\tilde{i}_{sd}-(\phi_{q2}+k\phi_{q4})\tilde{i}_{sq} \\
\frac{\mathrm{d}\tilde{i}_{sd}}{\mathrm{d}t} &= -k_e\tilde{i}_{sd}+z_d\phi_{d1}+z_q\phi_{d2}+(\alpha-k)\phi_{d3}\tilde{z}_d \\
&+(\alpha-k)\phi_{d4}\tilde{z}_q+k\phi_{d3}z_d+k\phi_{d4}z_q \\
&+(\phi_{d1}+\hat{\alpha}\beta M+k\phi_{d3})e_d \\
&+(\phi_{d2}+k\phi_{d4})e_q+\phi_{d5}\tilde{T}_L+\phi_{d6}\tilde{\alpha} \\
\frac{\mathrm{d}\tilde{i}_{sq}}{\mathrm{d}t} &= -k_e\tilde{i}_{sq}+z_d\phi_{q1}+z_q\phi_{q2}+(\alpha-k)\phi_{q3}\tilde{z}_d \\
&+(\alpha-k)\phi_{q4}\tilde{z}_q+k\phi_{q3}z_d+k\phi_{q4}z_q \\
&-\gamma_1\mu\psi^*\tilde{\omega}+(\phi_{q1}+k\phi_{q3})e_d \\
&+(\phi_{q2}+k\phi_{q4})e_q+\phi_{q5}\tilde{T}_L+\phi_{q6}\tilde{\alpha} \\
\frac{\mathrm{d}z_d}{\mathrm{d}t} &= -ke_d+\omega_0 z_q-\frac{\gamma_1\mu}{\beta}i_{sq}\tilde{\omega}+\omega e_q-(\phi_{d1}+k\phi_{d3})\tilde{i}_{sd} \\
&-(\phi_{q1}+k\phi_{q3})\tilde{i}_{sq} \\
\frac{\mathrm{d}z_q}{\mathrm{d}t} &= -ke_q-\omega_0 z_d+\frac{\gamma_1\mu}{\beta}i_{sd}\tilde{\omega}-\omega e_d-(\phi_{d2}+k\phi_{d4})\tilde{i}_{sd} \\
&-(\phi_{q2}+k\phi_{q4})\tilde{i}_{sq} \\
\frac{\mathrm{d}\tilde{z}_d}{\mathrm{d}t} &= \omega_0\tilde{z}_q-\frac{1}{\gamma_0}(e_d+\phi_{d3}\tilde{i}_{sd}+\phi_{q3}\tilde{i}_{sq}) \\
\frac{\mathrm{d}\tilde{z}_q}{\mathrm{d}t} &= -\omega_0\tilde{z}_d-\frac{1}{\gamma_0}(e_q+\phi_{d4}\tilde{i}_{sd}+\phi_{q4}\tilde{i}_{sq}) \\
\frac{\mathrm{d}\tilde{\alpha}}{\mathrm{d}t} &= -\mathrm{Proj}\Bigg(\frac{1}{\gamma_4}[(\beta\psi^*-\beta M i_{sd}+\hat{z}_d)e_d+(\hat{z}_q-\beta M i_{sq})e_q \\
&+\phi_{d6}\tilde{i}_{sd}+\phi_{q6}\tilde{i}_{sq}],\hat{\alpha}\Bigg) \\
\frac{\mathrm{d}\tilde{T}_L}{\mathrm{d}t} &= \frac{1}{\gamma_3}\left(\frac{\gamma_1}{J}\tilde{\omega}-\phi_{d5}\tilde{i}_{sd}-\phi_{q5}\tilde{i}_{sq}\right).
\end{aligned} \tag{4.105}
$$

The equations (4.105) may be rewritten as

$$
\begin{aligned}
\dot{x}(t) &= A(t)x(t)+B(t)z(t) \\
\dot{z}(t) &= D(t)x(t)
\end{aligned} \tag{4.106}
$$

with $x=[\tilde{\omega},e_d,e_q,\tilde{i}_{sd},\tilde{i}_{sq}]^{\mathrm{T}}$, $z=[z_a,z_b,\tilde{z}_a,\tilde{z}_b,\tilde{\alpha},\tilde{T}_L]^{\mathrm{T}}$, $A(t)$, $B(t)$, $D(t)$ suitable matrices and

$$
\begin{bmatrix} z_a \\ z_b \end{bmatrix} = \begin{bmatrix} \cos\varepsilon_0 & -\sin\varepsilon_0 \\ \sin\varepsilon_0 & \cos\varepsilon_0 \end{bmatrix} \begin{bmatrix} z_d \\ z_q \end{bmatrix}
$$

$$\begin{bmatrix} \tilde{z}_a \\ \tilde{z}_b \end{bmatrix} = \begin{bmatrix} \cos \varepsilon_0 & -\sin \varepsilon_0 \\ \sin \varepsilon_0 & \cos \varepsilon_0 \end{bmatrix} \begin{bmatrix} \tilde{z}_d \\ \tilde{z}_q \end{bmatrix} .$$

The radially unbounded function (4.96) may be written as

$$V = \frac{1}{2}(x^{\mathrm{T}} P x + z^{\mathrm{T}} \Lambda^{-1} z)$$

with

$$\begin{aligned} \Lambda &= \mathrm{diag}\left[1, 1, \frac{1}{(\alpha - k)\gamma_0}, \frac{1}{(\alpha - k)\gamma_0}, \frac{1}{\gamma_4}, \frac{1}{\gamma_3}\right] \\ P &= \mathrm{diag}[\gamma_1, 1, 1, 1, 1] . \end{aligned}$$

Its time derivative (4.104) satisfies

$$\dot{V} \leq -x^{\mathrm{T}} Q x$$

with

$$Q = \mathrm{diag}[\gamma_1 k_\omega, k_i + \alpha, k_i + \alpha, k_e, k_e] .$$

By virtue of Lemma A.3 in Appendix A, we can establish that if persistency of excitation conditions are satisfied, *i.e.* there exist two positive constants T and k_T such that

$$\int_t^{t+T} B^{\mathrm{T}}(\tau) B(\tau) \mathrm{d}\tau \geq k_T I > 0 , \quad \forall t \geq 0 , \tag{4.107}$$

then the state vectors $x(t)$ and $z(t)$ of system (4.106) exponentially converge to zero for any initial condition. In summary, if (4.107) is satisfied then, in addition to asymptotic speed tracking:

1. since both $(z_d(t), z_q(t))$ and $(e_d(t), e_q(t))$ exponentially tend to zero, from (4.88) it follows that exponential rotor flux tracking is achieved and, in addition, the rotor flux vector is exponentially oriented with respect to the (d,q) frame so that

$$\begin{aligned} \lim_{t \to \infty} [\psi_{rd}(t) - \psi^*(t)] &= 0 \\ \lim_{t \to \infty} \psi_{rq}(t) &= 0 ; \end{aligned}$$

2. since $(\tilde{\alpha}(t), \tilde{T}_L(t))$ exponentially tend to zero, both the rotor resistance R_r and the load torque T_L are exponentially estimated so that

$$\begin{aligned} \lim_{t \to \infty} [\alpha - \hat{\alpha}(t)] &= 0 \\ \lim_{t \to \infty} [T_L - \hat{T}_L(t)] &= 0 ; \end{aligned}$$

3. since $(\tilde{z}_d(t), \tilde{z}_q(t))$ exponentially tend to zero, the rotor flux vector is exponentially estimated: defining, according to (4.88),

$$\begin{aligned} \hat{\psi}_{rd} &= \psi^* + \frac{1}{\beta}(\hat{z}_d - e_d) \\ \hat{\psi}_{rq} &= \frac{1}{\beta}(\hat{z}_q - e_q) \end{aligned} \tag{4.108}$$

we have

$$\begin{aligned} &\lim_{t\to\infty}[\psi_{rd}(t) - \hat{\psi}_{rd}(t)] = 0 \\ &\lim_{t\to\infty}[\psi_{rq}(t) - \hat{\psi}_{rq}(t)] = 0\,. \end{aligned}$$

In conclusion, the seventh order dynamic *adaptive output feedback control*

$$\begin{aligned} \begin{bmatrix} u_{sa} \\ u_{sb} \end{bmatrix} &= \begin{bmatrix} \cos\varepsilon_0 & -\sin\varepsilon_0 \\ \sin\varepsilon_0 & \cos\varepsilon_0 \end{bmatrix} \begin{bmatrix} u_{sd} \\ u_{sq} \end{bmatrix} \\ u_{sq} &= \sigma[-\phi_{q0} - \gamma_1\mu\psi^*\tilde{\omega} + k\hat{z}_d\phi_{q3} + k\hat{z}_q\phi_{q4} \\ &\quad + e_d(\phi_{q1} + k\phi_{q3}) + e_q(\phi_{q2} + k\phi_{q4}) - k_e\tilde{i}_{sq}] \\ u_{sd} &= \sigma\Big[-\frac{2\gamma_1\mu\tilde{\omega}}{\sigma\beta^2\hat{\alpha}M}u_{sq} - \phi_{d0} + k\hat{z}_d\phi_{d3} + k\hat{z}_q\phi_{d4} \\ &\quad + e_d(\phi_{d1} + \hat{\alpha}\beta M + k\phi_{d3}) + e_q(\phi_{d2} + k\phi_{d4}) \\ &\quad - k_e\tilde{i}_{sd} - \frac{k_i - k}{\beta\hat{\alpha}M}w_{d2}\Big] \\ \frac{d\varepsilon_0}{dt} &= \omega + \frac{\hat{\alpha}Mi_{sq}}{\psi^*} - \frac{\eta_q}{\beta\psi^*} \\ \frac{d\hat{i}_{sd}}{dt} &= -\frac{R_s}{\sigma}i_{sd} + \omega_0 i_{sq} + \hat{\alpha}\beta\psi^* - \hat{\alpha}M\beta i_{sd} + \frac{1}{\sigma}u_{sd} \\ &\quad + k_i e_d + w_{d1} + w_{d2} \\ \frac{d\hat{i}_{sq}}{dt} &= -\frac{R_s}{\sigma}i_{sq} - \omega_0 i_{sd} - \beta\omega\psi^* - \hat{\alpha}M\beta i_{sq} + \frac{1}{\sigma}u_{sq} \\ &\quad + k_i e_q + w_{q1} + w_{q2} \\ \frac{d\hat{z}_d}{dt} &= -ke_d + \omega_0\hat{z}_q - \gamma_1\frac{\mu}{\beta}\tilde{\omega}i_{sq} + \omega e_q \\ &\quad - [\tilde{i}_{sd}\phi_{d1} + \tilde{i}_{sq}\phi_{q1} + k(\tilde{i}_{sd}\phi_{d3} + \tilde{i}_{sq}\phi_{q3})] \\ &\quad + \frac{1}{\gamma_0}(e_d + \tilde{i}_{sd}\phi_{d3} + \tilde{i}_{sq}\phi_{q3}) \\ \frac{d\hat{z}_q}{dt} &= -ke_q - \omega_0\hat{z}_d + \gamma_1\frac{\mu}{\beta}\tilde{\omega}i_{sd} - \omega e_d \end{aligned}$$

$$
\begin{aligned}
&\quad -[\tilde{i}_{sd}\phi_{d2}+\tilde{i}_{sq}\phi_{q2}+k(\tilde{i}_{sd}\phi_{d4}+\tilde{i}_{sq}\phi_{q4})] \\
&\quad +\frac{1}{\gamma_0}(e_q+\tilde{i}_{sd}\phi_{d4}+\tilde{i}_{sq}\phi_{q4}) \\
\frac{d\hat{T}_L}{dt} &= \frac{1}{\gamma_3}\left[-\frac{\gamma_1}{J}\tilde{\omega}+\tilde{i}_{sd}\phi_{d5}+\tilde{i}_{sq}\phi_{q5}\right] \\
\frac{d\hat{\alpha}}{dt} &= \mathrm{Proj}\left(\frac{1}{\gamma_4}[-\beta e_d(Mi_{sd}-\psi^*)-\beta e_q M i_{sq}+e_d\hat{z}_d\right. \\
&\quad \left. +e_q\hat{z}_q+\tilde{i}_{sd}\phi_{d6}+\tilde{i}_{sq}\phi_{q6}],\hat{\alpha}\right) \\
\begin{bmatrix} i_{sd} \\ i_{sq} \end{bmatrix} &= \begin{bmatrix} \cos\varepsilon_0 & \sin\varepsilon_0 \\ -\sin\varepsilon_0 & \cos\varepsilon_0 \end{bmatrix}\begin{bmatrix} i_{sa} \\ i_{sb} \end{bmatrix}
\end{aligned}
\tag{4.109}
$$

with η_q, w_{d1}, w_{q1}, w_{d2}, w_{q2} given in (4.90), (4.98), and ϕ_{di}, ϕ_{qi}, $0 \le i \le 6$, given in (4.93), has the following properties provided that $\hat{\alpha}(0) \ge \alpha_m - \delta$:

1. The rotor speed tracking error $[\omega(t)-\omega^*(t)]$ tends to zero, along with the stator current estimation errors $(e_d(t), e_q(t))$ and the stator current tracking errors $(\tilde{i}_{sd}, \tilde{i}_{sq})$, for any motor initial condition.
2. If the persistency of excitation condition (4.107) is satisfied, then the parameter estimation errors, the flux tracking errors and the flux estimation errors exponentially tend to zero:

$$
\begin{aligned}
&\lim_{t\to\infty}[T_L-\hat{T}_L(t)]=0 \\
&\lim_{t\to\infty}[\psi_{rd}(t)-\psi^*(t)]=0 \\
&\lim_{t\to\infty}\psi_{rq}(t)=0 \\
&\lim_{t\to\infty}\left\{\psi^*(t)+\frac{1}{\beta}[\hat{z}_d(t)-i_{sd}(t)+\hat{i}_{sd}(t)]-\psi_{rd}(t)\right\}=0 \\
&\lim_{t\to\infty}\left\{\frac{1}{\beta}[\hat{z}_q(t)-i_{sq}(t)+\hat{i}_{sq}(t)]-\psi_{rq}(t)\right\}=0\,.
\end{aligned}
$$

Remarks

1. The dynamic compensator (4.109) contains eight control parameters (k_ω, k_i, k_e, k, γ_0, γ_1, γ_3, γ_4) whose role may be evaluated by examining both the closed-loop error equations (4.105) and the corresponding function (4.96) with time derivative (4.104). The parameters (k_ω, k_i, k_e) determine the rate of decay of $\dot{V}$ in (4.104) and directly affect (see (4.105)) the dynamics of speed tracking error $\tilde{\omega}$, current

estimation errors (e_d, e_q), and current tracking errors $(\tilde{i}_{sd}, \tilde{i}_{sq})$, respectively. The parameter γ_1 determines the influence of speed tracking errors on the other errors and is typically chosen much smaller than one. The parameters k_ω and γ_1 may be separately tuned using the equations obtained from (4.85) and the first two equations in (4.90) in ideal conditions, *i.e.* by setting $e_d = e_q = 0$, $\tilde{\alpha} = 0$, $\tilde{i}_{sd} = \tilde{i}_{sq} = 0$, and $\hat{z}_d = z_d = \hat{z}_q = z_q = 0$, so that $\tilde{\omega}$ has the desired transients. The parameters $1/\gamma_3$ and $1/\gamma_4$ are the adaptation gains for $\hat{T}_L$ and $\hat{\alpha}$, respectively: the smaller they are chosen, the slower the adaptations for $\hat{T}_L$ and $\hat{\alpha}$ result. The parameter $\gamma_2 = (\alpha - k)\gamma_0 > 0$ is a weighting factor in the function (4.96): the choice of k depends on the interval of variation of the uncertain parameter α. The parameter k_e should be the last parameter to be tuned and should be chosen sufficiently large so that the current error dynamics are much faster than speed error dynamics while voltages are within saturation limits. Of course, different tuning strategies should be followed depending on sensor noise features, discretization strategies and other implementation issues.

2. For induction motors which allow for high gain current loops, the control may be greatly simplified as follows

$$
\begin{aligned}
u_{sa}(t) &= -k_P[i_{sa}(t) - i_{sa}^*(t)] - k_I \int_0^t (i_{sa}(\tau) - i_{sa}^*(\tau))\mathrm{d}\tau \\
u_{sb}(t) &= -k_P[i_{sb}(t) - i_{sb}^*(t)] - k_I \int_0^t (i_{sb}(\tau) - i_{sb}^*(\tau))\mathrm{d}\tau \\
\begin{bmatrix} i_{sa}^* \\ i_{sb}^* \end{bmatrix} &= \begin{bmatrix} \cos\varepsilon_0 & -\sin\varepsilon_0 \\ \sin\varepsilon_0 & \cos\varepsilon_0 \end{bmatrix} \begin{bmatrix} i_{sd}^* \\ i_{sq}^* \end{bmatrix}
\end{aligned}
\tag{4.110}
$$

with

$$
\begin{aligned}
i_{sd}^* &= \frac{\psi^*}{M} + \frac{\dot{\psi}^*}{\hat{\alpha}M} + \frac{1}{\beta\hat{\alpha}M}\Bigg[(k_i - k)e_d \\
&\quad -2\frac{\gamma_1}{\beta\psi^*}\tilde{\omega}\left(-k_\omega\tilde{\omega} + \frac{\hat{T}_L}{J} + \dot{\omega}^*\right) + (\hat{\alpha} - k)\hat{z}_d\Bigg] \\
i_{sq}^* &= \frac{1}{\mu\psi^*}\left(-k_\omega\tilde{\omega} + \frac{\hat{T}_L}{J} + \dot{\omega}^*\right) \\
\frac{\mathrm{d}\varepsilon_0}{\mathrm{d}t} &= \omega_0 = \omega + \frac{\hat{\alpha}Mi_{sq}^*}{\psi^*} - \frac{1}{\beta\psi^*}\left[(k_i - k)e_q + 2\frac{\gamma_1\mu}{\beta}\tilde{\omega}i_{sd}^* + (\hat{\alpha} - k)\hat{z}_q\right] \\
\frac{\mathrm{d}\hat{z}_d}{\mathrm{d}t} &= -ke_d + \omega_0\hat{z}_q - \frac{\gamma_1\mu}{\beta}\tilde{\omega}i_{sq} + \omega e_q + \frac{1}{\gamma_0}e_d \\
\frac{\mathrm{d}\hat{z}_q}{\mathrm{d}t} &= -ke_q - \omega_0\hat{z}_d + \frac{\gamma_1\mu}{\beta}\tilde{\omega}i_{sd} - \omega e_d + \frac{1}{\gamma_0}e_q \\
\frac{\mathrm{d}\hat{i}_{sd}}{\mathrm{d}t} &= -\frac{R_s}{\sigma}i_{sd} + \omega_0 i_{sq} + \hat{\alpha}\beta\psi^* - \hat{\alpha}M\beta i_{sd} + \frac{1}{\sigma}u_{sd} + k_i e_d \\
&\quad -\frac{\gamma_1\mu}{\beta}\tilde{\omega}i_{sq} + (\hat{\alpha} - k)\hat{z}_d - (\omega_0 + \omega)e_q
\end{aligned}
$$

$$\begin{aligned}
\frac{d\hat{i}_{sq}}{dt} &= -\frac{R_s}{\sigma} i_{sq} - \omega_0 i_{sd} - \beta\omega\psi^* - \hat{\alpha}M\beta i_{sq} + \frac{1}{\sigma}u_{sq} + k_i e_q \\
&\quad + \frac{\gamma_1\mu}{\beta}\tilde{\omega} i_{sd} + (\hat{\alpha} - k)\hat{z}_q + (\omega_0 + \omega)\, e_d \\
\frac{d\hat{\alpha}}{dt} &= \mathrm{Proj}\left(\frac{1}{\gamma_4}[-\beta e_d(Mi_{sd} - \psi^*) - \beta e_q M i_{sq} + e_d\hat{z}_d + e_q\hat{z}_q], \hat{\alpha}\right) \\
\frac{d\hat{T}_L}{dt} &= -\frac{\gamma_1}{\gamma_3 J}\tilde{\omega}
\end{aligned} \tag{4.111}$$

obtained by setting $\tilde{i}_{sd} = \tilde{i}_{sq} = 0$ in (4.81), (4.82), (4.86), (4.90), (4.98), (4.100), (4.103).

3. If rotor flux measurements are available, a simplified version of the adaptive algorithm (4.109) may be designed since $\hat{z}_d$ and $\hat{z}_q$ are not needed. A global state feedback adaptive algorithm is obtained which constitutes an adaptive version of the algorithm (2.113).

Illustrative Simulations

We tested the adaptive output feedback control by simulations for the three-phase single pole pair 0.6-kW induction motor whose parameters have been reported in Chapter 1. All the motor initial conditions have been set to zero except for $\psi_{ra}(0) = \psi_{rb}(0) = 0.1\,\mathrm{Wb}$. The control algorithm has been tested with control parameters (all values are in SI units) $k_e = 800$, $k_i = 50.1$, $k = 20.4$, $k_\omega = 450$, $\gamma_0 = -0.00105$, $\gamma_1 = 0.005$, $\gamma_3 = 5.2$, $\gamma_4 = 2.5$. All controller initial conditions have been set to zero except for the initial condition for the α-estimate which has been set to $\hat{\alpha}(0) = 13.2\,\mathrm{s}^{-1}$ which is 50% greater than the true parameter value $\alpha = 8.8\,\mathrm{s}^{-1}$. The references for the rotor speed and flux modulus along with the applied load torque are reported in Figures 4.22–4.24. The rotor flux modulus reference signal starts from 0.05 Wb at $t = 0\,\mathrm{s}$ and grows up to the constant value 1.16 Wb. The speed reference is zero until $t = 0.32\,\mathrm{s}$ and grows up to the constant value 100 rad/s; at $t = 1.5\,\mathrm{s}$ the speed is required to go up to the value 200 rad/s, while the reference for the flux modulus is reduced to 0.5 Wb. A 5.8-Nm load torque is applied to the motor and is reduced to 0.5 Nm. Figures 4.22 and 4.23 show the time histories of rotor speed and flux modulus along with the corresponding tracking errors. The rotor speed tracks its reference tightly even though load torque sharply changes since the load torque estimate quickly recovers the applied unknown load torque (see Figure 4.24). Also the α-estimate, according to Figure 4.25, converges to the true value. The rotor flux modulus tracks its reference: there is a coupling with rotor speed tracking at $t = 0.5\,\mathrm{s}$ and at $t = 1\,\mathrm{s}$ when the rotor speed is perturbed by the unknown load torque. Finally, the stator current and voltage profiles (which are within the physical saturation limits) are reported in Figures 4.26 and 4.27.

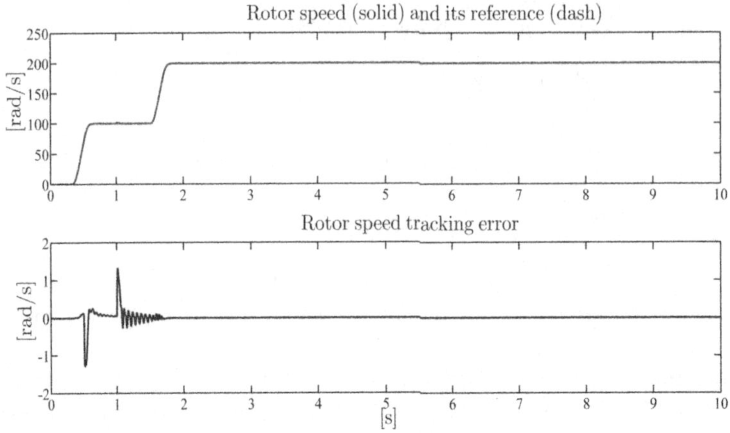

Fig. 4.22 Adaptive output feedback control: rotor speed ω and its reference ω^*; rotor speed tracking error

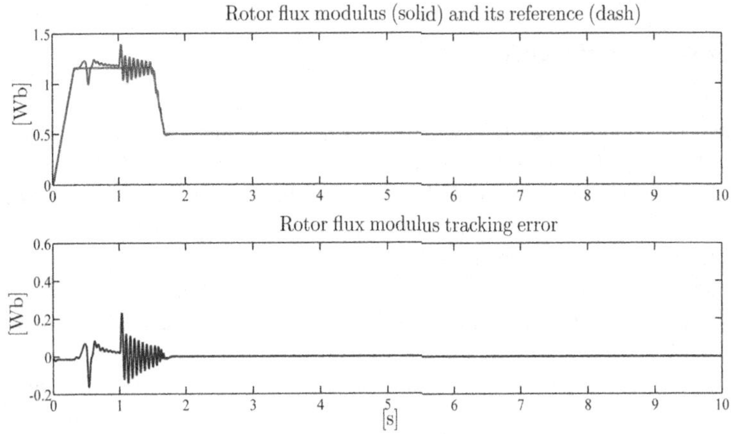

Fig. 4.23 Adaptive output feedback control: rotor flux modulus $\sqrt{\psi_{ra}^2 + \psi_{rb}^2}$ and its reference ψ^*; rotor flux modulus tracking error

4.5 Experimental Results

The simplified control algorithm (4.110)-(4.111) was tested experimentally with the control parameters values $k_i = 80$, $k = 40$, $\gamma_1 = 0.2$, $\beta/\gamma_4 = 1.7$, $\gamma_1/(\gamma_3 J) = 187$, $\gamma_0 = -0.2$, $k_\omega = 300$; the gains of the PI controllers (4.110) are chosen so that a unit step reference is tracked with a settling time of about 2.5 ms; all initial conditions of the controller are set to zero except $\hat{\alpha}(0)$. The following typical operating conditions were experimentally tested: the unloaded motor is required to reach the rated speed 100 rad/s with acceleration 1000 rad/s^2 in 140 ms starting from 0.5 s; during the initial time interval $[0, 0.31]$ s, the motor flux modulus is driven from the initial

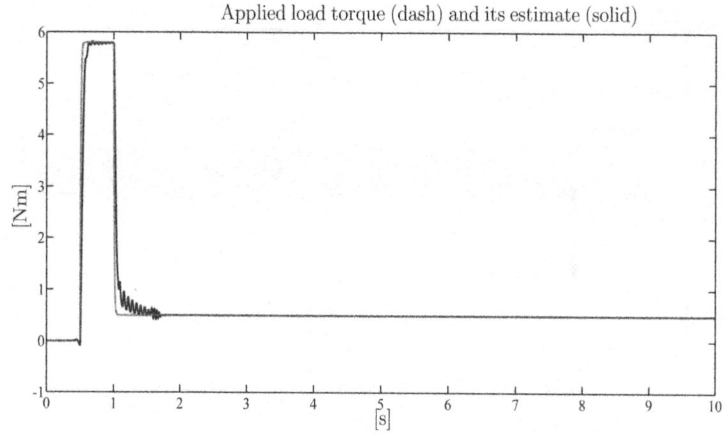

Fig. 4.24 Adaptive output feedback control: applied load torque T_L and its estimate $\hat{T}_L$

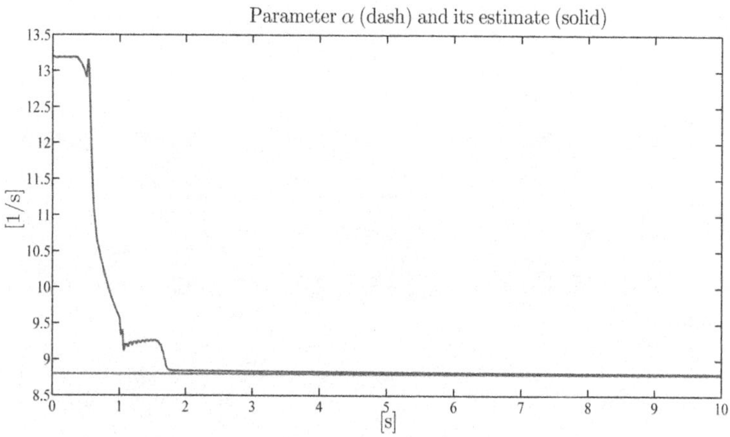

Fig. 4.25 Adaptive output feedback control: uncertain parameter α and its estimate $\hat{\alpha}$

value 10^{-4} Wb to its rated value 1.16 Wb, with flux speed 3.87 Wb/s; both speed and flux reference signals (given in Figure 4.28) are twice differentiable with bounded second order derivatives (the bounds are $2 \cdot 10^5$ rad/s^3 and 38.7 Wb/s^2, respectively); after start-up a constant load torque, equal to the rated value (5.8 Nm), is applied. We performed two experiments: in the first one (which is illustrated by Figures 4.29 and 4.30) $\hat{\alpha}(0)$ underestimates the correct value α, *i.e.* $\hat{\alpha}(0)/\alpha = 0.7$, while in the second one (which is reported in Figures 4.31 and 4.32) α is overestimated, *i.e.* $\hat{\alpha}(0)/\alpha = 1.5$. The closed-loop performance is documented in Figures 4.29–4.32, in which the speed error $\tilde{\omega}(t)$, the current estimation errors $e_d(t)$, $e_q(t)$, the estimated flux modulus $\sqrt{\hat{\psi}_{rd}^2(t) + \hat{\psi}_{rq}^2(t)}$ with $(\hat{\psi}_{rd}(t), \hat{\psi}_{rq}(t))$ given by (4.108), the stator currents $(i_{sd}(t), i_{sq}(t))$, the stator voltage $u_{sq}(t)$, and the normalized estimate

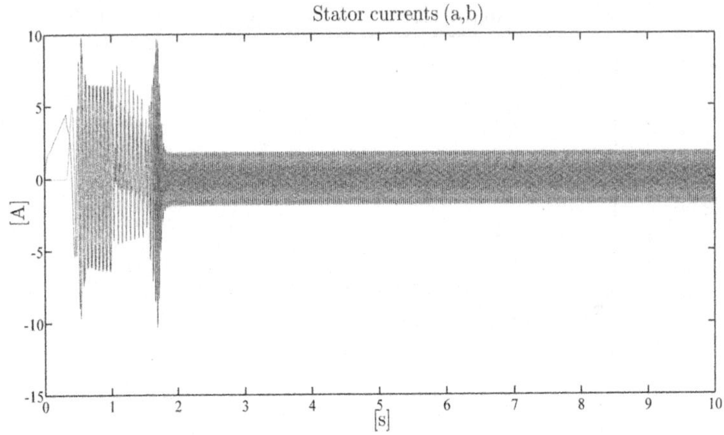

Fig. 4.26 Adaptive output feedback control: stator current vector (a,b)-components (i_{sa}, i_{sb})

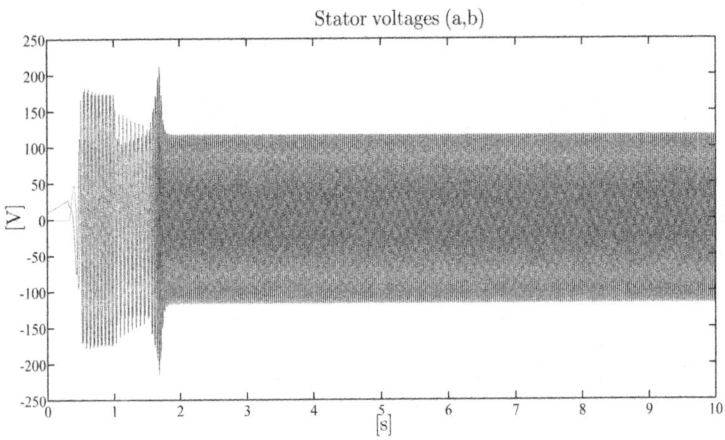

Fig. 4.27 Adaptive output feedback control: stator voltage vector (a,b)-components (u_{sa}, u_{sb})

$\hat{\alpha}(t)/\alpha$ are given. In both cases: the speed errors are compatible with a high performance drive; the estimated flux modulus converges to the reference value and $\hat{\alpha}(t)/\alpha$ converges to 1; the estimates of the flux modulus and of α depend on the torque level, which may be evaluated from $i_{sq}(t)$.

4.6 Conclusions

In this chapter we have finalized the developments in Chapter 2, in which state feedback controls have been designed to obtain good performance from any initial condition, and the developments in Chapter 3, in which rotor flux observers

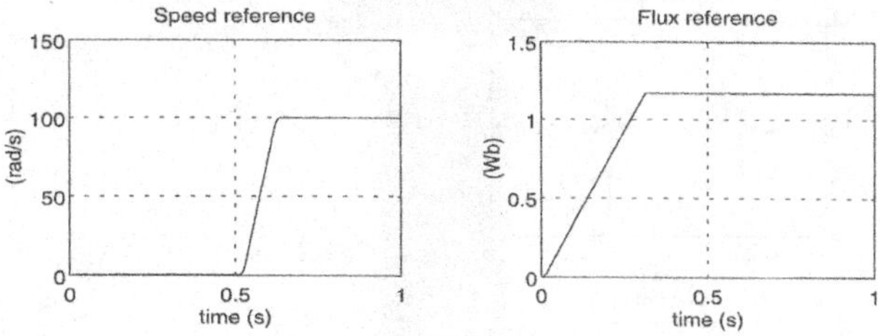

Fig. 4.28 Reference signals in the experiments

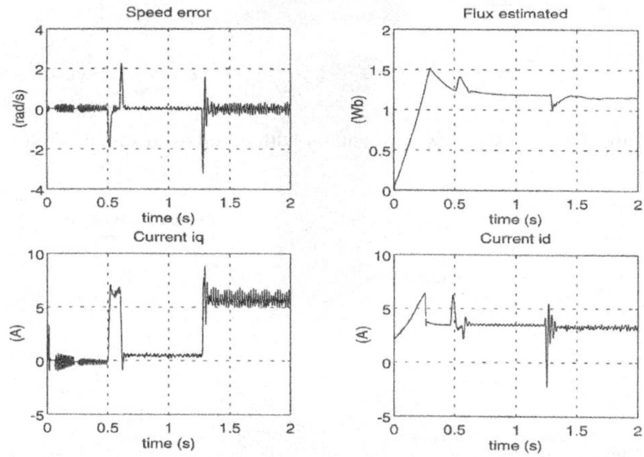

Fig. 4.29 Experimental $\tilde{\omega}$, $\sqrt{\hat{\psi}_{rd}^2 + \hat{\psi}_{rq}^2}$, i_{sq}, and i_{sd} with initial underestimated rotor resistance

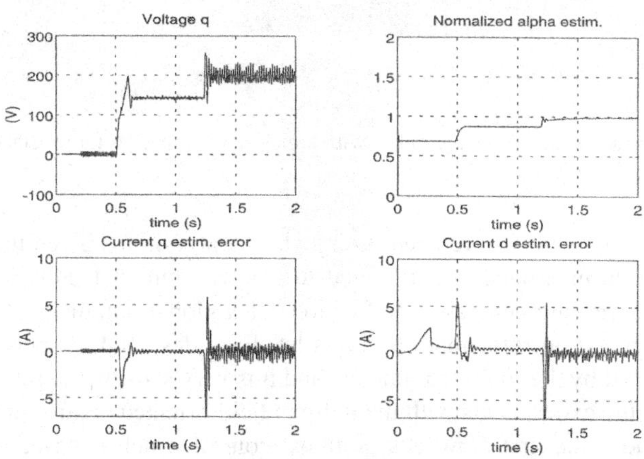

Fig. 4.30 Experimental u_{sq}, $\hat{\alpha}/\alpha$, e_q, and e_d with initial underestimated rotor resistance

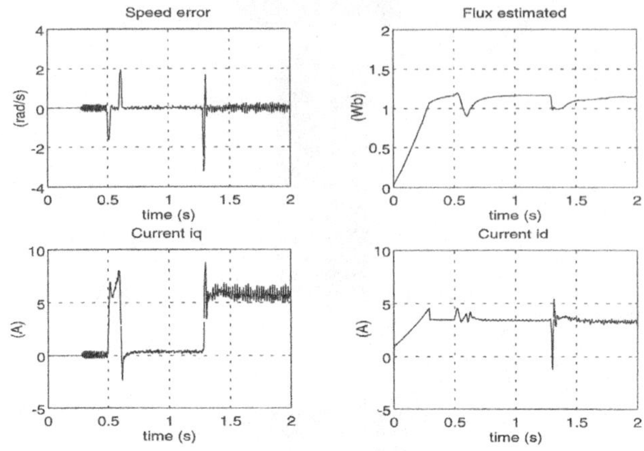

Fig. 4.31 Experimental $\tilde{\omega}$, $\sqrt{\hat{\psi}_{rd}^2 + \hat{\psi}_{rq}^2}$, i_{sq}, and i_{sd} with initial overestimated rotor resistance

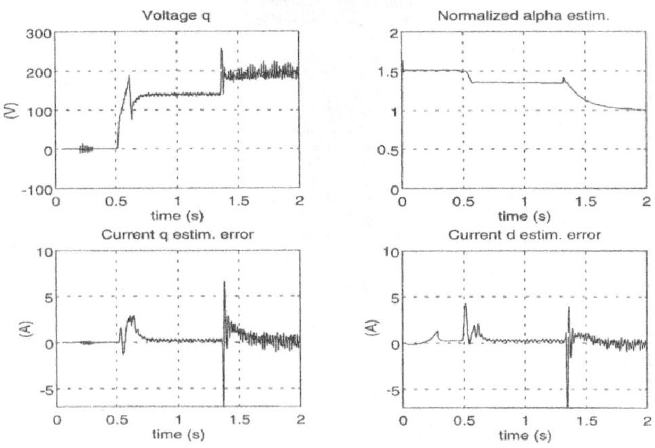

Fig. 4.32 Experimental u_{sq}, $\hat{\alpha}/\alpha$, e_q, and e_d with initial overestimated rotor resistance

and parameter estimators have been designed. The algorithms given in Chapters 2 and 3 have been integrated and modified to generate output feedback control algorithms of increasing generality. In Section 4.1 a global output feedback control which is not based on rotor flux observers has been designed: its performance is, however, limited by the motor parameters and it is very sensitive to rotor resistance variations, as the experiments with the indirect field-oriented control in Section 2.8 show. To remove the first drawback, a global control which is based on the rotor flux observers presented in Chapter 3 has been designed in Section 4.2; its adaptive version when the load torque is uncertain has been given in Section 4.3 on the basis of the load torque estimator obtained in Section 3.3. The second drawback is eliminated by incorporating the load torque and the rotor resistance estimators developed in Chapter 3 into the generalized indirect field-oriented control designed in

Section 4.1. The most general and most complex output feedback control, which is adaptive with respect to both unknown load torque and uncertain rotor resistance and allows for any motor initial condition, is presented in Section 4.4. Some experimental results obtained when the simplified algorithm (4.110), (4.111) given in Section 4.4 is implemented are finally reported in Section 4.5. While input–output linearization and decoupling were obtained in Chapter 2, no such result has been obtained in this chapter. Moreover, when both load torque and rotor resistance are uncertain, rotor flux modulus tracking is achieved in this chapter under persistency of excitation, while no persistency of excitation condition is required in Chapter 2 to achieve rotor flux modulus tracking. Hence, the lack of rotor flux measurements leads to a reduced closed-loop performance.

Problems

4.1. Consider the current-fed model

$$\begin{aligned}
\dot{\omega} &= \mu(\psi_{ra} i_{sb} - \psi_{rb} i_{sa}) - \frac{T_L}{J} \\
\dot{\psi}_{ra} &= -\alpha\psi_{ra} - \omega\psi_{rb} + \alpha M i_{sa} \\
\dot{\psi}_{rb} &= -\alpha\psi_{rb} + \omega\psi_{ra} + \alpha M i_{sb}\,;
\end{aligned}$$

show that the output feedback observer-based control (k_ψ, λ, and k_ω are positive reals while ω^* and ψ^* are the reference signals)

$$\begin{aligned}
\begin{bmatrix} i_{sa} \\ i_{sb} \end{bmatrix} &= \frac{1}{\sqrt{\hat{\psi}_{ra}^2 + \hat{\psi}_{rb}^2}} \begin{bmatrix} \hat{\psi}_{ra} & -\hat{\psi}_{rb} \\ \hat{\psi}_{rb} & \hat{\psi}_{ra} \end{bmatrix} \begin{bmatrix} \xi_1 \\ \xi_2 \end{bmatrix} \\
\dot{\hat{\psi}}_{ra} &= -\frac{R_r}{L_r}\hat{\psi}_{ra} - \omega\hat{\psi}_{rb} + \frac{R_r M}{L_r} i_{sa} + \lambda\mu i_{sb}\tilde{\omega} \\
\dot{\hat{\psi}}_{rb} &= -\frac{R_r}{L_r}\hat{\psi}_{rb} + \omega\hat{\psi}_{ra} + \frac{R_r M}{L_r} i_{sb} - \lambda\mu i_{sa}\tilde{\omega} \\
\xi_1 &= \frac{1}{2\alpha M}\Bigg[2\alpha\psi^{*2} + 2\psi^*\dot{\psi}^* - k_\psi(\hat{\psi}_{ra}^2 + \hat{\psi}_{rb}^2 - \psi^{*2}) \\
&\quad -2\lambda(\omega - \omega^*)\left(\frac{T_L}{J} + \dot{\omega}^* - k_\omega(\omega - \omega^*)\right)\Bigg] \\
\xi_2 &= \frac{1}{\mu}\left(\frac{T_L}{J} + \dot{\omega}^* - k_\omega(\omega - \omega^*)\right)
\end{aligned}$$

guarantees that $(\omega - \omega^*)$, $(\hat{\psi}_{ra}^2 + \hat{\psi}_{rb}^2 - \psi^{*2})$, $(\psi_{ra} - \hat{\psi}_{ra})$, and $(\psi_{rb} - \hat{\psi}_{rb})$ tend exponentially to zero, provided that the initial conditions of the observer are such that $\hat{\psi}_{ra}^2(0) + \hat{\psi}_{rb}^2(0) > \psi^{*2}(0)$. *Suggestion: use the function* $V = \frac{1}{2}\Big[(\omega - \omega^*)^2 + (\hat{\psi}_{ra}^2 + \hat{\psi}_{rb}^2 - \psi^{*2})^2 + \frac{1}{\lambda}(\psi_{ra} - \hat{\psi}_{ra})^2 + \frac{1}{\lambda}(\psi_{rb} - \hat{\psi}_{rb})^2\Big]$.

4.2. Consider the adaptive version of the control algorithm given in Problem 4.1 obtained by replacing T_L by its estimate $\hat{T}_L$ with adaptation dynamics (λ_T is a positive design parameter)

$$\dot{\hat{T}}_L = -\frac{\lambda_T}{J}(\omega - \omega^*);$$

show that the rotor speed tracking error $(\omega - \omega^*)$ converges asymptotically to zero by using the function $V_T = \frac{1}{2}\Big[(\omega - \omega^*)^2 + (\hat{\psi}_{ra}^2 + \hat{\psi}_{rb}^2 - \psi^{*2})^2 + \frac{1}{\lambda}(\psi_{ra} - \hat{\psi}_{ra})^2 + \frac{1}{\lambda}(\psi_{rb} - \hat{\psi}_{rb})^2 + \frac{1}{\lambda_T}(T_L - \hat{T}_L)^2\Big]$ and Barbalat's Lemma A.2 in Appendix A.

4.3. Consider the current-fed model in (d,q) arbitrarily rotating coordinates

$$\begin{aligned}
\dot{\omega} &= \mu(\psi_{rd} i_{sq} - \psi_{rq} i_{sd}) - \frac{T_L}{J} \\
\dot{\psi}_{rd} &= -\alpha\psi_{rd} + (\omega_0 - \omega)\psi_{rq} + \alpha M i_{sd} \\
\dot{\psi}_{rq} &= -\alpha\psi_{rq} - (\omega_0 - \omega)\psi_{rd} + \alpha M i_{sq};
\end{aligned}$$

show that the output feedback observer-based control (k_ψ and k_ω are positive reals)

$$\begin{aligned}
\dot{\hat{\psi}}_{rd} &= -\alpha\hat{\psi}_{rd} + \alpha M i_{sd} \\
\dot{\varepsilon}_0 &= \omega_0 = \omega + \frac{\alpha M i_{sq}}{\hat{\psi}_{rd}} \\
i_{sd} &= \frac{1}{M}\psi^* + \frac{1}{\alpha M}\dot{\psi}^* - \frac{1}{\alpha M}k_\psi(\hat{\psi}_{rd} - \psi^*) \\
i_{sq} &= \frac{1}{\mu\hat{\psi}_{rd}}\left[\frac{T_L}{J} + \dot{\omega}^* - k_\omega(\omega - \omega^*)\right]
\end{aligned}$$

guarantees that $(\omega - \omega^*)$, $(\hat{\psi}_{rd} - \psi^*)$, and $(\psi_{rd} - \hat{\psi}_{rd})$ tend exponentially to zero, provided that $\hat{\psi}_{rd}(0) > \psi^*(0)$, while $(\varepsilon_0 - \rho)$ tends to zero, with

$$\rho(t) = \arctan\frac{\psi_{rb}(t)}{\psi_{ra}(t)}.$$

Show that when $\hat{\psi}_{rd}(0) = \psi_{rd}(0)$ and $\varepsilon_0(0) = \rho(0)$, then $\hat{\psi}_{rd}(t) = \psi_{rd}(t)$ and $\varepsilon_0(t) = \rho(t)$ for every $t \geq 0$, so that the control coincides with the input–output feedback linearizing control (2.21).

4.4. Consider the induction motor rotating frame model and the output feedback, observer-based control (k_ψ, k_ω, k_{id} and k_{iq} are positive reals)

$$
\begin{aligned}
\dot{\hat{\psi}}_{rd} &= -\alpha\hat{\psi}_{rd} + \alpha M i_{sd} \\
\dot{\varepsilon}_0 &= \omega_0 = \omega + \frac{\alpha M i_{sq}}{\hat{\psi}_{rd}} \\
i_{sd}^* &= \frac{1}{M}\psi^* + \frac{1}{\alpha M}\dot{\psi}^* - \frac{1}{\alpha M}k_\psi(\hat{\psi}_{rd} - \psi^*) \\
i_{sq}^* &= \frac{1}{\mu\hat{\psi}_{rd}}\left[\frac{T_L}{J} + \dot{\omega}^* - k_\omega(\omega - \omega^*)\right] \\
u_{sd} &= \sigma\left[\gamma i_{sd}^* - \omega i_{sq} - \frac{\alpha M i_{sq}^2}{\hat{\psi}_{rd}} - \beta\alpha\hat{\psi}_{rd} + \frac{1}{M}\dot{\psi}^*\right. \\
&\quad \left. - \frac{1}{\alpha M}\ddot{\psi}^* - k_{id}(i_{sd} - i_{sd}^*)\right] \\
u_{sq} &= \sigma\left[\gamma i_{sq}^* + \omega i_{sd} + \frac{\alpha M i_{sq} i_{sd}}{\hat{\psi}_{rd}} + \beta\omega\hat{\psi}_{rd}\right. \\
&\quad + \frac{1}{\mu^2\hat{\psi}_{rd}^2}\dot{\hat{\psi}}_{rd}\left(\frac{T_L}{J} + \dot{\omega}^* - k_\omega(\omega - \omega^*)\right) - k_{iq}(i_{sq} - i_{sq}^*) \\
&\quad \left. - \frac{1}{\mu\hat{\psi}_{rd}}\left(-k_\omega\mu\hat{\psi}_{rd}i_{sq} + k_\omega\frac{T_L}{J}\right)\right].
\end{aligned}
$$

Compute the closed-loop error dynamics using the error variables $\tilde{\omega} = \omega - \omega^*$, $e_{\psi d} = \psi_{rd} - \hat{\psi}_{rd}$, $\tilde{\hat{\psi}}_{rd} = \hat{\psi}_{rd} - \psi^*$, $e_{\psi q} = \psi_{rq}$, $\tilde{i}_{sd} = i_{sd} - i_{sd}^*$, and $\tilde{i}_{sq} = i_{sq} - i_{sq}^*$, and investigate the asymptotic stability of the origin. Compare this control with the control given in Section 4.2.

4.5. Consider the input–output feedback linearizing control (2.64) with (ψ_{ra}, ψ_{rb}) replaced by $(\hat{\psi}_{ra}, \hat{\psi}_{rb})$ given by the reduced order rotor flux observer (3.8). Compute the closed-loop error dynamics using the error variables $\omega - \omega^*$, $\hat{a} - \dot{\omega}^*$ with $\hat{a} = \mu(\hat{\psi}_{ra}i_{sb} - \hat{\psi}_{rb}i_{sa}) - \frac{T_L}{J}$, $(\hat{\psi}_{ra}^2 + \hat{\psi}_{rb}^2 - \psi^{*2})$, $-2\alpha(\hat{\psi}_{ra}^2 + \hat{\psi}_{rb}^2) + 2\alpha M(\hat{\psi}_{ra}i_{sa} + \hat{\psi}_{rb}i_{sb}) - 2\psi^*\dot{\psi}^*$, $(\psi_{ra} - \hat{\psi}_{ra})$, and $(\psi_{rb} - \hat{\psi}_{rb})$, and investigate the asymptotic stability of the origin. Compare this control with the control given in Section 4.2. Show that if the rotor flux observer is initialized at the true flux value then the control coincides with (2.64).

4.6. Specialize the global control algorithm given in Section 4.4 to the case in which $\hat{\alpha} = \alpha$ and compare with the algorithm given in Section 4.3.

4.7. Specialize the global control algorithm given in Section 4.4 to the case in which $\hat{T}_L = T_L$ and $\hat{\alpha} = \alpha$, and compare with the algorithm given in Section 4.2.

4.8. Design an adaptive version of the generalized indirect field-oriented control given in Section 4.1 when the load torque T_L is an unknown parameter replacing T_L by its estimate $\hat{T}_L$, by using the function $V_T = \frac{1}{2}\Big[(\omega - \omega^*)^2 + (\hat{\psi}_{ra}^2 + \hat{\psi}_{rb}^2 - \psi^{*2})^2 + \frac{1}{\lambda}(\psi_{ra} - \hat{\psi}_{ra})^2 + \frac{1}{\lambda}(\psi_{rb} - \hat{\psi}_{rb})^2 + \frac{1}{\lambda_T}(T_L - \hat{T}_L)^2\Big]$ to design the adaptation law for $\hat{T}_L$.

4.9. Consider the adaptive input–output feedback linearizing control (2.85) with (ψ_{ra}, ψ_{rb}) replaced by the estimates $(\hat{\psi}_{ra}, \hat{\psi}_{rb})$ given by the adaptive observer (3.58). Compare the closed-loop performance of the resulting adaptive observer-based control with those obtained using the global adaptive output feedback control given in Section 4.4.

4.10. Consider the simplified version (4.110), (4.111) of the adaptive output feedback control (4.109). Simulate this algorithm with the control parameter values, the initial conditions, the reference signals, and the load torque considered in Section 4.5 and compare the results with those reported in Figures 4.29–4.32.

4.11. Consider the adaptive output feedback control (4.76) with $\hat{T}_L$ and $\hat{\alpha}$ provided by the adaptive observer (3.58) (along with their time derivatives $\dot{\hat{T}}_L$ and $\dot{\hat{\alpha}}$) and by the load torque identifier (3.84). Show that, for sufficiently small initial errors and sufficiently large k_i, exponential rotor speed and flux modulus tracking is guaranteed in closed-loop provided that persistency of excitation conditions are satisfied for the adaptive observer (3.58).

4.12. Consider the adaptive output feedback control (4.109): assuming that the rotor fluxes (ψ_{ra}, ψ_{rb}) are measured and that the parameters (T_L, α) are known, modify the control algorithm (4.109) in order to obtain the rotor flux tracking even when the condition (4.107) fails and compare it with the global control (2.114).

4.13. Consider the adaptive control

$$
\begin{aligned}
\begin{bmatrix} u_{sa} \\ u_{sb} \end{bmatrix} &= \begin{bmatrix} \cos\varepsilon_0 & -\sin\varepsilon_0 \\ \sin\varepsilon_0 & \cos\varepsilon_0 \end{bmatrix} \begin{bmatrix} u_{sd} \\ u_{sq} \end{bmatrix} \\
u_{sd} &= \sigma\left[-k_{id}\tilde{i}_{sd} + \gamma i_{sd}^* - \omega_0 i_{sq} - \alpha\beta\psi^* + \frac{1}{\alpha M}(\alpha\dot{\psi}^* + \ddot{\psi}^*)\right] \\
u_{sq} &= \sigma\left[-k_{iq}\tilde{i}_{sq} + \gamma i_{sq}^* + \omega_0 i_{sd} + \beta\omega\psi^*\right. \\
&\quad \left. + \frac{1}{\mu\psi^*}(k_\omega^2\tilde{\omega} - k_\omega\mu\psi^*\tilde{i}_{sq} + \dot{\hat{T}}_L + \ddot{\omega}^*) - \frac{\dot{\psi}^*}{\psi^*} i_{sq}^* - \frac{1}{\psi^*}(\dot{\psi}^*\tilde{i}_{sq} + \zeta)\right] \\
\dot{\varepsilon}_0 &= \omega_0 = \omega + \frac{\alpha M i_{sq}}{\psi^*} + \frac{\beta\omega\tilde{i}_{sd}}{\psi^*} \\
i_{sd}^* &= \frac{\psi^*}{M} + \frac{\dot{\psi}^*}{\alpha M} \\
i_{sq}^* &= \frac{1}{\mu\psi^*}(-k_\omega\tilde{\omega} + \hat{T}_L + \dot{\omega}^*) \\
\dot{\hat{T}}_L &= -k_{\omega i}\tilde{\omega} \\
\dot{\zeta} &= k_{\eta_i}\psi^*\tilde{i}_{sq} \\
\begin{bmatrix} i_{sd} \\ i_{sq} \end{bmatrix} &= \begin{bmatrix} \cos\varepsilon_0 & \sin\varepsilon_0 \\ -\sin\varepsilon_0 & \cos\varepsilon_0 \end{bmatrix} \begin{bmatrix} i_{sa} \\ i_{sb} \end{bmatrix} .
\end{aligned}
$$

Choose the control parameters $(k_{id}, k_{iq}, k_\omega, k_{\omega i}, k_{\eta i})$ to evaluate the performance of the controller and compare it with the adaptive observer-based control (4.75) (compare also with Problem 4.8).

4.14. Consider the dynamic control algorithm

$$
\begin{aligned}
\begin{bmatrix} u_{sa} \\ u_{sb} \end{bmatrix} &= \begin{bmatrix} \cos\varepsilon_0 & -\sin\varepsilon_0 \\ \sin\varepsilon_0 & \cos\varepsilon_0 \end{bmatrix} \begin{bmatrix} u_{sd} \\ u_{sq} \end{bmatrix} \\
u_{sd} &= \sigma\Bigg\{ -k_i\tilde{i}_{sd} - \hat{\eta}_d - k_{ii}x_d + \frac{1}{\alpha\eta}\Bigg[\ddot{\psi}^* + \alpha\dot{\psi}^* - \frac{\gamma_1}{\gamma_3}\mu i_{sq}(-k_\omega\tilde{\omega} + \mu\psi^*\tilde{i}_{sq}) \\
&\quad - \frac{\gamma_1}{\gamma_3}\mu\tilde{\omega}\left(\hat{\eta}_q - \gamma\tilde{i}_{sq} + \frac{u_{sq}}{\sigma}\right)\Bigg]\Bigg\} \\
u_{sq} &= \sigma\Bigg[-\hat{\eta}_q - k_i\tilde{i}_{sq} - k_{ii}x_q - \frac{\gamma_1}{\gamma_4}\mu\psi^*\tilde{\omega} \\
&\quad + \frac{1}{\mu\psi^*}\left(\dot{\hat{T}}_L + \ddot{\omega}^* + k_\omega^2\tilde{\omega} - k_\omega\mu\psi^*\tilde{i}_{sq}\right) - \frac{\dot{\psi}^*}{\psi^*} i_{sq}^* \Bigg] \\
\hat{\eta}_d &= -\gamma i_{sd}^* + \omega_0 i_{sq} + \alpha\beta\psi^* \\
\hat{\eta}_q &= -\gamma i_{sq}^* - \omega_0 i_{sd} - \beta\omega\psi^* \\
\dot{x}_d &= \tilde{i}_{sd} \\
\dot{x}_q &= \tilde{i}_{sq} \\
i_{sd}^* &= \frac{1}{\alpha M}\left(\alpha\psi^* + \dot{\psi}^* - \frac{\gamma_1}{\gamma_3}\mu\tilde{\omega} i_{sq}\right) \\
i_{sq}^* &= \frac{1}{\mu\psi^*}(-k_\omega\tilde{\omega} + \hat{T}_L + \dot{\omega}^*) \\
\dot{\varepsilon}_0 &= \omega_0 = \omega + \frac{\alpha M i_{sq}}{\psi^*} - \frac{\gamma_1}{\gamma_3}\frac{\mu i_{sd}\tilde{\omega}}{\psi^*} + \frac{\gamma_4}{\gamma_3\psi^*}(v_d\tilde{i}_{sd} + v_q\tilde{i}_{sq}) \\
\dot{\hat{T}}_L &= -\frac{\gamma_1}{\gamma_2}\tilde{\omega} - \frac{\gamma_1\gamma_4}{\gamma_2\gamma_3\alpha M}\mu i_{sd}\tilde{i}_{sq} - \frac{\gamma_4 k_\omega}{\gamma_2\mu\psi^*}\tilde{i}_{sq} \\
v_d &= \beta\omega - \frac{\gamma_1\mu}{\alpha M\gamma_3}(\mu i_{sd} i_{sq} - \alpha\beta\tilde{\omega}) \\
v_q &= \alpha\beta - \frac{k_\omega}{\psi^*} i_{sd} \\
\begin{bmatrix} i_{sd} \\ i_{sq} \end{bmatrix} &= \begin{bmatrix} \cos\varepsilon_0 & \sin\varepsilon_0 \\ -\sin\varepsilon_0 & \cos\varepsilon_0 \end{bmatrix} \begin{bmatrix} i_{sa} \\ i_{sb} \end{bmatrix}.
\end{aligned} \tag{4.112}
$$

Choose the control parameters $(k_\omega, k_i, k_{ii}, \gamma_1, \gamma_2, \gamma_3, \gamma_4)$, simulate (4.112), and compare the results with those obtained with the adaptive observer-based control (4.75) (see also Problem 4.13).

4.15. Consider the adaptive observer-based control (4.75); can we conclude using Lemma A.4 in Appendix A that the exponential rate of convergence can be made arbitrarily large by a proper choice of the control parameters?

Chapter 5
Speed-sensorless Feedback Control

Abstract In this chapter we address the design of feedback control algorithms for speed-sensorless induction motors, *i.e.* motors in which the measurement of rotor speed is not available due to sensor failures or on purpose to reduce costs and complexity. In Section 5.1 the reference signals for stator currents are used together with stator current measurements, which are the only measured variables, to generate a feedback control algorithm which is a generalization of the feedforward control and includes a PI feedback based on stator current tracking errors: it turns out that the desired steady-state operating condition may be unstable, depending on the load torque value and on the desired reference flux modulus. In Section 5.2 we address the control problem by assuming that rotor flux measurements are available and that all parameters are known: the aim is to explore at least the possibility of estimating the rotor speed from any motor initial condition within a closed-loop control algorithm and then to obtain a global control for any load torque and reference signals. The design is made adaptive in Section 5.3 in which we explore the design of adaptation with respect to load torque and rotor resistance under the assumption that flux measurements are available: it turns out that the reference flux signal must be time-varying in order to satisfy the persistency of excitation condition which implies that the rotor resistance estimate converges to the true value. On the basis of the results obtained in Sections 5.2 and 5.3, in Section 5.4 the realistic situation in which rotor flux measurements are not available is considered and an adaptive speed-sensorless control algorithm is designed when the load torque is uncertain: it relies on rotor speed and flux closed-loop estimators. This control design is then extended in Section 5.5 to the case in which the rotor resistance is also uncertain: a speed-sensorless control algorithm which is adaptive with respect to both load torque and rotor resistance is finally obtained.

5.1 PI Control from Stator Current Errors

In this section we explore the behavior of the simplest feedback control which can be conceived under the assumption that only the stator currents are available for feedback. The starting point is the inverse system (1.74) developed in Chapter 1 which can be used as a feedforward control in the absence of state variables measurements. Recall that in Chapter 2 certain steady-state conditions have been shown to be unstable depending on the load torque values when only the feedforward control is used. This drawback has been removed by using feedback controls in Chapters 2 and 4 in which the measurement of rotor speed plays a crucial role. The question addressed in this preliminary section is this: what can be done if only stator currents are measured and observers for rotor speed and rotor flux are not used?

On the basis of the steady-state stator currents

$$\begin{aligned} i_{sd}^* &= \frac{\psi^*}{M} + \frac{\dot{\psi}^*}{\alpha M} \\ i_{sq}^* &= \frac{\dot{\omega}^*}{\mu\psi^*} + \frac{T_L}{J\mu\psi^*}, \end{aligned} \tag{5.1}$$

depending on the reference signals (ω^*, ψ^*), and of the measured stator currents

$$\begin{aligned} \begin{bmatrix} i_{sd} \\ i_{sq} \end{bmatrix} &= \begin{bmatrix} \cos\varepsilon_0 & \sin\varepsilon_0 \\ -\sin\varepsilon_0 & \cos\varepsilon_0 \end{bmatrix} \begin{bmatrix} i_{sa} \\ i_{sb} \end{bmatrix} \\ \frac{\mathrm{d}\varepsilon_0}{\mathrm{d}t} &= \omega^* + \frac{\alpha M\dot{\omega}^*}{\mu\psi^{*2}} + \frac{\alpha M T_L}{J\mu\psi^{*2}}, \end{aligned} \tag{5.2}$$

the starting point of this section is to consider the simple PI feedback terms

$$\begin{aligned} v_d(t) &= -k_p(i_{sd}(t) - i_{sd}^*(t)) - k_I\int_0^t (i_{sd}(\tau) - i_{sd}^*(\tau))\mathrm{d}\tau \\ v_q(t) &= -k_p(i_{sq}(t) - i_{sq}^*(t)) - k_I\int_0^t (i_{sq}(\tau) - i_{sq}^*(\tau))\mathrm{d}\tau\,; \end{aligned} \tag{5.3}$$

they are added to the left inverse feedforward control (1.74), so that the following feedback control, which is based only on the stator current measurements (i_{sa}, i_{sb}), is obtained:

$$\begin{aligned} \begin{bmatrix} u_{sa} \\ u_{sb} \end{bmatrix} &= \begin{bmatrix} \cos\varepsilon_0 & -\sin\varepsilon_0 \\ \sin\varepsilon_0 & \cos\varepsilon_0 \end{bmatrix} \begin{bmatrix} u_{sd} \\ u_{sq} \end{bmatrix} \\ u_{sd} &= \sigma\left[\frac{\dot{\psi}^*}{M} + \frac{\ddot{\psi}^*}{\alpha M} + \frac{\gamma\psi^*}{M} + \frac{\gamma\dot{\psi}^*}{\alpha M} - \frac{\omega^*\dot{\omega}^*}{\mu\psi^*} - \frac{\omega^* T_L}{\mu J\psi^*} \right. \\ &\quad \left. - \frac{\alpha M}{\psi^*}\left(\frac{\dot{\omega}^*}{\mu\psi^*} + \frac{T_L}{J\mu\psi^*}\right)^2 - \beta\alpha\psi^* + v_d\right] \end{aligned}$$

$$
\begin{aligned}
u_{sq} &= \sigma \left[\frac{\ddot{\omega}^*}{\mu \psi^*} - \frac{\dot{\omega}^* \dot{\psi}^*}{\mu \psi^{*2}} - \frac{T_L \dot{\psi}^*}{J \mu \psi^{*2}} + \frac{\gamma \dot{\omega}^*}{\mu \psi^*} + \frac{\gamma T_L}{J \mu \psi^*} + \frac{\omega^* \psi^*}{M} \right. \\
&\quad \left. + \frac{\omega^* \dot{\psi}^*}{\alpha M} + \frac{\alpha M}{\psi^*} \left(\frac{\dot{\omega}^*}{\mu \psi^*} + \frac{T_L}{J \mu \psi^*} \right) \left(\frac{\psi^*}{M} + \frac{\dot{\psi}^*}{\alpha M} \right) + \beta \omega^* \psi^* + v_q \right] \\
v_d(t) &= -k_p (i_{sd}(t) - i_{sd}^*(t)) - k_I \int_0^t (i_{sd}(\tau) - i_{sd}^*(\tau)) \mathrm{d}\tau \\
v_q(t) &= -k_p (i_{sq}(t) - i_{sq}^*(t)) - k_I \int_0^t (i_{sq}(\tau) - i_{sq}^*(\tau)) \mathrm{d}\tau \\
\begin{bmatrix} i_{sd} \\ i_{sq} \end{bmatrix} &= \begin{bmatrix} \cos \varepsilon_0 & \sin \varepsilon_0 \\ -\sin \varepsilon_0 & \cos \varepsilon_0 \end{bmatrix} \begin{bmatrix} i_{sa} \\ i_{sb} \end{bmatrix} \\
i_{sd}^* &= \frac{\psi^*}{M} + \frac{\dot{\psi}^*}{\alpha M} \\
i_{sq}^* &= \frac{\dot{\omega}^*}{\mu \psi^*} + \frac{T_L}{J \mu \psi^*} \\
\dot{\varepsilon}_0 &= \omega^* + \frac{\alpha M \dot{\omega}^*}{\mu \psi^{*2}} + \frac{\alpha M T_L}{\mu J \psi^{*2}} \,.
\end{aligned}
\tag{5.4}
$$

Since the purpose of the additive feedback term (5.3) is to use sufficiently high gains (k_p, k_I) to drive the stator current tracking errors $(i_{sd} - i_{sd}^*)$ and $(i_{sq} - i_{sq}^*)$ quickly to zero, we analyze the motor behavior when this aim is achieved, *i.e.* when the stator current tracking errors $(i_{sd} - i_{sd}^*)$ and $(i_{sq} - i_{sq}^*)$ are kept equal to zero so that the tracking error dynamics (2.3) are given by

$$
\begin{aligned}
\frac{\mathrm{d}(\omega - \omega^*)}{\mathrm{d}t} &= \mu \left[(\psi_{rd} - \psi^*) i_{sq}^* - \psi_{rq} i_{sd}^* \right] \\
\frac{\mathrm{d}(\psi_{rd} - \psi^*)}{\mathrm{d}t} &= -\alpha (\psi_{rd} - \psi^*) - (\omega - \omega^*) \psi_{rq} + \omega_s^* \psi_{rq} \\
\frac{\mathrm{d}\psi_{rq}}{\mathrm{d}t} &= -\alpha \psi_{rq} + (\omega - \omega^*)(\psi_{rd} - \psi^*) \\
&\quad - \omega_s^* (\psi_{rd} - \psi^*) + (\omega - \omega^*) \psi^*
\end{aligned}
\tag{5.5}
$$

in which

$$
\omega_s^* = \frac{\alpha M i_{sq}^*}{\psi^*} \,.
$$

The reduced order tracking error dynamics (5.5), in the case of constant references (ω^*, ψ^*) and $T_L \neq 0$, have an additional equilibrium point besides the origin

$$
\left[(\omega - \omega^*), (\psi_{rd} - \psi^*), \psi_{rq} \right]^{\mathrm{T}} = 0
$$

which is given by

$$
\begin{aligned}
(\omega-\omega^*) &= -\frac{\alpha^2}{\omega_s^*}+\omega_s^* \\
&= \frac{R_r\left(-\psi^{*4}+T_L^2L_r^2\right)}{L_r^2\psi^{*2}T_L} \\
(\psi_{rd}-\psi^*) &= -\frac{\alpha\psi^*(-\frac{\alpha^2}{\omega_s^*}+\omega_s^*)}{\omega_s^*\left[-2\alpha+\frac{\alpha}{\omega_s^*}(-\frac{\alpha^2}{\omega_s^*}+\omega_s^*)\right]} \\
&= \frac{\psi^*\left(-\psi^{*4}+T_L^2L_r^2\right)}{T_L^2L_r^2+\psi^{*4}} \\
\psi_{rq} &= -\frac{\psi^*(-\frac{\alpha^2}{\omega_s^*}+\omega_s)}{\left[-2\alpha+\frac{\alpha}{\omega_s^*}(-\frac{\alpha^2}{\omega_s^*}+\omega_s^*)\right]} \\
&= \frac{L_rT_L\left(-\psi^{*4}+T_L^2L_r^2\right)}{\psi^*\left(T_L^2L_r^2+\psi^{*4}\right)}
\end{aligned} \tag{5.6}
$$

when $T_L^2 \neq \frac{\psi^{*4}}{L_r^2}$ and which collapses into the origin when $T_L^2 = \frac{\psi^{*4}}{L_r^2}$. Consequently, when

$$
T_L^2 \neq \frac{\psi^{*4}}{L_r^2}
$$

the origin is not a globally attractive equilibrium point for the error system (5.5) since an additional equilibrium point (5.6) exists. Furthermore, according to the linear approximation Theorem A.7 in Appendix A:

1. if $T_L^2 < \frac{\psi^{*4}}{L_r^2}$, then the origin is a locally (but not globally) exponentially stable equilibrium point for the error system (5.5) (with a region of attraction and a local transient behavior depending on motor parameters) while the additional equilibrium point (5.6) is unstable;
2. if $T_L^2 > \frac{\psi^{*4}}{L_r^2}$, then the origin is an unstable equilibrium point for the error system (5.5) while the additional equilibrium point (5.6) is locally exponentially stable;
3. if $T_L^2 = \frac{\psi^{*4}}{L_r^2}$, then we are in the critical case in which a saddle-node bifurcation occurs and the linear approximation Theorem A.7 does not apply, since the linear approximation has one eigenvalue with zero real part.

Illustrative Simulations

First Simulation

We tested the feedforward control (5.1), (5.2) by simulations for the three-phase single pole pair 0.6-kW induction motor whose parameters have been reported in Chapter 1: stator currents dynamics have been neglected so that the stator currents (i_{sa}, i_{sb}) constitute the motor control inputs. The rotor speed initial condition has been set to zero while the rotor fluxes initial conditions have been set to $\psi_{ra}(0) = \psi_{rb}(0) = 0.1\,\text{Wb}$. The references for the speed and flux modulus along with the applied load torque are reported in Figures 5.1–5.3. The rotor flux modulus reference signal starts from 0.001 Wb at $t = 0\,\text{s}$ and grows up to the constant value 1.16 Wb. The speed reference is zero until $t = 0.32\,\text{s}$ and grows up to the constant value 100 rad/s; at $t = 1.5\,\text{s}$ the speed is required to go up to the value 200 rad/s, while the reference for the flux modulus is reduced to 0.5 Wb. A 5.8-Nm load torque is applied to the motor and is reduced to 1.8 Nm. Figures 5.2 and 5.3 show the time histories of rotor speed and flux modulus along with the corresponding tracking errors: the rotor speed and flux modulus track their references as long as the load torque satisfies the inequality

$$T_L < \frac{\psi^{*2}}{L_r}$$

while rotor speed and flux modulus tracking cannot be guaranteed when the load torque is greater than the critical value $\frac{\psi^{*2}}{L_r}$. In fact, for a load torque $T_L = 1.8\,\text{Nm}$ and a constant rotor flux reference $\psi^* = 0.5\,\text{Wb}$, the origin of the error system (5.5) turns out to be unstable while the computed additional equilibrium point for (5.5) becomes exponentially stable: as illustrated by Figure 5.4, both the equilibrium points for (5.5) (the origin and the additional one) correspond to limit cycles in the state space, attractive or repulsive depending on the equilibrium nature. Finally, the stator currents profiles (which are within physical saturation limits) are reported in Figure 5.5.

Second Simulation

We tested the feedback control (5.4) by simulations for the three-phase single pole pair 0.6-kW induction motor whose parameters have been reported in Chapter 1. All the motor and controller initial conditions have been set to zero except for $\psi_{ra}(0) = \psi_{rb}(0) = 0.1\,\text{Wb}$. High gain integral-proportional actions with $k_p = 1200$ and $k_I = 900$ (in SI units) have been used. The references for the speed and flux modulus along with the applied load torque are reported in Figures 5.6–5.8. The rotor flux modulus reference signal starts from 0.001 Wb at $t = 0\,\text{s}$ and grows up to the constant value 1.16 Wb. The speed reference is zero until $t = 0.32\,\text{s}$ and grows up to the constant value 100 rad/s; at $t = 1.5\,\text{s}$ the speed is required to go up to the value

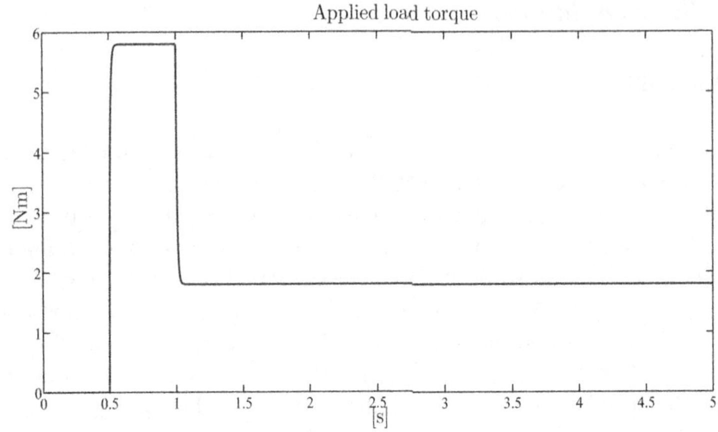

Fig. 5.1 Feedforward control for current-fed motors: applied load torque T_L

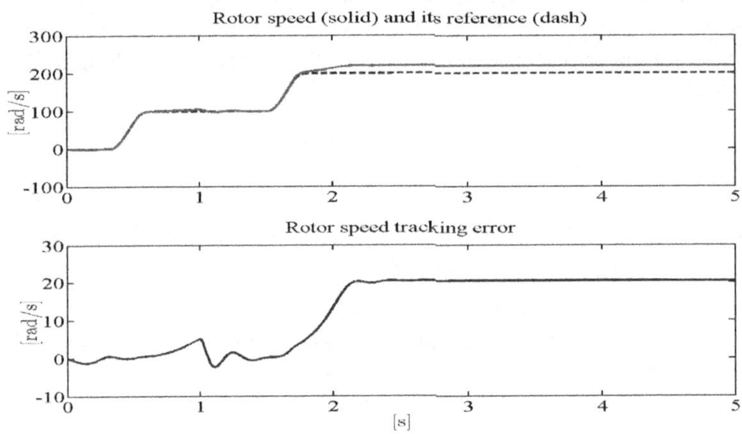

Fig. 5.2 Feedforward control for current-fed motors: rotor speed ω and its reference ω^*; rotor speed tracking error

200 rad/s, while the reference for the flux modulus is reduced to 0.5 Wb. A 5.8-Nm load torque is applied to the motor, is reduced to 0.5 Nm and then is increased to 2 Nm. Figures 5.7 and 5.8 show the time histories of rotor speed and flux modulus along with the corresponding tracking errors: the rotor speed and flux modulus track their references when a 0.5-Nm load torque satisfying the inequality

$$T_L < \frac{\psi^{*2}}{L_r}$$

is applied, while rotor speed and flux modulus tracking are not guaranteed when a 2 Nm load torque (greater than the value ψ^{*2}/L_r) is applied. Finally, the stator cur-

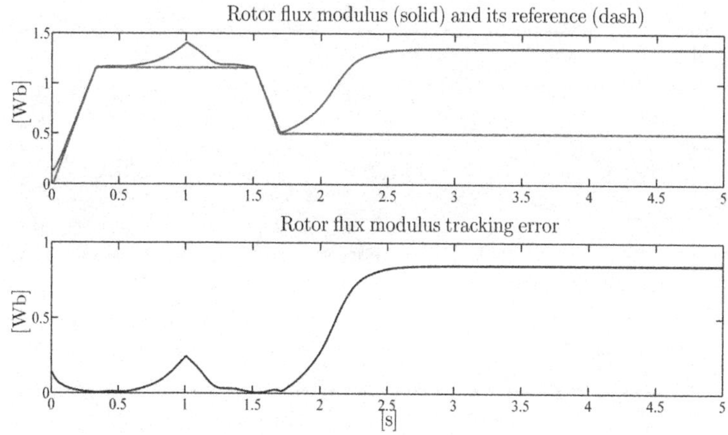

Fig. 5.3 Feedforward control for current-fed motors: rotor flux modulus $\sqrt{\psi_{ra}^2 + \psi_{rb}^2}$ and its reference ψ^*; rotor flux modulus tracking error

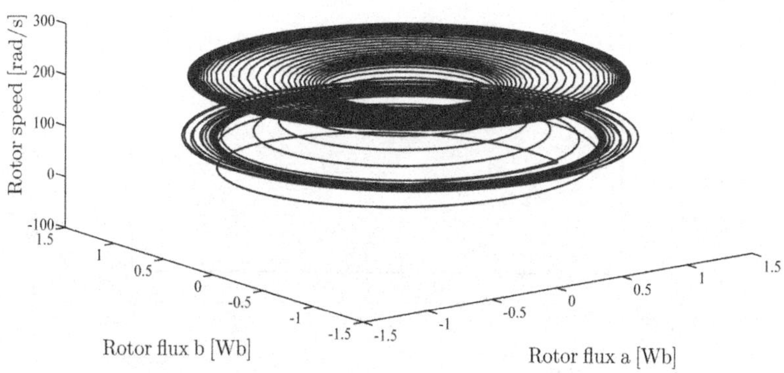

Fig. 5.4 Feedforward control for current-fed motors: $(\psi_{ra}, \psi_{rb}, \omega)$-trajectories

rent and voltages profiles (which are within physical saturation limits) are reported in Figures 5.9 and 5.10.

In the next sections we shall explore the possibility of improving the control algorithm analyzed in this section by using closed-loop rotor speed and flux observers.

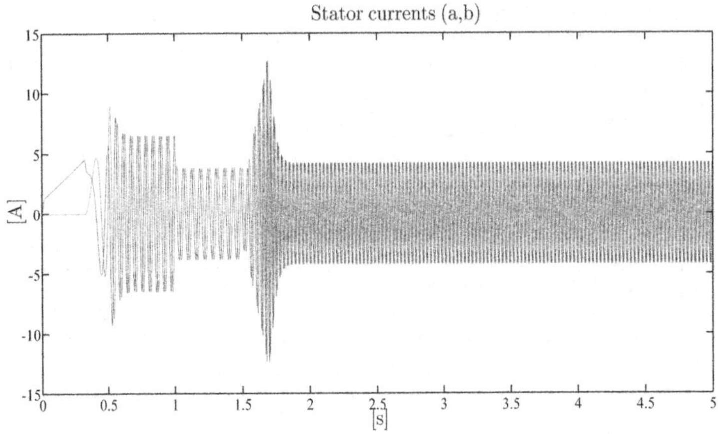

Fig. 5.5 Feedforward control for current-fed motors: stator current vector (a,b)-components

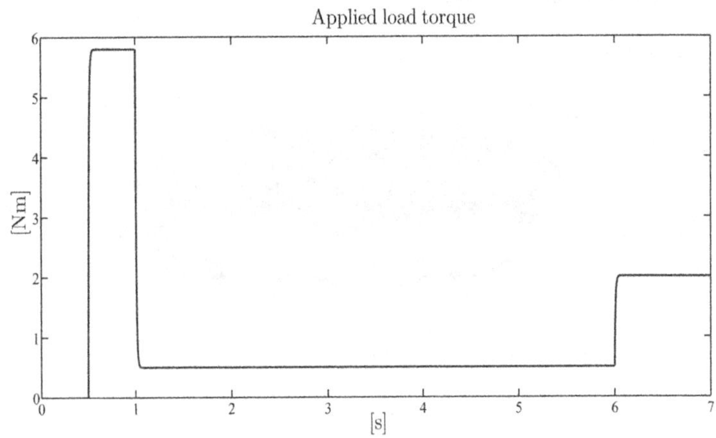

Fig. 5.6 Feedback control (5.4): applied load torque T_L

5.2 Global Control with Flux Measurements

The analysis carried out in Section 5.1 shows that the reference stator currents (5.1) do not guarantee the asymptotic stability of the steady-state solutions for any reference signal (ω^*, ψ^*) and any load torque value since the inequality

$$T_L^2 > \frac{\psi^{*4}}{L_r^2}$$

leads to instability. Hence, in this section we explore the possibility of adding feedback terms in the definition of the reference stator currents, on the basis of a closed-loop rotor speed observer. We then make the unrealistic assumptions that rotor flux

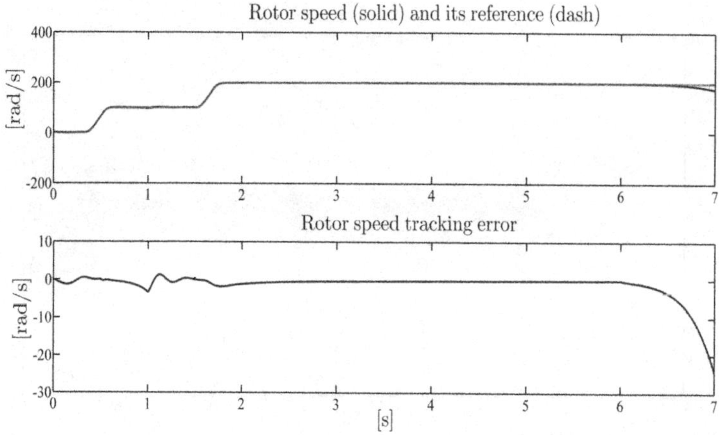

Fig. 5.7 Feedback control (5.4): rotor speed ω and its reference ω^*; rotor speed tracking error

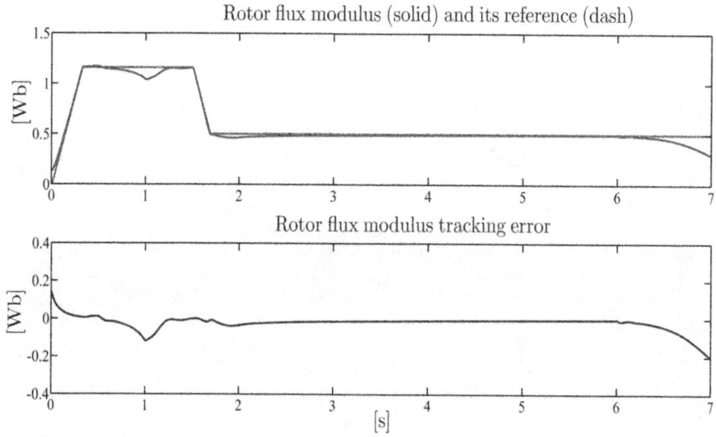

Fig. 5.8 Feedback control (5.4): rotor flux modulus $\sqrt{\psi_{ra}^2 + \psi_{rb}^2}$ and its reference ψ^*; rotor flux modulus tracking error

measurements are available and that all parameters are exactly known, to determine whether asymptotic rotor speed and flux modulus tracking of arbitrary reference signals can be obtained from any motor initial condition in the most favorable situation since, as we have seen, this property was not achieved by the control in the previous section. Those unrealistic assumptions (measured fluxes and known motor parameters) will then be relaxed in Sections 5.4 and 5.5 by using rotor flux observers and parameter estimators.

We define the reference signals for i_{sd} and i_{sq} as (k_ω is a positive control parameter)

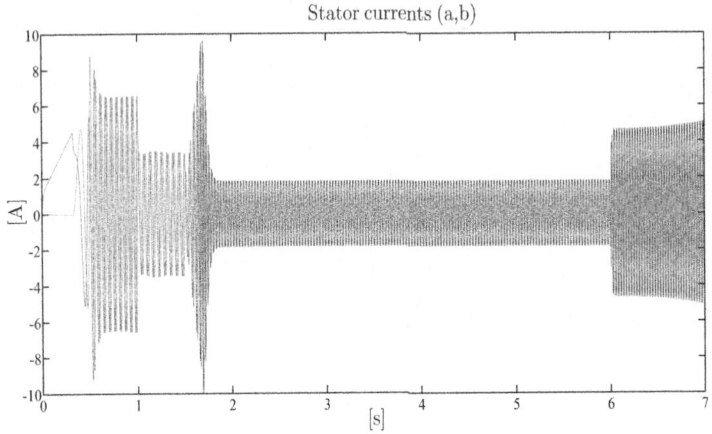

Fig. 5.9 Feedback control (5.4): stator current vector (a,b)-components (i_{sa}, i_{sb})

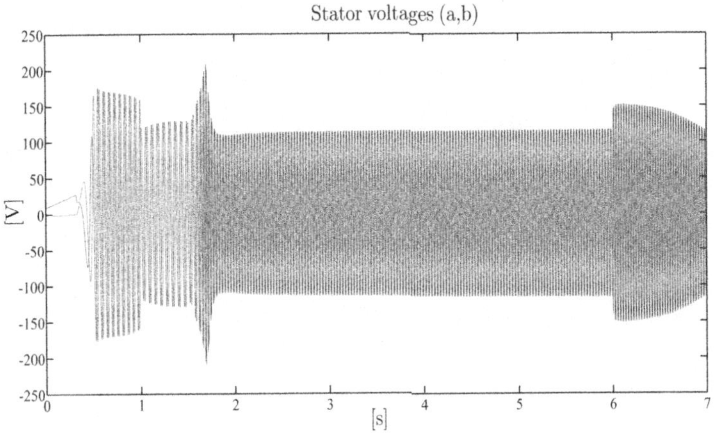

Fig. 5.10 Feedback control (5.4): stator voltage vector (a,b)-components (u_{sa}, u_{sb})

$$\begin{aligned} i_{sd}^* &= \frac{\psi^*}{M} + \frac{\dot{\psi}^*}{\alpha M} \\ i_{sq}^* &= \frac{1}{\mu\psi^*}\left[-k_\omega \mathrm{sat}(\hat{\omega} - \omega^*) + \frac{T_L}{J} + \dot{\omega}^*\right] \end{aligned} \tag{5.7}$$

by inserting the feedback saturating term $-k_\omega \mathrm{sat}(\hat{\omega} - \omega^*)$ in (5.4) which depends on the speed estimate $\hat{\omega}$, whose dynamics are to be designed: the saturation function $\mathrm{sat}(q)$ is a continuously differentiable odd function which is linear in a neighborhood of the origin $q = 0$ and has a finite limit as $|q|$ goes to infinity. The speed of the rotating (d,q) frame is defined from (5.4) by replacing ω^* by its estimate $\hat{\omega}$ and i_{sq}^* by the current measurement i_{sq} as follows:

$$\frac{d\varepsilon_0}{dt} = \omega_0 = \hat{\omega} + \frac{\alpha M i_{sq}}{\psi^*}. \tag{5.8}$$

The speed estimate $\hat{\omega}$ is provided by the rotor speed observer

$$\dot{\hat{\omega}} = \mu(\psi_{rd} i_{sq} - \psi_{rq} i_{sd}) - \frac{T_L}{J} + w \tag{5.9}$$

which is based on the measurements of (ψ_{rd}, ψ_{rq}), (i_{sd}, i_{sq}) and depends on the additional signal w which will be designed on the basis of the tracking errors. Note that the availability of rotor fluxes (ψ_{rd}, ψ_{rq}) is crucial in the design of the rotor speed observer (5.9). Introducing the tracking and estimation errors

$$\begin{aligned}
\tilde{\omega} &= \omega - \omega^* \\
\tilde{\psi}_{rd} &= \psi_{rd} - \psi^* \\
\tilde{\psi}_{rq} &= \psi_{rq} \\
e_\omega &= \hat{\omega} - \omega \\
\tilde{i}_{sd} &= i_{sd} - i_{sd}^* \\
\tilde{i}_{sq} &= i_{sq} - i_{sq}^*
\end{aligned} \tag{5.10}$$

from (1.31), (5.7), (5.8), (5.9), and (5.10), we obtain the tracking and estimation error dynamics

$$\begin{aligned}
\frac{\mathrm{d}\tilde{\omega}}{\mathrm{d}t} &= -k_\omega \mathrm{sat}(\tilde{\omega} + e_\omega) + \mu(\tilde{\psi}_{rd} i_{sq} - \tilde{\psi}_{rq} i_{sd}) + \mu \psi^* \tilde{i}_{sq} \\
\frac{\mathrm{d}\tilde{\psi}_{rd}}{\mathrm{d}t} &= -\alpha \tilde{\psi}_{rd} + (\omega_0 - \omega)\tilde{\psi}_{rq} + \alpha M \tilde{i}_{sd} \\
\frac{\mathrm{d}\tilde{\psi}_{rq}}{\mathrm{d}t} &= -\alpha \tilde{\psi}_{rq} - (\omega_0 - \omega)\tilde{\psi}_{rd} - e_\omega \psi^* \\
\frac{\mathrm{d}e_\omega}{\mathrm{d}t} &= w \\
\frac{\mathrm{d}\tilde{i}_{sd}}{\mathrm{d}t} &= \frac{u_{sd}}{\sigma} + \phi_{d0} + \phi_{d1} e_\omega \\
\frac{\mathrm{d}\tilde{i}_{sq}}{\mathrm{d}t} &= \frac{u_{sq}}{\sigma} + \phi_{q0} + \phi_{q1} e_\omega
\end{aligned} \tag{5.11}$$

in which the known terms

$$\begin{aligned}
\phi_{d0} &= -\gamma i_{sd} + \omega_0 i_{sq} + \alpha\beta\psi_{rd} + \beta\hat{\omega}\psi_{rq} - \frac{\dot{\psi}^*}{M} - \frac{\ddot{\psi}^*}{\alpha M} \\
\phi_{d1} &= -\beta\psi_{rq} \\
\phi_{q0} &= -\gamma i_{sq} - \omega_0 i_{sd} + \alpha\beta\psi_{rq} - \beta\hat{\omega}\psi_{rd} \\
&\quad + \frac{\dot{\psi}^*}{\mu\psi^{*2}}\left[-k_\omega \mathrm{sat}(\hat{\omega} - \omega^*) + \frac{T_L}{J} + \dot{\omega}^*\right]
\end{aligned}$$

$$
\begin{aligned}
& -\frac{1}{\mu\psi^*}\left[-k_\omega\frac{\mathrm{dsat}(\hat{\omega}-\omega^*)}{\mathrm{d}(\hat{\omega}-\omega^*)}(\dot{\hat{\omega}}-\dot{\omega}^*)+\ddot{\omega}^*\right] \\
\phi_{q1} &= \beta\psi_{rd}
\end{aligned} \tag{5.12}
$$

appear. In order to design the yet undetermined signal w in (5.9) and the feedback control inputs (u_{sd}, u_{sq}), consider the function (λ is a positive control parameter)

$$
V_0 = \frac{1}{2}(\tilde{\psi}_{rd}^2+\tilde{\psi}_{rq}^2)+\frac{1}{2}\lambda e_\omega^2+\frac{1}{2}(\tilde{i}_{sd}^2+\tilde{i}_{sq}^2) \tag{5.13}
$$

whose time derivative along the trajectories of the closed-loop system (5.11) is

$$
\begin{aligned}
\dot{V}_0 = & -\alpha(\tilde{\psi}_{rd}^2+\tilde{\psi}_{rq}^2)+(\lambda w-\psi^*\tilde{\psi}_{rq}+\phi_{d1}\tilde{i}_{sd}+\phi_{q1}\tilde{i}_{sq})e_\omega \\
& +\left(\frac{u_{sd}}{\sigma}+\phi_{d0}+\alpha M\tilde{\psi}_{rd}\right)\tilde{i}_{sd}+\left(\frac{u_{sq}}{\sigma}+\phi_{q0}\right)\tilde{i}_{sq}\,.
\end{aligned} \tag{5.14}
$$

If we define the feedback term w and the feedback control inputs (u_{sd}, u_{sq}) as (k_i is a positive control parameter)

$$
\begin{aligned}
w &= \frac{\psi^*\tilde{\psi}_{rq}}{\lambda}-\frac{\phi_{d1}\tilde{i}_{sd}}{\lambda}-\frac{\phi_{q1}\tilde{i}_{sq}}{\lambda} \\
u_{sd} &= \sigma(-\phi_{d0}-\alpha M\tilde{\psi}_{rd}-k_i\tilde{i}_{sd}) \\
u_{sq} &= \sigma(-\phi_{q0}-k_i\tilde{i}_{sq})
\end{aligned} \tag{5.15}
$$

then (5.14) becomes

$$
\dot{V}_0 = -\alpha(\tilde{\psi}_{rd}^2+\tilde{\psi}_{rq}^2)-k_i(\tilde{i}_{sd}^2+\tilde{i}_{sq}^2)\,. \tag{5.16}
$$

From (5.9), (5.12), and (5.15) the complete rotor speed observer becomes

$$
\dot{\hat{\omega}} = \mu(\psi_{rd}i_{sq}-\psi_{rq}i_{sd})-\frac{T_L}{J}+\frac{\psi^*\tilde{\psi}_{rq}}{\lambda}+\frac{\beta\psi_{rq}\tilde{i}_{sd}}{\lambda}-\frac{\beta\psi_{rd}\tilde{i}_{sq}}{\lambda}\,. \tag{5.17}
$$

From (5.13) and (5.16), it follows that $\tilde{\psi}_{rd}(t), \tilde{\psi}_{rq}(t), e_\omega(t), \tilde{i}_{sd}(t), \tilde{i}_{sq}(t)$ are bounded time functions on $[0,\infty)$. Since $\psi^*(t)$, $i_{sd}^*(t)$, $i_{sq}^*(t)$ are bounded time functions on $[0,\infty)$, $\psi_{rd}(t)$, $\psi_{rq}(t)$, $i_{sd}(t)$, $i_{sq}(t)$ are bounded time functions on $[0,\infty)$. Since $\omega_0(t)-\omega(t) = e_\omega(t)+\frac{\alpha M i_{sq}(t)}{\psi^*(t)}$ is a bounded time function on $[0,\infty)$, $\dot{\psi}_{rd}(t)$, $\dot{\psi}_{rq}(t)$ are bounded time functions on $[0,\infty)$. From (5.11) and (5.15), the closed-loop system is

$$
\begin{aligned}
\frac{\mathrm{d}\tilde{\omega}}{\mathrm{d}t} &= -k_\omega\mathrm{sat}(\tilde{\omega}+e_\omega)+\mu(\tilde{\psi}_{rd}i_{sq}-\tilde{\psi}_{rq}i_{sd})+\mu\psi^*\tilde{i}_{sq} \\
\frac{\mathrm{d}\tilde{\psi}_{rd}}{\mathrm{d}t} &= -\alpha\tilde{\psi}_{rd}+(\omega_0-\omega)\tilde{\psi}_{rq}+\alpha M\tilde{i}_{sd} \\
\frac{\mathrm{d}\tilde{\psi}_{rq}}{\mathrm{d}t} &= -\alpha\tilde{\psi}_{rq}-(\omega_0-\omega)\tilde{\psi}_{rd}-e_\omega\psi^*
\end{aligned}
$$

$$\begin{aligned}\frac{d\tilde{i}_{sd}}{dt} &= -k_i\tilde{i}_{sd} - \alpha M\tilde{\psi}_{rd} - \beta\psi_{rq}e_\omega \\ \frac{d\tilde{i}_{sq}}{dt} &= -k_i\tilde{i}_{sq} + \beta\psi_{rd}e_\omega \\ \frac{de_\omega}{dt} &= \frac{\psi^*\tilde{\psi}_{rq}}{\lambda} + \frac{\beta\psi_{rq}\tilde{i}_{sd}}{\lambda} - \frac{\beta\psi_{rd}\tilde{i}_{sq}}{\lambda}\,.\end{aligned} \tag{5.18}$$

The last five equations in (5.18) may be rewritten as

$$\begin{aligned}\dot{x}(t) &= A(t)x(t) + \Gamma^{\mathrm{T}}(t)y(t) \\ \dot{y}(t) &= -\frac{1}{\lambda}\Gamma(t)x(t)\end{aligned} \tag{5.19}$$

with $x(t) = [\tilde{\psi}_{rd}(t), \tilde{\psi}_{rq}(t), \tilde{i}_{sd}(t), \tilde{i}_{sq}(t)]^{\mathrm{T}}$, $y(t) = e_\omega(t)$ and

$$A(t) = \begin{bmatrix} -\alpha & (\omega_0(t) - \omega(t)) & \alpha M & 0 \\ -(\omega_0(t) - \omega(t)) & -\alpha & 0 & 0 \\ -\alpha M & 0 & -k_i & 0 \\ 0 & 0 & 0 & -k_i \end{bmatrix}$$

$$\Gamma^{\mathrm{T}}(t) = \begin{bmatrix} 0 \\ -\psi^*(t) \\ -\beta\psi_{rq}(t) \\ \beta\psi_{rd}(t) \end{bmatrix}.$$

By virtue of the Persistency of Excitation Lemma A.3 in Appendix A, we can establish that, since persistency of excitation conditions are satisfied ($\psi^*(t) \geq c_\psi > 0$ for all $t \geq 0$), *i.e.* there exist two positive constants T_p and c_p such that, for all $t \geq 0$,

$$\int_t^{t+T_p} \Gamma(\tau)\Gamma^{\mathrm{T}}(\tau)d\tau = \int_t^{t+T_p} \left[\psi^{*2}(\tau) + \beta^2\psi_{rd}^2(\tau) + \beta^2\psi_{rq}^2(\tau)\right] d\tau \geq c_p\,,$$

the components of vectors x and y exponentially decay to zero for any initial condition $x(0)$ and $y(0)$. Recalling the first equation in (5.18), since $i_{sd}(t), i_{sq}(t), \psi^*(t)$ are bounded time functions on $[0,\infty)$ and $\tilde{\psi}_{rd}(t), \tilde{\psi}_{rq}(t), e_\omega(t), \tilde{i}_{sq}(t)$ tend, due to persistency of excitation, exponentially to zero, $\tilde{\omega}(t)$ tends exponentially to zero from any initial condition $\tilde{\omega}(0)$ according to Lemma A.1 in Appendix A.

In conclusion, the *global speed-sensorless control*

$$\begin{aligned}\begin{bmatrix} u_{sa} \\ u_{sb} \end{bmatrix} &= \begin{bmatrix} \cos\varepsilon_0 & -\sin\varepsilon_0 \\ \sin\varepsilon_0 & \cos\varepsilon_0 \end{bmatrix} \begin{bmatrix} u_{sd} \\ u_{sq} \end{bmatrix} \\ u_{sd} &= \sigma(-\phi_{d0} - \alpha M\tilde{\psi}_{rd} - k_i\tilde{i}_{sd}) \\ u_{sq} &= \sigma(-\phi_{q0} - k_i\tilde{i}_{sq})\end{aligned}$$

$$
\begin{aligned}
i_{sd}^* &= \frac{\psi^*}{M} + \frac{\dot{\psi}^*}{\alpha M} \\
i_{sq}^* &= \frac{1}{\mu\psi^*}\left[-k_\omega \mathrm{sat}(\hat{\omega}-\omega^*) + \frac{T_L}{J} + \dot{\omega}^*\right] \\
\frac{d\varepsilon_0}{dt} &= \omega_0 = \hat{\omega} + \frac{\alpha M i_{sq}}{\psi^*} \\
\dot{\hat{\omega}} &= \mu(\psi_{rd} i_{sq} - \psi_{rq} i_{sd}) - \frac{T_L}{J} + \frac{\psi^* \tilde{\psi}_{rq}}{\lambda} + \frac{\beta \psi_{rq} \tilde{i}_{sd}}{\lambda} - \frac{\beta \psi_{rd} \tilde{i}_{sq}}{\lambda} \\
\begin{bmatrix} \psi_{rd} \\ \psi_{rq} \end{bmatrix} &= \begin{bmatrix} \cos\varepsilon_0 & \sin\varepsilon_0 \\ -\sin\varepsilon_0 & \cos\varepsilon_0 \end{bmatrix} \begin{bmatrix} \psi_{ra} \\ \psi_{rb} \end{bmatrix} \\
\begin{bmatrix} i_{sd} \\ i_{sq} \end{bmatrix} &= \begin{bmatrix} \cos\varepsilon_0 & \sin\varepsilon_0 \\ -\sin\varepsilon_0 & \cos\varepsilon_0 \end{bmatrix} \begin{bmatrix} i_{sa} \\ i_{sb} \end{bmatrix} \\
\tilde{i}_{sd} &= i_{sd} - i_{sd}^* \\
\tilde{i}_{sq} &= i_{sq} - i_{sq}^* \\
\tilde{\psi}_{rd} &= \psi_{rd} - \psi^* \\
\tilde{\psi}_{rq} &= \psi_{rq} \\
\phi_{d0} &= -\gamma i_{sd} + \omega_0 i_{sq} + \alpha\beta\psi_{rd} + \beta\hat{\omega}\psi_{rq} - \frac{\dot{\psi}^*}{M} - \frac{\ddot{\psi}^*}{\alpha M} \\
\phi_{q0} &= -\gamma i_{sq} - \omega_0 i_{sd} + \alpha\beta\psi_{rq} - \beta\hat{\omega}\psi_{rd} \\
&\quad + \frac{\dot{\psi}^*}{\mu\psi^{*2}}\left[-k_\omega \mathrm{sat}(\hat{\omega}-\omega^*) + \frac{T_L}{J} + \dot{\omega}^*\right] \\
&\quad - \frac{1}{\mu\psi^*}\left[-k_\omega \frac{\mathrm{dsat}(\hat{\omega}-\omega^*)}{\mathrm{d}(\hat{\omega}-\omega^*)}(\dot{\hat{\omega}}-\dot{\omega}^*) + \ddot{\omega}^*\right]
\end{aligned} \tag{5.20}
$$

is a second order dynamic control algorithm which is based on a closed-loop rotor speed observer and on the measurements of the state variables $(\psi_{ra}, \psi_{rb}, i_{sa}, i_{sb})$, on the reference signals (ω^*, ψ^*), on the positive control parameters k_ω, λ, k_i, on the load torque T_L, and on the machine parameters M, R_r, L_r, J, R_s, L_s since $\mu = \frac{M}{JL_r}$, $\alpha = \frac{R_r}{L_r}$, $\sigma = L_s\left(1 - \frac{M^2}{L_s L_r}\right)$, $\beta = \frac{M}{\sigma L_r}$, $\gamma = \frac{R_s}{\sigma} + \alpha\beta M$; the origin $(\tilde{\omega}, \tilde{\psi}_{rd}, \tilde{\psi}_{rq}, \tilde{i}_{sd}, \tilde{i}_{sq}, e_\omega) = 0$ of the closed-loop system (5.18) is globally uniformly asymptotically stable and locally exponentially stable: in particular, for every motor initial condition $(\omega(0), \psi_{ra}(0), \psi_{rb}(0), i_{sa}(0), i_{sb}(0))$ and any controller initial condition $(\hat{\omega}(0), \varepsilon_0(0))$, exponential rotor speed and flux modulus tracking is achieved along with exponential rotor speed estimation

$$\lim_{t\to\infty} [\omega(t) - \omega^*(t)] = 0$$

$$\lim_{t\to\infty} [\hat{\omega}(t) - \omega(t)] = 0\,.$$

$$\lim_{t\to\infty} \left[\sqrt{\psi_{ra}^2(t) + \psi_{rb}^2(t)} - \psi^*(t) \right] = 0$$

Remarks

1. Note that the controller contains a globally convergent closed-loop rotor speed estimator since, in closed-loop,
$$\lim_{t\to\infty} [\hat{\omega}(t) - \omega(t)] = 0\,.$$
Hence (5.17) is a rotor speed observer when it operates along with the controller in the closed-loop system.
2. The steady-state operating condition of the controller coincides with the inverse control (1.74).

Illustrative Simulations

We tested the global speed-sensorless controller (5.20) by simulations for the three-phase single pole pair 0.6-kW induction motor whose parameters have been reported in Chapter 1. All initial conditions of the motor are set to zero. The simulation test involves the following operating sequences: for $0 \le t < 0.36$ s the motor is driven by the inverse control (1.74) ($\varepsilon_0^*(0) = 0$), while for $t \ge 0.36$ s the motor is controlled by the proposed algorithm ($\varepsilon_0(0.36) = \varepsilon_0^*(0.36)$) with control parameters (the values are in SI units) $k_\omega = 20, \lambda = 0.001, k_i = 2000$. At $t = 1.4$ s, a smooth increase up to 10% in both rotor and stator resistances is simulated in the motor equations, while the controller maintains the nominal values. The references for rotor speed and flux modulus and the applied load torque are reported in Figure 5.11. The flux reference starts from 0.001 Wb at $t = 0$ and grows up to the rated constant value 1.16 Wb. The speed reference is zero until $t = 0.32$ s and grows up to the constant value 100 rad/s. The closed-loop control is inserted at $t = 0.36$ s and not at $t = 0.32$ s, so that there is an initial speed estimation error $e_\omega(0.36) = -2.426$ rad/s. A constant load torque (5.8 Nm, the rated value) is applied at $t = 0.9$ s. Figures 5.12–5.14 show the rotor speed, the rotor flux modulus, and the rotor speed estimate along with the corresponding tracking and estimation errors, while Figures 5.15 and 5.16 show the stator current and the stator voltage vectors (a,b)-components. Figures 5.12–5.14 show a fast speed estimation and a good tracking performance in nominal conditions while steady-state tracking and estimation errors appear when both the stator and the rotor resistances, for $t \ge 1.4$ s, are different from their nominal values.

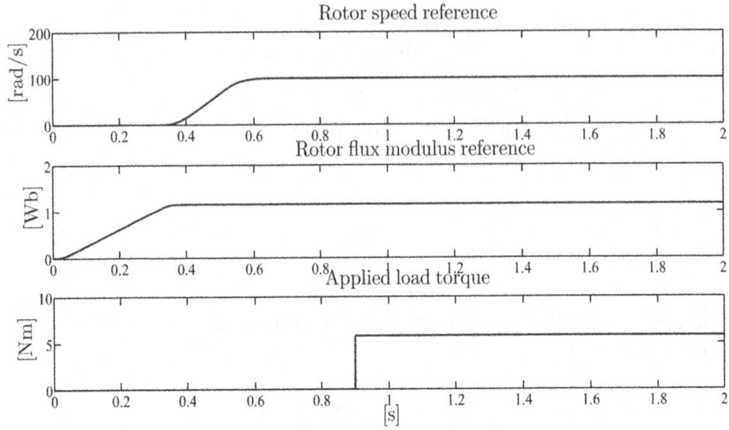

Fig. 5.11 Global speed-sensorless control (5.20): rotor speed and flux modulus reference signals and applied load torque

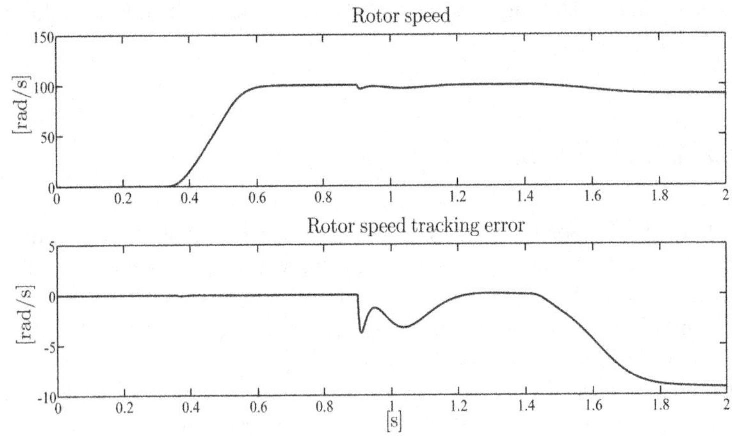

Fig. 5.12 Global speed-sensorless control (5.20): rotor speed and rotor speed tracking error

5.3 Adaptive Control with Flux Measurements

In this section we design an adaptive version of the global control presented in the previous section. While in the previous section all motor parameters were supposed to be known, in this section we shall assume that both the critical parameters load torque T_L and rotor resistance R_r (which affects the parameter $\alpha = R_r/L_r$) are uncertain. We still maintain the assumption that the rotor fluxes are available for feedback: this unrealistic assumption will be removed in the following sections by introducing rotor flux observers. Note that the control algorithm designed in the previous section depends on the parameters T_L and α: in particular the speed of the rotating (d,q) frame depends on α and the rotor speed estimation dynamics depends on T_L. We

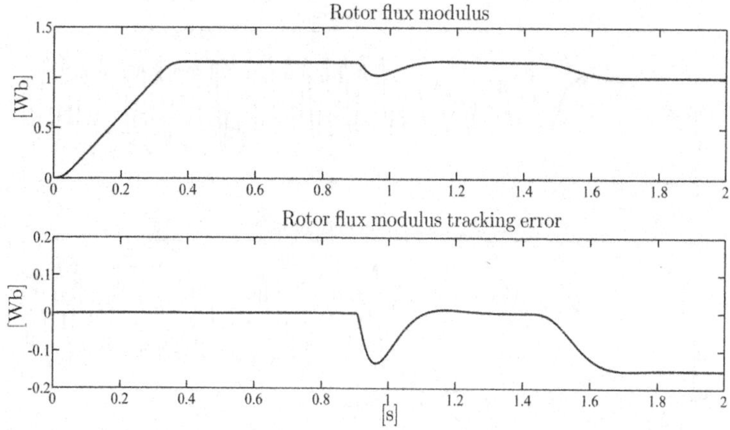

Fig. 5.13 Global speed-sensorless control (5.20): rotor flux modulus and rotor flux modulus tracking error

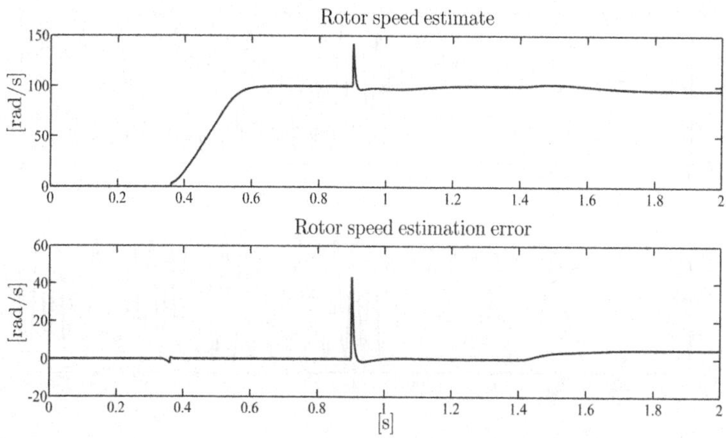

Fig. 5.14 Global speed-sensorless control (5.20): rotor speed estimate and rotor speed estimation error

introduce, as in (5.7), a saturation function $\text{sat}(q)$, a continuously differentiable odd function which is linear in a neighborhood of the origin $q = 0$ and has a finite limit as $|q|$ goes to infinity. We modify the reference current signals for i_{sd} and i_{sq} in (5.7) as (k_ω is a positive control parameter)

$$
\begin{aligned}
i_{sd}^* &= \frac{\psi^*}{M} + \frac{\dot{\psi}^*}{\hat{\alpha} M} \\
i_{sq}^* &= \frac{1}{\mu \psi^*}\left[-k_\omega \text{sat}(\hat{\omega} - \omega^*) + \frac{\hat{T}_L}{J} + \dot{\omega}^* \right]
\end{aligned} \tag{5.21}
$$

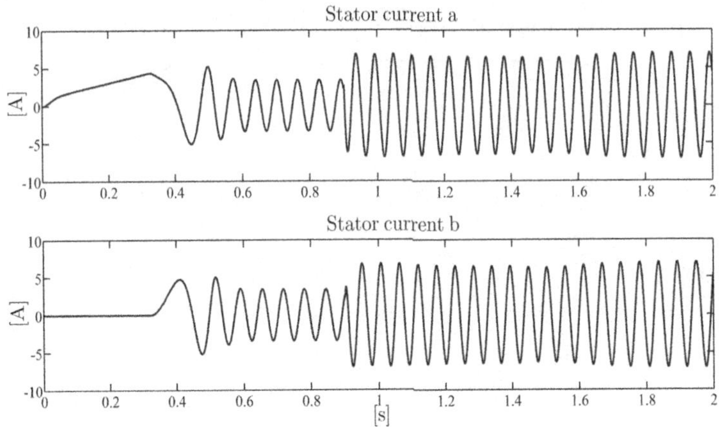

Fig. 5.15 Global speed-sensorless control (5.20): stator current vector (a,b)-components

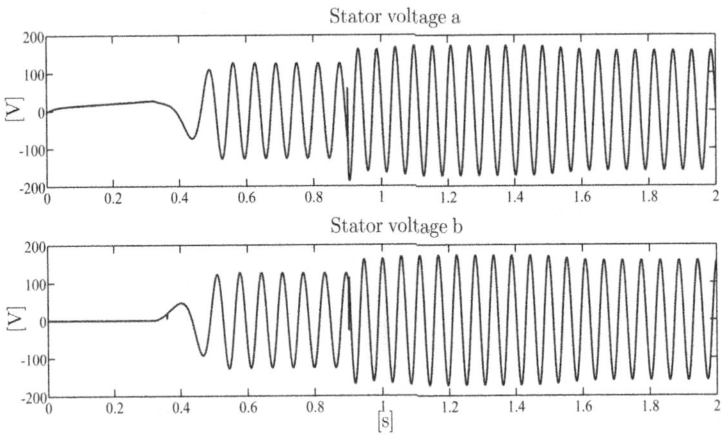

Fig. 5.16 Global speed-sensorless control (5.20): stator voltage vector (a,b)-components

and the speed of the rotating (d,q) frame in (5.8) as

$$\dot{\varepsilon}_0 = \omega_0 = \hat{\omega} + \frac{\hat{\alpha} M i_{sq}^*}{\psi^*} \tag{5.22}$$

where $\hat{\alpha}$ and $\hat{T}_L$ are the estimates of the uncertain parameters α and T_L, respectively, and $\hat{\omega}$ is the speed estimate provided by the rotor speed observer [a modification of (5.9)]

$$\dot{\hat{\omega}} = \mu(\psi_{rd} i_{sq} - \psi_{rq} i_{sd}) - \frac{\hat{T}_L}{J} + \eta \tag{5.23}$$

in which the additional term η will be determined later on the basis of the tracking errors. If we compare (5.7), (5.8), (5.9) with (5.21), (5.22), (5.23) we note that they are the same with T_L and α replaced by their estimates $\hat{T}_L$ and $\hat{\alpha}$ and with i_{sq}^* in (5.22) replacing i_{sq} in (5.8). As we shall see, the adaptation for $\hat{\alpha}$ will be designed using a projection algorithm guaranteeing that $\hat{\alpha}(t) \geq c_\alpha > 0$ for all $t \geq 0$ so that i_{sd}^*, which is divided by $\hat{\alpha}$, is well defined. Our goal is to design the feedback term η in (5.23), the feedback control inputs (u_{sd}, u_{sq}), and the adaptation laws for the estimates $\hat{T}_L$ and $\hat{\alpha}$, so that the tracking and the estimation errors asymptotically tend to zero. To this end, we introduce the tracking and the estimation errors

$$\begin{aligned}
\tilde{\omega} &= \omega - \omega^* \\
\tilde{\psi}_{rd} &= \psi_{rd} - \psi^* \\
\tilde{\psi}_{rq} &= \psi_{rq} \\
e_\omega &= \hat{\omega} - \omega \\
\tilde{i}_{sd} &= i_{sd} - i_{sd}^* \\
\tilde{i}_{sq} &= i_{sq} - i_{sq}^* \\
\tilde{\alpha} &= \alpha - \hat{\alpha} \\
\tilde{T}_L &= T_L - \hat{T}_L\,;
\end{aligned} \tag{5.24}$$

from (1.31), (5.21), (5.22), (5.23), (5.24), we can write the tracking and the estimation error dynamics as

$$\begin{aligned}
\frac{d\omega}{dt} &= -k_\omega \mathrm{sat}(\tilde{\omega} + e_\omega) + \mu(\tilde{\psi}_{rd} i_{sq} - \tilde{\psi}_{rq} i_{sd}) + \mu\psi^* \tilde{i}_{sq} - \frac{\tilde{T}_L}{J} \\
\frac{d\tilde{\psi}_{rd}}{dt} &= -\alpha\tilde{\psi}_{rd} + (\omega_0 - \omega)\tilde{\psi}_{rq} + (Mi_{sd}^* - \psi^*)\tilde{\alpha} + \alpha M\tilde{i}_{sd} \\
\frac{d\tilde{\psi}_{rq}}{dt} &= -\alpha\tilde{\psi}_{rq} - (\omega_0 - \omega)\tilde{\psi}_{rd} - e_\omega\psi^* + Mi_{sq}^*\tilde{\alpha} + \alpha M\tilde{i}_{sq} \\
\frac{de_\omega}{dt} &= \eta + \frac{\tilde{T}_L}{J} \\
\frac{d\tilde{i}_{sd}}{dt} &= \frac{u_{sd}}{\sigma} + \phi_{d0} + \beta\alpha\tilde{\psi}_{rd} - \beta(\omega_0 - \omega)\tilde{\psi}_{rq} - \beta(Mi_{sd}^* - \psi^*)\tilde{\alpha} - \beta\alpha M\tilde{i}_{sd} \\
\frac{d\tilde{i}_{sq}}{dt} &= \frac{u_{sq}}{\sigma} + \phi_{q0} + \beta\alpha\tilde{\psi}_{rq} + \beta(\omega_0 - \omega)\tilde{\psi}_{rd} + \beta e_\omega\psi^* - \beta Mi_{sq}^*\tilde{\alpha} - \beta\alpha M\tilde{i}_{sq}
\end{aligned} \tag{5.25}$$

in which

$$\begin{aligned}
\phi_{d0} &= -\frac{R_s}{\sigma} i_{sd} + \omega_0 i_{sq} + \beta\hat{\alpha}\psi^* + \beta\omega_0\psi_{rq} - \beta\hat{\alpha}Mi_{sd}^* \\
&\quad - \frac{\dot{\psi}^*}{M} - \frac{\ddot{\psi}^*}{\hat{\alpha}M} + \frac{\dot{\psi}^*\dot{\hat{\alpha}}}{\hat{\alpha}^2 M} \\
\phi_{q0} &= -\frac{R_s}{\sigma} i_{sq} - \omega_0 i_{sd} - \beta\hat{\omega}\psi^* - \beta\hat{\alpha}Mi_{sq}^* - \beta\omega_0\tilde{\psi}_{rd}
\end{aligned}$$

$$
\begin{aligned}
&+\frac{\dot{\psi}^*}{\mu\psi^{*2}}\left[-k_\omega \mathrm{sat}(\hat{\omega}-\omega^*)+\frac{\hat{T}_L}{J}+\dot{\omega}^*\right]\\
&-\frac{1}{\mu\psi^*}\left[-k_\omega\frac{\mathrm{d}[\mathrm{sat}(\hat{\omega}-\omega^*)]}{\mathrm{d}[\hat{\omega}-\omega^*]}(\dot{\hat{\omega}}-\dot{\omega}^*)+\frac{\dot{\hat{T}}_L}{J}+\ddot{\omega}^*\right].
\end{aligned}
\tag{5.26}
$$

We introduce the new error variables

$$
\begin{aligned}
z_d &= \tilde{i}_{sd}+\beta\tilde{\psi}_{rd}\\
z_q &= \tilde{i}_{sq}+\beta\tilde{\psi}_{rq}
\end{aligned}
\tag{5.27}
$$

so that the error equations (5.25), expressed in the new coordinates, become

$$
\begin{aligned}
\dot{\tilde{\omega}} &= -k_\omega \mathrm{sat}(\tilde{\omega}+e_\omega)+\mu(\tilde{\psi}_{rd}i_{sq}-\tilde{\psi}_{rq}i_{sd})+\mu\psi^* z_q\\
&\quad -\mu\psi^*\beta\tilde{\psi}_{rq}-\frac{\tilde{T}_L}{J}\\
\dot{\tilde{\psi}}_{rd} &= -(\alpha+\alpha M\beta)\tilde{\psi}_{rd}+(\omega_0-\omega)\tilde{\psi}_{rq}\\
&\quad +(Mi^*_{sd}-\psi^*)\tilde{\alpha}+\alpha M z_d\\
\dot{\tilde{\psi}}_{rq} &= -(\alpha+\alpha M\beta)\tilde{\psi}_{rq}-(\omega_0-\omega)\tilde{\psi}_{rd}-e_\omega\psi^*\\
&\quad +Mi^*_{sq}\tilde{\alpha}+\alpha M z_q\\
\dot{e}_\omega &= \eta+\frac{\tilde{T}_L}{J}\\
\dot{z}_d &= \frac{u_{sd}}{\sigma}+\phi_{d0}\\
\dot{z}_q &= \frac{u_{sq}}{\sigma}+\phi_{q0}\,.
\end{aligned}
\tag{5.28}
$$

Consider the positive definite function ($\lambda_1, \lambda_2, \lambda_3$ are positive control parameters)

$$
\begin{aligned}
V &= \frac{1}{2}\left(\tilde{\psi}_{rd}^2+\tilde{\psi}_{rq}^2+z_d{}^2+z_q{}^2\right)+\frac{\lambda_1}{2}e_\omega^2+\frac{\lambda_2{}^2}{\lambda_1 J^2}\tilde{T}_L^2\\
&\quad -\frac{\lambda_2}{J}e_\omega\tilde{T}_L+\frac{\lambda_3}{2}\tilde{\alpha}^2
\end{aligned}
\tag{5.29}
$$

whose time derivative along the trajectories of the closed-loop system (5.28) is

$$
\begin{aligned}
\dot{V} &= -(\alpha+\alpha M\beta)\left(\tilde{\psi}_{rd}^2+\tilde{\psi}_{rq}^2\right)+\alpha M\left(z_d\tilde{\psi}_{rd}+z_q\tilde{\psi}_{rq}\right)\\
&\quad +\left[\frac{u_{sd}}{\sigma}+\phi_{d0}\right]z_d+\left[\frac{u_{sq}}{\sigma}+\phi_{q0}\right]z_q\\
&\quad +\frac{\lambda_1}{J}e_\omega\tilde{T}_L-\frac{\lambda_2}{J^2}\tilde{T}_L^2+\left[\lambda_1\eta-\psi^*\tilde{\psi}_{rq}\right]e_\omega-\frac{\lambda_2}{J}e_\omega\dot{\tilde{T}}_L\\
&\quad +\left(\frac{2\lambda_2{}^2}{\lambda_1 J^2}\dot{\tilde{T}}_L-\frac{\lambda_2}{J}\eta\right)\tilde{T}_L
\end{aligned}
$$

$$+\left[(Mi_{sd}^* - \psi^*)\tilde{\psi}_{rd} + Mi_{sq}^*\tilde{\psi}_{rq} + \lambda_3\dot{\hat{\alpha}}\right]\tilde{\alpha}\,. \tag{5.30}$$

We design the adaptation law for $\hat{\alpha}$ as

$$\dot{\hat{\alpha}} = \mathrm{Proj}_\alpha\left(\frac{1}{\lambda_3}\left[(Mi_{sd}^* - \psi^*)\tilde{\psi}_{rd} + Mi_{sq}^*\tilde{\psi}_{rq}\right], \hat{\alpha}\right), \quad \hat{\alpha}(0) = \hat{\alpha}_0 \tag{5.31}$$

where $\mathrm{Proj}_\alpha(\xi, \hat{\alpha})$ is the smooth projection algorithm given by (A.25) in Appendix A and defined in our case by

$$\mathrm{Proj}_\alpha(\xi, \hat{\alpha}) = \begin{cases} \xi & \text{if } \alpha_m \le \hat{\alpha} \le \alpha_M \\ \xi & \text{if } \hat{\alpha} < \alpha_m \text{ and } \xi \ge 0 \\ \xi & \text{if } \hat{\alpha} > \alpha_M \text{ and } \xi \le 0 \\ \xi\left[1 - \frac{\alpha_m{}^2 - \hat{\alpha}^2}{\alpha_m{}^2 - (\alpha_m - \varepsilon_1)^2}\right] & \text{if } \hat{\alpha} < \alpha_m \text{ and } \xi < 0 \\ \xi\left[1 - \frac{\hat{\alpha}^2 - \alpha_M{}^2}{(\alpha_M + \varepsilon_1)^2 - \alpha_M{}^2}\right] & \text{if } \hat{\alpha} > \alpha_M \text{ and } \xi > 0 \end{cases} \tag{5.32}$$

with $\alpha_m = R_{rm}/L_r$ and $\alpha_M = R_{rM}/L_r$ the (known) lower and upper bounds of α, respectively, and ε_1 an arbitrary positive constant such that $\alpha_m - \varepsilon_1 > 0$. The initial condition $\hat{\alpha}_0$ in (5.31) is chosen so that $\alpha_m \le \hat{\alpha}_0 \le \alpha_M$. Property (iii) of the projection algorithm (see Appendix A) implies that substituting (5.31) in (5.30) we obtain the inequality

$$\begin{aligned} \dot{V} \le & -(\alpha + \alpha M\beta)\left(\tilde{\psi}_{rd}^2 + \tilde{\psi}_{rq}^2\right) + \alpha M\left(z_d\tilde{\psi}_{rd} + z_q\tilde{\psi}_{rq}\right) \\ & + \left[\frac{u_{sd}}{\sigma} + \phi_{d0}\right]z_d + \left[\frac{u_{sq}}{\sigma} + \phi_{q0}\right]z_q \\ & + \frac{\lambda_1}{J}e_\omega\tilde{T}_L - \frac{\lambda_2}{J^2}\tilde{T}_L^2 + \left[\lambda_1\eta - \psi^*\tilde{\psi}_{rq}\right]e_\omega \\ & - \frac{\lambda_2}{J}e_\omega\dot{\hat{T}}_L + \left(\frac{2\lambda_2{}^2}{\lambda_1 J^2}\dot{\hat{T}}_L - \frac{\lambda_2}{J}\eta\right)\tilde{T}_L\,. \end{aligned} \tag{5.33}$$

We define the additive feedback term η in (5.23) and the adaptation dynamics for $\hat{T}_L$ as

$$\eta = \begin{cases} \frac{2\psi^*}{\lambda_1}\tilde{\psi}_{rq} & \text{if } p_\eta(\hat{T}_L) \le 0 \\ \frac{2\psi^*}{\lambda_1}\tilde{\psi}_{rq} & \text{if } p_\eta(\hat{T}_L) > 0 \text{ and } -\frac{dp_\eta(\hat{T}_L)}{d\hat{T}_L}\tilde{\psi}_{rq} \le 0 \\ \frac{2\psi^*}{\lambda_1[1 + p_\eta(\hat{T}_L)]}\tilde{\psi}_{rq} & \text{otherwise.} \end{cases}$$

$$\dot{\hat{T}}_L = \begin{cases} -\frac{J\psi^*}{\lambda_2}\tilde{\psi}_{rq} & \text{if } p_\eta(\hat{T}_L) \le 0 \\ -\frac{J\psi^*}{\lambda_2}\tilde{\psi}_{rq} & \text{if } p_\eta(\hat{T}_L) > 0 \text{ and } -\frac{dp_\eta(\hat{T}_L)}{d\hat{T}_L}\tilde{\psi}_{rq} \le 0 \\ -\frac{J\psi^*[1 - p_\eta(\hat{T}_L)]}{\lambda_2[1 + p_\eta(\hat{T}_L)]}\tilde{\psi}_{rq} & \text{otherwise} \end{cases}$$

$$\hat{T}_L(0) = \hat{T}_{L0} \tag{5.34}$$

where $p_\eta(\hat{T}_L) = \frac{\hat{T}_L^2 - T_{LM}{}^2}{2\varepsilon_2 T_{LM} + \varepsilon_2^2}$ and $|\hat{T}_{L0}| \leq T_{LM}$, with T_{LM} a known upper bound on $|T_L|$ and $\varepsilon_2 > 0$. Since

$$\text{sgn}(\tilde{\psi}_{rq}) = \text{sgn}(\eta) \tag{5.35}$$

from (5.34) we can write

$$\dot{\hat{T}}_L = \text{Proj}_T\left(-\frac{\lambda_1 J}{2\lambda_2}\eta, \hat{T}_L\right), \quad \hat{T}_L(0) = \hat{T}_{L0} \tag{5.36}$$

where $\text{Proj}_T(\varphi, \hat{T}_L)$ is the smooth projection algorithm given by (A.25) in Appendix A and defined in our case by

$$\text{Proj}_T[\varphi, \hat{T}_L] = \begin{cases} \varphi & \text{if } p_\eta(\hat{T}_L) \leq 0 \\ \varphi & \text{if } p_\eta(\hat{T}_L) > 0 \text{ and } \frac{dp_\eta(\hat{T}_L)}{d\hat{T}_L}\varphi \leq 0 \\ [1 - p_\eta(\hat{T}_L)]\varphi & \text{otherwise.} \end{cases}$$

Denoting by

$$\begin{aligned} \chi = &-(\alpha + \alpha M\beta)\left(\tilde{\psi}_{rd}^2 + \tilde{\psi}_{rq}^2\right) + \alpha M\left(z_d\tilde{\psi}_{rd} + z_q\tilde{\psi}_{rq}\right) \\ &+ \left[\frac{u_{sd}}{\sigma} + \phi_{d0}\right] z_d + \left[\frac{u_{sq}}{\sigma} + \phi_{q0}\right] z_q + \frac{\lambda_1}{J} e_\omega \tilde{T}_L - \frac{\lambda_2}{J^2}\tilde{T}_L^2 \end{aligned}$$

by virtue of Property (iii) of the projection algorithm (see Appendix A), from (5.33), (5.34), (5.35), and (5.36), we can write

$$\begin{aligned} \dot{V} &\leq \chi + \left[\lambda_1\eta - \psi^*\tilde{\psi}_{rq}\right] e_\omega + \frac{\lambda_2}{J} e_\omega \dot{\hat{T}}_L \\ &\quad + \left(-\frac{2\lambda_2{}^2}{\lambda_1 J^2}\text{Proj}_T\left[-\frac{\lambda_1 J}{2\lambda_2}\eta, \hat{T}_L\right] - \frac{\lambda_2}{J}\eta\right)\tilde{T}_L \\ &\leq \chi + \left[\lambda_1\eta - \psi^*\tilde{\psi}_{rq}\right] e_\omega + \frac{\lambda_2}{J} e_\omega \dot{\hat{T}}_L = \chi \end{aligned} \tag{5.37}$$

and, completing the squares, we finally obtain

$$\begin{aligned} \dot{V} \leq &-\alpha\left(\tilde{\psi}_{rd}^2 + \tilde{\psi}_{rq}^2\right) + \frac{\alpha M}{4\beta}\left(z_d^2 + z_q^2\right) \\ &+ \left[\frac{u_{sd}}{\sigma} + \phi_{d0}\right] z_d + \left[\frac{u_{sq}}{\sigma} + \phi_{q0}\right] z_q + \frac{\lambda_1}{J} e_\omega \tilde{T}_L - \frac{\lambda_2}{J^2}\tilde{T}_L^2 . \end{aligned} \tag{5.38}$$

We define the control inputs as (k_z is a positive control parameter)

$$
\begin{aligned}
u_{sd} &= \sigma\left[-\phi_{d0} - \left(k_z + \frac{\alpha_M M}{4\beta}\right) z_d\right] \\
u_{sq} &= \sigma\left[-\phi_{q0} - \left(k_z + \frac{\alpha_M M}{4\beta}\right) z_q\right]
\end{aligned}
\tag{5.39}
$$

so that (5.38) becomes

$$
\begin{aligned}
\dot{V} &\leq -\alpha\left(\tilde{\psi}_{rd}^2 + \tilde{\psi}_{rq}^2\right) - \left[k_z + \frac{(\alpha_M - \alpha)M}{4\beta}\right]\left(z_d^2 + z_q^2\right) + \frac{\lambda_1}{J} e_\omega \tilde{T}_L - \frac{\lambda_2}{J^2}\tilde{T}_L^2 \\
&\leq -\min\left\{\alpha_m, k_z\right\}\left(\tilde{\psi}_{rd}^2 + \tilde{\psi}_{rq}^2 + z_d^2 + z_q^2\right) + \frac{\lambda_1}{J} e_\omega \tilde{T}_L - \frac{\lambda_2}{J^2}\tilde{T}_L^2 \,.
\end{aligned}
\tag{5.40}
$$

The closed-loop system is

$$
\begin{aligned}
\dot{\tilde{\omega}} &= -k_\omega \mathrm{sat}(\tilde{\omega} + e_\omega) + \mu(\tilde{\psi}_{rd} i_{sq} - \tilde{\psi}_{rq} i_{sd}) + \mu\psi^* z_q \\
&\quad - \mu\psi^*\beta\tilde{\psi}_{rq} - \frac{\tilde{T}_L}{J} \\
\dot{\tilde{\psi}}_{rd} &= -(\alpha + \alpha M\beta)\tilde{\psi}_{rd} + (\omega_0 - \omega)\tilde{\psi}_{rq} + (Mi_{sd}^* - \psi^*)\tilde{\alpha} + \alpha M z_d \\
\dot{\tilde{\psi}}_{rq} &= -(\alpha + \alpha M\beta)\tilde{\psi}_{rq} - (\omega_0 - \omega)\tilde{\psi}_{rd} - e_\omega\psi^* + Mi_{sq}^*\tilde{\alpha} + \alpha M z_q \\
\dot{z}_d &= -\left(k_z + \frac{\alpha_M M}{4\beta}\right) z_d \\
\dot{z}_q &= -\left(k_z + \frac{\alpha_M M}{4\beta}\right) z_q \\
\dot{e}_\omega &= \frac{\tilde{T}_L}{J} + \begin{cases} \frac{2\psi^*}{\lambda_1}\tilde{\psi}_{rq} & \text{if } p_\eta(\hat{T}_L) \leq 0 \\ \frac{2\psi^*}{\lambda_1}\tilde{\psi}_{rq} & \text{if } p_\eta(\hat{T}_L) > 0 \text{ and } -\frac{dp_\eta(\hat{T}_L)}{d\hat{T}_L}\tilde{\psi}_{rq} \leq 0 \\ \frac{2\psi^*}{\lambda_1[1 + p_\eta(\hat{T}_L)]}\tilde{\psi}_{rq} & \text{otherwise} \end{cases} \\
\dot{\tilde{\alpha}} &= -\mathrm{Proj}_\alpha\left(\frac{1}{\lambda_3}\left[(Mi_{sd}^* - \psi^*)\tilde{\psi}_{rd} + Mi_{sq}^*\tilde{\psi}_{rq}\right], \hat{\alpha}\right) \\
\dot{\tilde{T}}_L &= \begin{cases} \frac{J\psi^*}{\lambda_2}\tilde{\psi}_{rq} & \text{if } p_\eta(\hat{T}_L) \leq 0 \\ \frac{J\psi^*}{\lambda_2}\tilde{\psi}_{rq} & \text{if } p_\eta(\hat{T}_L) > 0 \text{ and } -\frac{dp_\eta(\hat{T}_L)}{d\hat{T}_L}\tilde{\psi}_{rq} \leq 0 \\ \frac{J\psi^*[1 - p_\eta(\hat{T}_L)]}{\lambda_2[1 + p_\eta(\hat{T}_L)]}\tilde{\psi}_{rq} & \text{otherwise.} \end{cases}
\end{aligned}
\tag{5.41}
$$

The last seven equations in (5.41) may be rewritten as

$$
\begin{aligned}
\dot{x}(t) &= \left[A(t) + S(t,y)\right]x(t) + B(t)y(t) \\
\dot{y}(t) &= D(t)x(t) + Ew(t) \\
\dot{w}(t) &= F(t)x(t)
\end{aligned}
\tag{5.42}
$$

with $x = [\tilde{\psi}_{rd}, \tilde{\psi}_{rq}, z_d, z_q]^{\mathrm{T}}$, $y = [e_\omega, \tilde{\alpha}]^{\mathrm{T}}$, $w = \tilde{T}_L$, and

$$A = \begin{bmatrix} -(\alpha+\alpha M\beta) & \frac{\hat{\alpha} M i^*_{sq}}{\psi^*} & \alpha M & 0 \\ -\frac{\hat{\alpha} M i^*_{sq}}{\psi^*} & -(\alpha+\alpha M\beta) & 0 & \alpha M \\ 0 & 0 & -\left(k_z + \frac{\alpha_M M}{4\beta}\right) & 0 \\ 0 & 0 & 0 & -\left(k_z + \frac{\alpha_M M}{4\beta}\right) \end{bmatrix}$$

$$S = \begin{bmatrix} 0 & e_\omega & 0 & 0 \\ -e_\omega & 0 & 0 & 0 \\ 0 & 0 & 0 & 0 \\ 0 & 0 & 0 & 0 \end{bmatrix}$$

$$B = \begin{bmatrix} 0 & M i^*_{sd} - \psi^* \\ -\psi^* & M i^*_{sq} \\ 0 & 0 \\ 0 & 0 \end{bmatrix}$$

$$D = \begin{bmatrix} 0 & g_1 \frac{2\psi^*}{\lambda_1} & 0 & 0 \\ -g_2 \frac{\dot{\psi}^*}{\lambda_3 \hat{\alpha}} & -g_2 \frac{M i^*_{sq}}{\lambda_3} & 0 & 0 \end{bmatrix}$$

$$E = \begin{bmatrix} 1/J \\ 0 \end{bmatrix}$$

$$F = \begin{bmatrix} 0 & \frac{J\psi^*}{\lambda_2} g_3 & 0 & 0 \end{bmatrix} \tag{5.43}$$

with $g_i(t)$, $i = 1,2,3$, being suitable time functions such that $0 \le g_i(t) \le 1$ for all $t \ge 0$. Consider the positive definite function

$$U = \frac{1}{2}\left(z_d^2 + z_q^2 + \psi_{rd}^2 + \psi_{rq}^2\right) \tag{5.44}$$

whose time derivative is

$$\dot{U} = -\left(k_z + \frac{\alpha_M M}{4\beta}\right)\left(z_d^2 + z_q^2\right) - \alpha\left(\psi_{rd}^2 + \psi_{rq}^2\right) + \alpha M i_{sd} \psi_{rd} + \alpha M i_{sq} \psi_{rq}\,. \tag{5.45}$$

Substituting in (5.45) the expressions of (i_{sd}, i_{sq}) obtained from (5.27), we have

$$\begin{aligned} \dot{U} = & -\left(k_z + \frac{\alpha_M M}{4\beta}\right)\left(z_d^2 + z_q^2\right) - (\alpha + \alpha M\beta)\left(\psi_{rd}^2 + \psi_{rq}^2\right) \\ & + \alpha M\left[i^*_{sd} + z_d + \beta\psi^*\right]\psi_{rd} + \alpha M\left[i^*_{sq} + z_q\right]\psi_{rq} \\ \le & -\min\left\{k_z, \alpha_m\right\}\left(z_d^2 + z_q^2 + \psi_{rd}^2 + \psi_{rq}^2\right) \end{aligned}$$

$$+\alpha M\left[i_{sd}^{*}+\beta\psi^{*}\right]\psi_{rd}+\alpha M i_{sq}^{*}\psi_{rq}\,. \tag{5.46}$$

Since $i_{sd}^{*}(t)$ and $i_{sq}^{*}(t)$ in (5.21) are bounded time functions on $[0,\infty)$ [recall that the adaptation for $\hat{\alpha}$ and $\hat{T}_L$ has been designed using a projection algorithm], from (5.44) and (5.46) it follows that $z_d(t)$, $z_q(t)$, $\psi_{rd}(t)$, and $\psi_{rq}(t)$ are bounded time functions on $[0,\infty)$ and, consequently, $\eta(t)$, $\dot{\hat{\alpha}}(t)$, and $\dot{\hat{T}}_L(t)$ are bounded time functions on $[0,\infty)$. From (5.21), since $\dot{\hat{\alpha}}(t)$ is a bounded time function on $[0,\infty)$, $\frac{\mathrm{d}i_{sd}^{*}(t)}{\mathrm{d}t}$ is a bounded time function on $[0,\infty)$. Since $\psi_{rd}(t)$ and $\psi_{rq}(t)$ are bounded time functions on $[0,\infty)$ and since, from (5.27)

$$\begin{aligned} i_{sd} &= i_{sd}^{*}+z_d-\beta\psi_{rd}+\beta\psi^{*} \\ i_{sq} &= i_{sq}^{*}+z_q-\beta\psi_{rq} \end{aligned}$$

$i_{sd}(t)$ and $i_{sq}(t)$ are bounded time functions on $[0,\infty)$, it follows from (5.23) that $\dot{\hat{\omega}}(t)$ is a bounded time function on $[0,\infty)$. Since $\dot{\hat{T}}_L(t)$ and $\dot{\hat{\omega}}(t)$ are bounded time functions on $[0,\infty)$, it follows that $\frac{\mathrm{d}i_{sq}^{*}(t)}{\mathrm{d}t}$ is a bounded time function on $[0,\infty)$. Therefore we can establish that $\|A(t)\|\le A_M$, $\|B(t)\|\le B_M$, $\|D(t)\|\le D_M$, $\|\dot{B}(t)\|\le \dot{B}_M$ for all $t\ge 0$, with $A_M, B_M, D_M, \dot{B}_M$ known positive reals. If we define $\psi^{*}(t)$ and the positive reals T_α, k_α in order to satisfy

$$\int_t^{t+T_\alpha}\left[\frac{\mathrm{d}\psi^{*}(\tau)}{\mathrm{d}\tau}\right]^2\mathrm{d}\tau\ge k_\alpha>0,\quad \forall t\ge 0 \tag{5.47}$$

so that persistency of excitation condition holds, *i.e.* a positive real k exists such that

$$\int_t^{t+T_\alpha} B^{\mathrm{T}}(\tau)B(\tau)\mathrm{d}\tau\ge kI,\quad \forall t\ge 0$$

and if we choose the positive control parameter λ_2 in order to satisfy the inequalities

$$\lambda_2>\frac{\lambda^2 e^{4T}}{pk^2}+8pB_M^4 \tag{5.48}$$

$$p<\frac{\min\left\{\alpha_m,k_z\right\}}{16(B_M^4D_M^2+B_M^2A_M^2+\dot{B}_M^2+B_M^2)+B_M^2}\,, \tag{5.49}$$

then system (5.42) complies with the hypotheses of the Persistency of Excitation Lemma A.3 in Appendix A: the equilibrium point $(\tilde{\psi}_{rd},\tilde{\psi}_{rq},z_d,z_q,e_\omega,\tilde{\alpha},\tilde{T}_L)=0$ is locally exponentially stable. Recalling the first equation in (5.41), since $i_{sd}(t)$ and $i_{sq}(t)$ are bounded time functions on $[0,\infty)$ and $\tilde{\psi}_{rd}$, $\tilde{\psi}_{rq}$, z_q, e_ω, and $\tilde{T}_L$ tend exponentially to zero as t goes to infinity, $\tilde{\omega}$ tends exponentially to zero for any sufficiently small initial conditions. In summary, if λ_2 is chosen so that (5.48) and (5.49) are satisfied then, for any sufficiently small initial condition:

1. since $\tilde{\omega}(t)$ tends exponentially to zero, exponential rotor speed tracking is achieved;

2. since $\tilde{\psi}_{rd}(t)$ and $\tilde{\psi}_{rq}(t)$ tend exponentially to zero, exponential rotor flux tracking is achieved and the rotor flux vector is exponentially oriented as the d-axis of the (d,q) frame;
3. since $\tilde{\psi}_{rd}(t)$, $\tilde{\psi}_{rq}(t)$, $z_d(t)$ and $z_q(t)$ tend exponentially to zero, from (5.27) it follows that $\tilde{i}_{sd}(t)$ and $\tilde{i}_{sq}(t)$ tend exponentially to zero, *i.e.* the stator currents tend exponentially to the corresponding reference signals;
4. since $e_\omega(t)$ tends exponentially to zero, the rotor speed is exponentially estimated;
5. since $\tilde{\alpha}(t)$ and $\tilde{T}_L(t)$ tend exponentially to zero, the rotor resistance R_r and the load torque T_L are exponentially estimated.

In conclusion, the *adaptive speed-sensorless control*

$$\begin{bmatrix} u_{sa} \\ u_{sb} \end{bmatrix} = \begin{bmatrix} \cos\varepsilon_0 & -\sin\varepsilon_0 \\ \sin\varepsilon_0 & \cos\varepsilon_0 \end{bmatrix} \begin{bmatrix} u_{sd} \\ u_{sq} \end{bmatrix}$$

$$u_{sd} = \sigma\left[-\phi_{d0} - \left(k_z + \frac{\alpha_M M}{4\beta}\right) z_d\right]$$

$$u_{sq} = \sigma\left[-\phi_{q0} - \left(k_z + \frac{\alpha_M M}{4\beta}\right) z_q\right]$$

$$i_{sd}^* = \frac{\psi^*}{M} + \frac{\dot{\psi}^*}{\hat{\alpha} M}$$

$$i_{sq}^* = \frac{1}{\mu\psi^*}\left[-k_\omega \mathrm{sat}(\hat{\omega} - \omega^*) + \frac{\hat{T}_L}{J} + \dot{\omega}^*\right]$$

$$\dot{\varepsilon}_0 = \omega_0 = \hat{\omega} + \frac{\hat{\alpha} M i_{sq}^*}{\psi^*}$$

$$\dot{\hat{\alpha}} = \mathrm{Proj}_\alpha\left(\frac{1}{\lambda_3}\left[(M i_{sd}^* - \psi^*)\tilde{\psi}_{rd} + M i_{sq}^* \tilde{\psi}_{rq}\right], \hat{\alpha}\right)$$

$$\mathrm{Proj}_\alpha(\xi, \hat{\alpha}) = \begin{cases} \xi & \text{if } \alpha_m \le \hat{\alpha} \le \alpha_M \\ \xi & \text{if } \hat{\alpha} < \alpha_m \text{ and } \xi \ge 0 \\ \xi & \text{if } \hat{\alpha} > \alpha_M \text{ and } \xi \le 0 \\ \xi\left[1 - \frac{\alpha_m^2 - \hat{\alpha}^2}{\alpha_m^2 - (\alpha_m - \varepsilon_1)^2}\right] & \text{if } \hat{\alpha} < \alpha_m \text{ and } \xi < 0 \\ \xi\left[1 - \frac{\hat{\alpha}^2 - \alpha_M^2}{(\alpha_M + \varepsilon_1)^2 - \alpha_M^2}\right] & \text{if } \hat{\alpha} > \alpha_M \text{ and } \xi > 0 \end{cases}$$

$$\dot{\hat{\omega}} = \mu(\psi_{rd} i_{sq} - \psi_{rq} i_{sd}) - \frac{\hat{T}_L}{J} + \eta$$

$$\eta = \begin{cases} \frac{2\psi^*}{\lambda_1}\tilde{\psi}_{rq} & \text{if } p_\eta(\hat{T}_L) \le 0 \\ \frac{2\psi^*}{\lambda_1}\tilde{\psi}_{rq} & \text{if } p_\eta(\hat{T}_L) > 0 \text{ and } -\frac{dp_\eta(\hat{T}_L)}{d\hat{T}_L}\tilde{\psi}_{rq} \le 0 \\ \frac{2\psi^*}{\lambda_1[1 + p_\eta(\hat{T}_L)]}\tilde{\psi}_{rq} & \text{otherwise.} \end{cases}$$

$$
\begin{aligned}
\dot{\hat{T}}_L &= \begin{cases} -\frac{J\psi^*}{\lambda_2}\tilde{\psi}_{rq} & \text{if } p_\eta(\hat{T}_L) \le 0 \\ -\frac{J\psi^*}{\lambda_2}\tilde{\psi}_{rq} & \text{if } p_\eta(\hat{T}_L) > 0 \text{ and } -\frac{dp_\eta(\hat{T}_L)}{d\hat{T}_L}\tilde{\psi}_{rq} \le 0 \\ -\frac{J\psi^*[1-p_\eta(\hat{T}_L)]}{\lambda_2[1+p_\eta(\hat{T}_L)]}\tilde{\psi}_{rq} & \text{otherwise} \end{cases} \\
p_\eta(\hat{T}_L) &= \frac{\hat{T}_L^2 - {T_{LM}}^2}{2\varepsilon_2 T_{LM} + \varepsilon_2^2} \\
\tilde{\psi}_{rd} &= \psi_{rd} - \psi^* \\
\tilde{\psi}_{rq} &= \psi_{rq} \\
\tilde{i}_{sd} &= i_{sd} - i_{sd}^* \\
\tilde{i}_{sq} &= i_{sq} - i_{sq}^* \\
z_d &= \tilde{i}_{sd} + \beta\tilde{\psi}_{rd} \\
z_q &= \tilde{i}_{sq} + \beta\tilde{\psi}_{rq} \\
\begin{bmatrix} \psi_{rd} \\ \psi_{rq} \end{bmatrix} &= \begin{bmatrix} \cos\varepsilon_0 & \sin\varepsilon_0 \\ -\sin\varepsilon_0 & \cos\varepsilon_0 \end{bmatrix} \begin{bmatrix} \psi_{ra} \\ \psi_{rb} \end{bmatrix} \\
\begin{bmatrix} i_{sd} \\ i_{sq} \end{bmatrix} &= \begin{bmatrix} \cos\varepsilon_0 & \sin\varepsilon_0 \\ -\sin\varepsilon_0 & \cos\varepsilon_0 \end{bmatrix} \begin{bmatrix} i_{sa} \\ i_{sb} \end{bmatrix} \\
\phi_{d0} &= -\frac{R_s}{\sigma}i_{sd} + \omega_0 i_{sq} + \beta\hat{\alpha}\psi^* + \beta\omega_0\psi_{rq} - \beta\hat{\alpha}Mi_{sd}^* \\
&\quad -\frac{\dot{\psi}^*}{M} - \frac{\ddot{\psi}^*}{\hat{\alpha}M} + \frac{\psi^*\dot{\hat{\alpha}}}{\hat{\alpha}^2 M} \\
\phi_{q0} &= -\frac{R_s}{\sigma}i_{sq} - \omega_0 i_{sd} - \beta\hat{\omega}\psi^* - \beta\hat{\alpha}Mi_{sq}^* - \beta\omega_0\tilde{\psi}_{rd} \\
&\quad + \frac{\dot{\psi}^*}{\mu{\psi^*}^2}\left[-k_\omega \text{sat}(\hat{\omega}-\omega^*) + \frac{\hat{T}_L}{J} + \dot{\omega}^*\right] \\
&\quad - \frac{1}{\mu\psi^*}\left[-k_\omega \frac{\mathrm{d}[\text{sat}(\hat{\omega}-\omega^*)]}{\mathrm{d}[\hat{\omega}-\omega^*]}(\dot{\hat{\omega}} - \dot{\omega}^*) + \frac{\dot{\hat{T}}_L}{J} + \ddot{\omega}^*\right]
\end{aligned} \tag{5.50}
$$

is a fourth order dynamic control algorithm which depends on the measurements of the state variables $(\psi_{ra}, \psi_{rb}, i_{sa}, i_{sb})$, on the reference signals (ω^*, ψ^*), on the positive control parameters k_ω, k_z, λ_1, λ_2, λ_3, α_m, α_M, T_{LM}, and on the machine parameters M, L_r, J, R_s, L_s since $\mu = \frac{M}{JL_r}$, $\sigma = L_s\left(1 - \frac{M^2}{L_s L_r}\right)$, $\beta = \frac{M}{\sigma L_r}$; the origin $(\tilde{\omega}, \tilde{z}_d, \tilde{z}_q, \tilde{i}_{sd}, \tilde{i}_{sq}, e_\omega, \tilde{\alpha}, \tilde{T}_L) = 0$ of the closed-loop system (5.41) is globally uniformly asymptotically stable and locally exponentially stable: in particular for every motor initial condition $(\omega(0), \psi_{ra}(0), \psi_{rb}(0), i_{sa}(0), i_{sb}(0))$ and any controller initial condition $(\hat{\omega}(0), \varepsilon_0(0))$, exponential rotor speed and flux modulus tracking is achieved along with exponential estimation of rotor speed, load torque and parameter

α, provided that the reference signal ψ^* is chosen so that the persistency of excitation (5.47) is satisfied and that the control parameter λ_2 is chosen to satisfy the inequality

$$\lambda_2 > \frac{\lambda^2 e^{4T}}{pk^2} + 8pB_M^4$$

with

$$p < \frac{\min\left\{\alpha_m, k_z\right\}}{16(B_M^4 D_M^2 + B_M^2 A_M^2 + \dot{B}_M^2 + B_M^2) + B_M^2}.$$

Remarks

1. The dynamic compensator contains five parameters $(k_\omega, \lambda_1, \lambda_2, \lambda_3, k_z)$ whose role may be evaluated by examining both the closed-loop error equations (5.41) and the corresponding stability analysis. The parameters k_z and k_ω directly affect the dynamics of the error variables (z_d, z_q) and the speed tracking error $\tilde{\omega}$, respectively; the parameter λ_1 determines the influence of the rotor flux vector quadrature component tracking error $\tilde{\psi}_{rq}$ on the rotor speed estimation error e_ω; the parameters λ_2^{-1} and λ_3^{-1} are the adaptation gains for $\hat{T}_L$ and $\hat{\alpha}$, respectively, and the smaller they are chosen, the slower the adaptations for $\hat{T}_L$ and $\hat{\alpha}$ result.
2. From (5.43) it follows that $\dot{B}_M$ depends on $\frac{di_{sq}^*(t)}{dt}$ which, in turn, depends on λ_2^{-1} (through $\dot{\hat{T}}_L$). More precisely, $\dot{B}_M$ either decreases or remains constant when λ_2 increases, so that (5.49) may be satisfied.
3. If $\|[\dot{\psi}^*(t),\, \ddot{\omega}^*(t)]\| = 0$, then the persistency of excitation condition, recalling the structure of $B(t)$, may not be satisfied. This fact can be clearly deduced considering the closed-loop system (5.41): since, from (5.21), $(Mi_{sd}^* - \psi^*) = \frac{\dot{\psi}^*}{\alpha}$, if $\|[\dot{\psi}^*(t), \ddot{\omega}^*(t)]\| = 0$, we have that all the points $(\tilde{\omega}, \tilde{\psi}_{rd}, \tilde{\psi}_{rq}, z_d, z_q, e_\omega, \tilde{\alpha}, \tilde{T}_L) = (-g\tilde{\alpha}, 0, 0, 0, 0, g\tilde{\alpha}, \tilde{\alpha}, 0)$, with $g = \frac{M}{\mu\psi^{*2}}\left(\dot{\omega}^* + \frac{T_L}{J}\right)$, are equilibrium points for system (5.41). Moreover, if $\|[\dot{\omega}^*(t), T_L(t)]\|$=0, then $g = 0$ and all the points $(\tilde{\omega}, \tilde{\psi}_{rd}, \tilde{\psi}_{rq}, z_d, z_q, e_\omega, \tilde{\alpha}, \tilde{T}_L) = (0, 0, 0, 0, 0, 0, \tilde{\alpha}, 0)$ are equilibrium points for system (5.41).
4. If T_L is known, we may set

$$i_{sq}^* = \frac{1}{\mu\psi^*}\left[-k_\omega \text{sat}(\hat{\omega} - \omega^*) + \frac{T_L}{J} + \dot{\omega}^*\right]$$

$$
\begin{aligned}
\dot{\hat{\omega}} &= \mu(\psi_{rd} i_{sq} - \psi_{rq} i_{sd}) - \frac{T_L}{J} + \frac{2\psi^*}{\lambda_1}\tilde{\psi}_{rq} \\
\phi_{q0} &= -\frac{R_s}{\sigma} i_{sq} - \omega_0 i_{sd} - \beta\hat{\omega}\psi^* - \beta\hat{\alpha}M i_{sq}^* - \beta\omega_0\tilde{\psi}_{rd} \\
&\quad + \frac{\dot{\psi}^*}{\mu\psi^{*2}}\left[-k_\omega \mathrm{sat}(\hat{\omega} - \omega^*) + \frac{T_L}{J} + \dot{\omega}^*\right] \\
&\quad - \frac{1}{\mu\psi^*}\left[-k_\omega \frac{\mathrm{d}[\mathrm{sat}(\hat{\omega} - \omega^*)]}{\mathrm{d}[\hat{\omega} - \omega^*]}(\dot{\hat{\omega}} - \dot{\omega}^*) + \ddot{\omega}^*\right].
\end{aligned}
$$

The closed-loop system becomes

$$
\begin{aligned}
\dot{\tilde{\omega}} &= -k_\omega \mathrm{sat}(\tilde{\omega} + e_\omega) + \mu(\tilde{\psi}_{rd} i_{sq} - \tilde{\psi}_{rq} i_{sd}) + \mu\psi^* z_q - \mu\psi^*\beta\tilde{\psi}_{rq} \\
\dot{\tilde{\psi}}_{rd} &= -(\alpha + \alpha M\beta)\tilde{\psi}_{rd} + (\omega_0 - \omega)\tilde{\psi}_{rq} + (M i_{sd}^* - \psi^*)\tilde{\alpha} + \alpha M z_d \\
\dot{\tilde{\psi}}_{rq} &= -(\alpha + \alpha M\beta)\tilde{\psi}_{rq} - (\omega_0 - \omega)\tilde{\psi}_{rd} - e_\omega\psi^* + M i_{sq}^*\tilde{\alpha} + \alpha M z_q \\
\dot{z}_d &= -\left(k_z + \frac{\alpha_M M}{4\beta}\right) z_d \\
\dot{z}_q &= -\left(k_z + \frac{\alpha_M M}{4\beta}\right) z_q \\
\dot{e}_\omega &= \frac{2\psi^*}{\lambda_1}\tilde{\psi}_{rq} \\
\dot{\tilde{\alpha}} &= -\mathrm{Proj}_\alpha\left(\frac{1}{\lambda_3}\left[(M i_{sd}^* - \psi^*)\tilde{\psi}_{rd} + M i_{sq}^*\tilde{\psi}_{rq}\right], \hat{\alpha}\right).
\end{aligned}
\tag{5.51}
$$

Consider the positive definite function

$$
V_n = \frac{1}{2}\left(\tilde{\psi}_{rd}^2 + \tilde{\psi}_{rq}^2 + z_d{}^2 + z_q{}^2\right) + \frac{1}{4}\lambda_1 e_\omega^2 + \frac{\lambda_3}{2}\tilde{\alpha}^2 \tag{5.52}
$$

whose time derivative along the trajectories of the closed-loop system (5.51) satisfies the inequality

$$
\dot{V}_n \leq -\min\left\{\alpha_m, k_z\right\}\left(\tilde{\psi}_{rd}^2 + \tilde{\psi}_{rq}^2 + z_d^2 + z_q^2\right). \tag{5.53}
$$

From (5.52) and (5.53) it follows that $\tilde{\psi}_{rd}(t)$, $\tilde{\psi}_{rq}(t)$, $z_d(t)$, $z_q(t)$, $e_\omega(t)$, and $\tilde{\alpha}(t)$ are bounded time functions on $[0, \infty)$. Since $\tilde{\psi}_{rd}(t)$, $\tilde{\psi}_{rq}(t)$, $z_d(t)$, $z_q(t)$, $i_{sd}^*(t)$, and $i_{sq}^*(t)$ are bounded time functions on $[0, \infty)$, it follows that $i_{sd}(t)$ and $i_{sq}(t)$ are bounded time functions on $[0, \infty)$ and, consequently, $\frac{\mathrm{d}i_{sd}^*(t)}{\mathrm{d}t}$ and $\frac{\mathrm{d}i_{sq}^*(t)}{\mathrm{d}t}$ are bounded time functions on $[0, \infty)$. The last six equations in (5.51) may be rewritten as ($x = [\tilde{\psi}_{rd}, \tilde{\psi}_{rq}, z_d, z_q]^{\mathrm{T}}$, $y = [e_\omega, \tilde{\alpha}]^{\mathrm{T}}$)

$$
\dot{x}(t) = \bar{A}(t)x(t) + B(t)y(t)
$$

$$\dot{y}(t) = \bar{D}(t)x(t) \tag{5.54}$$

with $\bar{A}(t) = A(t) + S(t,y)$ and

$$\bar{D} = \begin{bmatrix} 0 & \frac{2\psi^*}{\lambda_1} & 0 & 0 \\ -\bar{g}\frac{\dot{\psi}^*}{\lambda_3\hat{\alpha}} & -\bar{g}\frac{M i_{sq}^*}{\lambda_3} & 0 & 0 \end{bmatrix}$$

with $\bar{g}(t)$ a suitable bounded time function such that $0 \leq \bar{g}(t) \leq 1$ for all $t \geq 0$. The reference signal $\psi^*(t)$ is chosen so that there exists positive reals T_α, k_α satisfying

$$\int_t^{t+T_\alpha} \left[\frac{\mathrm{d}\psi^*(\tau)}{\mathrm{d}\tau}\right]^2 \mathrm{d}\tau \geq k_\alpha > 0, \quad \forall t \geq 0$$

so that the persistency of excitation condition holds, *i.e.* a positive k exists such that

$$\int_t^{t+T_\alpha} B^{\mathrm{T}}(\tau)B(\tau)\mathrm{d}\tau \geq kI, \quad \forall t \geq 0\,;$$

then system (5.54) complies with the hypotheses of the Persistency of Excitation Lemma A.3 in Appendix A: the point $(\tilde{\psi}_{rd}, \tilde{\psi}_{rq}, z_d, z_q, e_\omega, \tilde{\alpha}) = 0$ is globally uniformly asymptotically stable and exponentially stable for initial conditions in any compact set. Recalling the first equation in (5.51), since $i_{sd}(t)$ and $i_{sq}(t)$ are bounded time functions on $[0,\infty)$ and $\tilde{\psi}_{rd}(t)$, $\tilde{\psi}_{rq}(t)$, $z_q(t)$, and $e_\omega(t)$ tend exponentially to zero, $\tilde{\omega}(t)$ tends exponentially to zero for any initial condition: in particular, the origin is a globally uniformly asymptotically and locally exponentially stable equilibrium point for system (5.51).

Illustrative Simulations

We tested the adaptive speed-sensorless control (5.50) by simulations for the three-phase single pole pair 0.6-kW induction motor whose parameters have been reported in Chapter 1. The control parameters are (the values are in SI units): $k_\omega = 60$, $\lambda_1^{-1} = 6{,}000$, $\lambda_2^{-1} = 180/J$, $\lambda_3^{-1} = 75$, $k_z + \frac{\alpha_M M}{4\beta} = 700$, $\alpha_m = 3.5$, $\alpha_M = 18$, $\varepsilon_1 = 0.5$, $T_{LM} = 8$, $\varepsilon_2 = 1$. Due to the conservative nature of Lyapunov analysis, the values of λ_2 satisfying (5.48) and (5.49) may be larger than those achieving the control objective: as a matter of fact, in order to improve the performance, the simulation was carried out by setting $\lambda_2{}^{-1} = 180/J$ which does not satisfy (5.48) and (5.49). All initial conditions of the motor and of the controller are set to zero except $\hat{\alpha}(0) = 13.2\ \mathrm{s}^{-1}$, which is 50% greater than the true parameter value $\alpha = 8.8\,\mathrm{s}^{-1}$. The references for rotor speed and flux modulus along with the applied torque are reported in Figure 5.17. The rotor flux modulus reference starts from 0.001 Wb at $t = 0$ s, grows up to the rated constant value 1.16 Wb and, at $t = 0.45$ s, a persistently exciting signal is added in order to avoid the identifiability problems. The speed reference is zero until $t = 0.32$ s and grows up to the constant value 100 rad/s. A constant load torque (5.8 Nm) is applied at $t = 0.9$ s. Figures 5.18–5.22 show the rotor speed, the rotor flux modulus, the rotor speed estimate, the load torque estimate, and the α estimate, along with the corresponding tracking and estimation errors, while Figures 5.23 and 5.24 show the stator current and the stator voltage vectors (a,b) components. Under persistency of excitation, fast estimation and good tracking performance are obtained.

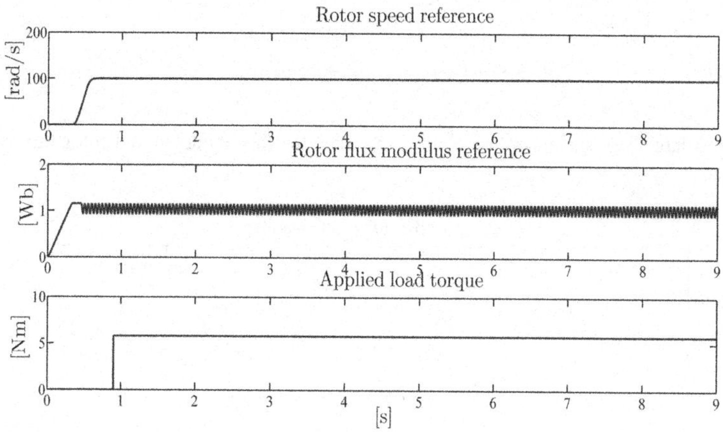

Fig. 5.17 Adaptive speed-sensorless control (5.50): rotor speed and flux modulus reference signals and applied load torque

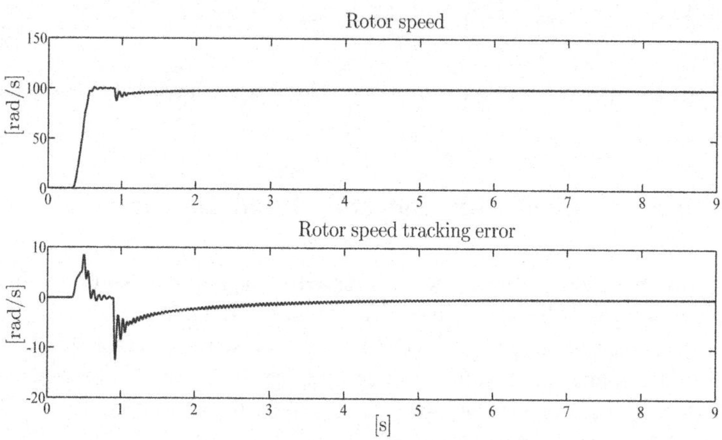

Fig. 5.18 Adaptive speed-sensorless control (5.50): rotor speed and rotor speed tracking error

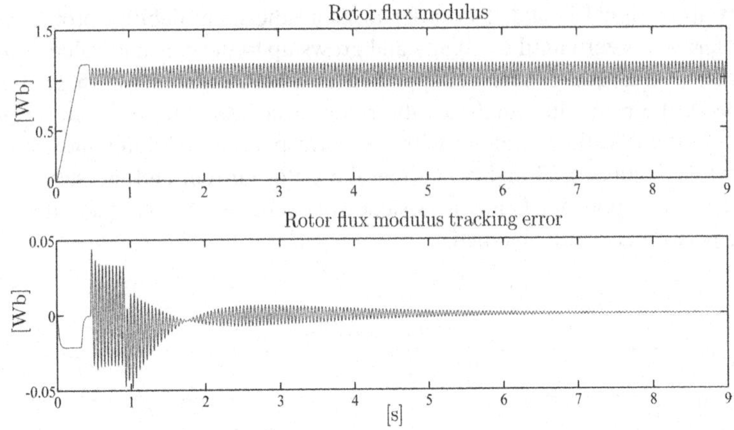

Fig. 5.19 Adaptive speed-sensorless control (5.50): rotor flux modulus and rotor flux modulus tracking error

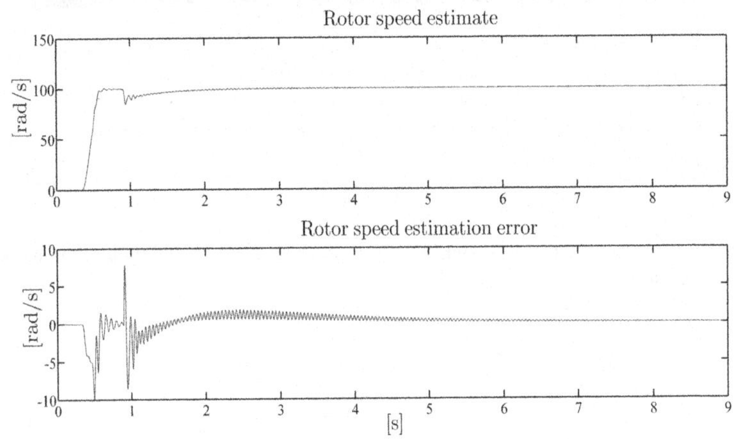

Fig. 5.20 Adaptive speed-sensorless control (5.50): rotor speed estimate and rotor speed estimation error

5.4 Adaptive Control with Uncertain Load Torque

In this section we address the problem discussed in the previous section without the unrealistic assumption that rotor fluxes are available for feedback but assuming that the rotor resistance is known. We will introduce closed-loop rotor flux observers and modify the control algorithm obtained in the previous section by replacing the rotor fluxes by the corresponding estimates given by the observers. To this end, we design the stator current control loop as

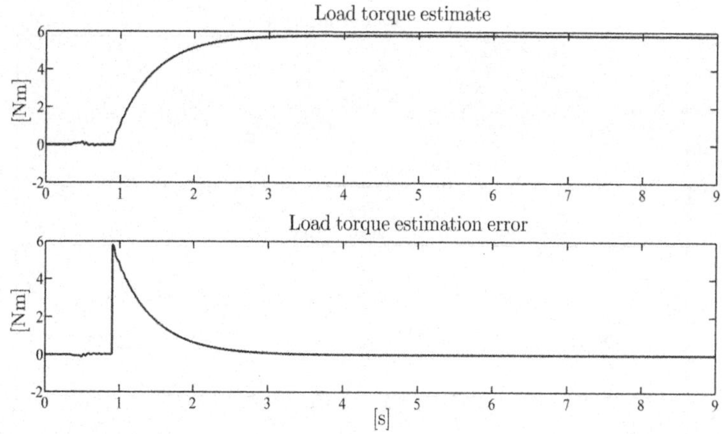

Fig. 5.21 Adaptive speed-sensorless control (5.50): load torque estimate and load torque estimation error

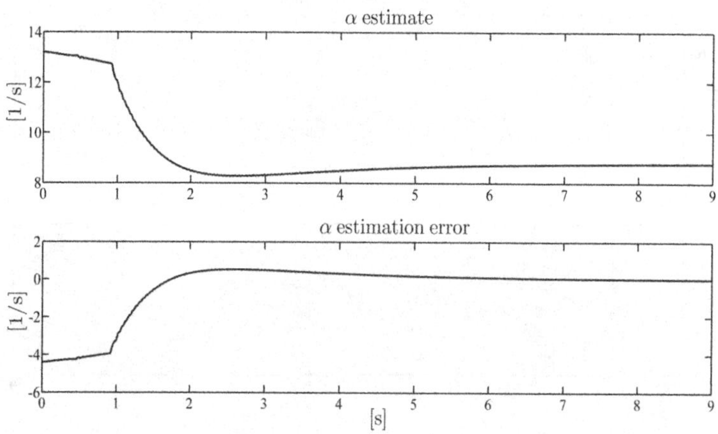

Fig. 5.22 Adaptive speed-sensorless control (5.50): α estimate and α estimation error

$$
\begin{aligned}
\begin{bmatrix} u_{sa} \\ u_{sb} \end{bmatrix} &= \begin{bmatrix} \cos\varepsilon_0 & -\sin\varepsilon_0 \\ \sin\varepsilon_0 & \cos\varepsilon_0 \end{bmatrix} \begin{bmatrix} u_{sd} \\ u_{sq} \end{bmatrix} \\
u_{sd} &= \sigma\beta\left[-\frac{g_d}{\beta} - \alpha(\hat{\psi}_{rd} - \psi^*) - \frac{k_i}{\gamma_f\mu}\tilde{i}_{sd}\right] \\
u_{sq} &= \sigma\beta\left[-\frac{g_q}{\beta} - \psi^*(\hat{\omega} - \omega^*) - \frac{k_i}{\gamma_f\mu}\tilde{i}_{sq}\right] \\
\tilde{i}_{sd} &= i_{sd} - i_{sd}^* \\
\tilde{i}_{sq} &= i_{sq} - i_{sq}^*
\end{aligned}
$$

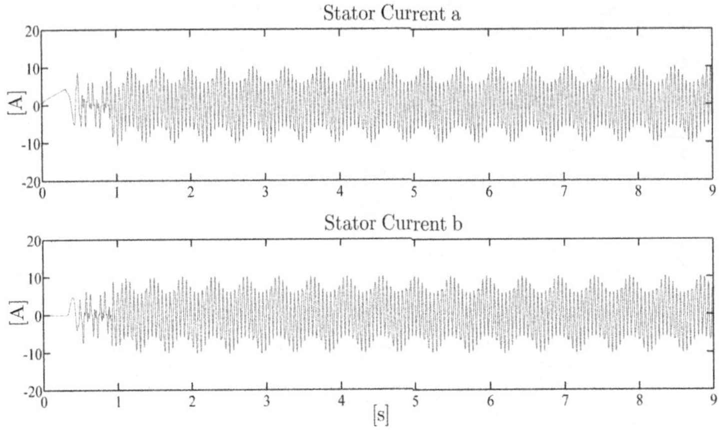

Fig. 5.23 Adaptive speed-sensorless control (5.50): stator current vector (a,b)-components

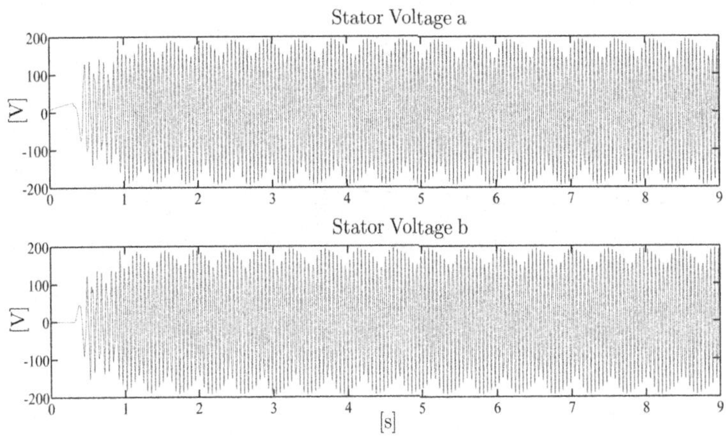

Fig. 5.24 Adaptive speed-sensorless control (5.50): stator voltage vector (a,b)-components

$$\begin{bmatrix} i_{sd} \\ i_{sq} \end{bmatrix} = \begin{bmatrix} \cos\varepsilon_0 & \sin\varepsilon_0 \\ -\sin\varepsilon_0 & \cos\varepsilon_0 \end{bmatrix} \begin{bmatrix} i_{sa} \\ i_{sb} \end{bmatrix} \tag{5.55}$$

in which the reference signals (i_{sd}^*, i_{sq}^*) for (i_{sd}, i_{sq}) and the speed of the rotating frame ω_0 (which, as in the field-oriented control strategy, are responsible for rotor speed and flux modulus tracking) are designed as (compare with the corresponding signals in the control algorithm presented in Section 2.7)

$$\begin{aligned} i_{sd}^* = & \frac{\psi^*}{M} + \frac{\dot{\psi}^*}{\alpha M} + \frac{k_\psi}{\alpha\mu\gamma_f}(\psi^* - \hat{\psi}_{rd}) \\ & - \frac{1}{\alpha\mu\psi^*}\left(\frac{T_{Ln}}{J} + \frac{\mathrm{sat}(\hat{\theta})}{J} + \dot{\omega}^*\right)(\hat{\omega} - \omega^*) \end{aligned}$$

$$
\begin{aligned}
i_{sq}^{*} &= \frac{1}{\mu\psi^{*}}\left[-k_{\omega}(\hat{\omega}-\omega^{*})+\frac{T_{Ln}}{J}+\frac{\mathrm{sat}(\hat{\theta})}{J}+\dot{\omega}^{*}\right] \\
\dot{\varepsilon}_{0} &= \omega_{0} = \hat{\omega}+\frac{\alpha M i_{sq}}{\psi^{*}}+\frac{k_{\psi}M\hat{\psi}_{rq}}{\gamma_{f}\mu\psi^{*}} \\
&\quad -\frac{M}{\psi^{*}}\left[\frac{\psi^{*}}{M}+\frac{\dot{\psi}^{*}}{\alpha M}\right](\hat{\omega}-\omega^{*})\,;
\end{aligned}
\tag{5.56}
$$

the estimates $(\hat{\omega}, \hat{\psi}_{rd}, \hat{\psi}_{rq})$ for the unmeasured state variables $(\omega, \psi_{rd}, \psi_{rq})$ appearing in (5.55) and (5.56) are given by

$$
\begin{aligned}
\hat{\omega} &= \xi_{1}-i_{sq}\zeta+i_{sd}\eta \\
\hat{\psi}_{rd} &= \xi_{2}+\left(M-\frac{L_{r}L_{s}}{M}\right)i_{sd} \\
\hat{\psi}_{rq} &= \xi_{3}+\left(M-\frac{L_{r}L_{s}}{M}\right)i_{sq}
\end{aligned}
$$

and depend on the controller internal signals (ξ_1, ξ_2, ξ_3) which satisfy the dynamic equations

$$
\begin{aligned}
\dot{\xi}_{1} &= -k_{\omega}\xi_{1}+i_{sq}\dot{\zeta}+k_{\omega}i_{sq}\zeta-i_{sd}\dot{\eta}-k_{\omega}i_{sd}\eta \\
&\quad +\mu\psi^{*}i_{sq}-\frac{T_{Ln}}{J}-\frac{\mathrm{sat}(\hat{\theta})}{J}+\mu\hat{\psi}_{rq}\tilde{i}_{sd}+k_{\omega}\omega^{*} \\
&\quad +\varphi_{1}-\zeta\left[-\frac{1}{\sigma}u_{sq}+\frac{R_{s}}{\sigma}i_{sq}+\omega_{0}i_{sd}-\alpha\beta\hat{\psi}_{rq}+\alpha M\beta i_{sq}\right. \\
&\quad \left.+\beta(\xi_{1}-i_{sq}\zeta+i_{sd}\eta)\hat{\psi}_{rd}\right] \\
&\quad +\eta\left[-\frac{1}{\sigma}u_{sd}+\frac{R_{s}}{\sigma}i_{sd}-\omega_{0}i_{sq}-\alpha\beta\hat{\psi}_{rd}\right. \\
&\quad \left.+\alpha M\beta i_{sd}-\beta(\xi_{1}-i_{sq}\zeta+i_{sd}\eta)\hat{\psi}_{rq}\right] \\
\dot{\xi}_{2} &= -\frac{k_{\psi}}{\lambda}\xi_{2}-\frac{k_{\psi}}{\lambda}\left(M-\frac{L_{r}L_{s}}{M}\right)i_{sd}+\frac{k_{\psi}}{\lambda}\psi^{*} \\
&\quad +\frac{\gamma_{f}\mu}{\lambda}\left(2\alpha\tilde{i}_{sd}-\hat{\omega}\tilde{i}_{sq}\right)+\frac{\varphi_{2}}{\lambda}+\frac{L_{r}}{M}u_{sd} \\
&\quad -\frac{L_{r}R_{s}}{M}i_{sd}+\omega_{0}\hat{\psi}_{rq}-\left(M-\frac{L_{r}L_{s}}{M}\right)\omega_{0}i_{sq} \\
\dot{\xi}_{3} &= -\frac{k_{\psi}}{\lambda}\xi_{3}-\frac{k_{\psi}}{\lambda}\left(M-\frac{L_{r}L_{s}}{M}\right)i_{sq} \\
&\quad +\frac{\gamma_{f}\mu}{\lambda}\left(\alpha\tilde{i}_{sq}+\hat{\omega}\tilde{i}_{sd}\right)+\frac{\varphi_{3}}{\lambda}+\frac{L_{r}}{M}u_{sq}
\end{aligned}
$$

$$-\frac{L_r R_s}{M} i_{sq} - \omega_0 \hat{\psi}_{rd} + \left(M - \frac{L_r L_s}{M}\right) \omega_0 i_{sd} \tag{5.57}$$

with initial conditions

$$\begin{aligned}
\xi_1(0) &= \hat{\omega}(0) + i_{sq}(0)\zeta(0) - i_{sd}(0)\eta(0) \\
\xi_2(0) &= \hat{\psi}_{rd}(0) - \left(M - \frac{L_r L_s}{M}\right) i_{sd}(0) \\
\xi_3(0) &= \hat{\psi}_{rq}(0) - \left(M - \frac{L_r L_s}{M}\right) i_{sq}(0) .
\end{aligned}$$

The terms $(g_d, g_q, \varphi_1, \varphi_2, \varphi_3, \zeta, \eta)$ appearing several times in (5.55), (5.57) and containing feedforward and feedback actions are given by

$$\begin{aligned}
g_d = &-\left(\frac{R_s}{\sigma} + \alpha M \beta\right) i_{sd} + \omega_0 i_{sq} + \alpha\beta\hat{\psi}_{rd} + \beta\hat{\omega}\hat{\psi}_{rq} - \frac{\dot{\psi}^*}{M} - \frac{\ddot{\psi}^*}{\alpha M} - \frac{k_\psi}{\alpha\mu\gamma_f}\dot{\psi}^* \\
&+ \frac{k_\psi}{\alpha\mu\gamma_f}\left\{\frac{\gamma_f\mu}{\lambda}\left(2\alpha\tilde{i}_{sd} - \hat{\omega}\tilde{i}_{sq}\right) - \frac{L_r R_s}{M} i_{sd} + \omega_0\hat{\psi}_{rq} + \frac{\varphi_2}{\lambda} + \frac{k_\psi}{\lambda}\left(\psi^* - \hat{\psi}_{rd}\right)\right. \\
&\left. + \left(M - \frac{L_r L_s}{M}\right)\left(-\frac{R_s}{\sigma} i_{sd} + \alpha\beta\hat{\psi}_{rd} - \alpha M \beta i_{sd} + \beta\hat{\omega}\hat{\psi}_{rq}\right)\right\} \\
&+ \frac{1}{\alpha\mu\psi^*}\left[\ddot{\omega}^* + \frac{\mathrm{dsat}(\hat{\theta})}{\mathrm{d}\hat{\theta}}\frac{\dot{\hat{\theta}}}{J} - \left(\frac{T_{Ln}}{J} + \frac{\mathrm{sat}(\hat{\theta})}{J} + \dot{\omega}^*\right)\frac{\dot{\psi}^*}{\psi^*}\right](\hat{\omega} - \omega^*) \\
&+ \frac{1}{\alpha\mu\psi^*}\left(\frac{T_{Ln}}{J} + \frac{\mathrm{sat}(\hat{\theta})}{J} + \dot{\omega}^*\right)\left[\mu\psi^* i_{sq} - \frac{T_{Ln}}{J} - \frac{\mathrm{sat}(\hat{\theta})}{J} + \mu\hat{\psi}_{rq}\tilde{i}_{sd}\right. \\
&\left. - k_\omega\left(\hat{\omega} - \omega^*\right) + \varphi_1 - \dot{\omega}^*\right] \\
g_q = &-\left(\frac{R_s}{\sigma} + \alpha M \beta\right) i_{sq} - \omega_0 i_{sd} + \alpha\beta\hat{\psi}_{rq} - \beta\hat{\omega}\hat{\psi}_{rd} + \frac{\dot{\psi}^*}{\mu\psi^{*2}}\left(-k_\omega(\hat{\omega} - \omega^*)\right. \\
&\left. + \frac{T_{Ln}}{J} + \frac{\mathrm{sat}(\hat{\theta})}{J} + \dot{\omega}^*\right) - \frac{1}{\mu\psi^*}\left(\ddot{\omega}^* + \frac{\mathrm{dsat}(\hat{\theta})}{\mathrm{d}\hat{\theta}}\frac{\dot{\hat{\theta}}}{J} + k_\omega\dot{\omega}^*\right) + \frac{k_\omega}{\mu\psi^*}\left[\mu\psi^* i_{sq}\right. \\
&\left. - \frac{T_{Ln}}{J} - \frac{\mathrm{sat}(\hat{\theta})}{J} + \mu\hat{\psi}_{rq}\tilde{i}_{sd} - k_\omega\left(\hat{\omega} - \omega^*\right) + \varphi_1\right] \\
\varphi_1 = &\frac{\mu}{\beta}\left\{\frac{k_\psi}{\alpha\mu\gamma_f}\left(M - \frac{L_r L_s}{M}\right)\beta\hat{\psi}_{rq}\tilde{i}_{sd} + \frac{\beta}{\alpha\mu\psi^*}\left(\frac{T_{Ln}}{J} + \frac{\mathrm{sat}(\hat{\theta})}{J} + \dot{\omega}^*\right)\left(\zeta\hat{\psi}_{rd}\tilde{i}_{sd}\right.\right. \\
&\left.\left. + \eta\hat{\psi}_{rq}\tilde{i}_{sd}\right) + \frac{k_\omega\beta}{\mu\psi^*}\left(\zeta\hat{\psi}_{rd}\tilde{i}_{sq} + \eta\hat{\psi}_{rq}\tilde{i}_{sq}\right)\right\} + \zeta\beta(\xi_1 - i_{sq}\zeta + i_{sd}\eta)\hat{\psi}_{rd} \\
&- \zeta\alpha\beta\hat{\psi}_{rq} + \eta\beta(\xi_1 - i_{sq}\zeta + i_{sd}\eta)\hat{\psi}_{rq} - \eta\alpha\beta\left(\psi^* - \hat{\psi}_{rd}\right) - \zeta\beta\psi^*\omega^*
\end{aligned}$$

$$\varphi_2 = \frac{\gamma_f \mu}{\beta}\left\{\frac{k_\psi}{\mu\gamma_f}\left(M - \frac{L_r L_s}{M}\right)\beta\tilde{i}_{sd} + \frac{\beta}{\alpha\mu\psi^*}\left(\frac{T_{Ln}}{J} + \frac{\mathrm{sat}(\hat{\theta})}{J} + \dot{\omega}^*\right)\left(\zeta\hat{\omega}\tilde{i}_{sd}\right.\right.$$
$$\left.\left.+\eta\alpha\tilde{i}_{sd}\right) + \frac{k_\omega\beta}{\mu\psi^*}\left(\zeta\hat{\omega}\tilde{i}_{sq} + \eta\alpha\tilde{i}_{sq}\right)\right\}$$

$$\varphi_3 = \frac{\gamma_f \mu}{\beta}\left\{\frac{k_\psi}{\alpha\mu\gamma_f}\left(M - \frac{L_r L_s}{M}\right)\beta\hat{\omega}\tilde{i}_{sd} - \frac{\beta}{\alpha\mu\psi^*}\left(\frac{T_{Ln}}{J} + \frac{\mathrm{sat}(\hat{\theta})}{J} + \dot{\omega}^*\right)\left(\zeta\alpha\tilde{i}_{sd}\right.\right.$$
$$\left.\left.-\eta\hat{\omega}\tilde{i}_{sd}\right) - \frac{k_\omega\beta}{\mu\psi^*}\left(\zeta\alpha\tilde{i}_{sq} - \eta\hat{\omega}\tilde{i}_{sq}\right)\right\}$$

$$\zeta = \frac{2L_r\left(T_{Ln} + \mathrm{sat}(\hat{\theta}) + J\dot{\omega}^*\right)\omega^* + \alpha\psi^{*2} + 2\psi^*\dot{\psi}^*}{\beta J\psi^* L_r\left(\alpha^2 + \omega^{*2}\right)}$$

$$\eta = \left[\beta\alpha M\left(\alpha^2 + \omega^{*2}\right)\psi^* J\right]^{-1}\left[-\psi^{*2}\omega^*\alpha\mu J - 2\mu J\psi^*\dot{\psi}^*\omega^*\right.$$
$$\left.+2\alpha^2 M\left(T_{Ln} + \mathrm{sat}(\hat{\theta}) + J\dot{\omega}^*\right)\right].$$

The load torque uncertainty saturated estimate appearing in (5.56) and (5.57) is defined as

$$\mathrm{sat}(\hat{\theta}) = \begin{cases} \hat{\theta} & \text{if } 0 \le \hat{\theta} \le \theta_m \\ \sum_{i=0}^{3} l_i\hat{\theta}^i & \text{if } \theta_m < \hat{\theta} < \theta_m + \varepsilon \\ \theta_m + \varepsilon & \text{if } \hat{\theta} \ge \theta_m + \varepsilon \end{cases}$$
$$l_0 = \frac{\theta_m^2(\theta_m + \varepsilon)}{\varepsilon^2}$$
$$l_1 = \frac{-2\theta_m\varepsilon - 3\theta_m^2 + \varepsilon^2}{\varepsilon^2}$$
$$l_2 = \frac{\varepsilon + 3\theta_m}{\varepsilon^2}$$
$$l_3 = -\frac{1}{\varepsilon^2}$$

in which $\mathrm{sat}(x)$ is a continuously differentiable odd function which is linear on the closed interval $[-\theta_m, \theta_m]$ and satisfies $|\mathrm{sat}(x)| \le \theta_m + \varepsilon$ for all $x \in \mathbb{R}$. The estimate $\hat{\theta}$ is provided by the sixth order closed-loop observer

$$\frac{\mathrm{d}\hat{i}_{sq}}{\mathrm{d}t} = -\left(\frac{R_s}{\sigma} + \beta\alpha M\right)i_{sq} + \frac{1}{\sigma}u_{sq} - \omega_0 i_{sd} - \beta\hat{\omega}_a\psi^* + (\lambda_1 + \lambda_2 + \lambda_3)(i_{sq} - \hat{i}_{sq})$$
$$-\frac{\dot{\psi}^*}{\psi^*}(i_{sq} - \hat{i}_{sq}) - \beta[\omega^*(\hat{\psi}_{da} - \psi^*) - \alpha\hat{\psi}_{qa}]$$

$$
\begin{aligned}
\frac{\mathrm{d}\hat{\omega}_a}{\mathrm{d}t} &= \mu\psi^* i_{sq} - \frac{(\lambda_1\lambda_3+\lambda_1\lambda_2+\lambda_2\lambda_3)}{\beta\psi^*}(i_{sq}-\hat{i}_{sq}) \\
&\quad -\frac{T_{Ln}}{J} - \frac{\hat{\theta}}{J} + \frac{1}{\psi^*}\left[\frac{T_{Ln}}{J} + \frac{\mathrm{sat}(\hat{\theta})}{J} + \dot{\omega}^*\right][\hat{\psi}_{da}-\psi^*] - \mu\left[\frac{\psi^*}{M}+\frac{\dot{\psi}^*}{\alpha M}\right]\hat{\psi}_{qa} \\
\frac{\mathrm{d}\hat{\theta}}{\mathrm{d}t} &= \frac{J\lambda_1\lambda_2\lambda_3}{\beta\psi^*}(i_{sq}-\hat{i}_{sq}) \\
\frac{\mathrm{d}\hat{i}_{sd}}{\mathrm{d}t} &= -\left(\frac{R_s}{\sigma}+\alpha M\beta\right)i_{sd} + \frac{1}{\sigma}u_{sd} + \omega_0 i_{sq} + \left(\alpha\cos\varepsilon_0 - \omega^*\sin\varepsilon_0\right)(\hat{z}_a - i_{sa}) \\
&\quad + \left(\alpha\sin\varepsilon_0 + \omega^*\cos\varepsilon_0\right)(\hat{z}_b - i_{sb}) + k_e(i_{sd}-\hat{i}_{sd}) \\
\frac{\mathrm{d}\hat{z}_a}{\mathrm{d}t} &= \begin{cases} \Phi_{za} & \text{if } -\bar{z}_M \le \hat{z}_a \le \bar{z}_M \\ \Phi_{za} & \text{if } \hat{z}_a < -\bar{z}_M \text{ and } \Phi_{za} \ge 0 \\ \Phi_{za} & \text{if } \hat{z}_a > \bar{z}_M \text{ and } \Phi_{za} \le 0 \\ \varepsilon_{am}\Phi_{za} & \text{if } \hat{z}_a < -\bar{z}_M \text{ and } \Phi_{za} < 0 \\ \varepsilon_{aM}\Phi_{za} & \text{if } \hat{z}_a > \bar{z}_M \text{ and } \Phi_{za} > 0 \end{cases} \\
\hat{z}_a(0) &\in \left[-\bar{z}_M, \bar{z}_M\right] \\
\frac{\mathrm{d}\hat{z}_b}{\mathrm{d}t} &= \begin{cases} \Phi_{zb} & \text{if } -\bar{z}_M \le \hat{z}_b \le \bar{z}_M \\ \Phi_{zb} & \text{if } \hat{z}_b < -\bar{z}_M \text{ and } \Phi_{zb} \ge 0 \\ \Phi_{zb} & \text{if } \hat{z}_b > \bar{z}_M \text{ and } \Phi_{zb} \le 0 \\ \varepsilon_{bm}\Phi_{zb} & \text{if } \hat{z}_b < -\bar{z}_M \text{ and } \Phi_{zb} < 0 \\ \varepsilon_{bM}\Phi_{zb} & \text{if } \hat{z}_b > \bar{z}_M \text{ and } \Phi_{zb} > 0 \end{cases} \\
\hat{z}_b(0) &\in \left[-\bar{z}_M, \bar{z}_M\right] \\
\begin{bmatrix} \hat{\psi}_{da} \\ \hat{\psi}_{qa} \end{bmatrix} &= \begin{bmatrix} \cos\varepsilon_0 & \sin\varepsilon_0 \\ -\sin\varepsilon_0 & \cos\varepsilon_0 \end{bmatrix} \begin{bmatrix} \frac{(\hat{z}_a - i_{sa})}{\beta} \\ \frac{(\hat{z}_b - i_{sb})}{\beta} \end{bmatrix} \\
\Phi_{za} &= -\frac{R_s}{\sigma}i_{sa} + \frac{1}{\sigma}u_{sa} + \frac{1}{\gamma_a}\left[\alpha\cos\varepsilon_0 - \omega^*\sin\varepsilon_0\right](i_{sd}-\hat{i}_{sd}) \\
\varepsilon_{am} &= 1 - \frac{\bar{z}_M^2 - \hat{z}_a^2}{\bar{z}_M^2 - (-\bar{z}_M - \varepsilon_{za})^2} \\
\varepsilon_{aM} &= 1 - \frac{\hat{z}_a^2 - \bar{z}_M^2}{(\bar{z}_M + \varepsilon_{za})^2 - \bar{z}_M^2} \\
\Phi_{zb} &= -\frac{R_s}{\sigma}i_{sb} + \frac{1}{\sigma}u_{sb} + \frac{1}{\gamma_b}\left[\alpha\sin\varepsilon_0 + \omega^*\cos\varepsilon_0\right](i_{sd}-\hat{i}_{sd}) \\
\varepsilon_{bm} &= 1 - \frac{\bar{z}_M^2 - \hat{z}_b^2}{\bar{z}_M^2 - (-\bar{z}_M - \varepsilon_{zb})^2} \\
\varepsilon_{bM} &= 1 - \frac{\hat{z}_b^2 - \bar{z}_M^2}{(\bar{z}_M + \varepsilon_{zb})^2 - \bar{z}_M^2} \\
\bar{z}_M &= z_M + \gamma_\varepsilon
\end{aligned} \tag{5.58}
$$

which is designed, by means of a suitable Lyapunov function, for the system in $(i_{sq}, \omega, i_{sd}, z_a, z_b)$ coordinates

$$\begin{aligned}
\frac{\mathrm{d}i_{sq}}{\mathrm{d}t} &= -\left(\frac{R_s}{\sigma}+\beta\alpha M\right)i_{sq}+\frac{1}{\sigma}u_{sq}-\omega_0 i_{sd} \\
&\quad +\alpha[-\sin\varepsilon_0(z_a-i_{sa})+\cos\varepsilon_0(z_b-i_{sb})] \\
&\quad -\omega[\cos\varepsilon_0(z_a-i_{sa})+\sin\varepsilon_0(z_b-i_{sb})] \\
\frac{\mathrm{d}\omega}{\mathrm{d}t} &= \mu\left[\left(\cos\varepsilon_0\frac{(z_a-i_{sa})}{\beta}+\sin\varepsilon_0\frac{(z_b-i_{sb})}{\beta}\right)i_{sq}\right. \\
&\quad \left.-\left(-\sin\varepsilon_0\frac{(z_a-i_{sa})}{\beta}+\cos\varepsilon_0\frac{(z_b-i_{sb})}{\beta}\right)i_{sd}\right]-\frac{T_{Ln}+\theta}{J} \\
\frac{\mathrm{d}i_{sd}}{\mathrm{d}t} &= -\left(\frac{R_s}{\sigma}+\beta\alpha M\right)i_{sd}+\frac{1}{\sigma}u_{sd}+\omega_0 i_{sq}+\alpha[\cos\varepsilon_0(z_a-i_{sa})+\sin\varepsilon_0(z_b-i_{sb})] \\
&\quad +\omega[-\sin\varepsilon_0(z_a-i_{sa})+\cos\varepsilon_0(z_b-i_{sb})] \\
\frac{\mathrm{d}z_a}{\mathrm{d}t} &= -\frac{R_s}{\sigma}i_{sa}+\frac{1}{\sigma}u_{sa} \\
\frac{\mathrm{d}z_b}{\mathrm{d}t} &= -\frac{R_s}{\sigma}i_{sb}+\frac{1}{\sigma}u_{sb}\,.
\end{aligned}$$

The advantage of using the variables $z_a = i_{sa}+\beta\psi_{ra}$ and $z_b = i_{sb}+\beta\psi_{rb}$ instead of the rotor fluxes ψ_{ra} and ψ_{rb} relies on the fact that their dynamics no longer depend on the rotor speed ω. The control algorithm (5.55), (5.56), (5.57), (5.58), as we shall see, guarantees bounded closed-loop signals and exponential rotor speed and flux modulus tracking under the technical assumptions stated below. The proposed dynamic control algorithm is of tenth order and depends on: the measurements of the stator currents (i_{sa}, i_{sb}); the reference signals (ω^*, ψ^*); the known parameters T_{Ln}, J, R_r, R_s, L_r, L_s, M; the known nonnegative bound θ_m on the load torque uncertainty $T_L - T_{Ln}$; the positive control parameters k_ω, k_ψ, γ_f, λ, k_i, λ_1, λ_2, λ_3, k_e, γ_a, γ_b, z_M, ε_{za}, ε_{zb}, γ_ε, ε.

Let us denote by $y(t) = [y_1(t)^{\mathrm{T}}, y_2(t)^{\mathrm{T}}, y_3(t)^{\mathrm{T}}]^{\mathrm{T}}$ the solution of the closed-loop system (1.26), (5.55), (5.56), (5.57), (5.58), with

$$\begin{aligned}
y_1 &= [\omega-\omega^*, \psi_{rd}-\psi^*, \psi_{rq}, \hat{\omega}-\omega, \psi_{rd}-\hat{\psi}_{rd}, \psi_{rq}-\hat{\psi}_{rq}, \tilde{i}_{sd}, \tilde{i}_{sq}]^{\mathrm{T}} \\
y_2 &= \left[\frac{1}{\beta^2\psi^{*2}}(i_{sq}-\hat{i}_{sq}), \frac{1}{\beta\psi^*}(\hat{\omega}_a-\omega), \frac{1}{J\beta\psi^*}(\theta-\hat{\theta})\right]^{\mathrm{T}} \\
y_3 &= [i_{sd}-\hat{i}_{sd}, z_a-\hat{z}_a, z_b-\hat{z}_b]^{\mathrm{T}} \triangleq [e_d, \tilde{z}_a, \tilde{z}_b]^{\mathrm{T}}\,.
\end{aligned}$$

Such a decomposition of the closed-loop system solution is directly related to the Lyapunov functions which will be considered in the following. Let us define

$$\begin{aligned}
r_1 &= \min\left\{\frac{\gamma_f\beta}{2}\inf_{t\geq 0}\{\zeta(t)\psi^*(t)\}, \frac{\gamma_f\mu\alpha}{2M}, k_i, \frac{k_\psi\gamma_f\mu\alpha}{2k_\psi M+\gamma_f\mu\alpha}\right\} \\
r_2(\theta_m,\varepsilon) &= \sup_{t\geq 0}\left\{\|A(t)\|+\|B(t)\|\right\} \\
A &= A^{\mathrm{T}} = \left[a_{ij}\right]_{1\leq i,j\leq 8}
\end{aligned}$$

$$B = B^{\mathrm{T}} = \left[b_{ij}\right]_{1\leq i,j\leq 8}$$

with the nonzero elements of A and B given by

$$\begin{aligned}
a_{12} &= -b_{24} = -\frac{\gamma_f k_\omega}{2\psi^*} \\
a_{13} &= -b_{34} = \frac{\gamma_f}{2\alpha\psi^*}\left(\frac{T_{Ln}}{J} + \frac{\mathrm{sat}(\hat{\theta})}{J} + \dot{\omega}^*\right) \\
a_{23} &= -a_{35} = -b_{23} = b_{35} = \frac{k_\psi}{2\alpha} \\
a_{24} &= \frac{\gamma_f \zeta \beta}{2} \\
a_{28} &= -a_{37} = b_{37} = \frac{\gamma_f \mu}{2} \\
a_{34} &= \frac{\gamma_f \eta \beta}{2} \\
b_{57} &= -\frac{\gamma_f \zeta}{2\alpha\psi^*}\left(\frac{T_{Ln}}{J} + \frac{\mathrm{sat}(\hat{\theta})}{J} + \dot{\omega}^*\right) \\
b_{58} &= -\frac{\gamma_f k_\omega \zeta}{2\psi^*} \\
b_{67} &= -b_{37} + \frac{\eta}{\zeta} b_{57} - \left(M - \frac{L_r L_s}{M}\right) b_{35} \\
b_{68} &= \frac{\eta}{\zeta} b_{58}\,.
\end{aligned}$$

Let us define

$$\begin{aligned}
c_l &= \min\{\lambda_1, \lambda_2, \lambda_3\} + \frac{\dot{\psi}^*}{\psi^*} \\
c_d &= \frac{1}{2c_l}\left[\frac{m_p m_s}{\beta}\left((2z_M + \gamma_\varepsilon + \varepsilon_{za})^2 + (2z_M + \gamma_\varepsilon + \varepsilon_{zb})^2\right)^{1/2} + \frac{m_p m_q r_1^2}{\sqrt{2} r_2^2}\right]^2 \\
m_s &= \sup_{t\geq 0}\left\{\|S(t)\|\right\} \\
S &= \left[s_{ij}\right]_{1\leq i\leq 3, 1\leq j\leq 2}
\end{aligned}$$

with the nonzero elements of S given by

$$\begin{aligned}
s_{11} &= -\frac{\omega^*}{\beta\psi^{*2}} \\
s_{12} &= \frac{\alpha}{\beta\psi^{*2}}
\end{aligned}$$

$$s_{21} = -\frac{1}{\beta\psi^{*2}}\left(\frac{T_{Ln}}{J} + \frac{\text{sat}(\hat{\theta})}{J} + \dot{\omega}^*\right)$$
$$s_{22} = \frac{\mu}{\beta\psi^*}\left(\frac{\psi^*}{M} + \frac{\dot{\psi}^*}{\alpha M}\right).$$

Let us define

$$\begin{aligned}
m_p &= \|P\| \\
n_p &= \|P^{-1}\| \\
P^{-1} &= \left[p_{ij}\right]_{1\leq i,j\leq 3} \\
p_{11} &= p_{12} = p_{13} = 1 \\
p_{2k} &= \sum_{l=1,l\neq k}^{3} \lambda_l \\
p_{3k} &= \prod_{l=1,l\neq k}^{3} \lambda_l\,, \qquad 1\leq k\leq 3 \\
m_q &= \sup_{t\geq 0}\left\{\|Q(t)\|\right\} \\
Q &= \left[q_{ij}\right]_{1\leq i\leq 3,1\leq j\leq 8}
\end{aligned}$$

with the nonzero elements of Q given by

$$\begin{aligned}
q_{11} &= -\frac{1}{\beta\psi^{*2}} \\
q_{21} &= q_{22} = -k_\omega q_{11} \\
q_{23} &= \frac{k_\psi}{\gamma_f\alpha\beta\psi^*} \\
q_{24} &= -q_{23} \\
q_{25} &= q_{26} = -\frac{1}{\alpha\beta\psi^{*2}}\left(\frac{T_{Ln}}{J} + \frac{\text{sat}(\hat{\theta})}{J} + \dot{\omega}^*\right) \\
q_{27} &= -q_{28} = -\frac{\mu}{\beta\psi^*}\,.
\end{aligned}$$

Finally, let us define

$$l_{\gamma 1} = \frac{\beta\psi^*}{\mu}|s_{22}|$$
$$l_{\gamma 2} = \frac{\beta\psi^*}{\mu}|s_{21}|\,.$$

We now show that the trajectories of the y_1-subsystem remain bounded, provided that the perturbation $\theta_e = \theta - \text{sat}(\hat{\theta})$ as well as the initial state $y_1(0)$ are sufficiently small: the boundedness of the vector y_1, along with the use of projection algorithms in the design of $(\hat{z}_a, \hat{z}_b)$ dynamics, guarantee the boundedness of the vectors y_2 and y_3 and, therefore, of the control signals (u_{sa}, u_{sb}). Let us first rewrite the rotor flux dynamic equations in (1.31) so that the rotor speed ω does not explicitly appear. To this end, recalling that the rotor fluxes (ψ_{rd}, ψ_{rq}) are dynamically related to the stator fluxes (ψ_{sd}, ψ_{sq}) in the (d,q) reference frame by the following relations

$$\begin{aligned}
\psi_{rd} &= Mi_{sd} - \frac{L_r L_s}{M} i_{sd} + \frac{L_r}{M}\psi_{sd} \\
\psi_{rq} &= Mi_{sq} - \frac{L_r L_s}{M} i_{sq} + \frac{L_r}{M}\psi_{sq} \\
\dot{\psi}_{sd} &= u_{sd} - R_s i_{sd} + \omega_0 \psi_{sq} \\
\dot{\psi}_{sq} &= u_{sq} - R_s i_{sq} - \omega_0 \psi_{sd} ,
\end{aligned}$$

by direct computation we obtain

$$\begin{aligned}
\frac{\mathrm{d}\psi_{rd}}{\mathrm{d}t} &= \left(M - \frac{L_r L_s}{M}\right)\frac{\mathrm{d}i_{sd}}{\mathrm{d}t} + \frac{L_r}{M}u_{sd} - \frac{L_r R_s}{M} i_{sd} + \omega_0 \psi_{rq} \\
&\quad - \left(M - \frac{L_r L_s}{M}\right)\omega_0 i_{sq} \\
\frac{\mathrm{d}\psi_{rq}}{\mathrm{d}t} &= \left(M - \frac{L_r L_s}{M}\right)\frac{\mathrm{d}i_{sq}}{\mathrm{d}t} + \frac{L_r}{M}u_{sq} - \frac{L_r R_s}{M} i_{sq} - \omega_0 \psi_{rd} \\
&\quad + \left(M - \frac{L_r L_s}{M}\right)\omega_0 i_{sd} .
\end{aligned} \tag{5.59}$$

Denoting by $\tilde{\omega} = \omega - \omega^*$, $\tilde{\psi}_{rd} = \psi_{rd} - \psi^*$, $\tilde{\psi}_{rq} = \psi_{rq}$, $e_\omega = \hat{\omega} - \omega^*$, $e_{\psi d} = \hat{\psi}_{rd} - \psi^*$, $e_{\psi q} = \hat{\psi}_{rq}$ (so that $y_1 = [\tilde{\omega}, \tilde{\psi}_{rd}, \tilde{\psi}_{rq}, e_\omega - \tilde{\omega}, \tilde{\psi}_{rd} - e_{\psi d}, \tilde{\psi}_{rq} - e_{\psi q}, \tilde{i}_{sd}, \tilde{i}_{sq}]^{\mathrm{T}}$) and by $e_q = \frac{1}{\beta^2 \psi^{*2}}(i_{sq} - \hat{i}_{sq})$, $\tilde{e}_\omega = \frac{1}{\beta\psi^*}(\hat{\omega}_a - \omega)$, $\tilde{\theta} = \frac{1}{J\beta\psi^*}(\theta - \hat{\theta})$ (so that $y_2 = [e_q, \tilde{e}_\omega, \tilde{\theta}]^{\mathrm{T}}$), from (1.26), (5.55)–(5.59) we obtain the closed-loop error dynamics. Restrict the analysis to the time-varying smooth bounded reference signals $\omega^*(t)$ and $\psi^*(t) \geq c_\psi > 0$ with bounded first and second order time derivatives satisfying the condition (for all $t \geq 0$)

$$\left[2JL_r(\alpha^2 + \omega^*(t)^2)\right]^{-1} h(t) \geq 2c_p > 0 \tag{5.60}$$

with

$$h(t) = 2L_r\left(T_{Ln} + J\dot{\omega}^*(t)\right)\omega^*(t) + \alpha\psi^*(t)^2 + 2\psi^*(t)\dot{\psi}^*(t)$$

and c_p any positive real. Let k_ψ, γ_f, λ, k_i, k_e, γ_a, γ_b, ε_{za}, ε_{zb}, γ_ε be any positive control parameters and choose the positive control parameter k_ω such that

$$k_\omega > \sup_{t\geq 0}\left\{\left[2JL_r(\alpha^2 + \omega^*(t)^2)\right]^{-1}\left[h(t) + 2L_r(\theta_m + \varepsilon)|\omega^*(t)|\right]\right\} .$$

Choose λ_1, λ_2, λ_3 such that $\lambda_1 \neq \lambda_2 \neq \lambda_3$ and

$$\min\{\lambda_1,\lambda_2,\lambda_3\}+\inf_{t\geq 0}\left\{\frac{\dot{\psi}^*(t)}{\psi^*(t)}\right\}\geq c_{\psi d}>0$$

with $c_{\psi d}$ any positive real, and choose the positive control parameter z_M such that

$$z_M \geq \sup_{t\geq 0}\left\{1+\beta+\frac{k_\psi}{\alpha\mu\gamma_f}+\frac{l_{\gamma 2}(t)}{\alpha}+\frac{k_\omega}{\mu\psi^*(t)}\right\}\frac{r_1}{r_2}$$
$$+\sup_{t\geq 0}\left\{l_{\gamma 1}(t)+l_{\gamma 2}(t)+\beta|\psi^*(t)|\right\}.$$

Let θ_m,ε be sufficiently small and restrict the initial conditions $y_1(0)$ such that

$$(\theta_m+\varepsilon)|\omega^*(t)|\leq J[\alpha^2+\omega^{*2}(t)]c_p$$
$$\|y_1(0)\|\leq\left[\frac{\min\{\gamma_f,\frac{\gamma_f\mu}{M},\lambda,\frac{\gamma_f\mu}{\beta}\}r_1^2}{4\max\{\gamma_f,\frac{\gamma_f\mu}{M},\lambda,\frac{\gamma_f\mu}{\beta}\}r_2^2}-\frac{32\gamma_f^2}{r_1^2J^2}(2\theta_m+\varepsilon)^2\right]^{1/2}$$

where c_p has been previously defined. Consider the quadratic positive definite function

$$V_1(y_1)=\frac{1}{2}y_1^{\mathrm{T}}\mathrm{diag}\left[\gamma_f,\frac{\gamma_f\mu}{M},\frac{\gamma_f\mu}{M},\gamma_f,\lambda,\lambda,\frac{\gamma_f\mu}{\beta},\frac{\gamma_f\mu}{\beta}\right]y_1\,. \tag{5.61}$$

Compute the time derivative of the function V_1 along the trajectories of the closed-loop y_1-subsystem and complete the squares so that

$$\dot{V}_1\leq -r_1\|y_1\|^2+r_2\|y_1\|^3+\frac{2\gamma_f}{J}(2\theta_m+\varepsilon)\|y_1\|\,. \tag{5.62}$$

Equation (5.61) and inequality (5.62) imply the inequality

$$\|y_1(t)\|\leq\frac{r_1}{2r_2}\,.$$

By observing that $|z_s|\leq z_M$ and $|\hat{z}_s|\leq(\bar{z}_M+\varepsilon_{zs})$, $s=a,b$, we have

$$|\tilde{z}_a(t)|\leq 2z_M+\gamma_\varepsilon+\varepsilon_{za}$$
$$|\tilde{z}_b(t)|\leq 2z_M+\gamma_\varepsilon+\varepsilon_{zb}\,.$$

Since

$$\dot{e}_d=-k_e e_d+(\alpha\cos\varepsilon_0-\omega^*\sin\varepsilon_0)\tilde{z}_a+(\alpha\sin\varepsilon_0+\omega^*\cos\varepsilon_0)\tilde{z}_b+\beta\tilde{\omega}\tilde{\psi}_{rq}$$

the inequality

$$|e_d(t)|\leq|e_d(0)|+\frac{1}{k_e}\left[(4z_M+2\gamma_\varepsilon+\varepsilon_{za}+\varepsilon_{zb})\left(\alpha+\sup_{t\geq 0}\left\{|\omega^*(t)|\right\}\right)+\frac{\beta r_1^2}{4r_2^2}\right]$$

is guaranteed. Define $\tilde{y}_2 = Py_2$ and consider the quadratic positive definite function

$$V_2(\tilde{y}_2) = \frac{1}{2}\|\tilde{y}_2\|^2 . \tag{5.63}$$

By computing the time derivative of the function V_2 along the trajectories of the closed-loop $\tilde{y}_2$-subsystem and by completing the squares we obtain

$$\dot{V}_2 \leq -\frac{c_l}{2}\|\tilde{y}_2\|^2 + c_d . \tag{5.64}$$

Equation (5.63) and inequality (5.64) imply the inequality

$$\|y_2(t)\| \leq n_p \left[m_p^2 \|y_2(0)\|^2 + \frac{2c_d}{c_l} \right]^{1/2} .$$

We now determine sufficient conditions for the exponential stability of the equilibrium point $y = 0$ of the closed-loop system (1.26), (5.55)–(5.58). Define

$$\Gamma^{\mathrm{T}}(t) = \begin{bmatrix} \alpha \cos \varepsilon_0(t) - \omega^*(t) \sin \varepsilon_0(t) \\ \alpha \sin \varepsilon_0(t) + \omega^*(t) \cos \varepsilon_0(t) \end{bmatrix} .$$

Assume that there exist two positive reals T_p and k_p such that the persistency of excitation condition (I is the identity matrix)

$$\int_t^{t+T_p} \Gamma^{\mathrm{T}}(\tau)\Gamma(\tau)d\tau \geq k_p I, \qquad \forall t \geq 0 \tag{5.65}$$

is satisfied. Consider the candidate Lyapunov function

$$\begin{aligned} W(t,y) = V_1(y_1) + \varepsilon_{s0} V_2(Py_2) + \varepsilon_{s1} \Bigg[& \frac{1}{2}\left(e_d^2 + \gamma_a \tilde{z}_a^2 + \gamma_b \tilde{z}_b^2 \right) \\ & + p \left\| Q_p(t)[\tilde{z}_a, \tilde{z}_b]^{\mathrm{T}} - \Gamma^{\mathrm{T}}(t) e_d \right\|^2 \Bigg] \end{aligned}$$

in which $V_1(\cdot)$ and $V_2(\cdot)$ have been previously defined, while the matrix $Q_p(t)$ is the solution of the linear matrix differential equation

$$\dot{Q}_p(t) = -Q_p(t) + \Gamma^{\mathrm{T}}(t)\Gamma(t) , \quad Q_p(0) = \mathrm{e}^{-T_p} k_p I .$$

Arguments similar to those adopted in the proof of the Persistency of Excitation Lemma A.3 in Appendix A guarantee the existence of suitable positive reals ε_{s0}, ε_{s1}, p (depending on T_p and k_p) such that the function W has a locally negative definite time derivative along the trajectories of the closed-loop system (1.26), (5.55)–(5.58) so that the origin of the closed-loop system is locally exponentially stable.

In conclusion, the *adaptive control from current measurements* (5.55)–(5.58) guarantees that the following inequalities hold:

$$\|y_1(t)\| \leq \frac{r_1}{2r_2}$$

$$\|y_2(t)\| \leq n_p \left[m_p^2 \|y_2(0)\|^2 + \frac{2c_d}{c_l} \right]^{1/2}$$

and $y_3(t) = [e_d(t), \tilde{z}_a(t), \tilde{z}_b(t)]^{\mathrm{T}}$ is such that

$$|\tilde{z}_a(t)| \leq 2z_M + \gamma_\varepsilon + \varepsilon_{za}$$
$$|\tilde{z}_b(t)| \leq 2z_M + \gamma_\varepsilon + \varepsilon_{zb}$$
$$|e_d(t)| \leq |e_d(0)| + \frac{1}{k_e} \left[(4z_M + 2\gamma_\varepsilon + \varepsilon_{za} + \varepsilon_{zb}) \left(\alpha + \sup_{t \geq 0} \left\{ |\omega^*(t)| \right\} \right) + \frac{\beta r_1^2}{4r_2^2} \right]$$

- for any time-varying smooth bounded reference signals $\omega^*(t)$ and $\psi^*(t) \geq c_\psi > 0$ with bounded first and second order time derivatives satisfying the condition (for all $t \geq 0$)

$$\left[2JL_r(\alpha^2 + \omega^*(t)^2) \right]^{-1} h(t) \geq 2c_p > 0$$

  with

$$h(t) = 2L_r \left(T_{Ln} + J\dot{\omega}^*(t) \right) \omega^*(t) + \alpha \psi^*(t)^2 + 2\psi^*(t)\dot{\psi}^*(t)$$

  and c_p any positive real;
- for any positive control parameters k_ψ, γ_f, λ, k_i, k_e, γ_a, γ_b, ε_{za}, ε_{zb}, γ_ε and for any positive control parameter k_ω such that

$$k_\omega > \sup_{t \geq 0} \left\{ \left[2JL_r(\alpha^2 + \omega^*(t)^2) \right]^{-1} \left[h(t) + 2L_r(\theta_m + \varepsilon)|\omega^*(t)| \right] \right\};$$

- for any positive control parameters λ_1, λ_2, λ_3 such that $\lambda_1 \neq \lambda_2 \neq \lambda_3$ and

$$\min\{\lambda_1, \lambda_2, \lambda_3\} + \inf_{t \geq 0} \left\{ \frac{\dot{\psi}^*(t)}{\psi^*(t)} \right\} \geq c_{\psi d} > 0$$

  with $c_{\psi d}$ any positive real;
- for any positive control parameter z_M such that

$$z_M \geq \sup_{t \geq 0} \left\{ 1 + \beta + \frac{k_\psi}{\alpha \mu \gamma_f} + \frac{l_{\gamma 2}(t)}{\alpha} + \frac{k_\omega}{\mu \psi^*(t)} \right\} \frac{r_1}{r_2}$$

$$+\sup_{t\geq 0}\left\{l_{\gamma 1}(t)+l_{\gamma 2}(t)+\beta|\psi^*(t)|\right\};$$

- for sufficiently small θ_m, ε and restricted initial conditions $y_1(0)$ such that

$$(\theta_m+\varepsilon)|\omega^*(t)| \leq J\,[\alpha^2+\omega^{*2}(t)]c_p$$

$$\|y_1(0)\| \leq \left[\frac{\min\{\gamma_f, \frac{\gamma_f\mu}{M}, \lambda, \frac{\gamma_f\mu}{\beta}\}r_1^2}{4\max\{\gamma_f, \frac{\gamma_f\mu}{M}, \lambda, \frac{\gamma_f\mu}{\beta}\}r_2^2} - \frac{32\gamma_f^2}{r_1^2 J^2}(2\theta_m+\varepsilon)^2\right]^{1/2}.$$

If two positive reals T_p and k_p exist such that the persistency of excitation condition (I is the identity matrix)

$$\int_t^{t+T_p} \Gamma^{\mathrm{T}}(\tau)\Gamma(\tau)d\tau \geq k_p I, \qquad \forall t \geq 0 \tag{5.66}$$

is satisfied with

$$\Gamma^{\mathrm{T}}(t) = \begin{bmatrix} \alpha\cos\varepsilon_0(t) - \omega^*(t)\sin\varepsilon_0(t) \\ \alpha\sin\varepsilon_0(t) + \omega^*(t)\cos\varepsilon_0(t) \end{bmatrix},$$

then the equilibrium point $y = [y_1^{\mathrm{T}}, y_2^{\mathrm{T}}, y_3^{\mathrm{T}}]^{\mathrm{T}} = [\omega-\omega^*, \psi_{rd}-\psi^*, \psi_{rq}, \hat{\omega}-\omega, \psi_{rd}-\hat{\psi}_d, \psi_{rq}-\hat{\psi}_{rq}, \tilde{i}_{sd}, \tilde{i}_{sq}, \frac{1}{\beta^2\psi^{*2}}(i_{sq}-\hat{i}_{sq}), \frac{1}{\beta\psi^*}(\hat{\omega}_a-\omega), \frac{1}{J\beta\psi^*}(\theta-\hat{\theta}), i_{sd}-\hat{i}_{sd}, z_a-\hat{z}_a, z_b-\hat{z}_b]^{\mathrm{T}} = 0$ of the closed-loop system (1.26), (5.55)–(5.58) is locally exponentially stable.

Remarks

1. Condition (5.60) represents a constraint on the time-varying reference signals $\omega^*(t)$ and $\psi^*(t)$, which may limit the applicability of the proposed solution: however, it is only a sufficient condition. In any case, if braking actions are not considered, every pair of constant reference signals ω^* and $\psi^* \geq c_\psi > 0$ satisfies condition (5.60) so that closed-loop boundedness may be guaranteed for any motor parameters and any constant operating condition. Note that if $\hat{\theta} = \theta$, the closed-loop observer (5.58) and the persistency of excitation condition (5.66) are no longer needed so that the controller is no longer observer-based and the internal variables $\hat{\omega}$, $\hat{\psi}_{rd}$ and $\hat{\psi}_{rq}$ converge to the references ω^*, ψ^*, and zero, respectively.
2. Under persistency of excitation, exponential rotor speed and flux modulus tracking may be guaranteed despite constant load torque uncertainty even in the presence of uncertain flux initial values: this improves the adaptive control algorithms in Sections 5.2 and 5.3 which make use of flux measurements.

3. A more useful persistency of excitation condition, which is related to motor observability (see Section 1.5), can be derived. In fact, inequality (5.65) is implied, for sufficiently small initial conditions $\|y(0)\|$ and for sufficiently small θ_m, ε by

$$\int_t^{t+T_p} \Gamma^{*\mathrm{T}}(\tau)\Gamma^*(\tau)d\tau \geq 4k_p I, \qquad \forall t \geq 0 \tag{5.67}$$

where

$$\Gamma^{*\mathrm{T}}(t) = \begin{bmatrix} \alpha \cos \varepsilon_0^*(t) - \omega^*(t) \sin \varepsilon_0^*(t) \\ \alpha \sin \varepsilon_0^*(t) + \omega^*(t) \cos \varepsilon_0^*(t) \end{bmatrix}$$

$$\dot{\varepsilon}_0^*(t) = \omega^*(t) + \frac{\alpha M}{\mu \psi^*(t)^2} \left[\frac{T_L}{J} + \dot{\omega}^*(t) \right]$$

$$\varepsilon_0^*(0) = \varepsilon_0(0)\,.$$

According to Schwarz's inequality, a necessary conditions for (5.67) to be satisfied is ($\tau \in [t, t+T_p]$, $t \geq 0$): $\dot{\varepsilon}_0^*(\tau) \neq 0$ in the case of constant rotor speed reference signal $\omega^*(\tau)$, which is also sufficient when the rotor flux modulus reference signal ψ^* is constant.

Illustrative Simulations

We tested the adaptive control from current measurements (5.55)–(5.58) by simulations for the three-phase single pole pair 0.6-kW induction motor whose parameters have been reported in Chapter 1. The control parameters are (the values are in SI units): $k_\omega = 60$, $k_i = 300$, $k_\psi = 12\,\beta/M$, $\gamma_f = \beta/\mu$, $\lambda = \beta/M$, $\lambda_1 = 3$, $\lambda_2 = 4$, $\lambda_3 = 5$, $k_e = 24$, $\gamma_a = \gamma_b = 0.3^{-1}$, $z_M = 42$, $\varepsilon_{za} = \varepsilon_{zb} = 0.9$, $\gamma_\varepsilon = 3$, $\varepsilon = 0.1$ ($\theta_m = 3.6$). All initial conditions for the motor and for the controller are set to zero except $\psi_{ra}(0) = 0.1$ Wb. The references for rotor speed and flux modulus along with the applied torque are reported in Figure 5.25. The rotor flux reference starts from 0.001 Wb at $t = 0$ s and grows up to the rated constant value 1.16 Wb; field weakening starts at $t = 4.32$ s. The rotor speed reference is zero until $t = 0.32$ s and grows up to the constant value 100 rad/s; at $t = 4.32$ s the rotor speed is required to go up to the value 200 rad/s, while the reference for the rotor flux is reduced to 0.72 Wb. The load torque T_L (5.104 Nm) is applied at $t = 0.9$ s and is reduced to 0.44 Nm at $t = 4.2$ s. The uncertainty θ is -12% of the load torque nominal value T_{Ln}, *i.e.*, $T_{Ln} = T_L/0.88$. Note that the inequality (5.60) is satisfied. Figures 5.26–5.28 show the rotor speed, the rotor flux modulus, and the load torque uncertainty saturated estimate along with the corresponding tracking and estimation errors, while Figures 5.29 and 5.30 show the stator current and the stator voltage vectors (a, b) components. Under persistency of excitation, fast estimation and good tracking performance are obtained. The initial error (about 10%) between the rotor flux vector d-component ψ_{rd} and its reference ψ^* is compensated by the controller.

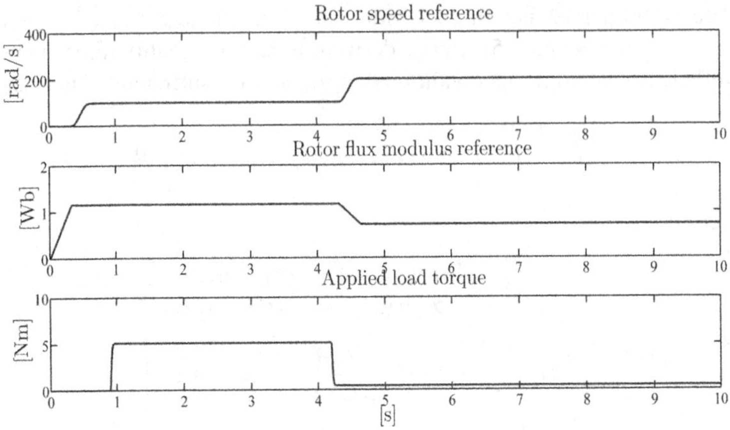

Fig. 5.25 Adaptive control from current measurements (5.55)–(5.58): rotor speed and flux modulus reference signals and applied load torque

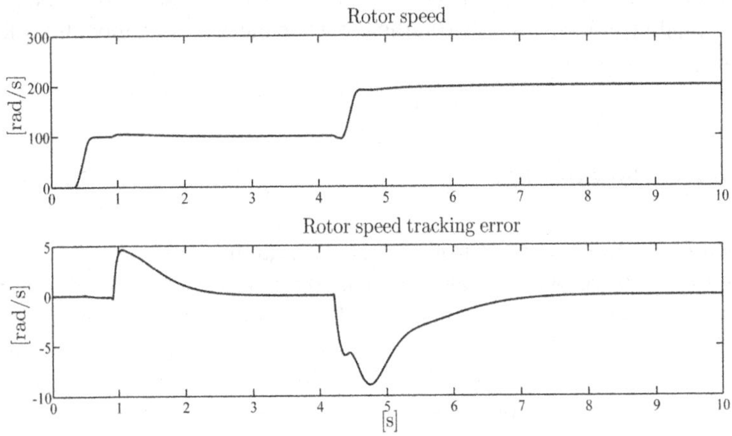

Fig. 5.26 Adaptive control from current measurements (5.55)–(5.58): rotor speed and rotor speed tracking error

5.5 Adaptive Control with Uncertain Load Torque and Rotor Resistance

The control algorithm presented in the previous section makes use of rotor resistance in a critical way: boundedness of closed-loop signals and convergence of tracking errors are guaranteed if the design parameters lie in a range depending on α, while the allowed reference signals for rotor speed and flux modulus are constrained by an inequality which also depends on α. The controller proposed in this section, which is adaptive with respect to α as well, is substantially different from the previous one: no constraints on the reference signals are to be imposed for tracking and estimation

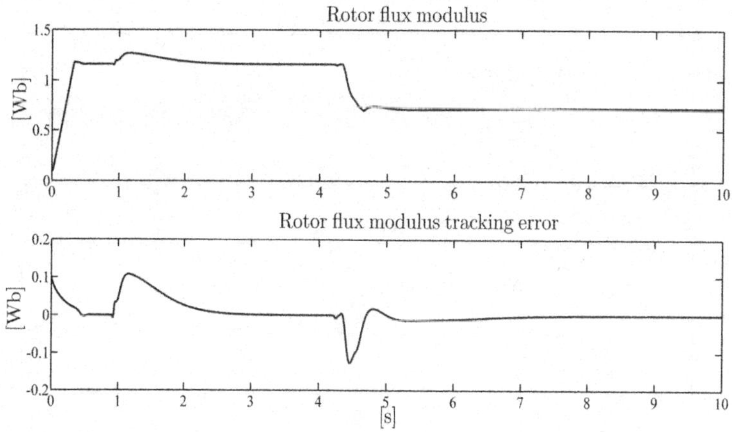

Fig. 5.27 Adaptive control from current measurements (5.55)–(5.58): rotor flux modulus and rotor flux modulus tracking error

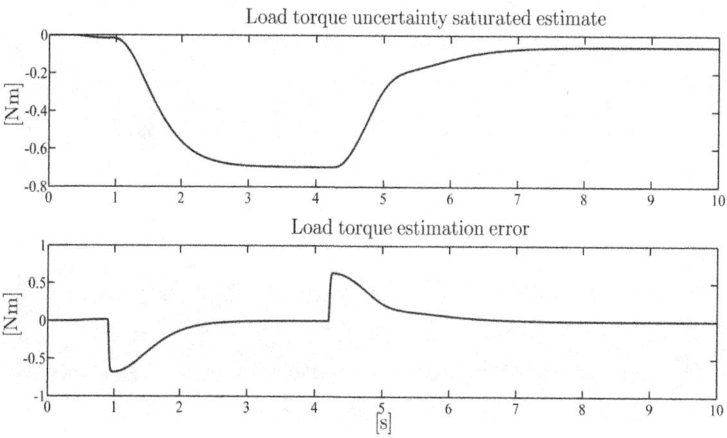

Fig. 5.28 Adaptive control from current measurements (5.55)–(5.58): load torque uncertainty saturated estimate and load torque estimation error

errors to converge to zero as long as they are persistently exciting. We propose the following control algorithm consisting of a stator current control loop (containing feedforward actions and suitable stabilizing and robust feedback terms):

$$\begin{bmatrix} u_{sa} \\ u_{sb} \end{bmatrix} = \begin{bmatrix} \cos\varepsilon_0 & -\sin\varepsilon_0 \\ \sin\varepsilon_0 & \cos\varepsilon_0 \end{bmatrix} \begin{bmatrix} u_{sd} \\ u_{sq} \end{bmatrix}$$

$$u_{sd} = \sigma\Bigg[\left(\frac{R_s}{\sigma} + \hat{\alpha}\beta M\right) i_{sd} - \omega_0 i_{sq} - \beta\hat{\alpha}\hat{\psi}_{rd} - \beta\omega^*\hat{\psi}_{rq} - k_i(i_{sd} - i_{sd}^*) + \frac{\mathrm{d}}{\mathrm{d}t} i_{sd}^* \\ -\frac{k}{4}(i_{sd} - i_{sd}^*)\beta^2\left(3 + \alpha_M^2 + \omega^{*2} + \frac{\dot{\psi}^{*2}}{\hat{\alpha}^2} + M^2(i_{sd} - i_{sd}^*)^2\right)\Bigg]$$

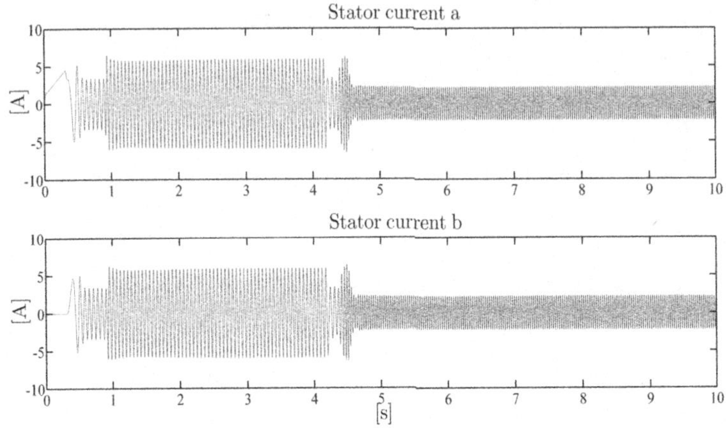

Fig. 5.29 Adaptive control from current measurements (5.55)–(5.58): stator current vector (a,b)-components

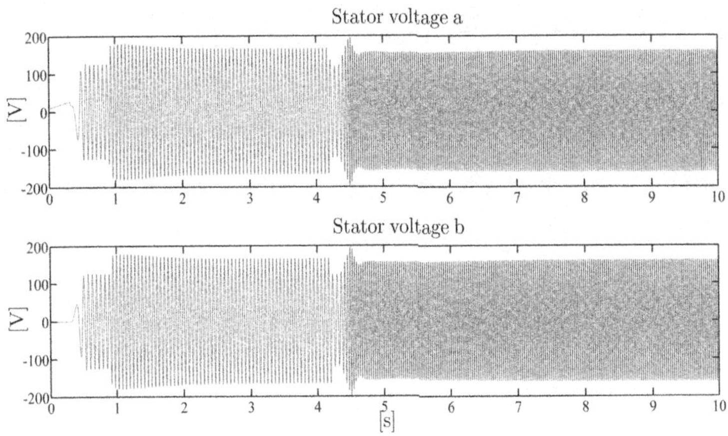

Fig. 5.30 Adaptive control from current measurements (5.55)–(5.58): stator voltage vector (a,b)-components

$$u_{sq} = \sigma\Big[\Big(\frac{R_s}{\sigma} + \hat{\alpha}\beta M\Big)i_{sq} + \omega_0 i_{sd} - \beta\hat{\alpha}\hat{\psi}_{rq} + \beta\hat{\omega}\hat{\psi}_{rd} - k_i(i_{sq} - i^*_{sq}) + \frac{\mathrm{d}}{\mathrm{d}t}i^*_{sq}$$

$$-\frac{k}{4}(i_{sq} - i^*_{sq})\beta^2\Big(M^2[i^{*2}_{sq} + (i_{sq} - i^*_{sq})^2] + \omega^{*2} + 5 + \alpha_M^2 + \psi^{*2}\Big)\Big]$$

$$\begin{bmatrix} i_{sd} \\ i_{sq} \end{bmatrix} = \begin{bmatrix} \cos\varepsilon_0 & \sin\varepsilon_0 \\ -\sin\varepsilon_0 & \cos\varepsilon_0 \end{bmatrix} \begin{bmatrix} i_{sa} \\ i_{sb} \end{bmatrix} \tag{5.68}$$

in which the reference signals (i^*_{sd}, i^*_{sq}) for (i_{sd}, i_{sq}) and the speed ω_0 of the (d,q) rotating frame (which, as in field-oriented control, are responsible for rotor speed and flux modulus tracking) are chosen as (compare with the corresponding signals in the control algorithm presented in Section 5.3)

$$\begin{aligned}
i_{sd}^* &= \frac{\psi^*}{M} + \frac{\dot{\psi}^*}{\hat{\alpha}M} \\
i_{sq}^* &= \frac{1}{\mu\psi^*}\left[-k_\omega(\hat{\omega}-\omega^*) + \frac{T_{Ln}}{J} + \frac{\text{sat}(\hat{\theta})}{J} + \dot{\omega}^*\right] \\
\dot{\varepsilon}_0 &= \omega_0 = \hat{\omega} + \frac{\hat{\alpha}Mi_{sq}^*}{\psi^*};
\end{aligned} \tag{5.69}$$

the estimates $(\hat{\omega}, \hat{\psi}_{rd}, \hat{\psi}_{rq}, \hat{\theta}, \hat{\alpha})$ for the unmeasured states $(\omega, \psi_{rd}, \psi_{rq})$ and for the uncertain constant parameters (θ, α) appearing in (5.68) and (5.69) are provided by the seventh order closed-loop adaptive observer which includes the auxiliary variables $(\hat{i}_{sd}, \hat{i}_{sq}, \hat{z}_d, \hat{z}_q)$

$$\begin{aligned}
\hat{\psi}_{rd} &= -\frac{1}{\beta}(i_{sd}-\hat{z}_d) \\
\hat{\psi}_{rq} &= -\frac{1}{\beta}(i_{sq}-\hat{z}_q) \\
\frac{d\hat{i}_{sq}}{dt} &= -\left(\frac{R_s}{\sigma}+\beta\hat{\alpha}M\right)i_{sq} + \frac{1}{\sigma}u_{sq} - \omega_0 i_{sd} \\
&\quad -\beta\hat{\omega}\psi^* + (\lambda_1+\lambda_2+\lambda_3)(i_{sq}-\hat{i}_{sq}) \\
&\quad -\frac{\dot{\psi}^*}{\psi^*}(i_{sq}-\hat{i}_{sq}) - \beta[\omega^*(\hat{\psi}_{rd}-\psi^*) - \hat{\alpha}\hat{\psi}_{rq}] \\
\frac{d\hat{\omega}}{dt} &= \mu\psi^* i_{sq} - \frac{(\lambda_1\lambda_3+\lambda_1\lambda_2+\lambda_2\lambda_3)}{\beta\psi^*}(i_{sq}-\hat{i}_{sq}) - \frac{T_{Ln}}{J} - \frac{\hat{\theta}}{J} \\
&\quad +\frac{1}{\psi^*}\left[\frac{T_{Ln}}{J} + \frac{\text{sat}(\hat{\theta})}{J} + \dot{\omega}^*\right][\hat{\psi}_{rd}-\psi^*] - \mu\left[\frac{\psi^*}{M}+\frac{\dot{\psi}^*}{\hat{\alpha}M}\right]\hat{\psi}_{rq} \\
\frac{d\hat{\theta}}{dt} &= \frac{J\lambda_1\lambda_2\lambda_3}{\beta\psi^*}(i_{sq}-\hat{i}_{sq}) \\
\frac{d\hat{i}_{sd}}{dt} &= -\left(\frac{R_s}{\sigma}+\beta\hat{\alpha}M\right)i_{sd} + \frac{1}{\sigma}u_{sd} + \omega_0 i_{sq} \\
&\quad -\omega^*(i_{sq}-\hat{z}_q) - \hat{\alpha}(i_{sd}-\hat{z}_d) + k_e(i_{sd}-\hat{i}_{sd}) \\
\frac{d\hat{z}_d}{dt} &= -\frac{R_s}{\sigma}i_{sd} + \frac{1}{\sigma}u_{sd} + \omega_0\hat{z}_q + \frac{\hat{\alpha}}{\gamma_1}(i_{sd}-\hat{i}_{sd}) \\
\frac{d\hat{z}_q}{dt} &= -\frac{R_s}{\sigma}i_{sq} + \frac{1}{\sigma}u_{sq} - \omega_0\hat{z}_d + \frac{\omega^*}{\gamma_1}(i_{sd}-\hat{i}_{sd}) \\
\frac{d\hat{\alpha}}{dt} &= \text{Proj}\left(-\frac{\beta\dot{\psi}^*}{\gamma_2\hat{\alpha}}(i_{sd}-\hat{i}_{sd}), \hat{\alpha}\right), \quad \hat{\alpha}(0) \in [\alpha_m, \alpha_M], \; 0 < \alpha_m - \varepsilon_\alpha
\end{aligned} \tag{5.70}$$

in which $\text{Proj}(\zeta, \hat{\alpha})$ is the smooth projection algorithm given by (A.25) in Appendix A and defined in our case by

$$\mathrm{Proj}(\zeta,\hat{\alpha}) = \begin{cases} \zeta & \text{if } \alpha_m \leq \hat{\alpha} \leq \alpha_M \\ \zeta & \text{if } \hat{\alpha} < \alpha_m \text{ and } \zeta \geq 0 \\ \zeta & \text{if } \hat{\alpha} > \alpha_M \text{ and } \zeta \leq 0 \\ \xi_{\zeta 1}\zeta & \text{if } \hat{\alpha} < \alpha_m \text{ and } \zeta < 0 \\ \xi_{\zeta 2}\zeta & \text{if } \hat{\alpha} > \alpha_M \text{ and } \zeta > 0 \end{cases}$$

$$\xi_{\zeta 1} = 1 - \frac{{\alpha_m}^2 - \hat{\alpha}^2}{{\alpha_m}^2 - (\alpha_m - \varepsilon_\alpha)^2}$$

$$\xi_{\zeta 2} = 1 - \frac{\hat{\alpha}^2 - {\alpha_M}^2}{(\alpha_M + \varepsilon_\alpha)^2 - {\alpha_M}^2}.$$

The load torque uncertainty saturated estimate appearing in (5.69) and (5.70) is defined as

$$\mathrm{sat}(\hat{\theta}) = \begin{cases} \hat{\theta} & \text{if } 0 \leq \hat{\theta} \leq \theta_m \\ \sum_{i=0}^{3} l_i \hat{\theta}^i & \text{if } \theta_m < \hat{\theta} < \theta_m + \varepsilon \\ \theta_m + \varepsilon & \text{if } \hat{\theta} \geq \theta_m + \varepsilon \end{cases}$$

$$l_0 = \frac{\theta_m^2(\theta_m + \varepsilon)}{\varepsilon^2}$$

$$l_1 = \frac{-2\theta_m\varepsilon - 3\theta_m^2 + \varepsilon^2}{\varepsilon^2}$$

$$l_2 = \frac{\varepsilon + 3\theta_m}{\varepsilon^2}$$

$$l_3 = -\frac{1}{\varepsilon^2}$$

in which $\mathrm{sat}(q)$ is a continuously differentiable odd function which is linear in the closed set $[-\theta_m, \theta_m]$ and satisfies $|\mathrm{sat}(q)| \leq \theta_m + \varepsilon$ for all $q \in \mathbb{R}$. The overall control algorithm (5.68), (5.69), (5.70) is of eighth order and depends on: the measurements of the stator currents (i_{sa}, i_{sb}); the reference signals (ω^*, ψ^*); the known parameters $T_{Ln}, J, R_s, L_r, L_s, M$; the known nonnegative bound θ_m on the load torque uncertainty $T_L - T_{Ln}$, the known bounds α_m, α_M such that $\alpha_m \leq \alpha \leq \alpha_M$; the positive control parameters k_ω, k_i, k, k_e, λ_1, λ_2, λ_3, γ_1, γ_2, ε_α, ε. Let us introduce the angle ε_0^* satisfying

$$\dot{\varepsilon}_0^*(t) = \omega^*(t) + \frac{\alpha M}{\mu \psi^{*2}(t)}\left[\frac{T_L}{J} + \dot{\omega}^*(t)\right]$$

$$\varepsilon_0^*(0) = \varepsilon_0(0)$$

depending on the uncertain parameters (α, T_L). Let us define the tracking and estimation errors: $\tilde{\omega} = \omega - \omega^*$, $\tilde{\psi}_{rd} = \psi_{rd} - \psi^*$, $\tilde{\psi}_{rq} = \psi_{rq}$, $\tilde{i}_{sd} = i_{sd} - i_{sd}^*$, $\tilde{i}_{sq} = i_{sq} - i_{sq}^*$, $e_d = i_{sd} - \hat{i}_{sd}$, $e_q = i_{sq} - \hat{i}_{sq}$, $e_\omega = \hat{\omega} - \omega$, $e_{\psi d} = \psi_{rd} - \hat{\psi}_{rd}$, $e_{\psi q} = \psi_{rq} - \hat{\psi}_{rq}$, $\tilde{\theta} = \theta - \hat{\theta}$, $\tilde{\alpha} = \alpha - \hat{\alpha}$. Let us introduce the changes of variables $[\lambda_1 \neq \lambda_2 \neq \lambda_3]$:

$$\begin{bmatrix} \tilde{z}_a \\ \tilde{z}_b \end{bmatrix} = \begin{bmatrix} \cos\varepsilon_0^* & -\sin\varepsilon_0^* \\ \sin\varepsilon_0^* & \cos\varepsilon_0^* \end{bmatrix} \begin{bmatrix} \beta e_{\psi d} \\ \beta e_{\psi q} \end{bmatrix}$$

$$\tilde{x}_1 = \frac{e_q}{\beta^2 \psi^{*2}}, \quad \tilde{x}_2 = \frac{e_\omega}{\beta \psi^*}, \quad \tilde{x}_3 = \frac{\tilde{\theta}}{J\beta\psi^*}$$

$$e = \begin{bmatrix} 1 & 1 & 1 \\ \lambda_2+\lambda_3 & \lambda_1+\lambda_3 & \lambda_1+\lambda_2 \\ \lambda_2\lambda_3 & \lambda_1\lambda_3 & \lambda_1\lambda_2 \end{bmatrix}^{-1} \begin{bmatrix} \tilde{x}_1 \\ \tilde{x}_2 \\ \tilde{x}_3 \end{bmatrix}.$$

Let us define the matrices

$$A_x = \begin{bmatrix} -(\lambda_1+\lambda_2+\lambda_3) & 1 & 0 \\ -(\lambda_2\lambda_3+\lambda_1\lambda_3+\lambda_1\lambda_2) & 0 & 1 \\ -\lambda_1\lambda_2\lambda_3 & 0 & 0 \end{bmatrix}$$

$$S_x = \begin{bmatrix} -\frac{\omega^*}{\beta\psi^{*2}} & \frac{\alpha}{\beta\psi^{*2}} & -\frac{M}{\mu\beta\psi^{*3}}\eta \\ -\frac{1}{\beta\psi^{*2}}\eta & \frac{\mu}{\beta\psi^*}i^*_{sd} & 0 \\ 0 & 0 & 0 \end{bmatrix}$$

$$R_x = \begin{bmatrix} \frac{1}{\beta\psi^{*2}}\tilde{\alpha}(\tilde{\psi}_{rq} - e_{\psi q} - M\tilde{i}_{sq}) - \frac{1}{\beta\psi^{*2}}\tilde{\omega}\tilde{\psi}_{rd} + \frac{k_\omega M}{\mu\beta\psi^{*3}}\tilde{\alpha}(\tilde{\omega}+e_\omega) \\ \frac{k_\omega}{\beta\psi^{*2}}\tilde{\psi}_{rd}(\tilde{\omega}+e_\omega) + \frac{\mu}{\beta\psi^*}(\tilde{\psi}_{rq}\tilde{i}_{sd} - \tilde{\psi}_{rd}\tilde{i}_{sq}) \\ 0 \end{bmatrix}$$

in which $\eta = \left[\frac{T_{Ln}}{J} + \frac{\text{sat}(\hat{\theta})}{J} + \dot{\omega}^*\right]$. Let $y = [\tilde{\omega}, \tilde{\psi}_{rd}, \tilde{\psi}_{rq}, \tilde{i}_{sd}, \tilde{i}_{sq}, e^{\mathrm{T}}, e_d, \tilde{z}_a, \tilde{z}_b, \tilde{\alpha}]^{\mathrm{T}} \in \mathbb{R}^{12}$. Choose the positive control parameters k_i, λ_1, λ_2, λ_3 in the control algorithm (5.68), (5.69), (5.70) so that $k_i > \alpha_M M^2$, $\lambda_1 \neq \lambda_2 \neq \lambda_3$, and

$$\min\{\lambda_1, \lambda_2, \lambda_3\} + \inf_{t\geq 0}\left\{\frac{\dot{\psi}^*(t)}{\psi^*(t)}\right\} \geq \tilde{c}_\psi > 0\,.$$

Assume that there exist two positive reals t_p and k_p such that the persistency of excitation condition (I is the identity matrix)

$$\int_t^{t+t_p} \Gamma^{\mathrm{T}}(\tau)\Gamma(\tau)d\tau \geq k_p I, \quad \forall t \geq 0 \tag{5.71}$$

holds with

$$\Gamma^{\mathrm{T}}(t) = \begin{bmatrix} \alpha\cos\varepsilon_0^*(t) - \omega^*(t)\sin\varepsilon_0^*(t) \\ \alpha\sin\varepsilon_0^*(t) + \omega^*(t)\cos\varepsilon_0^*(t) \\ -\frac{\beta\dot{\psi}^*(t)}{\alpha} \end{bmatrix}.$$

Introduce the auxiliary variables $z_d = i_{sd} + \beta\psi_{rd}$, $z_q = i_{sq} + \beta\psi_{rq}$, which are linear combinations of the motor state variables and whose dynamics

$$\begin{aligned} \dot{z}_d &= -\frac{R_s}{\sigma}i_{sd} + \frac{1}{\sigma}u_{sd} + \omega_0 z_q \\ \dot{z}_q &= -\frac{R_s}{\sigma}i_{sq} + \frac{1}{\sigma}u_{sq} - \omega_0 z_d \end{aligned}$$

do not depend on the unmeasured rotor speed ω and the uncertain parameter α. Since

$$\begin{aligned}
\dot{e}_{\psi d} &= \omega_0 e_{\psi q} - \frac{\hat{\alpha}}{\gamma_1 \beta} e_d \\
\dot{e}_{\psi q} &= -\omega_0 e_{\psi d} - \frac{\omega^*}{\gamma_1 \beta} e_d
\end{aligned}$$

the closed-loop system can be written as

$$\begin{aligned}
\frac{\mathrm{d}\tilde{\omega}}{\mathrm{d}t} &= -k_\omega(\tilde{\omega} + e_\omega) - \frac{1}{J}\left[\theta - \mathrm{sat}(\hat{\theta})\right] + \mu \tilde{\psi}_{rd} i_{sq}^* \\
&\quad - \mu \tilde{\psi}_{rq} i_{sd}^* + \mu \psi^* \tilde{i}_{sq} + \mu \tilde{\psi}_{rd} \tilde{i}_{sq} - \mu \tilde{\psi}_{rq} \tilde{i}_{sd} \\
\frac{\mathrm{d}\tilde{\psi}_{rd}}{\mathrm{d}t} &= -\alpha \tilde{\psi}_{rd} + (\omega_0 - \omega)\tilde{\psi}_{rq} + \alpha M \tilde{i}_{sd} + \frac{\dot{\psi}^*}{\hat{\alpha}} \tilde{\alpha} \\
\frac{\mathrm{d}\tilde{\psi}_{rq}}{\mathrm{d}t} &= -\alpha \tilde{\psi}_{rq} - (\omega_0 - \omega)\tilde{\psi}_{rd} - e_\omega \psi^* + \alpha M \tilde{i}_{sq} + \tilde{\alpha} M i_{sq}^* \\
\frac{\mathrm{d}\tilde{i}_{sd}}{\mathrm{d}t} &= -k_i \tilde{i}_{sd} + \beta \alpha e_{\psi d} + \beta \omega^* e_{\psi q} - \tilde{\alpha}\beta \frac{\dot{\psi}^*}{\hat{\alpha}} + \tilde{\alpha}\beta(\tilde{\psi}_{rd} - e_{\psi d}) - \tilde{\alpha}\beta M \tilde{i}_{sd} \\
&\quad + \beta \tilde{\omega} \tilde{\psi}_{rq} - \frac{k}{4}\beta^2\left(3 + \alpha_M^2 + \omega^{*2} + \frac{\dot{\psi}^{*2}}{\hat{\alpha}^2} + M^2 \tilde{i}_{sd}^2\right)\tilde{i}_{sd} \\
\frac{\mathrm{d}\tilde{i}_{sq}}{\mathrm{d}t} &= -k_i \tilde{i}_{sq} + \beta \alpha e_{\psi q} - \beta \omega^* e_{\psi d} - \tilde{\alpha}\beta M i_{sq}^* + \tilde{\alpha}\beta(\tilde{\psi}_{rq} - e_{\psi q}) \\
&\quad - \tilde{\alpha}\beta M \tilde{i}_{sq} - \beta \tilde{\omega} e_{\psi d} + \beta e_\omega \psi^* + \beta e_\omega(\tilde{\psi}_{rd} - e_{\psi d}) \\
&\quad - \frac{k}{4}\beta^2\left(M^2[i_{sq}^{*2} + \tilde{i}_{sq}^2] + \omega^{*2} + 5 + \alpha_M^2 + \psi^{*2}\right)\tilde{i}_{sq} \\
\begin{bmatrix} \dot{\tilde{x}}_1 \\ \dot{\tilde{x}}_2 \\ \dot{\tilde{x}}_3 \end{bmatrix} &= \left(A_x - \frac{\dot{\psi}^*}{\psi^*} I\right)\begin{bmatrix} \tilde{x}_1 \\ \tilde{x}_2 \\ \tilde{x}_3 \end{bmatrix} + S_x \begin{bmatrix} e_{\psi d} \\ e_{\psi q} \\ \tilde{\alpha} \end{bmatrix} + R_x \\
\frac{\mathrm{d}e_d}{\mathrm{d}t} &= -k_e e_d - \tilde{\alpha}\frac{\beta \dot{\psi}^*}{\alpha} + [\alpha \cos \varepsilon_0^* - \omega^* \sin \varepsilon_0^*]\tilde{z}_a \\
&\quad + [\alpha \sin \varepsilon_0^* + \omega^* \cos \varepsilon_0^*]\tilde{z}_b - \tilde{\alpha}\beta M \tilde{i}_{sd} \\
&\quad + \tilde{\alpha}\beta \tilde{\psi}_{rd} + \beta \tilde{\omega} \tilde{\psi}_{rq} - \cos \varepsilon_0^* \tilde{z}_a \tilde{\alpha} - \sin \varepsilon_0^* \tilde{z}_b \tilde{\alpha} - \frac{\beta \dot{\psi}^*}{\alpha \hat{\alpha}} \tilde{\alpha}^2 \\
\frac{\mathrm{d}\tilde{z}_a}{\mathrm{d}t} &= \tilde{\omega}_0 \tilde{z}_b - \frac{1}{\gamma_1}[\alpha \cos \varepsilon_0^* - \omega^* \sin \varepsilon_0^*]e_d + \frac{\cos \varepsilon_0^*}{\gamma_1}\tilde{\alpha} e_d \\
\frac{\mathrm{d}\tilde{z}_b}{\mathrm{d}t} &= -\tilde{\omega}_0 \tilde{z}_a - \frac{1}{\gamma_1}[\alpha \sin \varepsilon_0^* + \omega^* \cos \varepsilon_0^*]e_d + \frac{\sin \varepsilon_0^*}{\gamma_1}\tilde{\alpha} e_d \\
\frac{\mathrm{d}\tilde{\alpha}}{\mathrm{d}t} &= -\mathrm{Proj}\left(-\frac{\beta \dot{\psi}^*}{\gamma_2 \hat{\alpha}} e_d, \hat{\alpha}\right)
\end{aligned}$$

in which

$$\begin{aligned}\tilde{\omega}_0 = \tilde{\omega} + e_\omega - \frac{M}{\mu\psi^{*2}}\left[\frac{T_L}{J} + \dot{\omega}^*\right]\tilde{\alpha} + \frac{M}{\mu\psi^{*2}}(\alpha \\ -\tilde{\alpha})\left[-k_\omega(\tilde{\omega} + e_\omega) - \frac{(\theta - \text{sat}(\hat{\theta}))}{J}\right].\end{aligned}$$

Recalling that the projection algorithm $\text{Proj}(\zeta, \hat{\alpha})$ guarantees the properties in Appendix A, consider the positive definite function

$$\begin{aligned}W = \frac{1}{2}\Big[\tilde{\omega}^2 + s_\alpha(\tilde{\psi}_{rd}^2 + \tilde{\psi}_{rq}^2 + \tilde{i}_{sd}^2 + \tilde{i}_{sq}^2) \\ + s_\beta\|e\|^2 + s_\gamma\Big(e_d^2 + \gamma_1\tilde{z}_a^2 + \gamma_1\tilde{z}_b^2 + \gamma_2\tilde{\alpha}^2 \\ + 2p\left\|Q_p(t)[\tilde{z}_a, \tilde{z}_b, \tilde{\alpha}]^{\mathrm{T}} - \Gamma^{\mathrm{T}}(t)e_d\right\|^2\Big)\Big]\end{aligned}$$

in which $s_\alpha, s_\beta, s_\gamma, p \in \mathbb{R}^+$, and the matrix $Q_p(t)$ is the solution of the linear matrix differential equation

$$\begin{aligned}\dot{Q}_p(t) &= -Q_p(t) + \Gamma^{\mathrm{T}}(t)\Gamma(t) \\ Q_p(0) &= \mathrm{e}^{-t_p}k_p I\,.\end{aligned}$$

According to the Persistency of Excitation Lemma A.3 in Appendix A, there exist suitable positive reals s_α, s_β, s_γ, p (depending on t_p and k_p) for which the function W has, along the trajectories of the closed-loop system, a locally negative definite time derivative

$$\begin{aligned}\dot{W} &\le -a_1\|x\|^2 + a_2\|x\|^4 + a_3\|x\|^6 \\ &\le -\frac{a_1}{2}\|x\|^2 + \left(\frac{a_2^2}{2a_1} + a_3\right)\|x\|^6\end{aligned}$$

where $x = \left[y^{\mathrm{T}}, (Q_p(t)[\tilde{z}_a, \tilde{z}_b, \tilde{\alpha}]^{\mathrm{T}} - \Gamma^{\mathrm{T}}(t)e_d)^{\mathrm{T}}\right]^{\mathrm{T}}$ and a_i, $1 \le i \le 3$, are positive reals depending on t_p and k_p. Since $\nu_1\|x\|^2 \le W \le \nu_2\|x\|^2$ with $\nu_1 = \frac{1}{2}\min\{1, s_\alpha, s_\beta, s_\gamma, s_\gamma\gamma_1, s_\gamma\gamma_2, 2s_\gamma p\}$ and $\nu_2 = \frac{1}{2}\max\{1, s_\alpha, s_\beta, s_\gamma, s_\gamma\gamma_1, s_\gamma\gamma_2, 2s_\gamma p\}$, and since $\|x(t)\|^2 \le b\|y(t)\|^2$ with $b = 1 + 2\Gamma_M^2(\Gamma_M^2 + 1)$, and $\Gamma_M \ge \|\Gamma(t)\|$ for all $t \ge 0$, we can establish that there exists a positive real s such that

$$\|y(t)\|^2 \le \frac{\nu_2 b}{\nu_1}\|y(0)\|^2\mathrm{e}^{-st}, \quad \forall t \ge 0$$

for any initial condition $y(0)$ satisfying

$$\|y(0)\|^2 < \frac{\nu_1}{\nu_2 b}\sqrt{\frac{a_1}{2\left(\frac{a_2^2}{2a_1} + a_3\right)}}\,.$$

Therefore, the origin $y = 0$ of the closed-loop error variables system is locally exponentially stable so that exponential rotor speed and flux modulus tracking are achieved.

In conclusion, the *adaptive control from current measurements* (5.68)–(5.70) guarantees that the origin $y = 0$ of the closed-loop system is locally exponentially stable, so that exponential rotor speed and flux modulus tracking are achieved, under the following conditions:

1. the positive control parameters k_i, λ_1, λ_2, λ_3 are chosen so that $k_i > \alpha_M M^2$, $\lambda_1 \neq \lambda_2 \neq \lambda_3$, and

$$\min\{\lambda_1, \lambda_2, \lambda_3\} + \inf_{t \geq 0} \left\{ \frac{\dot{\psi}^*(t)}{\psi^*(t)} \right\} \geq \tilde{c}_\psi > 0\,;$$

2. there exist two positive reals t_p and k_p such that the persistency of excitation condition (I is the identity matrix)

$$\int_t^{t+t_p} \Gamma^{\mathrm{T}}(\tau)\Gamma(\tau)d\tau \geq k_p I, \quad \forall t \geq 0$$

holds with

$$\Gamma^{\mathrm{T}}(t) = \begin{bmatrix} \alpha \cos \varepsilon_0^*(t) - \omega^*(t) \sin \varepsilon_0^*(t) \\ \alpha \sin \varepsilon_0^*(t) + \omega^*(t) \cos \varepsilon_0^*(t) \\ -\frac{\beta \dot{\psi}^*(t)}{\alpha} \end{bmatrix}.$$

Remarks

1. According to the previous analysis, under persistency of excitation, exponential rotor speed and flux modulus tracking may be guaranteed despite constant load torque and rotor resistance uncertainties even in the presence of uncertain flux initial values: this improves the adaptive speed-sensorless control designed in Section 5.3 which makes use of measured rotor fluxes.
2. The dynamic compensator (5.68)–(5.70) contains 11 control parameters k_ω, k_i, k, k_e, λ_1, λ_2, λ_3, γ_1, γ_2, ε_α, ε whose role may be evaluated by examining both the closed-loop error equations and the corresponding stability analysis. The parameters k_ω, k_i, k_e, $(\lambda_1, \lambda_2, \lambda_3)$ directly affect the dynamics of the tracking and estimation errors $\tilde{\omega}$, $(\tilde{i}_{sd}, \tilde{i}_{sq})$, e_d, $(\tilde{x}_1, \tilde{x}_2, \tilde{x}_3)$, respectively, while the parameter γ_1 determines the influence of the estimation error e_d on the dynamics of the error variables $(\tilde{z}_a, \tilde{z}_b)$; the parameter γ_2^{-1} is the adaptation gain for $\hat{\alpha}$, while the pa-

rameters k, ε_α, ε characterize the robustifying terms in (u_{sd}, u_{sq}), the projection algorithm $\mathrm{Proj}(\cdot,\cdot)$, and the saturation function $\mathrm{sat}(\cdot)$, respectively.

3. The key inequality (5.71) may be physically interpreted in terms of motor observability and rotor resistance identifiability (see Sections 1.5 and 1.6). In fact, necessary conditions for (5.71) to be satisfied are ($\tau \in [t, t+T_p], t \geq 0$): (1) $\dot{\varepsilon}_0^*(\tau) \neq 0$, in the case of constant rotor speed reference signal $\omega^*(\tau)$; (2) $\dot{\psi}^*(\tau) \neq 0$. A sufficient condition for inequality (5.71) to be satisfied is to choose a time-varying rotor flux modulus reference signal $\psi^*(t)$ and a constant rotor speed reference value ω^* such that

$$\frac{1}{\psi^{*2}(t)} = \psi_c^{*2} + \varepsilon_* \cos\left(\frac{t}{\varepsilon_*}\right) > 0$$

$$0 \neq \omega^* + \frac{\alpha M T_L}{\mu J}\psi_c^{*2}$$

in which ψ_c^* is a positive constant and ε_* is a sufficiently small positive real.

4. As shown in Section 5.3, the rotor flux modulus reference signal is required to be time-varying: if $\|[\psi^*(t), \dot{\omega}^*(t)]\| = 0$ for all $t \geq 0$, then all the points ($\tilde{\omega}$, $\tilde{\psi}_{rd}$, $\tilde{\psi}_{rq}$, $\tilde{i}_{sd}$, $\tilde{i}_{sq}$, $\tilde{x}_1$, $\tilde{x}_2$, $\tilde{x}_3$, e_d, $\tilde{z}_a$, $\tilde{z}_b$, $\tilde{\alpha}$) = ($-g\tilde{\alpha}$, 0, 0, 0, 0, 0, $\frac{g}{\beta\psi^*}\tilde{\alpha}$, 0, 0, 0, 0, $\tilde{\alpha}$), with $g = \frac{L_r}{\psi^{*2}}(T_L + J\dot{\omega}^*)$, are equilibrium points for the closed-loop system so that, when both rotor speed and flux modulus reference signals are constant, local exponential rotor speed tracking may not be guaranteed by the control algorithm (5.68)–(5.70).

Illustrative Simulations

We tested the adaptive control from current measurements (5.68)–(5.70) by simulations for the three-phase single pole pair 0.6-kW induction motor whose parameters have been reported in Chapter 1. The control parameters are (the values are in SI units): $k_\omega = 120$, $k_i = 900$, $k = 0.01$, $k_e = 900$, $\lambda_1 = 120$, $\lambda_2 = 150$, $\lambda_3 = 180$, $\gamma_1 = 120^{-1}$, $\gamma_2 = 900^{-1}$, $\alpha_m = 4.5$, $\alpha_M = 13.5$, $\theta_m = 3.6$, $\varepsilon = 0.1$, $\varepsilon_\alpha = 0.9$. All initial conditions of the motor and of the controller are set to zero except for $\psi_{ra}(0) = 0.1\,\mathrm{Wb}$ and $\hat{\alpha}(0) = 11.1\,\mathrm{s}^{-1}$. The references for rotor speed and flux modulus along with the applied torque are reported in Figure 5.31. The load torque T_L (5.104 Nm) is applied at $t = 0.9\,\mathrm{s}$; the uncertainty θ is -12% of the load torque nominal value T_{Ln}. Figures 5.32–5.35 show the rotor speed, the rotor flux modulus, the load torque uncertainty saturated estimate, and the α estimate, along with the corresponding tracking and estimation errors, while Figures 5.36 and 5.37 show the stator current and the stator voltage vectors (a, b) components. Under persistency of excitation, fast estimation and good tracking performance are obtained. The initial error (about 10%) between the rotor flux vector d-component ψ_{rd} and its reference ψ^* is compensated by the controller.

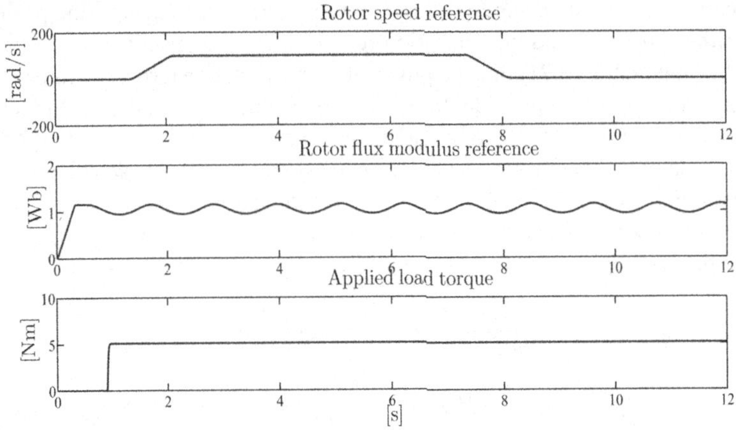

Fig. 5.31 Adaptive control from current measurements (5.68)–(5.70): rotor speed and flux modulus reference signals and applied load torque

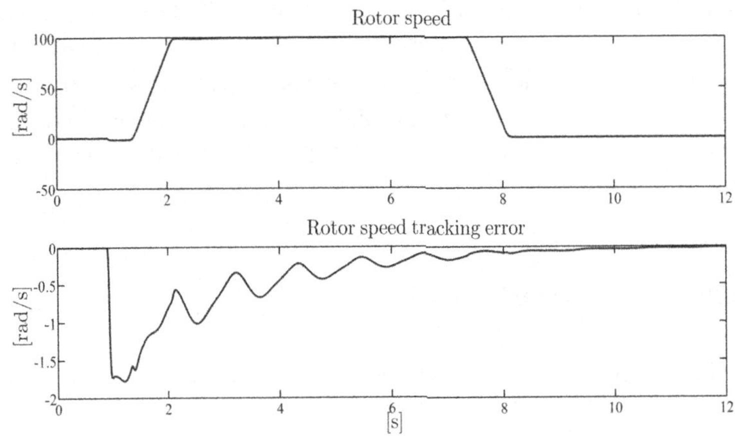

Fig. 5.32 Adaptive control from current measurements (5.68)–(5.70): rotor speed and rotor speed tracking error

5.6 Conclusions

In this chapter we have addressed the design of control algorithms for induction motors in which the rotor speed sensor is not available. The starting point is the feedforward control developed in Chapter 1 and analyzed in Chapter 2. In Section 5.1 a PI feedback term is added to the feedforward control which depends on the error between the measured stator currents and their steady-state profiles corresponding to given references for rotor speed and flux modulus. The desired steady-state operating condition may be unstable depending on the load torque value and the desired reference flux modulus. Hence, additional feedback terms are needed to remove this

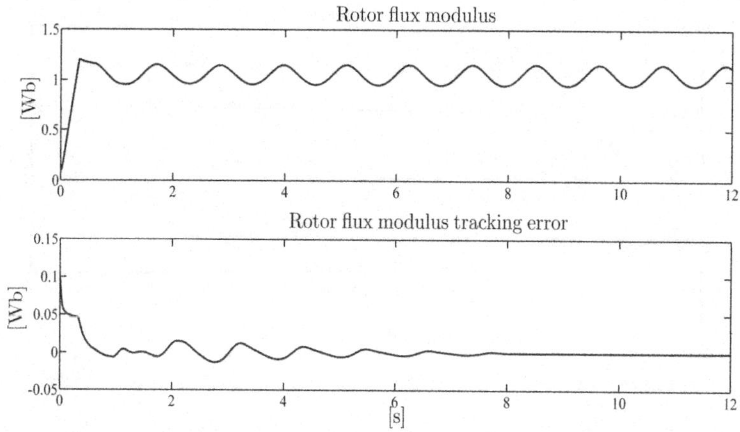

Fig. 5.33 Adaptive control from current measurements (5.68)–(5.70): rotor flux modulus and rotor flux modulus tracking error

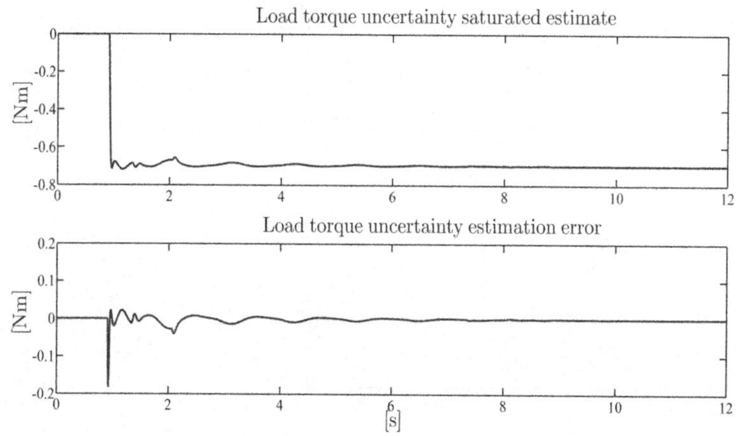

Fig. 5.34 Adaptive control from current measurements (5.68)–(5.70): load torque uncertainty saturated estimate and load torque estimation error

drawback. Since in Chapter 2 several state feedback controls have been obtained which have then been successfully integrated in Chapter 4 by rotor flux observers, in this chapter we have introduced observers both for rotor speed and rotor fluxes in the feedback control algorithm. As intermediate steps, in Sections 5.2 and 5.3 both nonadaptive and adaptive controls have been designed under the assumption that all state variables are available for feedback with the exception of rotor speed. This unrealistic assumption allows us to concentrate on the design of closed-loop rotor speed observers and to clarify the identifiability conditions which are required to estimate the rotor resistance in closed-loop: it turns out in fact that the rotor flux modulus reference has to be time-varying. The goal of designing observer-based controls

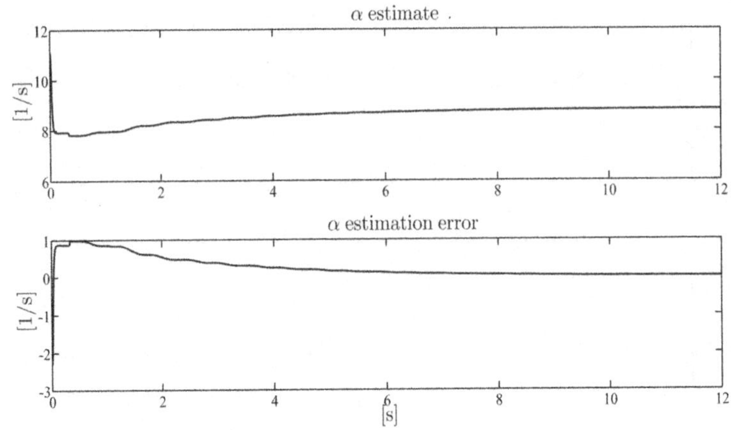

Fig. 5.35 Adaptive control from current measurements (5.68)–(5.70): α estimate and α estimation error

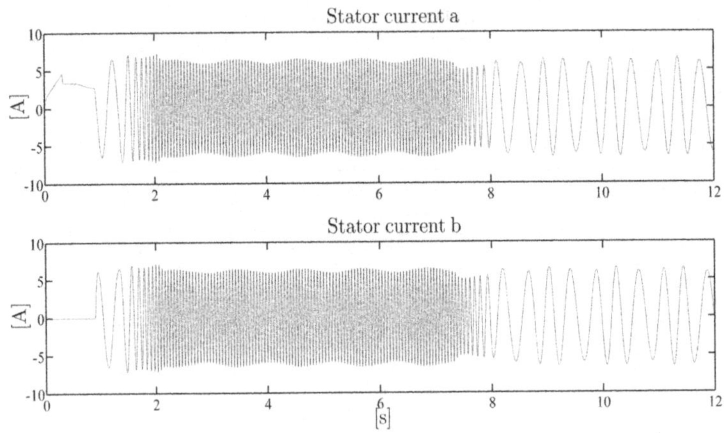

Fig. 5.36 Adaptive control from current measurements (5.68)–(5.70): stator current vector (a,b)-components

from stator current measurements has only been accomplished in Sections 5.4 and 5.5, in which adaptation with respect to load torque and rotor resistance is also obtained in closed-loop. The rotor speed tracking in the presence of critical parameters has been obtained only when time-varying references for the rotor flux modulus are imposed, while the rotor speed tracking with speed measurements has also been obtained for any initial condition in Chapter 4 when rotor flux modulus is constant. In other words, the rotor speed tracking has been achieved provided that the uncertain parameters are correctly estimated, while when rotor speed is measured the convergence to zero of the uncertain parameter estimation errors is not required to achieve rotor speed tracking from any initial condition.

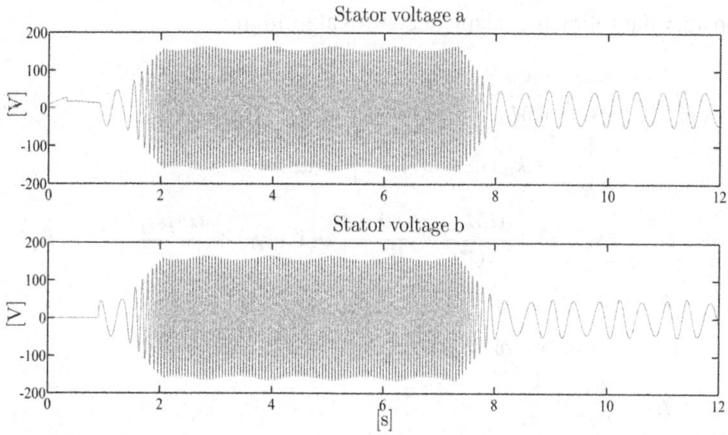

Fig. 5.37 Adaptive control from current measurements (5.68)–(5.70): stator voltage vector (a,b)-components

Problems

5.1. Consider the following control algorithm:

$$
\begin{aligned}
i_{sd}^* &= \frac{\psi^*}{M} + \frac{\dot{\psi}^*}{\alpha M} \\
i_{sq}^* &= \frac{1}{\mu \psi^*}\left[-k_\omega(\hat{\omega} - \omega^*) + \frac{T_L}{J} + \dot{\omega}^*\right] \\
\dot{\varepsilon}_0 = \omega_0 &= \hat{\omega} + \frac{\alpha M i_{sq}}{\psi^*} + \frac{1}{\beta \psi^*}\left[\hat{\omega}(1+\gamma_1) + \frac{\alpha M i_{sq}}{\psi^*}\right](i_{sd} - i_{sd}^*) \\
\dot{\hat{\omega}} &= -k_\omega(\hat{\omega} - \omega^*) + \dot{\omega}^* - k_{io}(i_{sq} - i_{sq}^*) \\
\gamma_1 &= \frac{R_s}{\alpha \sigma} + \frac{k_{id}}{\alpha} \\
\begin{bmatrix} i_{sd} \\ i_{sq} \end{bmatrix} &= \begin{bmatrix} \cos\varepsilon_0 & \sin\varepsilon_0 \\ -\sin\varepsilon_0 & \cos\varepsilon_0 \end{bmatrix}\begin{bmatrix} i_{sa} \\ i_{sb} \end{bmatrix} \\
u_{sd} &= \sigma\left[\gamma i_{sd}^* - \omega_0 i_{sq} - \alpha\beta\psi^* + \frac{\mathrm{d}i_{sd}^*}{\mathrm{d}t} - k_{id}(i_{sd} - i_{sd}^*)\right] \\
u_{sq} &= \sigma\left[\gamma i_{sq}^* + \omega_0 i_{sd} + \beta\hat{\omega}\psi^* + \frac{\mathrm{d}i_{sq}^*}{\mathrm{d}t} - k_{iq}(i_{sd} - i_{sd}^*)\right] \\
\begin{bmatrix} u_{sa} \\ u_{sb} \end{bmatrix} &= \begin{bmatrix} \cos\varepsilon_0 & -\sin\varepsilon_0 \\ \sin\varepsilon_0 & \cos\varepsilon_0 \end{bmatrix}\begin{bmatrix} u_{sd} \\ u_{sq} \end{bmatrix}.
\end{aligned}
$$

Choose the control parameters k_ω, k_{io}, k_{id}, k_{iq} and simulate the closed-loop performance. Compare with the controller given in Section 5.1 and with the controller given in Section 5.4 with $\hat{T}_L$ replaced by T_L.

5.2. Consider the following adaptive control algorithm:

$$
\begin{aligned}
i_{sd}^* &= \frac{\psi^*}{M} + \frac{\dot{\psi}^*}{\alpha M} \\
i_{sq}^* &= \frac{1}{\mu \psi^*}\left[-k_\omega(\hat{\omega} - \omega^*) + \frac{\hat{T}_L}{J} + \dot{\omega}^*\right] \\
\dot{\varepsilon}_0 = \omega_0 &= \hat{\omega} + \frac{\alpha M i_{sq}}{\psi^*} + \frac{1}{\beta \psi^*}\left[\hat{\omega}(1+\gamma_1) + \frac{\alpha M i_{sq}}{\psi^*}\right](i_{sd} - i_{sd}^*) \\
\dot{\hat{\omega}} &= \dot{\omega}^* - k_{io}(i_{sq} - i_{sq}^*) \\
\dot{\hat{T}}_L &= -k_T(\hat{\omega} - \omega^*) \\
\gamma_1 &= \frac{R_s}{\alpha\sigma} + \frac{k_{id}}{\alpha} \\
\begin{bmatrix} i_{sd} \\ i_{sq} \end{bmatrix} &= \begin{bmatrix} \cos\varepsilon_0 & \sin\varepsilon_0 \\ -\sin\varepsilon_0 & \cos\varepsilon_0 \end{bmatrix} \begin{bmatrix} i_{sa} \\ i_{sb} \end{bmatrix} \\
u_{sd} &= \sigma\left[\gamma i_{sd}^* - \omega_0 i_{sq} - \alpha\beta\psi^* + \frac{di_{sd}^*}{dt} - k_{id}(i_{sd} - i_{sd}^*)\right] \\
u_{sq} &= \sigma\left[\gamma i_{sq}^* + \omega_0 i_{sd} + \beta\hat{\omega}\psi^* + \frac{di_{sq}^*}{dt} - k_{iq}(i_{sd} - i_{sd}^*)\right] \\
\begin{bmatrix} u_{sa} \\ u_{sb} \end{bmatrix} &= \begin{bmatrix} \cos\varepsilon_0 & -\sin\varepsilon_0 \\ \sin\varepsilon_0 & \cos\varepsilon_0 \end{bmatrix} \begin{bmatrix} u_{sd} \\ u_{sq} \end{bmatrix} .
\end{aligned}
$$

Choose the control parameters k_ω, k_{io}, k_{id}, k_{iq}, k_T and simulate the closed-loop performance. Compare with the controller given in Section 5.4.

5.3. Consider the global speed-sensorless control (5.20): assuming that the rotor speed ω is measured, replace $\hat{\omega}$ by ω and compare the resulting control algorithm with the global control algorithm with arbitrary rate of convergence (2.114).

5.4. Consider the adaptive speed-sensorless control (5.50): assuming that the rotor speed ω is measured, modify the control algorithm so that the uncertain parameters are estimated even in the cases in which condition (5.47) fails and compare it with the adaptive input–output feedback linearizing control (2.85).

5.5. Consider the adaptive control from current measurements (5.68)–(5.70): modify the control algorithm assuming that the rotor speed ω is measured so that the uncertain parameters are estimated even in the case in which the reference for the rotor flux modulus is constant ($\dot{\psi}^* = 0$) and compare it with the control algorithms given in Section 4.4.

5.6. Redesign the adaptive speed-sensorless control (5.50) assuming that T_L is known while R_s is uncertain.

5.7. Design an adaptive version of the feedforward control (1.74) replacing T_L by $\hat{T}_L$ and assuming that only stator current measurements are available. *Suggestion: assume constant references ω^* and ψ^* and use the linear approximation.*

5.8. Design an adaptive version of the feedforward control (1.74) replacing T_L by $\hat{T}_L$ and $\gamma = \alpha\beta M + R_s/\sigma$ by $\hat{\gamma} = \alpha\beta M + \hat{R}_s/\sigma$ and assuming that only stator current measurements are available. *Suggestion: assume constant references* ω^* *and* ψ^* *and use the linear approximation.*

Chapter 6
Conclusions

In this book the problem of designing a control for an induction motor has been systematically studied and thoroughly discussed using tools from nonlinear system theory, both for analysis and control purposes. Adaptive control techniques have been used to identify online critical parameters.

The analysis has been carried out in Chapter 1 in which several nonlinear state space models are introduced for balanced unsaturated induction motors. The dynamical models are multivariable and highly nonlinear: they relate two independent stator voltage inputs to the output to be tracked, which is the rotor speed, via the unknown load torque and the stator and rotor electric state variables. The power balance is best understood using the energy model in which the rotor speed and the motor currents are the state variables: the input power from a voltage source is balanced by the power losses and by the output power which is required to match a load torque at a desired speed. The field-oriented model, which is expressed in a frame rotating in synchronism with the rotor flux vector, allows us to compute the steady-state operating conditions when sinusoidal input voltages with constant modulus and frequency are applied. The nonlinear analytical relation between the load torque and the rotor speed, which is usually called torque–speed characteristics, has been explicitly computed. It shows that, while for low load torques there is only one stable operating condition at high rotor speed, for higher load torques there might be two operating conditions depending on motor parameters such as the stator resistance. The operating condition at higher speed is exponentially stable while the operating condition at lower speed is unstable: they get closer and closer as the load torque increases giving rise to a saddle-node nonlinear bifurcation phenomenon. In particular, zero rotor speed is typically an unstable operating condition in which the rotor flux modulus has its minimum value while the stator current modulus achieves its maximum value. The inverse system has also been computed using the field-oriented model. It is a first order nonlinear system, whose state is the flux vector angle, which generates the stator voltage inputs required to track the desired time varying rotor speed reference signal; it is parameterized by a time varying rotor flux modulus which may also be viewed as an additional output to be tracked. The choice of the flux modulus reference aims at the power losses minimization or, more generally, at maintaining the motor within its physical limitations in those stable operating conditions in which the model used to design the control is reliable. When the desired rotor speed and rotor flux modulus are constant, limit cycles arise in the state space, whose attractivity is in general not global and not even guaranteed depending on the required flux modulus, on the load torque, and on the motor parameters. These features reveal the nonlinear nature of the induction motor dynamics and motivate the need of a feedback control action to improve the motor dynamics, particularly at low speed. The models have been experimentally validated for a low power induction motor by comparing the characteristic curves obtained theoretically and experimentally: a good match was obtained at higher speed while at low speed and high torque some discrepancies appear due to measurement inaccuracies and unmodeled effects such as magnetic saturation and power electronics dynamics. Transient responses have also been compared: the models are suitable for control design even though they are less reliable at low speed, low flux modulus, and high torque.

Before discussing the design of feedback controls, which is the topic of the book, structural properties of the induction motor, such as feedback linearizability, rotor flux observability, and parameter identifiability have been studied in Chapter 1. Several encouraging facts have been established. The induction motor can be made input–output decoupled and linear by static state feedback and can be transformed into a linear, controllable system by a first order dynamic state feedback control. Hence, the highly nonlinear induction motor can be effectively controlled if all the state space variables are available for feedback and all motor parameters are exactly known. Unfortunately the rotor variables, either fluxes or currents, cannot be easily measured and two critical parameters, the load torque and the rotor resistance, may vary widely during operations and are therefore uncertain. These two parameters play a crucial role in the induction motor dynamics. Nevertheless the rotor fluxes are observable and the load torque is identifiable for any voltage input from stator current and rotor speed measurements; in this case the rotor resistance is also identifiable when the electromagnetic torque generated by the motor is different from zero. When the rotor speed is not measured the observability and the identifiability issues are more involved: time-varying rotor flux modulus may be required in order to identify the rotor resistance.

Once good controllability, observability, and identifiability properties have been established, the design of feedback control algorithms has followed the classical path: the full potentiality of state feedback design has first been explored in Chapter 2; then the design of observers, parameter estimators and adaptive observers have been studied in Chapter 3; in Chapter 4 the output feedback design based on observers and estimators has been discussed. Finally, in Chapter 5 the control of induction motors without rotor speed sensors has been dealt with: this is the most difficult control task since the speed tracking error is no longer available for feedback.

In particular, since globally attractive operating conditions are not guaranteed by the feedforward control strategy, which may also lead to instabilities, depending on motor parameters and reference signals, six different feedback control schemes have been designed and discussed in Chapter 2, under the assumption that all the state variables are available for feedback. Since in practice the rotor state variables cannot be measured, the actual control algorithms that are implementable will have a more complex structure and will achieve inferior performance. Hence, in Chapter 2 the foundations of any feedback control algorithm are laid and the best achievable performance is explored. A very satisfactory control algorithm has been presented in Section 2.7 which achieves speed tracking with arbitrary rate of convergence from any initial condition and for any rotor speed reference: it represents an evolution of the indirect field-oriented control which is commonly used in practice and is illustrated in Section 2.3. Classical tools from Lyapunov stability theory have been used in the design of this controller. Superior closed-loop performance can be achieved if singularities are accepted which imply that the motor cannot be operated from any initial condition. In this case the controlled motor can be rendered linear as shown in Sections 2.4–2.6. Moreover, estimators for load torque and rotor resistance can also be designed when all state variables are measured: they allow the convergence to zero of the rotor speed and rotor flux tracking errors even when the parameters are not correctly estimated. However, when the electromagnetic torque is different from zero both load torque and rotor resistance are identified online. The indirect field-oriented control, which does not require rotor flux measurements, has been experimentally tested and its robustness with respect to rotor resistance variations has been evaluated: the experiments confirm that the rotor resistance is a very critical parameter in the control design, since its variations lead to poor control of the rotor flux and, consequently, to reduced power efficiency.

Chapter 3 has been devoted to the design of rotor flux observers and of rotor resistance estimators from rotor speed and stator current measurements. It has been shown that the two problems are closely related and can simultaneously be solved by an adaptive flux observer which has been presented in Section 3.2: the rotor flux estimation errors converge to zero only when the rotor resistance estimation errors also converge to zero; this happens in most operating conditions. Basic techniques from adaptive control theory such as the concept of persistency of excitation have been used. The preliminary Section 3.1 has shown that, when the rotor resistance is known, rotor flux observers with arbitrary rate of convergence can be designed. Exponentially convergent load torque estimators have been discussed in Section 3.3. Rotor flux observers which are adaptive with respect to rotor resistance variations have been experimentally tested: their robustness with respect to stator resistance variations has been evaluated.

In Chapter 4 physically implementable feedback control algorithms have been obtained on the basis of the results discussed in Chapters 2 and 3. The most general and most complex output feedback controller has been presented in Section 4.4: it is adaptive with respect to both unknown load torque and rotor resistance and allows for any motor initial condition and any speed reference. It achieves in any case the tracking of the rotor speed reference: the tracking of the desired rotor flux modulus, however, is achieved if persistency of excitation conditions are met so that the parameters are correctly estimated. The persistency of excitation conditions cannot be met when the load torque is zero and both the rotor speed and the rotor flux modulus are constant; in typical operating conditions, that is when the electromagnetic torque is different from zero, they are satisfied. Hence, the lack of rotor flux measurements only prevents us from achieving rotor flux tracking even when the rotor resistance is uncertain and is not correctly identified. However, when the electromagnetic torque is different from zero, the exact value of the rotor resistance is recovered online and the flux tracking errors exponentially tend to zero. In Chapter 4 it has also been clarified that even though output feedback control not based on observers can also be designed when rotor resistance is known, as shown in Section 4.1, observers can improve the closed-loop performance and motor efficiency and should be used when the rotor resistance is uncertain. An output feedback control algorithm, which is adaptive with respect to load torque and rotor resistance variations, has been experimentally tested: experiments confirm that the rotor resistance is estimated online.

The difficult problem of controlling the induction motor without rotor speed measurements has been studied in Chapter 5. Since no rotor speed tracking error is available for feedback, the main problem is to design rotor speed observers so that the control techniques developed in Chapter 4 can still be applied. In Section 5.1 it has been shown that a control which is based on the stator current errors only and does not attempt to estimate the rotor speed has severe limitations and may lead to instability. The possibility of observing the rotor speed, at least when a feedback control is applied and the rotor flux vector is measured, has been discussed in Section 5.2, which opens the way to the design of adaptive speed-sensorless controls in the realistic cases in which the rotor flux is not measured and both torque load and rotor resistance are uncertain. In these cases the obtained results, which are presented in Sections 5.4 and 5.5, are local, *i.e.* they hold for sufficiently small initial errors, while global results have been obtained in Chapter 4 when rotor speed measurements are available. Moreover, in order to identify the rotor resistance which is required for the rotor speed estimation and for the speed tracking, the rotor flux reference is required to be time-varying so that power efficiency is more problematic: this limitation is not present in the case in which the rotor speed is measured.

This book clearly illustrates that nonlinear adaptive control techniques can be successfully applied to the design of induction motor control. A nonlinear adaptive control design is required for the induction motor control since its dynamics are highly nonlinear and critical parameters should be identified online to improve motor efficiency and performance. All control and estimation algorithms presented in this book have been simulated for the same induction motor which has been used in the experiments, so that the corresponding performance can be compared and validated. Experimental tests confirm the simulation results: they show that critical parameters such as torque load and rotor resistance can be estimated online and that the presented control algorithms are implementable, achieve good performance, and improve power efficiency when the rotor speed is measured. Even in the cases in which only stator currents are measured while rotor speed and fluxes are not measured, rotor speed tracking can still be achieved by complex nonlinear adaptive feedback controls, provided that persistently exciting rotor fluxes are generated.

We may conclude that the two parameters load torque and rotor resistance have been the main characters of this book which specifically require adaptive control techniques; their variations may turn a stable operating condition into an unstable one and *vice versa* when the induction motor is operated by a feedforward control; they both contribute to determine the optimal flux modulus which minimizes the power losses; they determine, for a given flux modulus, the slip speed, that is the difference between the flux vector speed of rotation and the rotor speed. Consequently, since the quadrature component of the stator current vector, which is orthogonal to the flux vector, is responsible for the electromagnetic torque generation while the direct component of the stator current vector, which is parallel to the flux vector, is responsible for the flux tracking, unless the rotor flux is directly measured, the exact values of those two parameters are needed by any control algorithm to achieve high dynamic performance in speed control with high power efficiency. Nonlinear adaptive observers are then the key tools for induction motor control, since they can observe the rotor flux vector and, simultaneously, identify online the rotor resistance. Nonadaptive rotor flux observers are, in fact, very sensitive to rotor resistance variations. Even though there are operating conditions in which the rotor flux vector is not observable and the rotor resistance is not identifiable, in typical operating conditions, such as nonzero electromagnetic torque, the adaptive observer errors converge exponentially to zero and this is experimentally confirmed. Once the rotor flux estimation errors converge to zero, the online identification of the load torque is not difficult. The incorporation of adaptive observers into an output feedback control algorithm presents technical difficulties and leads to complex nonlinear feedback controls, which are designed on the basis of Lyapunov techniques. The most difficult problem has been to control the induction motor without rotor speed measurements. Adaptive observers for rotor speed and fluxes and online identification algorithms for the load torque and the rotor resistance are to be designed on the basis of stator current and stator voltage measurements only. This is, to a large extent, still an open problem: it is of special interest for traction applications in which induction motor controls are required to tolerate rotor speed sensor faults, especially at low speed, and to maintain at the same time high power efficiency at every speed.

Appendix A
Lyapunov Stability

Consider the ordinary differential equation

$$\dot{x} = f(x,t)\,, \qquad x(t_0) = x_0\,,\; x \in \mathbb{R}^n \tag{A.1}$$

with $f \in C^0(\mathbb{R}^n \times \mathbb{R}^+, \mathbb{R}^n)$. It is called autonomous when f does not depend on t, *i.e.*

$$\dot{x} = f(x)\,, \quad x(0) = x_0 \tag{A.2}$$

and nonautonomous otherwise. It is called linear if $f(x,t) = A(t)x$, with $A(t)$ an $n \times n$ time-varying matrix, and nonlinear otherwise. A solution of (A.1) over an interval $[t_0, t_0 + T]$ is denoted by $\varphi(t_0 + t, t_0, x_0) \in C^1(\mathbb{R}^+ \times \mathbb{R}^+ \times \mathbb{R}^n, \mathbb{R}^n)$, and satisfies

(i) $\varphi(t_0, t_0, x_0) = x_0$,
(ii) $\dfrac{\mathrm{d}\varphi(t_0+t, t_0, x_0)}{\mathrm{d}t} = f(\varphi(t_0+t, t_0, x_0), t)$, $\forall t \in [0, T]$.

A solution of (A.2) over an interval $[0, T]$ is denoted by $\varphi(t, x_0)$ with $x_0 = x(0)$.

Theorem A.1. *(Local Existence and Uniqueness) Suppose that there exist positive reals τ, α_τ, and β_τ such that*

(i) $\|f(x_1,t) - f(x_2,t)\| \le \alpha_\tau \|x_1 - x_2\|$, $\forall x_1, x_2 \in B_r$, $\forall t \in [t_0, \tau]$,
(ii) $\|f(x_0,t)\| \le \beta_\tau$, $\forall t \in [t_0, \tau]$,

with $B_r(x_0) = \{x \in \mathbb{R}^n : \|x - x_0\| \le r\}$; then (A.1) has only one solution over $[t_0, t_0 + T]$ for a sufficiently small positive real T such that $0 < T \le \tau$. □

Theorem A.2. *(Global Existence and Uniqueness) Suppose that for each $\tau \in [t_0, \infty)$ there exist positive reals α_τ and β_τ such that:*

(i) $\|f(x_1,t) - f(x_2,t)\| \le \alpha_\tau \|x_1 - x_2\|$, $\forall x_1, x_2 \in \mathbb{R}^n$, $\forall t \in [t_0, \tau]$,
(ii) $\|f(x_0,t)\| \le \beta_\tau$, $\forall t \in [t_0, \tau]$;

then (A.1) has only one solution over $[t_0, \infty)$. □

A point x_e is called an equilibrium point of (A.1) if $f(x_e,t) = 0$, $\forall t \geq 0$. An equilibrium point is said to be stable if for any $\varepsilon > 0$ and any $t_0 \in \mathbb{R}^+$ there exists $\delta(t_0,\varepsilon) > 0$ such that $\|\varphi(t+t_0,t_0,x_0) - x_e\| < \varepsilon$, $\forall t \geq 0$, for every x_0 satisfying $\|x_0 - x_e\| < \delta(t_0,\varepsilon)$. If δ can be chosen independently of t_0 the equilibrium point x_e is said to be uniformly stable. An equilibrium point x_e which is not stable is said to be unstable.

An equilibrium point x_e is said to be attractive if there exists $\gamma(t_0) > 0$ such that

$$\lim_{t\to\infty} \|\varphi(t+t_0,t_0,x_0) - x_e\| = 0 \tag{A.3}$$

for every x_0 satisfying $\|x_0 - x_e\| < \gamma(t_0)$. If γ can be chosen independently of t_0 and (A.3) holds uniformly in t_0 and x_0, the equilibrium point is said to be uniformly attractive. Let x_e be a uniformly attractive equilibrium point: its domain of attraction $D(x_e)$ is defined as

$$D(x_e) = \{x \in \mathbb{R}^n : \lim_{t\to\infty} \|\varphi(t,0,x) - x_e\| = 0\} .$$

An equilibrium point is said to be (uniformly) asymptotically stable if it is both (uniformly) stable and (uniformly) attractive. The equilibrium point is said to be exponentially stable if there exist positive constants c, α and r such that

$$\|\varphi(t+t_0,t_0,x_0) - x_e\| \leq c\,\|x_0 - x_e\| \mathrm{e}^{-\alpha(t-t_0)}, \quad \forall t \geq t_0 \tag{A.4}$$

for every x_0 such that $\|x_0 - x_e\| < r$; it is called globally exponentially stable if (A.4) holds for every $x_0 \in \mathbb{R}^n$. The constant α is called the rate of convergence while $1/\alpha$ is called the time constant. Exponential stability implies uniform asymptotic stability. In the case of linear systems the two properties are equivalent.

Theorem A.3. *If the origin is a uniformly asymptotically stable equilibrium point for the linear time-varying system*

$$\dot{x} = A(t)x\,, \quad x \in \mathbb{R}^n\,,$$

then it is globally exponentially stable. □

An equilibrium point is said to be globally uniformly asymptotically stable if:

(i) it is uniformly asymptotically stable,
(ii) for every arbitrarily small positive real ε and every arbitrarily large positive real M there exists a positive $T(\varepsilon,M)$ such that $\|\varphi(t+t_0,t_0,x_0) - x_e\| < \varepsilon$ for every $t > T(\varepsilon,M)$ and every x_0 satisfying $\|x_0 - x_e\| < M$.

A function $\phi : \mathbb{R}^+ \to \mathbb{R}^+$ is said to be of class $\mathscr{K}$ if it is continuous, strictly increasing, and $\phi(0) = 0$.

A function $V : \mathbb{R}^+ \times \mathbb{R}^n \to \mathbb{R}$ is said to be a positive definite function if it is continuous, $V(t,0) = 0$, $\forall t \geq t_0$, and there exist a constant $r > 0$ and a function ϕ of class $\mathscr{K}$ such that

$$\phi(\|x\|) \leq V(t,x)\,, \quad \forall t \geq t_0\,,\, \forall x \in B_r(0) \tag{A.5}$$

with $B_r(0)$ an open ball in $\mathbb{R}^n$ of radius r centered at the origin. A function V is negative definite if $-V$ is positive definite. A function V is said to be a radially unbounded function if for some class $\mathcal{K}$ function $\psi: \mathbb{R}^+ \to \mathbb{R}^+$ such that $\lim_{r\to\infty} \psi(r) = \infty$, we have

$$\psi(\|x\|) \leq V(t,x)\,, \quad \forall t \geq t_0\,,\, \forall x \in \mathbb{R}^n\,.$$

A function V is said to be decrescent if there exist a constant $r > 0$ and a function ψ of class $\mathcal{K}$ such that

$$V(t,x) \leq \psi(\|x\|)\,, \quad \forall t \geq t_0\,,\, \forall x \in B_r(0)\,.$$

An $n \times n$ real matrix P is said to be a positive definite matrix (positive semidefinite) if $x^{\mathrm{T}}Px > 0$ ($x^{\mathrm{T}}Px \geq 0$) for every $x \in \mathbb{R}^n$, $x \neq 0$. Matrix P is negative definite (semidefinite) if $-P$ is positive definite (semidefinite).

Without loss of generality we assume that the equilibrium point x_e is the origin in the statement of the following basic result.

Theorem A.4. *(Lyapunov) Let $x_\mathrm{e} = 0$ be an equilibrium point of (A.1). Let $V(t,x)$ be a C^1 positive definite function whose time derivative along the solutions of (A.1) is denoted by*

$$\dot{V}(t,x) = \frac{\partial V(t,x)}{\partial t} + \sum_{i=1}^{n} \frac{\partial V(t,x)}{\partial x_i} f_i(x,t)\,.$$

(i) *If $\dot{V}(t,x) \leq 0$, $\forall t \geq t_0$, $\forall x \in B_r$ for some $r > 0$, then $x_\mathrm{e} = 0$ is stable.*
(ii) *If $V(t,x)$ is positive definite and decrescent and $\dot{V}(t,x) \leq 0$, $\forall t \geq t_0$, $\forall x \in B_r(0)$ for some $r > 0$, then $x_\mathrm{e} = 0$ is uniformly stable.*
(iii) *If $V(t,x)$ is positive definite and decrescent and $\dot{V}(t,x)$ is negative definite then $x_\mathrm{e} = 0$ is uniformly asymptotically stable.*
(iv) *If $V(t,x)$ is positive definite, decrescent in $\mathbb{R}^n$ and radially unbounded and $\dot{V}(t,x)$ is negative definite in $\mathbb{R}^n$, then $x_\mathrm{e} = 0$ is globally uniformly asymptotically stable.* □

A function $V(t,x)$ satisfying condition (i) above is called a Lyapunov function for system (A.1).

Theorem A.5. *Let $x_\mathrm{e} = 0$ be an equilibrium point of (A.1). Suppose there exist positive constants a, b, c, $p \geq 1$ and a C^1 function $V: \mathbb{R}^+ \times \mathbb{R}^n \to \mathbb{R}^+$ such that for every $x \in B_r(0)$ for some positive r*

(i) $a\|x\|^p \leq V(t,x) \leq b\|x\|^p$, $\forall t \geq t_0$,
(ii) $\dot{V}(t,x) = \dfrac{\partial V(t,x)}{\partial t} + \sum_{i=1}^{n} \dfrac{\partial V(t,x)}{\partial x_i} f_i(x,t) \leq -c\|x\|^p$, $\forall t \geq t_0$.

Then $x_\mathrm{e} = 0$ is exponentially stable. If, in addition, (i) *and* (ii) *hold for every $x \in \mathbb{R}^n$ then $x_\mathrm{e} = 0$ is globally exponentially stable.* □

Theorem A.6. *Consider the linear autonomous system*

$$\dot{x} = Ax, \qquad x \in \mathbb{R}^n . \tag{A.6}$$

The following statements are equivalent:

(i) *the origin is an asymptotically stable equilibrium point;*
(ii) *the origin is globally exponentially stable;*
(iii) *A is a Hurwitz matrix, i.e. all the eigenvalues of A have negative real parts;*
(iv) *for each symmetric positive definite $n \times n$ matrix Q, the $n \times n$ symmetric positive definite matrix*

$$P = \int_0^\infty \mathrm{e}^{A^{\mathrm{T}}t} Q \mathrm{e}^{At}\, \mathrm{d}t$$

is the unique solution of the Lyapunov matrix equation

$$A^{\mathrm{T}}P + PA = -Q$$

and $V = x^{\mathrm{T}}Px$ is a Lyapunov function for (A.6). □

Theorem A.7. *(Linear Approximation Theorem) Consider the autonomous system (A.2) with $f(x)$ a C^2 function and $f(0) = 0$. The origin is an exponentially stable equilibrium point if and only if the linear approximation about the origin*

$$\dot{x} = Fx\,, \quad x \in \mathbb{R}^n$$

is asymptotically stable, with the $n \times n$ constant matrix F being the Jacobian of $f(x)$ evaluated at the origin. If the matrix F has at least one eigenvalue with positive real part then the origin is an unstable equilibrium point. □

When the matrix F has all its eigenvalues with nonpositive real parts and at least one eigenvalue with real part equal to zero we are in a critical case since Theorem A.7 does not apply.

Lemma A.1. *Consider the system ($x_1 \in \mathbb{R}^n$, $x_2 \in \mathbb{R}^m$)*

$$\begin{aligned} \dot{x}_1 &= f_1(t,x_1,x_2), \quad f_1(t,0,0) = 0 \\ \dot{x}_2 &= f_2(t,x_2), \quad f_2(t,0) = 0 \end{aligned} \tag{A.7}$$

and assume that:

(i) for the x_1-subsystem there exists a function $V_1(t,x_1)$ such that

$$\begin{aligned} c_1\|x_1\|^2 \leq V_1(t,x_1) &\leq c_2\|x_1\|^2 \\ \frac{\partial V_1}{\partial t} + \frac{\partial V_1}{\partial x_1} f_1(t,x_1,x_2) &\leq -c_3\|x_1\|^2 + c_4\|x_2\|^2 \end{aligned} \tag{A.8}$$

with $c_i > 0$, $1 \leq i \leq 4$;

(ii) the equilibrium point $x_2 = 0$ of the x_2-subsystem is globally exponentially stable.

Then, the equilibrium point $(x_1, x_2) = 0$ of the whole system (A.7) is globally exponentially stable.

Proof. By assumption (ii), for any initial condition $x_2(t_0)$, $x_2(t)$ satisfies the inequality

$$\|x_2(t)\| \leq c_5 \|x_2(t_0)\| \mathrm{e}^{-c_6(t-t_0)} \tag{A.9}$$

with $c_5 > 0$ and $c_6 > 0$. Let $y(t) \in \mathbb{R}$ be the solution of

$$\dot{y}(t) = -c_6 y(t) \,, \quad y(t_0) = \|x_2(t_0)\| \,. \tag{A.10}$$

Since $y(t) = \mathrm{e}^{-c_6(t-t_0)} y(t_0)$, we can write

$$\|x_2(t)\| \leq c_5 y(t) \,, \quad \forall t \geq t_0 \,. \tag{A.11}$$

Consider the function

$$V(t, x_1, y) = V_1(t, x_1) + \frac{1}{2}(c_4 + 1)\frac{c_5^2}{c_6} y^2 \,.$$

By virtue of (A.8), (A.10), and (A.11), we have, for every $t \geq t_0$,

$$c_1 \|x_1\|^2 + \frac{1}{2}(c_4 + 1)\frac{c_5^2}{c_6} y^2 \leq V \leq c_2 \|x_1\|^2 + \frac{1}{2}(c_4 + 1)\frac{c_5^2}{c_6} y^2$$

and

$$\begin{aligned}\dot{V} &\leq -c_3 \|x_1\|^2 + c_4 \|x_2\|^2 - (c_4 + 1) c_5^2 y^2 \\ &\leq -c_3 \|x_1\|^2 + c_4 c_5^2 y^2 - (c_4 + 1) c_5^2 y^2 = -c_3 \|x_1\|^2 - c_5^2 y^2\end{aligned}$$

which imply that

$$\left\| \begin{bmatrix} x_1(t) \\ y(t) \end{bmatrix} \right\| \leq c_7 \left\| \begin{bmatrix} x_1(t_0) \\ y(t_0) \end{bmatrix} \right\| \mathrm{e}^{-c_8(t-t_0)} \tag{A.12}$$

for suitable $c_7 > 0$ and $c_8 > 0$. The inequalities (A.9) and (A.12) prove the thesis. □

A function $\varphi : \mathbb{R}^+ \to \mathbb{R}$ is said to be uniformly continuous on $[0, \infty)$ if for every $\varepsilon > 0$ there exists a $\delta(\varepsilon) > 0$ such that

$$|\varphi(t_1) - \varphi(t_2)| < \varepsilon \,, \quad \forall t_1, t_2 \in \mathbb{R}^+ : \, |t_1 - t_2| < \delta \,.$$

A continuous, differentiable function φ with bounded derivative in $[0, \infty)$ is uniformly continuous on $[0, \infty)$.

Lemma A.2. *(Barbalat) If $\varphi(t)$ is a real function of the real variable t which is defined and uniformly continuous on $[t_0,\infty)$, $t_0 \in \mathbb{R}^+$, and if the limit of the integral $\int_{t_0}^{t} \varphi(\tau)\mathrm{d}\tau$ as t tends to infinity exists and is a finite number, then*

$$\lim_{t\to\infty} \varphi(t) = 0\,. \tag{A.13}$$

Proof. The proof proceeds by contradiction. If (A.13) is not satisfied there exists a positive real $\varepsilon > 0$ such that for every positive real $T > 0$ one can find $t_T \geq T$ with $|\varphi(t_T)| \geq \varepsilon$. Since φ is uniformly continuous, there exists a positive real δ_ε such that, for every $t \geq t_0$ and every τ in the interval $0 \leq \tau \leq \delta_\varepsilon$,

$$|\varphi(t) - \varphi(t+\tau)| \leq \frac{\varepsilon}{2}$$

and in particular, for $t = t_T$,

$$|\varphi(t_T) - \varphi(t_T+\tau)| \leq \frac{\varepsilon}{2}\,, \quad 0 \leq \tau \leq \delta_\varepsilon$$

i.e.

$$|\varphi(t_T) - \varphi(t)| \leq \frac{\varepsilon}{2}\,, \quad t_T \leq t \leq t_T + \delta_\varepsilon\,.$$

Since $|\varphi(t_T)| \geq \varepsilon$, we have

$$|\varphi(t)| \geq \frac{\varepsilon}{2}\,, \quad t_T \leq t \leq t_T + \delta_\varepsilon$$

and, therefore,

$$\int_{t_T}^{t_T+\delta_\varepsilon} |\varphi(\tau)|\mathrm{d}\tau \geq \frac{\varepsilon\delta_\varepsilon}{2}\,.$$

Hence, we can determine an infinite unbounded sequence $\{t_i\}$ such that

$$\int_{t_i}^{t_i+\delta_i} |\varphi(\tau)|\mathrm{d}\tau \geq \frac{\varepsilon_i\delta_i}{2}\,.$$

It follows that the integral $\int_{t_0}^{t} \varphi(\tau)\mathrm{d}\tau$ cannot tend to a finite limit as $t \to \infty$. □

Example A.1. The function

$$\varphi(t) = \mathrm{e}^{-t}\sin(\mathrm{e}^{3t}-1) - 3\mathrm{e}^{2t}\cos(\mathrm{e}^{3t}-1)$$

is not uniformly continuous. Its integral

$$\int_0^t \varphi(\tau)\,d\tau = -\mathrm{e}^{-t}\sin(\mathrm{e}^{3t}-1)$$

tends to zero as t goes to infinity but the function itself does not tend to zero. □

Corollary A.1. *If* $\psi(t): [t_0,\infty) \to \mathbb{R}^n$ *is such that*

(i) $\lim_{t\to\infty} \int_{t_0}^{t} \psi^{\mathrm{T}}(\tau)\psi(\tau)\, d\tau < \infty$
(ii) $\|\psi(t)\| \le M_1,\ \forall t \ge t_0$
(iii) $\left\|\frac{\mathrm{d}\psi(t)}{\mathrm{d}t}\right\| \le M_2,\ \forall t \ge t_0$

with M_1, M_2 *positive reals, then*

$$\lim_{t\to\infty} \|\psi(t)\| = 0\,.$$

Proof. Let $\varphi(t) = \psi^{\mathrm{T}}(t)\psi(t)$. Conditions (ii) and (iii) imply that $\varphi(t)$ and $\dot{\varphi}(t)$ are bounded. Therefore, φ is uniformly continuous and we can apply Barbalat's Lemma A.2 to the function $\varphi(t)$ which guarantees that $\lim_{t\to\infty} \varphi(t) = 0$ and, consequently, the corollary is proved. □

Lemma A.3. *(Persistency of Excitation) Consider the system*

$$\begin{aligned} \dot{x} &= f_x(t,x,z) = A(t)x + B(t)z\,, \quad x \in \mathbb{R}^n \\ \dot{z} &= f_z(t,x) = D(t)x\,, \quad z \in \mathbb{R}^m \end{aligned} \tag{A.14}$$

in which f_x *and* f_z *are continuous and* $B(t)$ *is continuously differentiable for* $t \ge t_0$. *Assume that, for any* $t \ge t_0$:

(i) $\|A(t)\| \le A_M$, $\|B(t)\| \le B_M$, $\|D(t)\| \le D_M$ *and* $\|\dot{B}(t)\| \le \dot{B}_M$;
(ii) there exist two positive reals T *and* k_T *such that the persistency of excitation condition*

$$\int_t^{t+T} B^{\mathrm{T}}(\tau)B(\tau)\mathrm{d}\tau \ge k_T I \tag{A.15}$$

is satisfied;
(iii) there exists a smooth function $V(x,z,t)$ *and suitable positive reals* a_i, $1 \le i \le 3$ *such that*

$$\begin{aligned} a_1(\|x\|^2 + \|z\|^2) \le V(x,z,t) &\le a_2(\|x\|^2 + \|z\|^2) \\ \dot{V}(x,z,t) &\le -a_3\|x\|^2\,. \end{aligned} \tag{A.16}$$

Then, the equilibrium point $(x,z) = 0$ *of system (A.14) is globally exponentially stable.*

Proof. Consider the class of radially unbounded functions

$$W(x,z,t) = V(x,z,t) + p\|Qz - B^{\mathrm{T}}x\|^2$$

where p is a positive scalar parameter to be defined later, and $Q(t)$ is an $m \times m$ matrix generated by the matrix differential equation

$$\dot{Q}(t) = -Q(t) + B^{\mathrm{T}}(t)B(t)$$

with $Q(t_0) = \mathrm{e}^{-T} k_T I$. By virtue of assumption (ii),

$$Q(t+T) \geq \mathrm{e}^{-T} \int_t^{t+T} B^{\mathrm{T}}(\tau) B(\tau) \mathrm{d}\tau \geq \mathrm{e}^{-T} k_T I > 0 .$$

Since $\|B(t)\| \leq B_M$, it follows that

$$B_M^2 I \geq Q(t) > k_T \mathrm{e}^{-2T} I .$$

In view of hypothesis (iii),

$$\begin{aligned} W(x,z,t) &\geq a_1(\|x\|^2 + \|z\|^2) + p\|Qz - B^{\mathrm{T}} x\|^2 \\ W(x,z,t) &\leq a_2(\|x\|^2 + \|z\|^2) + p\|Qz - B^{\mathrm{T}} x\|^2 . \end{aligned} \tag{A.17}$$

By virtue of assumption (iii), we have

$$\dot{W} \leq -a_3\|x\|^2 + 2p(Qz - B^{\mathrm{T}} x)^{\mathrm{T}} (QDx - Qz - B^{\mathrm{T}} Ax - \dot{B}^{\mathrm{T}} x)$$

so that by adding and subtracting $2p(Qz - B^{\mathrm{T}} x)^{\mathrm{T}} B^{\mathrm{T}} x$, we obtain

$$\dot{W} \leq -a_3\|x\|^2 - 2p\|Qz - B^{\mathrm{T}} x\|^2 + 2p(Qz - B^{\mathrm{T}} x)^{\mathrm{T}} (QD - B^{\mathrm{T}} A - \dot{B}^{\mathrm{T}} - B^{\mathrm{T}}) x . \tag{A.18}$$

By hypothesis (i), there exists a positive real $a_4 \in \Re^+$, such that $a_4 \geq \|QD - B^{\mathrm{T}} A - \dot{B}^{\mathrm{T}} - B^{\mathrm{T}}\|^2$. By completing the squares, we have

$$\begin{aligned} 2p(Qz - B^{\mathrm{T}} x)^{\mathrm{T}} (QD - B^{\mathrm{T}} A - \dot{B}^{\mathrm{T}} - B^{\mathrm{T}}) x &\leq \frac{p}{4}\|Qz - B^{\mathrm{T}} x\|^2 + 4pa_4\|x\|^2 \\ -p\|Qz - B^{\mathrm{T}} x\|^2 &\leq p\left(\|B^{\mathrm{T}} x\|^2 - \frac{1}{2}\|Qz\|^2\right) . \end{aligned} \tag{A.19}$$

By substituting (A.19) into (A.18) and rearranging terms, we obtain

$$\dot{W} \leq -\left[a_3 - p\left(4a_4 + B_M^2\right)\right] \|x\|^2 - \frac{p}{2} k_T^2 \mathrm{e}^{-4T} \|z\|^2 - \frac{3}{4} p\|Qz - B^{\mathrm{T}} x\|^2 . \tag{A.20}$$

By choosing p such that $p < a_3 / \left(4a_4 + B_M^2\right)$, the right-hand side of the inequality (A.20) is negative definite and, therefore, from (A.17), it follows that the origin of system (A.14) is globally exponentially stable. □

Example A.2. Consider the system

$$\begin{aligned} \dot{\tilde{x}} &= -\tilde{x} + b_1(t)\tilde{\theta}_1 + b_2(t)\tilde{\theta}_2 \\ \dot{\tilde{\theta}}_1 &= -b_1(t)\tilde{x} \\ \dot{\tilde{\theta}}_2 &= -b_2(t)\tilde{x} \end{aligned} \tag{A.21}$$

in which $b_1(t)$ and $b_2(t)$ are bounded inputs with bounded time derivatives satisfying condition (ii) of Lemma A.3; since the positive definite function

$$V(\tilde{x},\tilde{\theta}_1,\tilde{\theta}_2) = \frac{1}{2}(\tilde{x}^2 + \tilde{\theta}_1^2 + \tilde{\theta}_2^2)$$

has the following time derivative, along the system solutions:

$$\dot{V}(x,\tilde{\theta}_1,\tilde{\theta}_2) = -\tilde{x}^2$$

so that condition (iii) of Lemma A.3 is satisfied, then, according to Lemma A.3, the origin $(\tilde{x},\tilde{\theta}_1,\tilde{\theta}_2) = 0$ is a globally exponentially stable equilibrium point for system (A.21). The above example (A.21) may arise from the adaptive control of the system

$$\dot{x} = u(t) + b_1(t)\theta_1 + b_2(t)\theta_2$$

in which θ_1 and θ_2 are unknown constant parameters and the following adaptive control

$$\begin{aligned} u &= -[x - x_r(t)] + \dot{x}_r(t) - b_1(t)\hat{\theta}_1 - b_2(t)\hat{\theta}_2 \\ \dot{\hat{\theta}}_1 &= b_1(t)(x - x_r) \\ \dot{\hat{\theta}}_2 &= b_2(t)(x - x_r) \end{aligned}$$

is designed to track the reference signal $x_r(t)$ by adjusting the parameter estimates $\hat{\theta}_1(t)$ and $\hat{\theta}_2(t)$. Defining the tracking error $\tilde{x} = x - x_r$ and the estimation errors $\tilde{\theta}_1 = \theta_1 - \hat{\theta}_1$, $\tilde{\theta}_2 = \theta_2 - \hat{\theta}_2$, the error dynamics are given by (A.21). On the other hand, the example (A.21) may arise from the design of an adaptive observer for the system

$$\begin{aligned} \dot{x} &= b_1(t)\theta_1 + b_2(t)\theta_2 \\ y &= x \end{aligned}$$

in which y is the measured variable and θ_1, θ_2 are unknown constant parameters. In fact, design the adaptive observer

$$\begin{aligned} \dot{\hat{x}} &= -\hat{x} + y + b_1(t)\hat{\theta}_1 + b_2(t)\hat{\theta}_2 \\ \dot{\hat{\theta}}_1 &= b_1(t)(y - \hat{x}) \\ \dot{\hat{\theta}}_2 &= b_2(t)(y - \hat{x}) \end{aligned}$$

in which $\hat{x}$ is the state variable estimate and $\hat{\theta}_1$, $\hat{\theta}_2$ are the parameter estimates. If we denote by $\tilde{x} = x - \hat{x}$ the observation error and by $\tilde{\theta}_1 = \theta_1 - \hat{\theta}_1$, $\tilde{\theta}_2 = \theta_2 - \hat{\theta}_2$ the parameter estimation errors, the error dynamics are given by (A.21). Note that, even though the persistency of excitation condition (A.15) fails, we can still apply Corollary A.1 to the function $\psi(t) = \tilde{x}^2(t)$ and conclude that $\lim_{t\to\infty}\tilde{x}(t) = 0$ for any initial condition $\tilde{x}(t_0)$, $\hat{\theta}_1(t_0)$, and $\hat{\theta}_2(t_0)$. Furthermore, it is interesting to note that if b_1 and b_2 are constant so that the persistency of excitation condition (A.15) fails, defining

$$\tilde{\theta} = b_1\tilde{\theta}_1 + b_2\tilde{\theta}_2$$

we obtain

$$\begin{aligned}\dot{\tilde{x}} &= -\tilde{x} + \tilde{\theta} \\ \dot{\tilde{\theta}} &= -(b_1^2 + b_2^2)\tilde{x}\,;\end{aligned} \tag{A.22}$$

considering the positive definite function

$$V = \frac{1}{2}\left(\tilde{x}^2 + \frac{\tilde{\theta}^2}{b_1^2 + b_2^2}\right)$$

whose time derivative along the system solution is

$$\dot{V} = -\tilde{x}^2$$

we can conclude, according to Lemma A.3, that the origin $(\tilde{x},\, \tilde{\theta}) = 0$ is a globally exponentially stable equilibrium point for system (A.22): in particular, even though $\tilde{\theta}_1$ and $\tilde{\theta}_2$ may not converge to zero, $b_1\tilde{\theta}_1 + b_2\tilde{\theta}_2$ exponentially converges to zero for any initial condition. □

Lemma A.4. *Consider the differential scalar inequality*

$$\dot{V}(t) = -c_1V(t) + W(t) \tag{A.23}$$

in which

$$W(t) \leq c_w\mathrm{e}^{-c_2t} \tag{A.24}$$

with $c_1 > 0$, $c_2 > 0$, and $c_w \geq 0$. Then, the solution $V(t)$ satisfies the following inequality:

$$V(t) \leq V(0)\mathrm{e}^{-c_1t} + c_w\frac{\mathrm{e}^{-c_2t} - \mathrm{e}^{-c_1t}}{c_1 - c_2}\,.$$

Proof. By integrating (A.23) and taking (A.24) into account, we obtain

$$\begin{aligned}V(t) &\leq V(0)\mathrm{e}^{-c_1t} + \int_0^t \mathrm{e}^{-c_1(t-\tau)}W(\tau)\mathrm{d}\tau \\ &= V(0)\mathrm{e}^{-c_1t} + c_w\mathrm{e}^{-c_1t}\left[\frac{\mathrm{e}^{\tau(c_1-c_2)}}{c_1 - c_2}\right]_0^t\end{aligned}$$

which implies the thesis. □

We define in the following the projection operator $\text{Proj}(f,\hat{\theta})$ which is used by the adaptation algorithm

$$\dot{\hat{\theta}} = \text{Proj}(f,\hat{\theta}), \quad \hat{\theta}(0) = \hat{\theta}_0$$

for the estimate $\hat{\theta}$ of the uncertain parameter $\theta \in [\theta_m, \theta_M]$, with known lower and upper bounds θ_m and θ_M, respectively. The initial condition $\hat{\theta}_0$ is such that $\theta_m \leq \hat{\theta}_0 \leq \theta_M$. The projection operator $\text{Proj}(f,\hat{\theta}) : \mathbb{R} \times \mathbb{R} \to \mathbb{R}$ is defined as

$$\text{Proj}(f,\hat{\theta}) = \begin{cases} f & \text{if } \theta_m \leq \hat{\theta} \leq \theta_M \\ f & \text{if } (\hat{\theta} < \theta_m \text{ and } f \geq 0) \\ f & \text{if } (\hat{\theta} > \theta_M \text{ and } f \leq 0) \\ f\left[1 - \frac{\theta_m{}^2 - \hat{\theta}^2}{\theta_m{}^2 - (\theta_m - \varepsilon)^2}\right] & \text{if } (\hat{\theta} < \theta_m \text{ and } f < 0) \\ f\left[1 - \frac{\hat{\theta}^2 - \theta_M{}^2}{(\theta_M + \varepsilon)^2 - \theta_M{}^2}\right] & \text{if } (\hat{\theta} > \theta_M \text{ and } f > 0) \end{cases} \tag{A.25}$$

with f a continuous scalar time function and $\varepsilon > 0$ satisfying $\theta_m(\theta_m - \varepsilon) \geq 0$ and $\theta_M(\theta_M + \varepsilon) \geq 0$. The projection algorithm $\text{Proj}(\cdot,\cdot)$ has the following properties:

(i) $\text{Proj}(f,\hat{\theta})$ is Lipschitz continuous;
(ii) $|\text{Proj}(f,\hat{\theta})| \leq |f|$;
(iii) $(\theta - \hat{\theta})\text{Proj}(f,\hat{\theta}) \geq (\theta - \hat{\theta})f$;
(iv) $\theta_m - \varepsilon \leq \hat{\theta}(t) \leq \theta_M + \varepsilon$, for any $t \geq 0$.

Appendix B
Nonlinear Control Theory

Let p be a point in $\mathbb{E}^n$, n-dimensional Euclidean space, and U a neighborhood of p. Let $\varphi(q) = (x_1(q),\ldots,x_n(q)) : U \to V \subset \mathbb{R}^n$ be a homeomorphism, that is one-to-one and onto (*i.e.* a bijection), with φ, φ^{-1} continuous and $\mathbb{R}^n = \mathbb{R} \times \ldots \times \mathbb{R}$, $\mathbb{R}$ the real numbers. (U, φ) is called a coordinate neighborhood or coordinate chart and the real numbers $x_1(q),\ldots,x_n(q)$, which vary continuously, are the local coordinates of $q \in \mathbb{E}^n$: $x_i(q)$ is called the ith coordinate function. If both φ and φ^{-1} are smooth maps, φ is called a diffeomorphism. If both φ and φ^{-1} are defined in $\mathbb{R}^n$ and are smooth maps, φ is called a global diffeomorphism.

Given two coordinate neighborhoods (U, φ), (W, ψ) with $U \cap W \neq 0$ and $\varphi(q) = (x_1(q),\ldots,x_n(q))$, $\psi(q) = (z_1(q),\ldots,z_n(q))$, the homeomorphism

$$\psi \circ \varphi^{-1} : \mathbb{R}^n \to \mathbb{R}^n$$

is a coordinate transformation in $U \cap W$, *i.e.*

$$z(x) = \psi \circ \varphi^{-1}(x) .$$

The inverse mapping is

$$x(z) = \varphi \circ \psi^{-1}(z) .$$

If x and z are represented by vectors with n components, namely

$$x = \begin{bmatrix} x_1 \\ \vdots \\ x_n \end{bmatrix}, \qquad z = \begin{bmatrix} z_1 \\ \vdots \\ z_n \end{bmatrix},$$

the coordinate transformations are expressed by n real valued continuous functions defined in $\mathbb{R}^n$, *i.e.*

$$x(z) = \begin{bmatrix} x_1(z_1,\ldots,z_n) \\ \vdots \\ x_n(z_1,\ldots,z_n) \end{bmatrix}, \qquad z(x) = \begin{bmatrix} z_1(x_1,\ldots,x_n) \\ \vdots \\ z_n(x_1,\ldots,x_n) \end{bmatrix}.$$

If both homeomorphisms $z(x)$ and $x(z)$ are smooth maps, the coordinate transformation is a diffeomorphism. If both homeomorphisms $z(x)$ and $x(z)$ are smooth maps defined in $\mathbb{R}^n$, the coordinate transformation is a global diffeomorphism.

We now recall a well known result from calculus which provides sufficient conditions for a map to be a diffeomorphism.

Theorem B.1. *(Inverse Function) Let U be an open subset of $\mathbb{R}^n$ and let $\varphi = (\varphi_1,\ldots,\varphi_n) : U \to \mathbb{R}^n$ be a smooth map. If the Jacobian matrix*

$$\frac{\mathrm{d}\varphi}{\mathrm{d}x} = \begin{bmatrix} \frac{\partial \varphi_1}{\partial x_1} & \cdots & \frac{\partial \varphi_1}{\partial x_n} \\ \vdots & \ddots & \vdots \\ \frac{\partial \varphi_n}{\partial x_1} & \cdots & \frac{\partial \varphi_n}{\partial x_n} \end{bmatrix}$$

is nonsingular at some point $p \in U$ then there exists a neighborhood $V \subset U$ of p such that $\varphi : V \to \varphi(V)$ is a diffeomorphism. □

Let $h : U \subset \mathbb{E}^n \to \mathbb{R}$ be a real valued function defined on U. Depending on the coordinate neighborhoods (U,φ) chosen, the function h is expressed in local coordinates as

$$h_\varphi = h \circ \varphi^{-1} : \mathbb{R}^n \to \mathbb{R}.$$

The expression h_φ depends on the chosen local coordinates.

The differential of a smooth function $h : U \subset \mathbb{R}^n \to \mathbb{R}$ is defined in local coordinates as

$$\mathrm{d}h = \frac{\partial h}{\partial x_1}\mathrm{d}x_1 + \cdots + \frac{\partial h}{\partial x_n}\mathrm{d}x_n$$

and may be denoted as a row gradient vector

$$\mathrm{grad}h = \begin{bmatrix} \frac{\partial h}{\partial x_1} & \cdots & \frac{\partial h}{\partial x_n} \end{bmatrix}.$$

Given r smooth real valued functions $\varphi_1, \ldots, \varphi_r$ in $U \subset \mathbb{R}^n$, $\mathrm{Rank}\{\mathrm{d}\varphi_1,\ldots,\mathrm{d}\varphi_r\} = r$ in $p \in U$ (in $U \subset \mathbb{R}^n$) is equivalent to

$$\text{Rank}\begin{bmatrix} \frac{\partial \varphi_1(x)}{\partial x_1} & \cdots & \frac{\partial \varphi_1(x)}{\partial x_n} \\ \vdots & \vdots & \vdots \\ \frac{\partial \varphi_r(x)}{\partial x_1} & \cdots & \frac{\partial \varphi_r(x)}{\partial x_n} \end{bmatrix} = r$$

for $x = p$ (for every $x \in U$). The inverse function theorem may also be stated as follows.

Theorem B.2. *If* $\text{Rank}\{\mathrm{d}\varphi_1, \ldots, \mathrm{d}\varphi_n\} = n$ *at some point* $p \in U$*, an open subset of* $\mathbb{R}^n$*, then there exists a neighborhood* $V \subset U$ *of* p *such that* $\varphi : V \longrightarrow \varphi(V)$ *is a diffeomorphism.* □

Theorem B.3. *Given* $n+1$ *real valued smooth functions* ψ, φ_1, …, φ_n *in* $U \subset \mathbb{R}^n$*, if in* $p \in U$ $\text{Rank}\{\mathrm{d}\varphi_i,\ 1 \le i \le n\} = n$ *and* $\mathrm{d}\psi \in \text{Span}\{\mathrm{d}\varphi_i,\ 1 \le i \le j\}$ *for some* $j \le n$*, then in local coordinates* $z_i = \varphi_i(x)$*,* $1 \le i \le n$*,* $\partial\psi(z)/\partial z_i = 0$*,* $j+1 \le i \le n$*, i.e.* $\psi = \psi(z_1, \ldots, z_j)$ *in a neighborhood* $V \subset U$ *of* p*.* □

Let $C^\infty(p)$ be the set of all C^∞ functions defined in U, a neighborhood of p. They form an algebra over the field $\mathbb{R}$. A tangent vector v at p has three properties:

(i) it maps $C^\infty(p)$ into $\mathbb{R}^n$;
(ii) it is a linear operator, that is

$$v(a_1h_1 + a_2h_2) = a_1v(h_1) + a_2v(h_2), \quad \forall h_i \in C^\infty(p),\ a_i \in \mathbb{R}\,;$$

(iii) it satisfies Leibniz's rule

$$v(h_1h_2) = v(h_1)h_2 + h_1v(h_2)\,.$$

All tangent vectors v at p constitute a vector space over the field $\mathbb{R}$ since

$$(a_1v_1 + a_2v_2)(h) = a_1v_1(h) + a_2v_2(h)\,.$$

If (U, φ) is a coordinate neighborhood and $x_1(p), \ldots, x_n(p)$ are local coordinates, a vector may be expressed as

$$v = v_1\frac{\partial}{\partial x_1} + \cdots + v_n\frac{\partial}{\partial x_n}$$

with $v_i \in \mathbb{R}$, $1 \le i \le n$, being the components of v with respect to $\partial/\partial x_1, \ldots, \partial/\partial x_n$ which is a basis for the vector space over $\mathbb{R}$ consisting of all tangent vectors v at p.

A vector field f on an open subset $U \subset \mathbb{R}^n$ is a function which assigns to each point $p \in U$ a vector f_p.

If (U, φ) is a coordinate neighborhood and $x_1(p)$, …, $x_n(p)$, $p \in U$, are local coordinates, a C^∞ (smooth) vector field is expressed as

$$f = f_1(x)\frac{\partial}{\partial x_1} + \ldots + f_n(x)\frac{\partial}{\partial x_n}$$

with $f_i \in C^\infty(p)$ depending on the chosen local coordinates. A vector field may also be expressed as a column vector

$$f(x) = \begin{bmatrix} f_1(x) \\ \vdots \\ f_n(x) \end{bmatrix}.$$

If $z(x)$ denotes a coordinate transformation whose inverse is $x(z)$, the vector field f is expressed in new coordinates as

$$\tilde{f} = \tilde{f}_1(z)\frac{\partial}{\partial z_1} + \cdots + \tilde{f}_n(z)\frac{\partial}{\partial z_n}$$

with

$$\tilde{f}_j(z) = \sum_{i=1}^{n} \frac{\partial z_j(x(z))}{\partial x_i} f_i(x(z)), \quad 1 \leq j \leq n.$$

Equivalently, in vector notation we can write

$$\begin{bmatrix} \tilde{f}_1 \\ \vdots \\ \tilde{f}_n \end{bmatrix}(z) = \frac{\mathrm{d}z}{\mathrm{d}x} \begin{bmatrix} f_1 \\ \vdots \\ f_n \end{bmatrix} \circ x(z)$$

with

$$\frac{\mathrm{d}z}{\mathrm{d}x} = \begin{bmatrix} \frac{\partial z_1}{\partial x_1} & \cdots & \frac{\partial z_1}{\partial x_n} \\ \vdots & \cdots & \vdots \\ \frac{\partial z_n}{\partial x_1} & \cdots & \frac{\partial z_n}{\partial x_n} \end{bmatrix}$$

denoting the Jacobian matrix of the coordinate transformation.

A dynamical system is a C^1 map $\phi_t(p): \mathbb{R} \times U \to U$ where U is an open set in the Euclidean space which satisfies:

(i) $\phi_0(p) = p$,
(ii) $\phi_t \circ \phi_s = \phi_{t+s}$ for each $t, s \in \mathbb{R}$,

with $\phi_t(p)$ mapping $U \to U$ for each t. This definition implies that the map $\phi_t(p)$ has a C^1 inverse $\phi_{-t}(p)$. A dynamical system defines a vector field

$$f(p) = \frac{\mathrm{d}}{\mathrm{d}t}\phi_t(p)\bigg|_{t=0}$$

which is the tangent vector to the curve $t \to \phi_t(p)$ at $t = 0$. Given a coordinate chart (U, φ), the curve $t \to \phi_t(p)$ may be expressed as $t \to (x_1(t), \ldots, x_n(t))$ and

$$\begin{bmatrix} f_1(x(t)) \\ \vdots \\ f_n(x(t)) \end{bmatrix} = \begin{bmatrix} \frac{\mathrm{d}}{\mathrm{d}t}x_1(t) \\ \vdots \\ \frac{\mathrm{d}}{\mathrm{d}t}x_n(t) \end{bmatrix}.$$

Hence $x_1(t),\ldots,x_n(t)$ is the solution of the differential equation $\dot{x} = f(x)$, with initial condition $x(0)$. If $z(x)$ denotes a coordinate transformation with inverse $x(z)$, the differential equation is expressed in new coordinates as

$$\dot{z} = \left(\frac{\mathrm{d}z}{\mathrm{d}x}f\right) \circ x(z) \triangleq \tilde{f}(z)\,.$$

Conversely, a vector field defines a unique dynamical system provided that f satisfies the local Lipschitz condition (see condition (i) in Theorem A.2). The converse process to the above is: given a differential equation

$$\dot{x} = f(x) \tag{B.1}$$

i.e. a vector field f, with initial condition x_0, determine an integral curve $t \to (x_1(t),\ldots,x_n(t))$ which is a solution to (B.1) and such that $x(0) = x_0$.

If f is a smooth vector field on U and h a smooth function on U then $f(h)$ is the smooth function on U defined by

$$f(h)(p) = \sum_{i=1}^{n} f_i(p)\left(\frac{\partial h}{\partial x_i}\right)(p)\,.$$

A vector field may be interpreted as an operator mapping the function h into the function $f(h)$. The function $f(h)$ is called the Lie derivative of the function h along the vector field f; it is usually denoted as $L_f h$ which is a more convenient notation in the case of repeated operations:

$$L_{f_1}L_{f_2}L_{f_3}\ldots L_{f_i}h = f_1(f_2(f_3(\ldots f_i(h)\ldots)))\,.$$

Repeated Lie derivatives along the same vector field f are denoted as

$$L_f^i h = L_f(L_f^{i-1}h)\,, \qquad L_f^1 h = L_f h\,, \qquad L_f^0 h = h\,.$$

The Lie derivative $L_f h$ of a smooth function h along a vector field f is also denoted by $\langle dh, f\rangle$. The set of all C^∞ vector fields on U is a vector space over $\mathbb{R}$ and over $C^\infty(p)$, *i.e.*

$$L_{\alpha f_1+\beta f_2}h = \alpha L_{f_1}h + \beta L_{f_2}h, \quad \alpha,\beta \in C^\infty(p)\,;$$

if we define the Lie bracket

$$[f,g](h) = f(g(h)) - g(f(h)) = L_f L_g h - L_g L_f h$$

they form a noncommutative Lie algebra over $C^\infty(p)$. The Lie bracket $[f,g]$ is a vector field since:

(i) it satisfies Leibniz's rule

$$\begin{aligned}[f,g](h_1h_2) &= L_fL_g(h_1h_2) - L_gL_f(h_1h_2)\\ &= L_f\big((L_gh_1)h_2 + h_1(L_gh_2)\big) - L_g\big((L_fh_1)h_2 + h_1(L_fh_2)\big)\\ &= (L_fL_gh_1 - L_gL_fh_1)h_2 + h_1(L_fL_gh_2 - L_gL_fh_2) + (L_gh_1)(L_fh_2)\\ &\quad +(L_fh_1)(L_gh_2) - (L_fh_1)(L_gh_2) - (L_gh_1)(L_fh_2)\\ &= [f,g](h_1)h_2 + h_1[f,g](h_2)\,;\end{aligned}$$

(ii) it is a linear operator (a_1, $a_2 \in \mathbb{R}$)

$$\begin{aligned}[f,g](a_1h_1 + a_2h_2) &= L_fL_g(a_1h_1 + a_2h_2) - L_gL_f(a_1h_1 + a_2h_2)\\ &= a_1[f,g](h_1) + a_2[f,g](h_2)\,.\end{aligned}$$

The Lie bracket $[f,g]$ of two vector fields f and g is also denoted by ad_fg or by L_fg. Repeated Lie brackets are denoted as

$$ad_f^ig = ad_f(ad_f^{i-1}g)\,, \qquad ad_f^1g = ad_fg\,, \qquad ad_f^0g = g\,.$$

Given two smooth functions α, β and two smooth vector fields f, g, the following formula holds:

$$[\alpha f, \beta g] = \alpha\beta[f,g] + \alpha(L_f\beta)g - \beta(L_g\alpha)f\,, \tag{B.2}$$

which allows us to compute a Lie bracket in local coordinates $(x_1,\ldots,x_n)$. Given

$$f = \sum_{i=1}^n f_i\frac{\partial}{\partial x_i}\,, \qquad g = \sum_{j=1}^n g_j\frac{\partial}{\partial x_j}$$

their Lie bracket is

$$\begin{aligned}[f,g] &= \left[\sum_{i=1}^n f_i\frac{\partial}{\partial x_i}, \sum_{j=1}^n g_j\frac{\partial}{\partial x_j}\right]\\ &= \sum_{i=1}^n\sum_{j=1}^n\left(f_i\left(\frac{\partial}{\partial x_i}g_j\right)\frac{\partial}{\partial x_j} - g_j\left(\frac{\partial}{\partial x_j}f_i\right)\frac{\partial}{\partial x_i} + f_ig_j\left[\frac{\partial}{\partial x_i}, \frac{\partial}{\partial x_j}\right]\right)\,.\end{aligned}$$

By definition

$$\left[\frac{\partial}{\partial x_i}, \frac{\partial}{\partial x_j}\right](h) = \frac{\partial}{\partial x_i}\left(\frac{\partial}{\partial x_j}h\right) - \frac{\partial}{\partial x_j}\left(\frac{\partial}{\partial x_i}h\right) = 0\,,$$

hence

$$\left[\frac{\partial}{\partial x_i}, \frac{\partial}{\partial x_j}\right] = 0\,.$$

In conclusion, the formula for computing a Lie bracket, given a coordinate neighborhood, is

$$[f,g] = \sum_{i=1}^{n} \left(\sum_{j=1}^{n} f_j \left(\frac{\partial}{\partial x_j} g_i \right) - g_j \left(\frac{\partial}{\partial x_j} f_i \right) \right) \frac{\partial}{\partial x_i}\,. \tag{B.3}$$

Denoting by $\mathrm{d}f/\mathrm{d}x$ or $J(f)$ the Jacobian matrix of a vector field $f(x) = [f_1(x), \ldots, f_n(x)]^{\mathrm{T}}$, defined as

$$\frac{\mathrm{d}f}{\mathrm{d}x} = \begin{bmatrix} \frac{\partial f_1}{\partial x_1} & \cdots & \frac{\partial f_1}{\partial x_n} \\ \vdots & \ddots & \vdots \\ \frac{\partial f_n}{\partial x_1} & \cdots & \frac{\partial f_n}{\partial x_n} \end{bmatrix},$$

formula (B.3) may be rewritten in vector notation as

$$[f,g] = \frac{\mathrm{d}g}{\mathrm{d}x} f - \frac{\mathrm{d}f}{\mathrm{d}x} g\,.$$

Given two smooth vector fields f, g, and a smooth function h, the definition of Lie bracket $[f,g]$ gives Leibniz's formula

$$L_{[f,g]} h = L_f L_g h - L_g L_f h\,.$$

Theorem B.4. *(Implicit Function) Let U be an open subset of $\mathbb{R}^n \times \mathbb{R}^r$ and let $\varphi : U \to \mathbb{R}^r$ be a smooth map such that $\varphi(p,q) = 0$ for some point $(p,q) \in U$. Let (x,y) be local coordinates in U, $x \in \mathbb{R}^n$, $y \in \mathbb{R}^r$. If the $r \times r$ Jacobian matrix*

$$\frac{\mathrm{d}\varphi}{\mathrm{d}y} = \begin{bmatrix} \frac{\partial \varphi_1}{\partial y_1} & \cdots & \frac{\partial \varphi_1}{\partial y_r} \\ \vdots & \cdots & \vdots \\ \frac{\partial \varphi_r}{\partial y_1} & \cdots & \frac{\partial \varphi_r}{\partial y_r} \end{bmatrix} (x,y)$$

is nonsingular at $x = p$, $y = q$, then there exist a neighborhood $V \subset U$ of (p,q), a neighborhood $W \subset \mathbb{R}^n$ of p, and a unique smooth map $\psi : W \to \mathbb{R}^r$ such that $\psi(p) = q$, $\varphi(x, \psi(x)) = 0$, $\forall x \in W$ and

$$\{(x,y) \in V : \varphi(x,y) = 0\} = \{(x, \psi(x)) : x \in W\}\,.$$

□

A subset $M \subset \mathbb{R}^n$ is an r-dimensional submanifold ($r < n$) of $\mathbb{R}^n$ if for each $\bar{x} \in M$ there exist an open set U, with $\bar{x} \in U$, and smooth functions $h_{r+1}(x), \ldots, h_n(x)$ such that $\{\mathrm{d}h_{r+1}(x), \ldots, \mathrm{d}h_n(x)\}$ is a linearly independent set of row vectors for all $x \in U$ and $U \cap M = \{x \in U : h_i(x) = 0, r+1 \leq i \leq n\}$. Note that the smooth functions $h_{r+1}, \ldots, h_n$ in the above definition are not unique.

Let M be an r-dimensional submanifold of $\mathbb{R}^n$. The tangent space of M at $x \in M$ is the r-dimensional subspace of $\mathbb{R}^n$

$$TM_x = \{v \in \mathbb{R}^n : \langle \mathrm{d}h_i(x), v \rangle = 0, \ r+1 \leq i \leq n\}.$$

A vector field f is said to be tangent to the submanifold M of $\mathbb{R}^n$ at $x \in M$ if $f(x) \in TM_x$.

An r-dimensional distribution $\mathscr{D}$ on W, an open connected subset of $\mathbb{R}^n$, is a map which assigns to each $p \in W$ an r-dimensional subspace of $\mathbb{R}^n$ such that for each $p_0 \in W$ there exist a neighborhood U of p_0 and r smooth vector fields $f_1, \ldots, f_r$ with the properties:

(i) $f_1(p), \ldots, f_r(p)$ are linearly independent for every $p \in U$;
(ii) $\mathscr{D}(p) = \mathrm{span}\{f_1(p), \ldots, f_r(p)\}, \forall p \in U$.

Given a distribution $\mathscr{D}$ and a vector field f in $U \subset \mathbb{R}^n$, we say that f belongs to $\mathscr{D}$ if

$$f(p) \in \mathscr{D}(p), \quad \forall p \in U.$$

From the above definitions it follows that, given any vector field $f \in \mathscr{D}$, there exist r smooth functions $\alpha_1(p), \ldots, \alpha_r(p)$ in U such that

$$f(p) = \sum_{i=1}^{r} \alpha_i(p) f_i(p), \qquad \forall p \in U.$$

A submanifold M of $\mathbb{R}^n$ is an integral manifold of the distribution $\mathscr{D}$ on $\mathbb{R}^n$ if for every $x \in M$

$$TM_x = \mathscr{D}(x).$$

A distribution $\mathscr{D}$ on $\mathbb{R}^n$ is integrable if through any point p of $\mathbb{R}^n$ there passes an integral manifold of $\mathscr{D}$.

A distribution $\mathscr{D}$ is called involutive if, given any two vector fields f and g belonging to $\mathscr{D}$, their Lie bracket $[f, g]$ also belongs to $\mathscr{D}$.

Given two distributions $\mathscr{D}_1$ and $\mathscr{D}_2$ defined in $U \subset \mathbb{R}^n$, we say that $\mathscr{D}_1$ is contained in $\mathscr{D}_2$, *i.e.* $\mathscr{D}_1 \subset \mathscr{D}_2$, if every vector field f belonging to $\mathscr{D}_1$ also belongs to $\mathscr{D}_2$.

Given a distribution $\mathscr{D}$, its involutive closure, denoted by $\bar{\mathscr{D}}$ or inv.cl. $\mathscr{D}$, is defined as the smallest involutive distribution containing $\mathscr{D}$.

Theorem B.5. *(Frobenius) Let $\mathscr{D}$ be an r-dimensional distribution on W, an open connected subset of $\mathbb{R}^n$. Around any point $p \in W$ there exists a coordinate neighborhood $(U, x_1, \ldots, x_n)$, with U a neighborhood of p, such that for any $q \in U$*

$$\mathscr{D}(q) = \text{Span}\left\{\frac{\partial}{\partial x_1}, \ldots, \frac{\partial}{\partial x_r}\right\}$$

if, and only if, $\mathscr{D}$ is involutive. □

Consider a multi-input nonlinear system

$$\dot{x} = f(x) + \sum_{i=1}^{m} g_i(x)u_i \triangleq f(x) + G(x)u\,, \qquad x \in \mathbb{R}^n \tag{B.4}$$

in which f, g_1, ..., g_m are smooth vector fields in $\mathbb{R}^n$, $f(0) = 0$, the $n \times m$ matrix $G(x)$ has maximum rank at the origin, *i.e.* Rank $G(0) = m$, and $u = [u_1, \ldots, u_m]^{\mathrm{T}}$ is the m-dimensional input vector. Linear systems are special cases of (B.4) with $f(x)$ a linear vector field and constant vector fields g_i. Recall the definition of controllability indices for linear controllable systems.

Definition B.1. Consider the multi-input linear systems

$$\dot{x} = Fx + \sum_{i=1}^{m} g_i u_i = Fx + Gu \tag{B.5}$$

with Rank $G = m$. A set of controllability indices $\{k_1, \ldots, k_m\}$ is uniquely associated with any controllable system (B.5), *i.e.* such that

$$\text{Rank}[G, FG, \ldots, F^{n-1}G] = n\,,$$

as follows:

$$k_i = \text{Card}\{m_j \geq i : j \geq 0\}\,, \qquad 1 \leq i \leq m$$

with

$$\begin{aligned}
m_0 &= \text{Rank } G \\
m_1 &= \text{Rank}[G, FG] - \text{Rank } G \\
&\vdots \\
m_k &= \text{Rank}[G, \ldots, F^k G] - \text{Rank}[G, \ldots, F^{k-1}G] \\
&\vdots \\
m_{n-1} &= \text{Rank}[G, \ldots, F^{n-1}G] - \text{Rank}[G, \ldots, F^{n-2}G]\,.
\end{aligned}$$

By definition $k_1 \geq k_2 \geq \cdots \geq k_m$ and, since the system is controllable, $\sum_{i=1}^{m} k_i = n$. □

The controllability indices are invariant under linear state space change of coordinates

$$z = Tx, \qquad z \in \mathbb{R}^n \tag{B.6}$$

with T nonsingular, and under nonsingular state feedback transformations

$$u = Kx + \beta v\,, \qquad v \in \mathbb{R}^m \tag{B.7}$$

where β is a nonsingular $m \times m$ matrix and K is an $m \times n$ matrix.

The following is a well known result from linear systems theory.

Theorem B.6. *For any controllable system (B.5) with controllability indices* $\{k_1, \ldots, k_m\}$ *there exists a linear state space change of coordinates (B.6) and a nonsingular state feedback (B.7) transforming* (F,G) *into*

$$(T(F+GK)T^{-1}, TG\beta) = (A_c, B_c)$$

with (A_c, B_c) *in Brunovsky controller form:*

$$\begin{aligned} A_c &= \text{block diag}[A_1, \ldots, A_m] \\ B_c &= \text{block diag}[B_1, \ldots, B_m] \end{aligned} \tag{B.8}$$

in which, for $1 \le i \le m$,

$$A_i = \begin{bmatrix} 0\;1\;0 \ldots 0 \\ \vdots\;\vdots\;\vdots \ldots \vdots \\ 0\;0\;0 \ldots 1 \\ 0\;0\;0 \ldots 0 \end{bmatrix}_{k_i \times k_i}, \qquad B_i = \begin{bmatrix} 0 \\ \vdots \\ 0 \\ 1 \end{bmatrix}_{1 \times k_i}.$$

□

Theorem B.6 is generalized to nonlinear systems as follows.

Theorem B.7. *(Multi-Input Feedback Linearization) The nonlinear system (B.4) is locally feedback linearizable, i.e. locally transformable in* V_0*, a neighborhood of the origin contained in* U_0*, into a linear controllable system in Brunovsky controller form by means of:*

(i) *a nonsingular state feedback*

$$u = K(x) + \beta(x)v\,, \qquad K(0) = 0 \tag{B.9}$$

where $K(x)$ *is a smooth function from* V_0 *into* $\mathbb{R}^n$, $\beta(x)$ *is an* $m \times m$ *matrix with smooth entries, nonsingular in* V_0,

(ii) *a local diffeomorphism in* V_0

$$z = T(x)\,, \qquad T(0) = 0 \tag{B.10}$$

if, and only if, in U_0:

(i) $\mathscr{G}_\ell = \text{Span}\{ad_f^j g_i : 1 \le i \le m, 0 \le j \le \ell\}$, $0 \le \ell \le n-2$, *is involutive and of constant rank,*

(ii) Rank $\mathcal{G}_{n-1} = n$. □

The controllability indices $\{k_1, \ldots, k_m\}$ associated with those systems (B.4) which satisfy the conditions of Theorem B.7 are defined as

$$k_i = \text{Card}\{m_j \geq i : j \geq 0\}, \qquad 1 \leq i \leq m$$

with

$$\begin{aligned} m_0 &= \text{Rank } \mathcal{G}_0 \\ m_1 &= \text{Rank } \mathcal{G}_1 - \text{Rank } \mathcal{G}_0 \\ &\vdots \\ m_{n-1} &= \text{Rank } \mathcal{G}_{n-1} - \text{Rank } \mathcal{G}_{n-2} \end{aligned}$$

and are invariant under feedback transformations. Theorem B.7 may also be stated as follows.

Theorem B.8. *The nonlinear system (B.4) is locally transformable in a neighborhood of the origin, by a nonsingular state feedback transformation (which consists of a nonsingular state feedback (B.9) and a diffeomorphism (B.10)) into a linear controllable system in Brunovsky controller form with controllability indices $k_1 \geq \cdots \geq k_m$ if, and only if, in U_0, a neighborhood of the origin:*

(i) $\mathcal{G}_{k_i-2}$, $1 \leq i \leq m$, are involutive and of constant rank;
(ii) Rank $\mathcal{G}_{k_1-1} = n$. □

Conditions (i) and (ii) guarantee the existence of m smooth functions $\phi_1(x), \ldots, \phi_m(x)$ such that

$$\langle \mathrm{d}\phi_i, \mathcal{G}_{k_j-2} \rangle = 0, \qquad j \geq i$$

and the $m \times m$ matrix

$$\begin{bmatrix} \langle \mathrm{d}\phi_1, ad_f^{k_1-1} g_1 \rangle & \ldots & \langle \mathrm{d}\phi_1, ad_f^{k_1-1} g_m \rangle \\ \vdots & \ddots & \vdots \\ \langle \mathrm{d}\phi_m, ad_f^{k_m-1} g_1 \rangle & \ldots & \langle \mathrm{d}\phi_m, ad_f^{k_m-1} g_m \rangle \end{bmatrix}$$

is nonsingular in V_0. The linearizing nonsingular state feedback transformation and diffeomorphism are, respectively,

$$v = \begin{bmatrix} L_f^{k_1} \phi_1 \\ \vdots \\ L_f^{k_m} \phi_m \end{bmatrix} + \begin{bmatrix} L_{g_1} L_f^{k_1-1} \phi_1 & \ldots & L_{g_m} L_f^{k_1-1} \phi_1 \\ \vdots & \ldots & \vdots \\ L_{g_1} L_f^{k_m-1} \phi_m & \ldots & L_{g_m} L_f^{k_m-1} \phi_m \end{bmatrix} u$$

and

$$z = \begin{bmatrix} \phi_1 \\ \vdots \\ L_f^{k_1-1}\phi_1 \\ \vdots \\ \phi_m \\ \vdots \\ L_f^{k_m-1}\phi_m \end{bmatrix}.$$

The controllability indices may be computed using the linear approximation about the origin.

Note that even though system (B.4) is not feedback linearizable, according to Theorem B.7, its dynamic extension

$$\begin{aligned} \dot{x} &= f(x) + \sum_{i=1}^{m} g_i(x)u_i\,, \qquad x \in \mathbb{R}^n \\ \frac{\mathrm{d}^{\lambda_i}u_i}{\mathrm{d}t^{\lambda_i}} &= v_i\,, \qquad 1 \le i \le m \end{aligned} \tag{B.11}$$

may be feedback linearizable for a suitable choice of the indices $(\lambda_1, \ldots, \lambda_m)$ from the new inputs v_i, $1 \le i \le m$.

Definition B.2. System (B.4) is said to be locally dynamically state feedback linearizable if there exist indices $(\lambda_1, \ldots, \lambda_m)$ such that the extended system (B.11) is locally state feedback linearizable. □

Definition B.3. System (B.4) is said to be locally partially state feedback linearizable with indices $\{k_1, \ldots, k_m\}$ if there exist a nonsingular state feedback transformation (B.9) and a local diffeomorphism (B.10) transforming (B.4) into

$$\begin{aligned} \dot{\xi} &= \varphi(\xi, z), \qquad \xi \in \mathbb{R}^{n-r} \\ \dot{z} &= A_c z + B_c v, \qquad z \in \mathbb{R}^r \end{aligned}$$

with (A_c, B_c) in Brunovsky controller form (B.8) with indices $\{k_1, \ldots, k_m\}$ such that $\sum_{i=1}^m k_i = r < n$. □

Define the distributions

$$\begin{aligned} \mathcal{G}_0 &= \mathrm{Span}\{g_1, \ldots, g_m\} \\ \mathcal{G}_j &= \mathrm{Span}\{\mathcal{G}_{j-1}, [f+g, \mathcal{G}_{j-1}] : \forall g \in \mathcal{G}_0\} \\ \mathcal{Q}_0 &= \mathcal{G}_0 \\ \mathcal{Q}_j &= \mathrm{Span}\{ad_f^j \mathcal{G}_0, \bar{\mathcal{G}}_{j-1}\}\,. \end{aligned}$$

Under the assumption that all distributions $\mathcal{G}_j$, $\mathcal{Q}_j$, and the involutive closure $\bar{\mathcal{G}}_j$ of $\mathcal{G}_j$, $j \ge 0$, have constant rank in U_0, we can define

$$\begin{aligned} m_0 &= \text{Rank}\, \mathscr{G}_0 \\ m_1 &= \text{Rank}\, \mathscr{Q}_1 - \text{Rank}\, \bar{\mathscr{G}}_0 \\ &\vdots \\ m_i &= \text{Rank}\, \mathscr{Q}_i - \text{Rank}\, \bar{\mathscr{G}}_{i-1} \end{aligned}$$

and

$$k_i^* = \text{Card}\{m_j \geq i : j \geq 0\}, \qquad 1 \leq i \leq m .$$

By definition

$$k_1^* \geq k_2^* \geq \cdots \geq k_m^* .$$

It can be shown that the integers $\{k_1^*, \ldots, k_m^*\}$ are invariant under nonsingular state feedback and local diffeomorphisms.

Theorem B.9. *(Multi-Input Partial Feedback Linearization) The system (B.4) is locally partially state feedback linearizable with indices* $\{k_1^*, \ldots, k_m^*\}$. □

Consider now multivariable nonlinear systems with as many inputs (m) as outputs

$$\begin{aligned} \dot{x} &= f(x) + \sum_{i=1}^{m} g_i(x) u_i \triangleq f(x) + G(x) u , \qquad x \in \mathbb{R}^n \\ y_j &= h_j(x) , \qquad 1 \leq j \leq m \end{aligned} \tag{B.12}$$

with f, $g_1, \ldots, g_m$ smooth vector fields, $h_1, \ldots, h_m$ smooth real valued functions, Rank $G(0) = m$ and Rank$\{dh_1(0), \ldots, dh_m(0)\} = m$.

Definition B.4. A set of m integers $\{\rho_1, \ldots, \rho_m\}$ called control characteristic indices are associated with system (B.12) in U_0, a neighborhood of the origin, as follows ($1 \leq i \leq m$):

$$\begin{aligned} L_{g_j} L_f^k h_i(x) &= 0 , \qquad 1 \leq j \leq m ,\ 0 \leq k \leq \rho_i - 2 ,\ \forall x \in U_0 \\ L_{g_j} L_f^{\rho_i - 1} h_i(x) &\neq 0 , \qquad \text{for some } j ,\ 1 \leq j \leq m ,\forall x \in U_0 . \end{aligned}$$

If

$$L_{g_j} L_f^k h_i(x) = 0 , \qquad \forall x \in U_0 ,\ \forall k \geq 0$$

we say that $\rho_i = \infty$. □

The above definition is given about the origin: it may be given around any point $\bar{x} \in \mathbb{R}^n$ such that Rank $G(\bar{x}) = m$ and Rank$\{dh_1(\bar{x}), \ldots, dh_m(\bar{x})\} = m$. If $\rho_i < \infty$, $1 \leq i \leq m$, then each ρ_i is equal to the least order of the time derivative of the output y_i which is directly affected at least by some input u_j, $1 \leq j \leq m$.

Definition B.5. If $\rho_i < \infty$, $1 \le i \le m$, an $m \times m$ matrix, called the decoupling matrix, is defined as

$$D(x) = \begin{bmatrix} L_{g_1} L_f^{\rho_1 - 1} h_1 & \dots & L_{g_m} L_f^{\rho_1 - 1} h_1 \\ \vdots & \vdots & \vdots \\ L_{g_1} L_f^{\rho_m - 1} h_m & \dots & L_{g_m} L_f^{\rho_m - 1} h_m \end{bmatrix} .$$

□

Note that the decoupling matrix may be nonsingular as the following example ($\rho_1 = 2$, $\rho_2 = 1$) shows:

$$\begin{aligned} \dot{x}_1 &= x_2 + x_3^2 \\ \dot{x}_2 &= x_1^2 + (1 + x_1^2) u_1 \\ \dot{x}_3 &= x_2^2 + u_2 \\ y_1 &= x_1 \\ y_2 &= x_3 . \end{aligned}$$

The decoupling matrix is

$$D(x) = \begin{bmatrix} 1 + x_1^2 & 2x_3 \\ 0 & 1 \end{bmatrix}$$

whose determinant $(1 + x_1^2)$ is always nonzero. The decoupling matrix may be singular as the following example ($\rho_1 = 2$, $\rho_2 = 1$) shows:

$$\begin{aligned} \dot{x}_1 &= x_2 + x_3 \\ \dot{x}_2 &= x_1^2 + (1 + x_1^2) u_1 + \frac{1}{2} u_2 \\ \dot{x}_3 &= x_2^2 + (1 + x_1^2) u_1 + \frac{1}{2} u_2 \\ y_1 &= x_1 \\ y_2 &= x_3 . \end{aligned}$$

The decoupling matrix is

$$D(x) = \begin{bmatrix} 2(1 + x_1^2) & 1 \\ 1 + x_1^2 & \frac{1}{2} \end{bmatrix}$$

whose determinant is equal to zero.

Lemma B.1. *If the decoupling matrix is nonsingular in U_0 then*

(i) Rank$\{dh_j, \dots, d(L_f^{\rho_j - 1} h_j),\ 1 \le j \le m\} = \sum_{j=1}^{m} \rho_j = \rho \le n$,
(ii) *there exist $n - \rho$ functions $\xi_i(x)$ such that, defining*

$$\begin{aligned}\xi &= [\xi_1,\ldots,\xi_{n-\rho}]^{\mathrm{T}}\\ z_j &= [h_j,\ldots,L_f^{\rho_j-1}h_j]^{\mathrm{T}}, \qquad 1\le j\le m\\ z &= [z_1,\ldots,z_m]^{\mathrm{T}},\end{aligned}$$

the coordinates (ξ,z) *form a local diffeomorphism about the origin. In local coordinates* (ξ,z) *system (B.12) is expressed in multivariable tracking form:*

$$\begin{aligned}y_j &= z_{j1}\\ \dot z_{ji} &= z_{j,i+1}, \qquad 1\le i\le \rho_j-1\\ \dot z_{j,\rho_j} &= L_f^{\rho_j}h_j(\xi,z)+\sum_{i=1}^{m}L_{g_i}L_f^{\rho_j-1}h_j(\xi,z)u_i, \qquad 1\le j\le m\\ \dot\xi &= \phi_0(\xi,z)+\Phi^{\mathrm{T}}(\xi,z)u\,.\end{aligned}$$

□

If the distribution $\mathscr{G}_0=\mathrm{Span}\{g_1,\ldots,g_m\}$ is involutive and of constant rank, then we can determine $n-\rho$ functions $\xi_i(x)$ such that $\langle d\xi_i,\mathscr{G}_0\rangle=0$ so that the statement in (ii) applies with the dynamics

$$\dot\xi=\phi_0(\xi,z)$$

unaffected by u, as happens in the single input case in which $\mathscr{G}_0=\mathrm{Span}\{g(x)\}$ is always an involutive distribution.

Theorem B.10. *(Multivariable Input–Output Feedback Linearization) If the decoupling matrix is nonsingular in U then the system is locally decouplable and input–output linearizable by state feedback: the state feedback*

$$\begin{bmatrix}L_f^{\rho_1}h_1\\ \vdots\\ L_f^{\rho_m}h_m\end{bmatrix}+D(x)u=v \tag{B.13}$$

makes the closed-loop system (B.12), (B.13) decoupled and input–output linear

$$\begin{aligned}y_j &= z_{j1}\\ \dot z_{ji} &= z_{j,i+1}, \qquad 1\le i\le\rho_j-1\\ \dot z_{j\rho_j} &= v_j, \qquad 1\le j\le m\\ \dot\xi &= \phi_0(\xi,z)+\Phi^{\mathrm{T}}(\xi,z)v\,, \quad \xi\in\mathbb{R}^{n-\rho}\,.\end{aligned}$$

□

Let $v_{\mathrm{r}}=[y_{\mathrm{r}1}^{(\rho_1)},\ldots,y_{\mathrm{r}m}^{(\rho_m)}]^{\mathrm{T}}$ be the open-loop reference input which allows y_j in (B.12), (B.13) to track $y_{\mathrm{r}j}$, $1\le j\le m$, in U. The dynamics

$$\dot{\xi}_{\rm r} = \phi_0(\xi_{\rm r}, Y_{\rm r1}, \ldots, Y_{{\rm r}m}) + \Phi^{\rm T}(\xi_{\rm r}, Y_{\rm r1}, \ldots, Y_{{\rm r}m}) v_{\rm r}$$

$$v_{\rm r} = \begin{bmatrix} y_{\rm r1}^{\rho_1} \\ \vdots \\ y_{{\rm r}m}^{\rho_m} \end{bmatrix}$$

with $Y_{{\rm r}j} = [y_{{\rm r}j}, \ldots, y_{{\rm r}j}^{(\rho_j - 1)}]^{\rm T}$, $1 \leq j \leq m$, are called the tracking dynamics. The inputs to the tracking dynamics are $Y_{\rm r1}, \ldots, Y_{{\rm r}m}, v_{\rm r}$.

Definition B.6. Consider the system (B.12) with nonsingular decoupling matrix in $U \in \mathbb{R}^n$; if $Y_{{\rm r}j} \in U_0$, $1 \leq j \leq m$, and $\xi(0) \in U_0$, then

$$\dot{\xi}_{\rm r} = \phi_0(\xi_{\rm r}, Y_{\rm r1}, \ldots, Y_{{\rm r}m}) + \Phi^{\rm T}(\xi_{\rm r}, Y_{\rm r1}, \ldots, Y_{{\rm r}m}) v_{\rm r}$$

$$u_{\rm r} = D^{-1}(\xi_{\rm r}, Y_{\rm r1}, \ldots, Y_{{\rm r}m}) \begin{bmatrix} y_{\rm r1}^{\rho_1} - L_f^{\rho_1} h_1 \\ \vdots \\ y_{{\rm r}m}^{\rho_m} - L_f^{\rho_m} h_m \end{bmatrix}$$

is the inverse system. □

Corollary B.1. *System (B.12) is simultaneously locally state feedback input–output linearizable and locally state feedback linearizable if, and only if,*

$$\sum_{j=1}^{m} \rho_j = n\,.$$

In this case the controllability indices are equal to the control characteristic indices. □

Consider nonlinear multi-input, multi-output systems with s outputs $y = [y_1, \ldots, y_s]^{\rm T}$, m inputs $u = [u_1, \ldots, u_m]^{\rm T}$, and p constant parameters $\theta = [\theta_1, \ldots, \theta_p]^{\rm T}$,

$$\dot{x} = f(x, \theta) + q(x, \theta, u))\,, \quad x \in \mathbb{R}^n\,,\ u \in \mathbb{R}^m\,,\ \theta \in \mathbb{R}^p$$

$$y = H(x) = \begin{bmatrix} h_1(x) \\ \vdots \\ h_s(x) \end{bmatrix}\,, \quad y \in \mathbb{R}^s \tag{B.14}$$

where $h_1, \ldots, h_s$ are smooth functions, $\mathrm{d}h_1, \ldots, \mathrm{d}h_s$ are linearly independent in $\mathbb{R}^n$, f is a smooth vector field, and $q(x, \theta, u)$ is a smooth vector field with $q(x, \theta, 0) = 0$, $\forall x \in \mathbb{R}^n$, $\forall \theta \in \mathbb{R}^p$.

Definition B.7. Two states x_1, $x_2 \in \mathbb{R}^n$ are said to be indistinguishable for (B.14) (denoted by $x_1 I x_2$) if for every admissible input function u the output function of the system for initial state $x(0) = x_1$, and the output function of the system for initial state $x(0) = x_2$, are identical on their common domain of definition. The system is called observable if $x_1 I x_2$ implies $x_1 = x_2$. □

Definition B.8. The system (B.14) is called locally observable at x_0 if there exists a neighborhood W of x_0 such that for every neighborhood $V \subset W$ of x_0 the relation $x_0 I x_1$ implies that $x_1 = x_0$. If the system is locally observable at each x_0 in U_0 then it is called locally observable in U_0. □

Roughly speaking, a system is locally observable if every state x_0 can be distinguished from its neighbors by using system trajectories remaining close to x_0.

Theorem B.11. *System (B.14) is locally observable at x_0 if*

$$\operatorname{rank}\{dh_i(x), \ldots, d(L_f^j h_i) : 1 \le i \le s,\ j \ge 0\} = n \tag{B.15}$$

for every $x \in U_0 \subset \mathbb{R}^n$. □

Observability indices may be defined for locally observable systems satisfying (B.15).

Definition B.9. A set of observability indices $\{k_1, \ldots, k_s\}$ is uniquely associated at x to system (B.14) with $u = 0$ and $\theta = 0$, that is

$$\begin{aligned} \dot{x} &= f(x) \\ y_i &= h_i(x)\,, \qquad 1 \le i \le s \end{aligned} \tag{B.16}$$

satisfying (B.15) as follows:

$$k_i = \operatorname{card}\{s_j \ge i :\ j \ge 0\}\,, \qquad 1 \le i \le s$$

with

$$\begin{aligned} s_0 &= \operatorname{rank}\{dh_i(x) :\ 1 \le i \le s\} \\ &\vdots \\ s_k &= \operatorname{rank}\{dh_i(x), \ldots, d(L_f^k h_i(x)) :\ 1 \le i \le s\} \\ &\quad -\operatorname{rank}\{dh_i(x), \ldots, d(L_f^{k-1} h_i(x)) :\ 1 \le i \le s\} \\ &\vdots \\ s_{n-1} &= \operatorname{rank}\{dh_i(x), \ldots, d(L_f^{n-1} h_i(x)) :\ 1 \le i \le s\} \\ &\quad -\operatorname{rank}\{dh_i(x), \ldots, d(L_f^{n-2} h_i(x)) :\ 1 \le i \le s\}\,. \end{aligned}$$

□

In the case of linear systems Definition B.9 takes the following form.

Definition B.10. A set of observability indices $\{k_1, \ldots, k_s\}$ is uniquely associated with any observable system

$$\begin{aligned} \dot{x} &= Fx \\ y &= Hx \end{aligned} \tag{B.17}$$

such that

$$\operatorname{rank}\begin{bmatrix} H \\ HF \\ \vdots \\ HF^{n-1} \end{bmatrix} = n$$

as follows:

$$k_i = \operatorname{card}\{s_j \geq i : \ j \geq 0\}, \quad 1 \leq i \leq s$$

with

$$\begin{aligned}
s_0 &= \operatorname{rank}[H] \\
&\vdots \\
s_k &= \operatorname{rank}\begin{bmatrix} H \\ \vdots \\ HF^k \end{bmatrix} - \operatorname{rank}\begin{bmatrix} H \\ \vdots \\ HF^{k-1} \end{bmatrix} \\
&\vdots \\
s_{n-1} &= \operatorname{rank}\begin{bmatrix} H \\ \vdots \\ HF^{n-1} \end{bmatrix} - \operatorname{rank}\begin{bmatrix} H \\ \vdots \\ HF^{n-2} \end{bmatrix}.
\end{aligned}$$

By definition, $k_1 \geq k_2 \geq \ldots \geq k_s$ and, since the system is observable, $\sum_{i=1}^{s} k_i = n$. □

The concept of local observability for nonlinear systems coincides with the well known concept of observability for linear systems.

Lemma B.2. *If, and only if, the linear system (B.17) is locally observable at* $x = 0$ *then it is observable.* □

Consider the extended system (B.14)

$$\begin{aligned}
\dot{x} &= f(x) + q(x, \theta, u) \\
\dot{\theta} &= 0 \\
y &= H(x)
\end{aligned} \tag{B.18}$$

in which $[x^{\mathrm{T}}, \theta^{\mathrm{T}}]^{\mathrm{T}}$ is the extended state vector.

Definition B.11. The system (B.14) is said to be locally observable and identifiable at (x_0, θ_0) if the extended system (B.18) is locally observable at (x_0, θ_0). □

Theorem B.11 may be applied to system (B.18): in this case it provides sufficient conditions for local observability of (B.18) and, consequently, for local observability and identifiability of system (B.14).

Example B.1. The condition of Theorem B.11 is only sufficient: consider the scalar system

$$\begin{aligned} \dot{x} &= 0 \\ y &= x^3 . \end{aligned}$$

Since $\mathrm{d}h = 3x^2\mathrm{d}x$ and rank $\mathrm{d}h(0) = 0$ Theorem B.11 does not apply, while this system is clearly observable since $x = (y)^{1/3}$. □

Example B.2. Local observability does not imply observability: consider the system

$$\begin{aligned} \dot{x} &= u \\ y_1 &= \sin x \\ y_2 &= \cos x . \end{aligned}$$

Since $\mathrm{rank}\{\mathrm{d}h_1, \mathrm{d}h_2\} = \mathrm{rank}\{\cos x\,\mathrm{d}x, -\sin x\,\mathrm{d}x\}$ is one for every $x \in \mathbb{R}$, this system is locally observable according to Theorem B.11 but it is not observable since the points x_1 and x_2 with x_1 and x_2 multiple of 2π are indistinguishable. □

Example B.3. Observability for nonlinear systems does not exclude that for a specific input the system is not observable: consider the system with constant input u

$$\begin{aligned} \dot{x}_1 &= \theta x_2 + u x_2 \\ \dot{x}_2 &= 0 \\ y &= x_1 . \end{aligned}$$

This system is not observable if $u = -\theta$ but it is observable, according to Definition B.7, since for every constant $u \neq -\theta$ the system is linear and observable. □

Example B.4. For the class of constant inputs the system

$$\begin{aligned} \dot{x} &= \theta_1 u_1 + \theta_2 u_2 \\ \dot{\theta}_1 &= 0 \\ \dot{\theta}_2 &= 0 \\ y &= x \end{aligned}$$

is not observable, but it is observable for the specific persistently exciting inputs $u_1 = \sin t$, $u_2 = -\cos t$: hence, this system is not observable for the class of constant inputs but it is observable if specific time-varying inputs are applied. Note that the inputs $u_1 = \sin t$, $u_2 = -\cos t$ satisfy the persistency of excitation condition (A.15) for any positive T; an adaptive observer is given by

$$\begin{aligned} \dot{\hat{x}} &= -\hat{x} + y + \hat{\theta}_1 u_1 + \hat{\theta}_2 u_2 \\ \dot{\hat{\theta}}_1 &= u_1 (y - \hat{x}) \\ \dot{\hat{\theta}}_2 &= u_2 (y - \hat{x}) \end{aligned}$$

since the persistency of excitation Lemma A.3 applies to the error dynamics ($\tilde{x} = y - \hat{x}$, $\tilde{\theta}_1 = \theta_1 - \hat{\theta}_1$, $\tilde{\theta}_2 = \theta_2 - \hat{\theta}_2$)

$$
\begin{aligned}
\dot{\tilde{x}} &= -\tilde{x} + \tilde{\theta}_1 u_1 + \tilde{\theta}_2 u_2 \\
\dot{\tilde{\theta}}_1 &= u_1 \tilde{x} \\
\dot{\tilde{\theta}}_2 &= u_2 \tilde{x}
\end{aligned}
$$

when $u_1 = \sin t$, $u_2 = -\cos t$, so that both the observation error $\tilde{x}$ and the estimation errors $\tilde{\theta}_1$ and $\tilde{\theta}_2$ converge exponentially to zero. □

Bibliographical Notes

Due to the large number of papers on induction motor control (more than four thousand at the time of writing) we list in the references only those books and journal papers which were actually consulted and used during the preparation of this book: the reported list is therefore by no means exhaustive and complete.

Chapter 1 and Appendices

The material in Chapter 1 and in the Appendices is, to a large extent, classical and may be found in the books listed in the references. The following books are on the modeling of electrical machines: [73, 130, 132, 179, 224, 225, 246, 259]. In particular [130] dedicates one chapter to the reference-frame theory and discusses the history of rotating frame models, including Park's transformations. The following books are specifically dedicated to the nonlinear control of induction motors and electric motors: [46, 51, 121, 196, 215, 242, 245, 247]. In particular [46] deals with modeling and control of both AC and DC machines (one chapter is dedicated to induction motor control); [51, 121] deal with nonlinear control of several types of motors, including brushless DC and stepper motors (two chapters are dedicated to induction motor control in [51, 121]); [196, 245] are entirely dedicated to the control of induction motors and synchronous motors using field orientation techniques while [242] is specifically dedicated to the field-oriented control of induction motors; [215, 247] address the sensorless control problem. The following books are on the control of electric drives including the power electronics: [29, 30, 61, 133, 147]. In particular [147] is the classical, most quoted reference on the control of electric drives while [29, 30] focus on the power electronics for electric drives; [61] is specifically dedicated to the use of digital signal processors for electric motor control; [133] is a comprehensive textbook on electric motor drives. The following books are on nonlinear control theory: [4, 58, 89, 106, 116, 135, 169, 191, 204]. In particular [89, 116] contain an extensive treatment of Lyapunov stability theory; [106, 191] are classical references for nonlinear systems analysis and control;

[4, 58, 169, 204] also describe specific applications; [135, 169] are focused on those nonlinear control algorithms which can be made adaptive. The following books are on adaptive control: [135, 169, 189, 222]. In particular [135, 169] are focused on the adaptive control of nonlinear systems while [189, 222] include identification techniques, robustness issues and are dedicated to linear systems. Magnetic saturations and iron losses are considered in [120, 195, 243]: the choice of the optimal flux modulus is discussed in [120, 195]. The determination of the rotor flux modulus which minimizes the power losses may be found in [248]. A comparison of standards for determining induction motor efficiency may be found in [218] while methods for motor flux and torque sensing are presented in [155, 214]. The torque control problem is specifically addressed in [7, 8, 26, 31, 33, 34, 53, 63, 64, 65, 67, 78, 96, 104, 126, 127, 180, 188, 192, 198, 201, 202, 203, 208, 232, 253, 260] (see [31] for a survey). Additional material on modeling, optimization, simulation, practical implementation, and nonlinear control theory applied to induction motors and general electric machines may be found in: [24, 27, 44, 84, 131, 146, 154, 167, 229, 237].

Chapter 2

The stability analysis of the feedforward control in Section 2.1 is taken from [170]. The direct field-oriented control was originally introduced in [20, 21]: the presentation in Section 2.2 follows the interpretation given in [168]. Sections 2.4 and 2.5 are adapted from [168] (see also the references therein). The indirect field-oriented control was introduced in [187] (see also [197] for the extension to general AC machines): the stability proof follows the original proofs given in [207] for constant references and in [81] for time-varying references (see also [55] for robustness analysis, [40] for a discussion on the tuning rules for the PI gains, and [206] for a discrete time implementation). The dynamic feedback linearizing control given in Section 2.6 is based on the flatness theory introduced in [74] (see also [24, 43, 45, 52]). The material in Section 2.7 is a simplification of the control algorithm given in [165]. The adaptation with respect to stator resistance is considered in [251] while the adaptation with respect to rotor and stator resistances, rotor inductance, and moment of inertia is considered in [15]. The robust tracking control problem in the presence of uncertainties in all model parameters is addressed in [102]. The experimental results presented in Section 2.9 concerning the influence of uncertain rotor resistance on the performance of the indirect field-oriented control are taken from [165]: in [9, 10, 83] this specific robustness issue is studied in detail via a bifurcation analysis. The problem 2.15 is adapted from [53]. Sliding mode controls are proposed in [140, 219]. Methods for achieving high motor efficiency may be found in [77, 123, 128, 139, 157, 158, 177, 185, 209, 235, 257]. Field-oriented controls are implemented in [75, 113, 119, 148] while the optimal induction motor control problem is addressed in [16, 62, 72, 186, 220, 221]. The effects of motor parameter variations in field orientation schemes are analyzed in [134, 193] while a stability analysis for an induction motor in closed-loop with an input–output feedback linearizing control in the presence of uncertain parameters may be found in [36]. Different methods for achieving observer-based field orientation are discussed in [111]. Speed regulation via a PI controller is achieved in [143]. Nonlinear predictive control techniques are used in [159].

Chapter 3

The first nonlinear observers for induction motors were independently obtained in [14, 59] (see also [60] for a reduced order nonlinear observer). The influence of parameter uncertainty on the performance of flux observers are investigated in [110]. Kalman-like and adaptive observers are presented in [238] and in [35], respectively, to estimate online and simultaneously the unmeasured states and the parameters while least square techniques are used in [49, 184, 231]. Gradient techniques are used in [68] for parameter estimation in induction motors at standstill. Sliding mode observers which are adaptive with respect to motor uncertain parameters are presented in [6, 93]. Discrete time observers are proposed in [5, 178, 258, 264]. Rotor resistance and mutual inductance are estimated in [233]. Rotor resistance identification at steady-state conditions is discussed in [39, 100] using the gradient technique; the rotor time constant is estimated in [239] by a simple online identification scheme; in [176] a current perturbation signal is injected for the estimation of rotor and stator resistances. Offline parameter estimations are treated in [13, 129, 241, 256]. Flux and speed observers for vehicular applications are designed in [217]. A review on induction motor parameter estimation techniques is given in [240]. The material in Section 3.1 on flux observers with arbitrary rate of convergence is adapted from [17, 250]; the adaptive flux observer with rotor resistance estimation is adapted from [161] while the simultaneous online identification of rotor and stator resistances is taken from [166]. The experimental results reported in Section 3.3 are taken from [161]. The problem 3.2 is taken from [175] in which an observer is also designed using a reparameterization of time. The problem 3.10 is taken from [238].

Chapter 4

The global control which is adaptive with respect to load torque and rotor resistance presented in Section 4.4 is taken from [165]. The experimental results reported in Section 4.4 are also taken from [165]. Related work is presented in [162, 163, 164, 212]. Problems 4.1 and 4.3 are taken from [163] while Problems 4.13 and 4.14 are taken from [212] and [213], respectively. In [211] the controller proposed in [212] is experimentally compared with an indirect field-oriented control while related work may be found in [213]. Speed tracking controls which are not based on observers are given in [11, 48, 71, 114, 248] while observer-based control are given in [57, 107, 122, 145, 216, 262]. A supervisory adaptive control is proposed in [41] while nonlinear predictive control techniques are used in [94]. Speed regulation with uncertain electrical parameters is achieved in [108] while an adaptive speed control is experimentally tested in [76]. Passivity based controls are proposed in [8, 37, 47, 63, 64, 65, 69, 79, 80, 82, 114, 124, 190, 194, 202, 203, 205]. Sliding mode controls are designed in [141, 230] while model reference adaptive controls may be found in [38, 150, 151, 153]. The problem of position control on the basis of rotor position measurements is addressed in [25, 103, 181] and, when some parameters are uncertain, in [3, 101, 249]. Magnetic saturations are considered in [12]. Discrete time control algorithms are proposed in [18, 156, 206, 228, 236, 252].

Chapter 5

The stability analysis in Section 5.1 is taken from [170]. The control algorithm in Section 5.2 was originally proposed in [170] while its adaptive version presented in Section 5.3 is taken from [171]. Section 5.4 is entirely adapted from [174] which is based on the nonadaptive version given in [172]. Section 5.5 presents the original algorithm given in [173]. Problems 5.1 and 5.2 are taken from [183] where experimental results are also reported. A review on sensorless control of induction motors may be found in [97]. The problem of motor parameter estimation from stator current measurements is dealt with in [32, 50, 54, 86, 87, 115, 226, 254, 255, 266]: in particular in [86] stator resistance, in [50] self and mutual inductances, in [32, 255] the rotor time constant, in [87, 115] both stator and rotor resistances, in [266] stator parameters are estimated; several parameters are estimated in [54, 254]. The problem of designing rotor speed and flux observers is addressed in [91, 92, 95, 118, 138, 149, 210, 223, 227, 263, 267]. The observability properties are analyzed in [105]. Simultaneous parameter and rotor speed estimation are carried out in [1, 42, 136, 137, 244, 265]. The speed-sensorless induction motor control is specifically addressed in [56, 70, 85, 98, 99, 125, 144, 152, 182, 183, 231, 234]. A sensorless solution for low-cost induction motor applications may be found in [28]. The performance limitations in sensorless controls for flux and speed estimators which are based on model reference adaptive system techniques are analyzed in [22, 23]. Discrete time controls are presented in [19, 160] while a sliding mode solution may be found in [261]. Experimental results for sensorless induction motor controls are reported in [88, 109, 112, 199] while a comparative analysis of three flux and speed observers is performed in [200]. Instability phenomena are reported in [90] when an estimated rotor speed is used for indirect field-oriented control. Additional work on parameter and state estimation for speed-sensorless induction motors may be found in [2, 111, 117, 142]. Automotive applications may be found in [56, 66, 217].

References

1. K. Akatsu and A. Kawamura. Sensorless very low-speed and zero-speed estimations with online rotor resistance estimation of induction motor without signal injection. *IEEE Transactions on Industry Applications*, 36(3):764–771, 2000.
2. B. Aloliwi, H.K. Khalil, and E.G. Strangas. Robust speed control of induction motors: application to a benchmark example. *International Journal of Adaptive Control and Signal Processing*, 14(2–3):157–170, 2000.
3. P. Aquino, M. Feemster, D.M. Dawson, and A. Behal. Adaptive partial state feedback control of the induction motor: elimination of rotor flux and rotor velocity measurements. *International Journal of Adaptive Control and Signal Processing*, 14(2–3):83–108, 2000.
4. A. Astolfi, D. Karagiannis, and R. Ortega. *Nonlinear and Adaptive Control with Applications*. Springer-Verlag, London, 2008.
5. D.J. Atkinson, P.P. Acarnley, and J.W. Finch. Observers for induction motor state and parameter estimation. *IEEE Transactions on Industry Applications*, 27(6):1119–1127, 1991.
6. G. Bartolini, A. Pisano, and P. Pisu. Simplified exponentially convergent rotor resistance estimation for induction motors. *IEEE Transactions on Automatic Control*, 48(2):325–330, 2003.
7. E. Bassi, F.P. Benzi, S. Bolognani, and G.S. Buja. A field orientation scheme for current-fed induction motor drives based on the torque angle closed-loop control. *IEEE Transactions on Industry Applications*, 28(5):1038–1044, 1992.
8. C. Batlle, A. Doria-Cerezo, G. Espinosa-Pérez, and R. Ortega. Simultaneous interconnection and damping assignment passivity-based control: the induction machine case study. *International Journal of Control*, 82(2):241–255, 2009.
9. A.S. Bazanella and R. Reginatto. Robustness margins for indirect field-oriented control of induction motors. *IEEE Transactions on Automatic Control*, 45(6):1226–1231, 2000.
10. A. S. Bazanella and R. Reginatto. Robust tuning of the speed loop in indirect field oriented control of induction motors. *Automatica*, 37(11):1811–1818, 2001.
11. A. Behal, M. Feemster, and D.M. Dawson. An improved indirect field-oriented controller for the induction motor. *IEEE Transactions on Control Systems Technology*, 11(2):248–252, 2003.
12. A. Behal, M. Feemster, D.M. Dawson, and A. Mangal. Partial state feedback control of induction motors with magnetic saturation: elimination of flux measurements. *Automatica*, 38(2):191–203, 2002.
13. A. Bellini, A. De Carli, and M. La Cava. Parameter identification for induction motor simulation. *Automatica*, 12(4):383–386, 1976.
14. A. Bellini and G. Figalli. A bilinear observer of the state of the induction machine. *Ricerche di Automatica*, 9(1):70–85, 1978.
15. A. Bellini and G. Figalli. Adaptive control with parameter estimation for induction motor drives. *Control Engineering Practice*, 3(2):181–188, 1995.
16. A. Bellini, G. Figalli, and G. Ulivi. A microcomputer-based optimal control system to reduce the effects of the parametric variations and speed measurements errors in induction motor drives. *IEEE Transactions on Industry Applications*, IA-22(1):42–50, 1986.

17. A. Bellini, G. Figalli, and G. Ulivi. Analysis and design of a microcomputer-based observer for an induction machine. *Automatica*, 24(4):549–555, 1988.
18. L. Ben-Brahim and A. Kawamura. Digital current regulation of field-oriented controlled induction motor based on predictive flux observer. *IEEE Transactions on Industry Applications*, 27(5):956–961, 1991.
19. L. Ben-Brahim and A. Kawamura. A fully digitized field-oriented controlled induction motor drive using only current sensors. *IEEE Transactions on Industrial Electronics*, 39(3):241–249, 1992.
20. F. Blaschke. Das prinzip der feldorientierung, die grundlage für die TRANSVEKTOR-regelung von asynchronmaschinen. *Siemens Zeitschrift*, 45:757–760, 1971.
21. F. Blaschke. The principle of field orientation as applied to the new TRANSVEKTOR closed loop control system for rotating field machines. *Siemens Review*, 39:217–220, 1972.
22. R. Blasco-Giménez, G.M. Asher, M. Sumner, and K.J. Bradley. Dynamic performance limitations for MRAS based sensorless induction motor drives. Part 1: stability analysis for the closed loop drive. *IEE Proceedings on Electric Power Applications*, 143(2):113–122, 1996.
23. R. Blasco-Giménez, G.M. Asher, M. Sumner, and K.J. Bradley. Dynamic performance limitations for MRAS based sensorless induction motor drives. Part 2: online parameter tuning and dynamic performance studies. *IEE Proceedings on Electric Power Applications*, 143(2):123–134, 1996.
24. M. Bodson and J. Chiasson. Differential-geometric methods for control of electric motors. *International Journal of Robust and Nonlinear Control*, 8(11):923–954, 1998.
25. M. Bodson, J. Chiasson, and R.T. Novotnak. High-performance induction motor control via input-output linearization. *IEEE Control Systems Magazine*, 14(4):25–33, 1994.
26. M. Bodson, J.N. Chiasson, and R.T. Novotnak. A systematic approach to selecting flux references for torque maximization in induction motors. *IEEE Transactions on Control Systems Technology*, 3(4):388–397, 1995.
27. M. Bodson, J. Chiasson, and R.T. Novotnak. Nonlinear speed observer for high-performance induction motor control. *IEEE Transactions on Industrial Electronics*, 42(4):337–343, 1995.
28. R. Bojoi, P. Guglielmi, and G.M. Pellegrino. Sensorless direct field-oriented control of three-phase induction motor drives for low-cost applications. *IEEE Transactions on Industry Applications*, 44(2):475–481, 2008.
29. B.K. Bose. *Power Electronics and AC Drives*. Prentice-Hall, New York, 1986.
30. B.K. Bose. *Power Electronics and Variable Frequency Drives*. IEEE Press, New York, 1997.
31. G.S. Buja and M.P. Kazmierkowski. Direct torque control of PWM inverter-fed AC motors — A survey. *IEEE Transactions on Industrial Electronics*, 51(4):744–757, 2004.
32. M.L. Campbell, J. Chiasson, M. Bodson, and L.M. Tolbert. Speed sensorless identification of the rotor time constant in induction machines. *IEEE Transactions on Automatic Control*, 52(4):758–763, 2007.
33. C. Canudas de Wit and S.I. Seleme. Robust torque control design for induction motors: the minimum energy approach. *Automatica*, 33(1):63–79, 1997.
34. C. Canudas de Wit and J. Ramirez. Optimal torque control for current-fed induction motors. *IEEE Transactions on Automatic Control*, 44(5):1084–1089, 1999.
35. P. Castaldi, W. Geri, M. Montanari, and A. Tilli. A new adaptive approach for on-line parameter and state estimation of induction motors. *Control Engineering Practice*, 13(1):81–94, 2005.
36. S. Cauët, L. Rambault, O. Bachelier, and D. Mehdi. Parameter-dependent Lyapunov functions applied to analysis of induction motor stability. *Control Engineering Practice*, 10(3):337–345, 2002.
37. C. Cecati and N. Rotondale. Torque and speed regulation of induction motors using the passivity theory approach. *IEEE Transactions on Industrial Electronics*, 46(1):119–127, 1999.
38. C.C. Chan, W.S. Leung, and C.W. Ng. Adaptive decoupling control of induction motor drives. *IEEE Transactions on Industrial Electronics*, 37(1):41–47, 1990.
39. C.C. Chan and H. Wang. An effective method for rotor resistance identification for high-performance induction motor vector control. *IEEE Transactions on Industrial Electronics*, 37(6):477–482, 1990.

40. G.W. Chang, G. Espinosa-Pérez, E. Mendes, and R. Ortega. Tuning rules for the PI gains of field-oriented controllers of induction motors. *IEEE Transactions on Industrial Electronics*, 47(3):592–602, 2000.
41. G.W. Chang, J.P. Hespanha, A.S. Morse, M.S. Netto, and R. Ortega. Supervisory field-oriented control of induction motors with uncertain rotor resistance. *International Journal of Adaptive Control and Signal Processing*, 15(3):353–375, 2001.
42. T.L. Chern, J. Chang, and K.L. Tsai. Integral-variable-structure-control-based adaptive speed estimator and resistance identifier for an induction motor. *International Journal of Control*, 69(1):31–47, 1998.
43. J. Chiasson. Dynamic feedback linearization of the induction motor. *IEEE Transactions on Automatic Control*, 38(10):1588–1594, 1993.
44. J. Chiasson. Nonlinear controllers for an induction motor. *Control Engineering Practice*, 4(7):977–990, 1996.
45. J. Chiasson. A new approach to dynamic feedback linearization control of an induction motor. *IEEE Transactions on Automatic Control*, 43(3):391–397, 1998.
46. J. Chiasson. *Modeling and High-Performance Control of Electric Machines*. Wiley-IEEE Press, Hoboken, 2005.
47. J. Chiasson and R.T. Novotnak. Comments on "A passivity-based method for induction motor control". *IEEE Transactions on Industrial Electronics*, 46(3):669–670, 1999.
48. C.S. Chiu, K.Y. Lian, C.Y. Hung, and P. Liu. Adaptive speed control for induction motors based on a semi-current-fed model. *International Journal of Control*, 81(2):307–316, 2008.
49. M. Cirrincione, M. Pucci, G. Cirrincione, and G.A. Capolino. A new experimental application of least-squares techniques for the estimation of the induction motor parameters. *IEEE Transactions on Industry Applications*, 39(5):1247–1256, 2003.
50. A. Consoli, G. Bottiglieri, G. Scarcella, and G. Scelba. Flux and voltage calculations of induction motors supplied by low- and high-frequency currents. *IEEE Transactions on Industry Applications*, 45(2):737–746, 2009.
51. D.M. Dawson, J. Hu, and T.C. Burg. *Nonlinear Control of Electric Machinery*. Marcel Dekker, New York, 1998.
52. E. Delaleau, J.P. Louis, and R. Ortega. Modeling and control of induction motors. *International Journal of Applied Mathematics and Computer Science*, 11(1):105–129, 2001.
53. A. De Luca and G. Ulivi. Design of an exact nonlinear controller for induction motors. *IEEE Transactions on Automatic Control*, 34(12):1304–1307, 1989.
54. L.A. De Souza Ribeiro, C.B. Jacobina, and A.M.N. Lima. Real-time estimation of the electric parameters of an induction machine using sinusoidal PWM voltage waveforms. *IEEE Transactions on Industry Applications*, 36(3):743–754, 2000.
55. P.A.S. De Wit, R. Ortega, and I. Mareels. Indirect field-oriented control of induction motors is robustly globally stable. *Automatica*, 32(10):1393–1402, 1996.
56. D. Diallo, M.E.H. Benbouzid, and A. Makouf. A fault-tolerant control architecture for induction motor drives in automotive applications. *IEEE Transactions on Vehicular Technology*, 53(6):1847–1855, 2004.
57. G. Ding, X. Wang, and Z. Han. H_∞ disturbance attenuation control of induction motor. *International Journal of Adaptive Control and Signal Processing*, 14(2–3):223–244, 2000.
58. W.E. Dixon, A. Behal, D.M. Dawson, and S.P. Nagarkatti. *Nonlinear Control of Engineering Systems*. Birkhäuser, Boston, 2003.
59. Y. Dote. Existence of limit cycle and stabilization of induction motor via new nonlinear state observer. *IEEE Transactions on Automatic Control*, AC-24(3):421–428, 1979.
60. Y. Dote. Stabilization of controlled current induction motor drive systems via new nonlinear state observer. *IEEE Transactions on Industrial Electronics and Control Instrumentation*, IECI-27(2):77–81, 1980.
61. Y. Dote. *Servo Motor and Motion Control Using Digital Signal Processors*. Prentice-Hall, New York, 1990.
62. M.A. El-Sharkawi and M. Akherraz. Tracking control technique for induction motors. *IEEE Transactions on Energy Conversion*, 4(1):81–87, 1989.

63. G. Espinosa and R. Ortega. State observers are unnecessary for induction motor control. *Systems and Control Letters*, 23(5):315–323, 1994.
64. G. Espinosa-Pérez and R. Ortega. An output feedback globally stable controller for induction motors. *IEEE Transactions on Automatic Control*, 40(1):138–143, 1995.
65. G. Espinosa-Pérez, R. Ortega, and P.J. Nicklasson. Torque and flux tracking of induction motors. *International Journal of Robust and Nonlinear Control*, 7(1):1–9, 1997.
66. J. Faiz, M.B.B. Sharifian, A. Keyhani, and A.B. Proca. Sensorless direct torque control of induction motors used in electric vehicle. *IEEE Transactions on Energy Conversion*, 18(1):1–10, 2003.
67. P. Famouri and J.J. Cathey. Loss minimization control of an induction motor drive. *IEEE Transactions on Industry Applications*, 27(1):32–37, 1991.
68. C.H. Fang, S.K. Lin, and S.J. Wang. On-line parameter estimator of an induction motor at standstill. *Control Engineering Practice*, 13(5):535–540, 2005.
69. H.A.A. Fattah, and K.A. Loparo. Speed control of electrical machines: unknown load torque case. *IEEE Transactions on Automatic Control*, 46(12):1979–1983, 2001.
70. M. Feemster, P. Aquino, D.M. Dawson, and A. Behal. Sensorless rotor velocity tracking control for induction motors. *IEEE Transactions on Control Systems Technology*, 9(4):645–653, 2001.
71. M. Feemster, P. Vedagarbha, D. Haste, and D.M. Dawson. Adaptive output-feedback control of induction motors. *International Journal of Systems Science*, 31(10):1195–1208, 2000.
72. G. Figalli, M. La Cava, and L. Tomasi. An optimal feedback control for a bilinear model of induction motor drives. *International Journal of Control*, 39(5):1007–1016, 1984.
73. A.E. Fitzgerald, C. Kingsley Jr, and S.D. Umans. *Electric Machinery*. McGraw-Hill, New York, 1983.
74. M. Fliess, J. Lévine, P. Martin, and P. Rouchon. Flatness and defect of non-linear systems: introductory theory and examples. *International Journal of Control*, 61(6):1327–1361, 1995
75. R. Gabriel, W. Leonhard, and C.J. Nordby. Field-oriented control of a standard AC motor using microprocessors. *IEEE Transactions on Industry Applications*, IA-16(2):186–192, 1980.
76. L.J. Garcés. Parameter adaption for the speed-controlled static AC drive with a squirrel-cage induction motor. *IEEE Transactions on Industry Applications*, IA-16(2):173–178, 1980.
77. G.O. Garcia, J.C.M. Luís, R.M. Stephan, and E.H. Watanabe. An efficient controller for an adjustable speed induction motor drive. *IEEE Transactions on Industrial Electronics*, 41(5):533–539, 1994.
78. D. Georges, C. Canudas de Wit, and J. Ramirez. Nonlinear H_2 and H_∞ optimal controllers for current-fed induction motors. *IEEE Transactions on Automatic Control*, 44(7):1430–1435, 1999.
79. L.U. Gökdere and M.A. Simaan. A passivity-based method for induction motor control. *IEEE Transactions on Industrial Electronics*, 44(5):688–695, 1997.
80. L.U. Gökdere and M.A. Simaan. Response to comments on "A passivity-based method for induction motor control" *IEEE Transactions on Industrial Electronics*, 46(3):670–671, 1999.
81. L.U. Gökdere, M.A. Simaan, and C.W. Brice. Global asymptotic stability of indirect field-oriented speed control of current-fed induction motors. *Automatica*, 34(1):133–135, 1998.
82. L.U. Gökdere, M.A. Simaan, and C.W. Brice. Passivity-based control of saturated induction motors. *IEEE Transactions on Industrial Electronics*, 48(4):870–872, 2001.
83. F. Gordillo, F. Salas, R. Ortega, and J. Aracil. Hopf bifurcation in indirect field-oriented control of induction motors. *Automatica*, 38(5):829–835, 2002.
84. F. Grognard and C. Canudas de Wit. Design of orbitally stable zero dynamics for a class of nonlinear systems. *Systems and Control Letters*, 51(2):89–103, 2004.
85. C. Guerrero, G. Espinosa-Pérez, J.A. Moreno, and R. Álvarez-Salas. On the globally defined sensorless control of induction motors. *International Journal of Robust and Nonlinear Control*, 19(2):117–134, 2009.
86. G. Guidi and H. Umida. A novel stator resistance estimation method for speed-sensorless induction motor drives. *IEEE Transactions on Industry Applications*, 36(6):1619–1627, 2000.
87. I.J. Ha and S.H. Lee. An online identification method for both stator and rotor resistances of induction motors without rotational transducers. *IEEE Transactions on Industrial Electronics*, 47(4):842–853, 2000.

88. J.I. Ha and S.K. Sul. Sensorless field-orientation control of an induction machine by high-frequency signal injection. *IEEE Transactions on Industry Applications*, 35(1):45–51, 1999.
89. W. Hahn. *Stability of Motion*. Springer-Verlag, Berlin, 1967.
90. L. Harnefors. Instability phenomena and remedies in sensorless indirect field oriented control. *IEEE Transactions on Power Electronics*, 15(4):733–743, 2000.
91. L. Harnefors. Design and analysis of general rotor-flux-oriented vector control systems. *IEEE Transactions on Industrial Electronics*, 48(2):383–390, 2001.
92. L. Harnefors. Globally stable speed-adaptive observers for sensorless induction motor drives. *IEEE Transactions on Industrial Electronics*, 54(2):1243–1245, 2007.
93. S.M.N. Hasan and I. Husain. A Luenberger-sliding mode observer for online parameter estimation and adaptation in high-performance induction motor drives. *IEEE Transactions on Industry Applications*, 45(2):772–781, 2009.
94. R. Hedjar, R. Toumi, P. Boucher, and D. Dumur. Cascaded nonlinear predictive control of induction motor. *European Journal of Control*, 10(1):65–80, 2004.
95. M. Hilairet, F. Auger, and E. Berthelot. Speed and rotor flux estimation of induction machines using a two-stage extended Kalman filter. *Automatica* 45(8):1819–1827, 2009.
96. E.Y.Y. Ho and P.C. Sen. Decoupling control of induction motor drives. *IEEE Transactions on Industrial Electronics*, 35(2):253–262, 1988.
97. J. Holtz. Sensorless control of induction motor drives. *Proceedings of the IEEE*, 90(8):1358–1394, 2002.
98. J. Holtz. Sensorless control of induction machines — with or without signal injection? *IEEE Transactions on Industrial Electronics*, 53(1):7–30, 2006.
99. J. Holtz and J. Quan. Sensorless vector control of induction motors at very low speed using a nonlinear inverter model and parameter identification. *IEEE Transactions on Industry Applications*, 38(4):1087–1095, 2002.
100. J. Holtz and T. Thimm. Identification of the machine parameters in a vector-controlled induction motor drive. *IEEE Transactions on Industry Applications*, 27(6):1111–1118, 1991.
101. J. Hu and D.M. Dawson. Adaptive control of induction motor systems despite rotor resistance uncertainty. *Automatica*, 32(8):1127–1143, 1996.
102. J. Hu and D.M. Dawson. Robust tracking control of an induction motor. *International Journal of Robust and Nonlinear Control*, 6(3):201–219, 1996.
103. J. Hu, D.M. Dawson, and Y. Qian. Position tracking control of an induction motor via partial state feedback. *Automatica*, 31(7):989–1000, 1995.
104. K.D. Hurst, T.G. Habetler, G. Griva, and F. Profumo. Zero-speed tacholess IM torque control: simply a matter of stator voltage integration. *IEEE Transactions on Industry Applications*, 34(4):790–795, 1998.
105. S. Ibarra-Rojas, J. Moreno, and G. Espinosa-Pérez. Global observability analysis of sensorless induction motors. *Automatica*, 40(6):1079–1085, 2004.
106. A. Isidori. *Nonlinear Control Systems*. Springer, Berlin, 1995.
107. M. Iwasaki and N. Matsui. Robust speed control of IM with torque feedforward control. *IEEE Transactions on Industrial Electronics*, 40(6):553–560, 1993.
108. F. Jadot, F. Malrait, J. Moreno-Valenzuela, and R. Sepulchre. Adaptive regulation of vector-controlled induction motors. *IEEE Transactions on Control Systems Technology*, 17(3):646–657, 2009.
109. F. Jadot, P. Martin, and P. Rouchon. Industrial sensorless control of induction motors. *In* A. Isidori, F. Lamnabhi-Lagarrigue, and W. Respondek (Eds), *Nonlinear Control in the Year 2000*, 1, 535–543, Springer-Verlag, London, 2001.
110. P.L. Jansen and R.D. Lorenz. A physically insightful approach to the design and accuracy assessment of flux observers for field oriented induction machine drives. *IEEE Transactions on Industry Applications*, 30(1):101–110, 1994.
111. P.L. Jansen, R.D. Lorenz and D.W. Novotny. Observer-based direct field orientation: analysis and comparison of alternative methods. *IEEE Transactions on Industry Applications*, 30(4):945–953, 1994.

112. R. Joetten and G. Maeder. Control methods for good dynamic performance induction motor drives based on current and voltage as measured quantities. *IEEE Transactions on Industry Applications*, IA-19(3):356–363, 1983.
113. M. Kaimoto, M. Hashii, T. Yanase, and T. Nakano. Performance improvement of current source inverter-fed induction motor drives. *IEEE Transactions on Industry Applications*, IA-18(6):703–711, 1982.
114. D. Karagiannis, A. Astolfi, R. Ortega, and M. Hilairet. A nonlinear tracking controller for voltage-fed induction motors with uncertain load torque. *IEEE Transactions on Control Systems Technology*, 17(3):608–619, 2009.
115. B. Karanayil, M.F. Rahman, and C. Grantham. Online stator and rotor resistance estimation scheme using artificial neural networks for vector controlled speed sensorless induction motor drive. *IEEE Transactions on Industrial Electronics*, 54(1):167–176, 2007.
116. H.K. Khalil. *Nonlinear Systems*. Prentice-Hall, Upper Saddle River, NJ, 1996.
117. H.K. Khalil and E.G. Strangas. Robust speed control of induction motors using position and current measurements. *IEEE Transactions on Automatic Control*, 41(8):1216–1220, 1996.
118. H.K. Khalil, E.G. Strangas, and S. Jurkovic. Speed observer and reduced nonlinear model for sensorless control of induction motors. *IEEE Transactions on Control Systems Technology*, 17(2):327–339, 2009.
119. A.M. Khambadkone and J. Holtz. Vector-controlled induction motor drive with a self-commissioning scheme. *IEEE Transactions on Industrial Electronics*, 38(5):322–327, 1991.
120. F.M.H. Khater, R.D. Lorenz, D.W. Novotny, and K. Tang. Selection of flux level in field-oriented induction machine controllers with consideration of magnetic saturation effects. *IEEE Transactions on Industry Applications*, IA-23(2):276–282, 1987.
121. F. Khorrami, P. Krishnamurthy, and H. Melkote. *Modeling and Adaptive Nonlinear Control of Electric Motors*. Springer-Verlag, Berlin, 2003.
122. D.I. Kim, I.J. Ha, and M.S. Ko. Control of induction motors via feedback linearization with input-output decoupling. *International Journal of Control*, 51(4):863–883, 1990.
123. G.S. Kim, I.J. Ha, and M.S. Ko. Control of induction motors for both high dynamic performance and high power efficiency. *IEEE Transactions on Industrial Electronics*, 39(4):323–333, 1992.
124. K.C. Kim, R. Ortega, A. Charara, and J.P. Vilain. Theoretical and experimental comparison of two nonlinear controllers for current-fed induction motors. *IEEE Transactions on Control Systems Technology*, 5(3):338–348, 1997.
125. Y.R. Kim, S.K. Sul, and M.H. Park. Speed sensorless vector control of induction motor using extended Kalman filter. *IEEE Transactions on Industry Applications*, 30(5):1225–1233, 1994.
126. D.S. Kirschen, D.W. Novotny, and T.A. Lipo. On-line efficiency optimization of a variable frequency induction motor drive. *IEEE Transactions on Industry Applications*, IA-21(4):610–616, 1985.
127. D.S. Kirschen, D.W. Novotny, and T.A. Lipo. Optimal efficiency control of an induction motor drive. *IEEE Transactions on Energy Conversion*, EC-2(1):70–75, 1987.
128. D.S. Kirschen, D.W. Novotny, and W. Suwanwisoot. Minimizing induction motor losses by excitation control in variable frequency drives. *IEEE Transactions on Industry Applications*, IA-20(5):1244–1250, 1984.
129. N.R. Klaes. Parameter identification of an induction machine with regard to dependencies on saturation. *IEEE Transactions on Industry Applications*, 29(6):1135–1140, 1993.
130. P.C. Krause. *Analysis of Electric Machinery*. McGraw-Hill, New York, 1986.
131. P.C. Krause and C.H. Thomas. Simulation of symmetrical induction machinery. *IEEE Transactions on Power Apparatus and Systems*, PAS-84(11):1038–1053, 1965.
132. P.C. Krause, O. Wasynczuk, and S.D. Sudhoff. *Analysis of Electric Machinery*. IEEE Press, New York, 1994.
133. R. Krishnan. *Electric Motor Drives: Modeling, Analysis and Control*. Prentice Hall, Englewood Cliffs, 2001.
134. R. Krishnan and F.C. Doran. Study of parameter sensitivity in high-performance inverter-fed induction motor drive systems. *IEEE Transactions on Industry Applications*, IA-23(4):623–635, 1987.

135. M. Krstic, I. Kanellakopoulos, and P.V. Kokotovic. *Nonlinear and Adaptive Control Design*. J. Wiley, New York, 1995.
136. H. Kubota and K. Matsuse. Speed sensorless field-oriented control of induction motor with rotor resistance adaptation. *IEEE Transactions on Industry Applications*, 30(5):1219–1224, 1994.
137. H. Kubota, K. Matsuse, and T. Nakano. DSP-based speed adaptive flux observer of induction motor. *IEEE Transactions on Industry Applications*, 29(2):344–348, 1993.
138. H. Kubota, I. Sato, Y. Tamura, K. Matsuse, H. Ohta, and Y. Hori. Regenerating-mode low-speed operation of sensorless induction motor drive with adaptive observer. *IEEE Transactions on Industry Applications*, 38(4):1081–1086, 2002.
139. A. Kusko and D. Galler. Control means for minimization of losses in AC and DC motor drives. *IEEE Transactions on Industry Applications*, IA-19(4):561–570, 1983.
140. C.M. Kwan. Robust adaptive control of induction motors. *International Journal of Control*, 67(4):539–552, 1997.
141. C.M. Kwan, F.L. Lewis, and K.S. Yeung. Adaptive control of induction motors without flux measurements. *Automatica*, 32(6):903–908, 1996.
142. C. Lascu, I. Boldea, and F. Blaabjerg. A modified direct torque control for induction motor sensorless drive. *IEEE Transactions on Industry Applications*, 36(1):122–130, 2000.
143. H.T. Lee, J.S. Chang, and L.C. Fu. Exponentially stable non-linear control for speed regulation of induction motor with field-oriented PI-controller. *International Journal of Adaptive Control and Signal Processing*, 14(2–3):297–312, 2000.
144. H.T. Lee, L.C. Fu, and H.S. Huang. Sensorless speed tracking of induction motor with unknown torque based on maximum power transfer. *IEEE Transactions on Industrial Electronics*, 49(4):911–924, 2002.
145. A.M. Lee, L.C. Fu, C.Y. Tsai, and Y.C. Lin. Nonlinear adaptive speed and torque control of induction motors with unknown rotor resistance. *IEEE Transactions on Industrial Electronics*, 48(2):391–401, 2001.
146. W. Leonhard. Microcomputer control of high dynamic performance AC-drives — a survey. *Automatica*, 22(1):1–19, 1986.
147. W. Leonhard. *Control of Electrical Drives*. Springer-Verlag, Berlin, 2001.
148. R. Lessmeier, W. Schumacher, and W. Leonhard. Microprocessor-controlled AC-servo drives with synchronous or induction motors: which is preferable? *IEEE Transactions on Industry Applications*, IA-22(5):812–819, 1986.
149. M. Li, J. Chiasson, M. Bodson, and L.M. Tolbert. A differential-algebraic approach to speed estimation in an induction motor. *IEEE Transactions on Automatic Control*, 51(7):1172–1177, 2006.
150. C.M. Liaw, C.T. Pan, and Y.C. Chen. An adaptive controller for current-fed induction motor. *IEEE Transactions on Aerospace and Electronic Systems*, 24(3):250–262, 1988.
151. C.M. Liaw, C.T. Pan, and Y.C. Chen. Design and implementation of an adaptive controller for current-fed induction motor. *IEEE Transactions on Industrial Electronics*, 35(3):393–401, 1988.
152. Y.C. Lin, L.C. Fu, and C.Y. Tsai. Non-linear sensorless indirect adaptive speed control of induction motor with unknown rotor resistance and load. *International Journal of Adaptive Control and Signal Processing*, 14(2–3):109–140, 2000.
153. F.J. Lin and C.M. Liaw. Reference model selection and adaptive control for induction motor drives. *IEEE Transactions on Automatic Control*, 38(10):1594–1600, 1993.
154. T.A. Lipo. Recent progress in the development of solid-state AC motor drives. *IEEE Transactions on Power Electronics*, 3(2):105–117, 1988.
155. T.A. Lipo and K.C. Chang. A new approach to flux and torque-sensing in induction machines. *IEEE Transactions on Industry Applications*, IA-22(4):731–737, 1986.
156. C.H. Liu, C.C. Hwu, and Y.F. Feng. Modeling and implementation of a microprocessor-based CSI-fed induction motor drive using field-oriented control. *IEEE Transactions on Industry Applications*, 25(4):588–597, 1989.

157. R.D. Lorenz and S.M. Yang. Efficiency-optimized flux trajectories for closed-cycle operation of field-orientation induction machine drives. *IEEE Transactions on Industry Applications*, 28(3):574–580, 1992.
158. R.D. Lorenz and S.M. Yang. AC induction servo sizing for motion control applications via loss minimizing real-time flux control. *IEEE Transactions on Industry Applications*, 28(3):589–593, 1992.
159. M.K. Maaziz, P. Boucher, and D. Dumur. A new control strategy for induction motor based on non-linear predictive control and feedback linearization. *International Journal of Adaptive Control and Signal Processing*, 14(2–3):313–329, 2000.
160. J. Maes and J.A. Melkebeek. Speed-sensorless direct torque control of induction motors using and adaptive flux observer. *IEEE Transactions on Industry Applications*, 36(3):778–785, 2000.
161. R. Marino, S. Peresada, and P. Tomei. Exponentially convergent rotor resistance estimation for induction motors. *IEEE Transactions on Industrial Electronics*, 42(5):508–515, 1995.
162. R. Marino, S. Peresada, and P. Tomei. Adaptive observer-based control of induction motors with unknown rotor resistance. *International Journal of Adaptive Control and Signal Processing*, 10(4–5):345–363, 1996.
163. R. Marino, S. Peresada, and P. Tomei. Output feedback control of current-fed induction motors with unknown rotor resistance. *IEEE Transactions on Control Systems Technology*, 4(4):336–347, 1996.
164. R. Marino, S. Peresada, and P. Tomei. Adaptive output feedback control of current-fed induction motors with uncertain rotor resistance and load torque. *Automatica*, 34(5):617–624, 1998.
165. R. Marino, S. Peresada, and P. Tomei. Global adaptive output feedback control of induction motors with uncertain rotor resistance. *IEEE Transactions on Automatic Control*, 44(5):967–983, 1999.
166. R. Marino, S. Peresada, and P. Tomei. On-line stator and rotor resistance estimation for induction motors. *IEEE Transactions on Control Systems Technology*, 8(3):570–579, 2000.
167. R. Marino, S. Peresada, and P. Valigi. Adaptive nonlinear control of induction motors via extended matching. *In* P.V. Kokotovic Ed., *Foundations of Adaptive Control* (Lecture Notes in Control and Information Sciences), 435–454, Springer-Verlag, Berlin, 1991.
168. R. Marino, S. Peresada, and P. Valigi. Adaptive input-output linearizing control of induction motors. *IEEE Transactions on Automatic Control*, 38(2):208–221, 1993.
169. R. Marino and P. Tomei. *Nonlinear Control Design — Geometric, Adaptive and Robust*. Prentice-Hall, London, 1995.
170. R. Marino, P. Tomei, and C.M. Verrelli. A global tracking control for speed-sensorless induction motors. *Automatica*, 40(6):1071–1077, 2004.
171. R. Marino, P. Tomei, and C.M. Verrelli. Adaptive control for speed-sensorless induction motors with uncertain load torque and rotor resistance. *International Journal of Adaptive Control and Signal Processing*, 19(9):661–685, 2005.
172. R. Marino, P. Tomei, and C.M. Verrelli. A nonlinear tracking control for sensorless induction motors. *Automatica*, 41(6):1071–1077, 2005.
173. R. Marino, P. Tomei, and C.M. Verrelli. An adaptive tracking control from current measurements for induction motors with uncertain load torque and rotor resistance. *Automatica*, 44(10):2593–2599, 2008.
174. R. Marino, P. Tomei, and C.M. Verrelli. A nonlinear tracking control for sensorless induction motors with uncertain load torque. *International Journal of Adaptive Control and Signal Processing*, 22(1):1–22, 2008.
175. P. Martin and P. Rouchon. Two simple flux observers for induction motors. *International Journal of Adaptive Control and Signal Processing*, 14(2–3):171–175, 2000.
176. T. Matsuo and T.A. Lipo. A rotor parameter identification scheme for vector-controlled induction motor drives. *IEEE Transactions on Industry Applications*, IA-21(4):624–632, 1985.
177. K. Matsuse, T. Yoshizumi, S. Katsuta, and S. Taniguchi. High-response flux control of direct-field-oriented induction motor with high efficiency taking core loss into account. *IEEE Transactions on Industry Applications*, 35(1):62–69, 1999.

178. M. Menaa, O. Touhami, R. Ibtiouen, and M. Fadel. Sensorless direct vector control of an induction motor. *Control Engineering Practice*, 16(1):67–77, 2008.
179. J. Meisel. *Principles of Electromechanical Energy Conversion*. McGraw-Hill, New York, 1966.
180. H. Miranda, P. Cortés, J.I. Yuz, and J. Rodríguez. Predictive torque control of induction machines based on state-space models. *IEEE Transactions on Industrial Electronics*, 56(6):1916–1924, 2009.
181. M. Montanari, S. Peresada, C. Rossi, and A. Tilli. Current sensorless position-flux tracking controller for induction motor drives. *Mechatronics*, 17(1):15–30, 2007.
182. M. Montanari, S. Peresada, C. Rossi, and A. Tilli. Speed sensorless control of induction motors based on a reduced-order adaptive observer. *IEEE Transactions on Control Systems Technology*, 15(6):1049–1064, 2007.
183. M. Montanari, S. Peresada, and A. Tilli. A speed-sensorless indirect field-oriented control for induction motors based on high gain speed estimation. *Automatica*, 42(10):1637–1650, 2006.
184. C. Moons and B. De Moor. Parameter identification of induction motor drives. *Automatica*, 31(8):1137–1147, 1995.
185. J.C. Moreira, T.A. Lipo, and V. Blasko. Simple efficiency maximizer for an adjustable frequency induction motor drive. *IEEE Transactions on Industry Applications*, 27(5):940–946, 1991.
186. T. Murata, T. Tsuchiya, and I. Takeda. Vector control for induction machine on the application of optimal control theory. *IEEE Transactions on Industrial Electronics*, 37(4):283–290, 1990.
187. A. Nabae, K. Otsuka, H. Uchino, and R. Kurosawa. An approach to flux control of induction motors operated with variable-frequency power supply. *IEEE Transactions on Industry Applications*, IA-16(3):342–350, 1980.
188. K. Nagata, T. Okuyama, H. Nemoto, and T. Katayama. A simple robust voltage control of high power sensorless induction motor drives with high start torque demand. *IEEE Transactions on Industry Applications*, 44(2):604–611, 2008.
189. K.S. Narendra and A.M. Annaswamy. *Stable Adaptive Systems*. Prentice Hall, Englewood Cliffs, NJ, 1989.
190. P.J. Nicklasson, R. Ortega, and G. Espinosa-Pérez. Passivity-based control of a class of Blondel-Park transformable electric machines. *IEEE Transactions on Automatic Control*, 42(5):629–647, 1997.
191. H. Nijmeijer and A.J. van der Schaft. *Nonlinear Dynamical Control Systems*. Springer-Verlag, Berlin, 1990.
192. T. Noguchi, S. Kondo, and I. Takahashi. Field-oriented control of an induction motor with robust on-line tuning of its parameters. *IEEE Transactions on Industry Applications*, 33(1):35–42, 1997.
193. K.B. Nordin, D.W. Novotny, and D.S. Zinger. The influence of motor parameter deviations in feedforward field orientation drive systems. *IEEE Transactions on Industry Applications*, IA-21(4):1009–1015, 1985.
194. R.T. Novotnak and J. Chiasson. Comments on "Passivity-based control of saturated induction motors". *IEEE Transactions on Industrial Electronics*, 50(4):820–821, 2003.
195. R.T. Novotnak, J. Chiasson, and M. Bodson. High-performance motion control of an induction motor with magnetic saturation. *IEEE Transactions on Control Systems Technology*, 7(3):315–327, 1999.
196. D.W. Novotny and T.A. Lipo. *Vector Control and Dynamics of AC Drives*. Oxford University Press, New York, 1996.
197. S. Ogasawara, H. Akagi, and A. Nabae. The generalized theory of indirect vector control for AC machines. *IEEE Transactions on Industry Applications*, 24(3):470–478, 1988.
198. K. Ohnishi, H. Suzuki, K. Miyachi, and M. Terashima. Decoupling control of secondary flux and secondary current in induction motor drive with controlled voltage source and its comparison with volts/hertz control. *IEEE Transactions on Industry Applications*, IA-21(1):241–247, 1985.
199. T. Ohtani, N. Takada, and K. Tanaka. Vector control of induction motor without shaft encoder. *IEEE Transactions on Industry Applications*, 28(1):157–164, 1992.

200. K. Ohyama, G.M. Asher, and M. Sumner. Comparative analysis of experimental performance and stability of sensorless induction motor drives. *IEEE Transactions on Industrial Electronics*, 53(1):178–186, 2006.
201. R. Ortega, N. Barabanov, and G.E. Valderrama. Direct torque control of induction motors: stability analysis and performance improvement. *IEEE Transactions on Automatic Control*, 46(8):1209–1222, 2001.
202. R. Ortega, C. Canudas De Wit, and S.I. Seleme. Nonlinear control of induction motors: torque tracking with unknown load disturbance. *IEEE Transactions on Automatic Control*, 38(11):1675–1680, 1993.
203. R. Ortega and G. Espinosa. Torque regulation of induction motors. *Automatica*, 29(3):621–633, 1993.
204. R. Ortega, A. Loría, P.J. Nicklasson, and H. Sira-Ramirez. *Passivity-based control of Euler-Lagrange Systems*. Springer-Verlag, London, 1998.
205. R. Ortega, P.J. Nicklasson, and G. Espinosa-Pérez. On speed control of induction motors. *Automatica*, 32(3):455–460, 1996.
206. R. Ortega and D. Taoutaou. A globally stable discrete-time controller for current-fed induction motors. *Systems and Control Letters*, 28(3):123–128, 1996.
207. R. Ortega and D. Taoutaou. Indirect field-oriented speed regulation for induction motors is globally stable. *IEEE Transactions on Industrial Electronics*, 43(2):340–341, 1996.
208. G. Papafotiou, J. Kley, K.G. Papadopoulos, P. Bohren, and M. Morari. Model predictive direct torque control — Part II: Implementation and experimental evaluation. *IEEE Transactions on Industrial Electronics*, 56(6):1906–1915, 2009.
209. M.H. Park and S.K. Sul. Microprocessor-based optimal-efficiency drive of an induction motor. *IEEE Transactions on Industrial Electronics*, IE-31(1):69–73, 1984.
210. F.Z. Peng and T. Fukao. Robust speed identification for speed-sensorless vector control of induction motors. *IEEE Transactions on Industry Applications*, 30(5):1234–1240, 1994.
211. S. Peresada, A. Tilli, and A. Tonielli. Theoretical and experimental comparison of indirect field-oriented controllers for induction motors. *IEEE Transactions on Power Electronics*, 18(1):151–163, 2003.
212. S. Peresada and A. Tonielli. High-performance robust speed-flux tracking controller for induction motor. *International Journal of Adaptive Control and Signal Processing*, 14(2–3):177–200, 2000.
213. S. Peresada, A. Tonielli, and R. Morici. High-performance indirect field-oriented output-feedback control of induction motors. *Automatica*, 35(6):1033–1047, 1999.
214. A.B. Plunkett. Direct flux and torque regulation in a PWM inverter-induction motor drive. *IEEE Transactions on Industry Applications*, IA-13(2):139–146, 1977.
215. K. Rajashekara, A. Kuwamura, and K. Matsuse. *Sensorless Control of AC Motor Drives: Speed and Position Sensorless Operation*. IEEE Press, New York, 1996.
216. M. Rashed, P.F.A. MacConnell, and A.F. Stronach. Nonlinear adaptive state-feedback speed control of a voltage-fed induction-motor with varying parameters. *IEEE Transactions on Industry Applications*, 42(3):723–732, 2006.
217. H.U. Rehman. An integrated starter-alternator and low-cost high-performance drive for vehicular applications. *IEEE Transactions on Vehicular Technology*, 57(3):1454–1465, 2008.
218. B. Renier, K. Hameyer, and R. Belmans. Comparison of standards for determining efficiency of three phase induction motors. *IEEE Transactions on Energy Conversion*, 14(3):512–517, 1999.
219. A. Sabanovic and D.B. Izosimov. Application of sliding modes to induction motor control. *IEEE Transactions on Industry Applications*, IA-17(1):41–49, 1981.
220. S. Sangwongwanich, M. Ishida, S. Okuma, and K. Iwata. Manipulation of rotor flux for time-optimal single-step velocity response of field-oriented induction machines. *IEEE Transactions on Industry Applications*, 24(2):262–270, 1988.
221. S. Sangwongwanich, M. Ishida, S. Okuma, Y. Uchikawa, and K. Iwata. Realization of time-optimal single-step velocity response control of field-oriented induction machines under the condition of nonsaturation of flux. *IEEE Transactions on Industry Applications*, 27(5):947–955, 1991.

222. S. Sastry and M. Bodson. *Adaptive Control — Stability, Convergence, and Robustness*. Prentice Hall, Englewood Cliffs, NJ, 1989.
223. C. Schauder. Adaptive speed identification for vector control of induction motors without rotational transducers. *IEEE Transactions on Industry Applications*, 28(5):1054–1061, 1992.
224. S. Seely. *Electromechanical Energy Conversion*. McGraw-Hill, New York, 1962.
225. P.C. Sen. *Principles of Electric Machines and Power Electronics*. Wiley, New York, 1989.
226. S.R. Shaw and S.B. Leeb. Identification of induction motor parameters from transient stator current measurements. *IEEE Transactions on Industrial Electronics*, 46(1):139–149, 1999.
227. K.L. Shi, T.F. Chan, Y.K. Wong, and S.L. Ho. Speed estimation of an induction motor drive using an optimized extended Kalman filter. *IEEE Transactions on Industrial Electronics*, 49(1):124–133, 2002.
228. S.L. Shishkin and D.C. Wunsch. Discrete-time method for robust global stabilization of induction motor. *International Journal of Adaptive Control and Signal Processing*, 14(2–3):141–156, 2000.
229. G.R. Slemon. Modelling of induction machines for electric drives. *IEEE Transactions on Industry Applications*, 25(6):1126–1131, 1989.
230. R. Soto and K.S. Yeung. Sliding-mode control of an induction motor without flux measurement. *IEEE Transactions on Industry Applications*, 31(4):744–751, 1995.
231. J. Stephan, M. Bodson, and J. Chiasson. Real-time estimation of the parameters and fluxes of induction motors. *IEEE Transactions on Industry Applications*, 30(3):746–759, 1994.
232. E.G. Strangas, H.K. Khalil, B.A. Oliwi, L. Laubinger, and J.M. Miller. A robust torque controller for induction motors without rotor position sensor: analysis and experimental results. *IEEE Transactions on Energy Conversion*, 14(4):1448–1458, 1999.
233. S.K. Sul. A novel technique of rotor resistance estimation considering variation of mutual inductance. *IEEE Transactions on Industry Applications*, 25(4):578–587, 1989.
234. H. Tajima and Y. Hori. Speed sensorless field-orientation control of the induction machine. *IEEE Transactions on Industry Applications*, 29(1):175–180, 1993.
235. I. Takahashi and T. Noguchi. A new quick-response and high-efficiency control strategy of an induction motor. *IEEE Transactions on Industry Applications*, IA-22(5):820–827, 1986.
236. D. Taoutaou, R. Puerto, R. Ortega, and L. Loron. A new field-oriented discrete-time controller for current-fed induction motors. *Control Engineering Practice*, 5(2):209–217, 1997.
237. D. Taylor. Nonlinear control of electric machines: an overview. *IEEE Control Systems Magazine*, 14(6):41–51, 1994.
238. A. Ţiclea and G. Besançon. Observer scheme for state and parameter estimation in asynchronous motors with application to speed control. *European Journal of Control*, 12(4):400–412, 2006.
239. H.A. Toliyat, M.S. Arefeen, K. M. Rahman, and D. Figoli. Rotor time constant updating scheme for a rotor flux-oriented induction motor drive. *IEEE Transactions on Power Electronics*, 14(5):850–857, 1999.
240. H.A. Toliyat, E. Levi, and M. Raina. A review of RFO induction motor parameter estimation techniques. *IEEE Transactions on Energy Conversion*, 18(2):271–283, 2003.
241. A. Trentin, P. Zanchetta, C. Gerada, J. Clare, and P.W. Wheeler. Optimized commissioning method for enhanced vector control of high-power induction motor drives. *IEEE Transactions on Industrial Electronics*, 56(5):1708–1717, 2009.
242. A.M. Trzynadlowski. *The Field Orientation Principle in Control of Induction Motors*. Kluwer, Norwell MA, 1994.
243. M.N. Uddin and S.W. Nam. Development and implementation of a nonlinear-controller-based IM drive incorporating iron loss with parameter uncertainties. *IEEE Transactions on Industrial Electronics*, 56(4):1263–1272, 2009.
244. P. Vaclavek and P. Blaha. Lyapunov-function-based flux and speed observer for AC induction motor sensorless control and parameters estimation. *IEEE Transactions on Industrial Electronics*, 53(1):138–145, 2006.
245. P. Vas. *Vector Control of AC Machines*. Oxford University Press, New York, 1990.
246. P. Vas. *Parameter Estimation, Condition Monitoring, and Diagnosis of Electrical Machines*. Oxford University Press, New York, 1993.

247. P. Vas. *Sensorless Vector and Direct Torque Control*. Oxford University Press, New York, 1998.
248. P. Vedagarbha, D.M. Dawson, and T. Burg. Rotor velocity/flux control of induction motors with improved efficiency. *Mechatronics*, 7(2):105–127, 1997.
249. P. Vedagarbha, M. Feemster, P. Aquino, and D.M. Dawson. Non-linear adaptive control of induction motors. *International Journal of Adaptive Control and Signal Processing*, 13(5):367–392, 1999.
250. G.C. Verghese and S.R. Sanders. Observers for flux estimation in induction machines. *IEEE Transactions on Industrial Electronics*, 35(1):85–94, 1988.
251. T. von Raumer, J.M. Dion, and L. Dugard. Adaptive non-linear speed control of induction motors. *International Journal of Adaptive Control and Signal Processing*, 7(5):435–455, 1993.
252. T. von Raumer, J.M. Dion, L. Dugard, and J.L. Thomas. Applied nonlinear control of an induction motor using digital signal processing. *IEEE Transactions on Control Systems Technology*, 2(4):327–335, 1994.
253. I.T. Wallace, D.W. Novotny, R.D. Lorenz, and D.M. Divan. Increasing the dynamic torque per ampere capability of induction machines. *IEEE Transactions on Industry Applications*, 30(1):146–153, 1994.
254. K. Wang, J. Chiasson, M. Bodson, and L.M. Tolbert. A nonlinear least-squares approach for identification of the induction motor parameters. *IEEE Transactions on Automatic Control*, 50(10):1622–1628, 2005.
255. K. Wang, J. Chiasson, M. Bodson, and L.M. Tolbert. An online rotor time constant estimator for the induction machine. *IEEE Transactions on Control Systems Technology*, 15(2):339–348, 2007.
256. C. Wang, D.W. Novotny, and T.A. Lipo. An automated rotor time constant measurement system for indirect field-oriented drives. *IEEE Transactions on Industry Applications*, 24(1):151–159, 1988.
257. O. Wasynczuk, S.D. Sudhoff, K.A. Corzine, J.L. Tichenor, P.C. Krause, I.G. Hansen, and L.M. Taylor. A maximum torque per ampere control strategy for induction motor drives. *IEEE Transactions on Energy Conversion*, 13(2):163–169, 1998.
258. N.T. West and R.D. Lorenz. Digital implementation of stator and rotor flux-linkage observers and a stator-current observer for deadbeat direct torque control of induction machines. *IEEE Transactions on Industry Applications*, 45(2):729–736, 2009.
259. H.H. Woodson and J.R. Melcher. *Electromechanical Dynamics*. Wiley, New York, 1968.
260. X. Xu and D.W. Novotny. Selection of the flux reference for induction machine drives in the field weakening region. *IEEE Transactions on Industry Applications*, 28(6):1353–1358, 1992.
261. Z. Yan, C. Jin, and V.I. Utkin. Sensorless sliding-mode control of induction motors. *IEEE Transactions on Industrial Electronics*, 47(6):1286–1297, 2000.
262. J.H. Yang, W.H. Yu, and L.C. Fu. Nonlinear observer-based adaptive tracking control for induction motors with unknown load. *IEEE Transactions on Industrial Electronics*, 42(6):579–586, 1995.
263. H.S. Yoo and I.J. Ha. A polar coordinate-oriented method of identifying rotor flux and speed of induction motors without rotational transducers. *IEEE Transactions on Control Systems Technology*, 4(3):230–243, 1996.
264. L.C. Zai, C.L. DeMarco, and T.A. Lipo. An extended Kalman filter approach to rotor time constant measurement in PWM induction motor drives. *IEEE Transactions on Industry Applications*, 28(1):96–104, 1992.
265. M.S. Zaky, M.M. Khater, S.S. Shokralla, and H.A. Yasin. Wide-speed-range estimation with online parameter identification schemes of sensorless induction motor drives. *IEEE Transactions on Industrial Electronics*, 56(5):1699–1707, 2009.
266. J.L. Zamora and A. García-Cerrada. Online estimation of the stator parameters in an induction motor using only voltage and current measurements. *IEEE Transactions on Industry Applications*, 36(3):805–816, 2000.
267. J.L. Zamora, A. García-Cerrada, and A. Zazo. Rotor-speed estimator for induction motors using voltage and current measurements. *Control Engineering Practice*, 6(3):369–383, 1998.

Index

Other titles published in this series (continued):

Soft Sensors for Monitoring and Control of Industrial Processes
Luigi Fortuna, Salvatore Graziani, Alessandro Rizzo and Maria G. Xibilia

Adaptive Voltage Control in Power Systems
Giuseppe Fusco and Mario Russo

Advanced Control of Industrial Processes
Piotr Tatjewski

Process Control Performance Assessment
Andrzej W. Ordys, Damien Uduehi and Michael A. Johnson (Eds.)

Modelling and Analysis of Hybrid Supervisory Systems
Emilia Villani, Paulo E. Miyagi and Robert Valette

Process Control
Jie Bao and Peter L. Lee

Distributed Embedded Control Systems
Matjaž Colnarič, Domen Verber and Wolfgang A. Halang

Precision Motion Control (2nd Ed.)
Tan Kok Kiong, Lee Tong Heng and Huang Sunan

Optimal Control of Wind Energy Systems
Iulian Munteanu, Antoneta Iuliana Bratcu, Nicolaos-Antonio Cutululis and Emil Ceangă

Identification of Continuous-time Models from Sampled Data
Hugues Garnier and Liuping Wang (Eds.)

Model-based Process Supervision
Arun K. Samantaray and Belkacem Bouamama

Diagnosis of Process Nonlinearities and Valve Stiction
M.A.A. Shoukat Choudhury, Sirish L. Shah, and Nina F. Thornhill

Magnetic Control of Tokamak Plasmas
Marco Ariola and Alfredo Pironti

Real-time Iterative Learning Control
Jian-Xin Xu, Sanjib K. Panda and Tong H. Lee

Deadlock Resolution in Automated Manufacturing Systems
ZhiWu Li and MengChu Zhou

Model Predictive Control Design and Implementation Using MATLAB®
Liuping Wang

Fault-tolerant Flight Control and Guidance Systems
Guillaume Ducard

Predictive Functional Control
Jacques Richalet and Donal O'Donovan

Fault-tolerant Control Systems
Hassan Noura, Didier Theilliol, Jean-Christophe Ponsart and Abbas Chamseddine

Control of Ships and Underwater Vehicles
Khac Duc Do and Jie Pan

Detection and Diagnosis of Stiction in Control Loops
Mohieddine Jelali and Biao Huang

Stochastic Distribution Control System Design
Lei Guo and Hong Wang

Dry Clutch Control for Automotive Applications
Pietro J. Dolcini, Carlos Canudas-de-Wit, and Hubert Béchart

Active Control of Flexible Structures
Alberto Cavallo, Giuseppe De Maria, Ciro Natale and Salvatore Pirozzi

Active Braking Control Design for Road Vehicles
Sergio M. Savaresi and Mara Tanelli
Publication due September 2010

Other titles published in this series (continued):

Made in the USA
Monee, IL
11 June 2026

53100301R00208